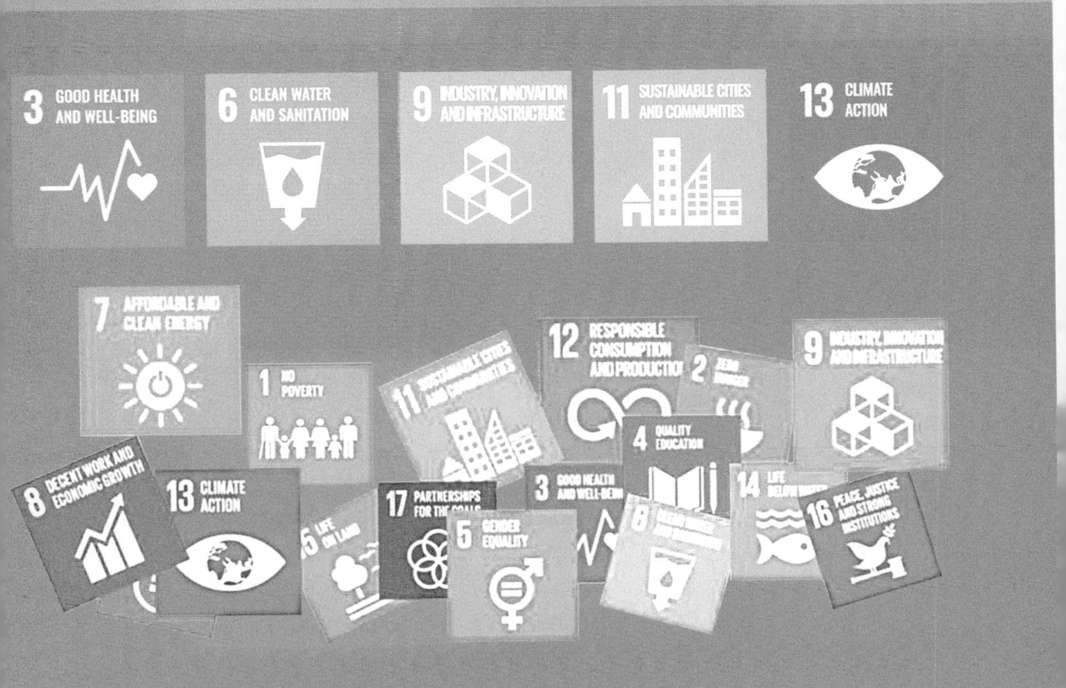

High-T_c Superconducting Technology

High-T_c Superconducting Technology

Towards Sustainable Development Goals

edited by

Muralidhar Miryala

JENNY STANFORD
PUBLISHING

Published by

Jenny Stanford Publishing Pte. Ltd.
Level 34, Centennial Tower
3 Temasek Avenue
Singapore 039190

Email: editorial@jennystanford.com
Web: www.jennystanford.com

British Library Cataloguing-in-Publication Data
A catalogue record for this book is available from the British Library.

High-T_c Superconducting Technology: Towards Sustainable Development Goals

ISBN 978-981-4877-65-7 (Hardcover)
ISBN 978-1-003-16468-5 (eBook)

Contents

Preface

With the discovery of the LaBaCuO superconductor in 1986 by Bednorz and Müller, a number of new superconducting compounds were found to exhibit superconductivity above the liquid nitrogen temperature, including rare-earth oxides and bismuth oxides, enabling the application of cheaper cooling methods. The new class of superconducting materials featuring high critical temperature (T_c above 77 K) and high upper critical magnetic field gave a new impetus to the research and development in superconductivity. Governments of many countries worldwide have encouraged scientists, engineers, and industry to develop high-temperature superconducting materials for practical use.

This volume covers the recent achievements, especially processing and applications of high-T_c superconductors and highlights the aim of the United Nations' Sustainable Goals. Economic growth, world peace, and future development of global community require the production of new materials that improve the quality of life. Superconductivity in general allows for 100% current transmission without losses. This makes it a super valuable resource for sustainability in many aspects. High-temperature superconducting materials (HTSC), which will be crucial for the daily-life applications, need to become more attractive for the sustainable development goals (SDGs). In medicine, diagnosis and treatments require sophisticated technology to overcome new challenges. Superconducting magnets can be used as compact high-field magnets that can be the hearts of a new class of mobile diagnostic devices such as MRI and NMR and can be used for magnetic drug delivery. Hunger can be partly avoided if there is sustainability in performing agriculture. This requires timely supply of water for irrigation. In remote areas, power losses are high so that the major part of power supply is lost by attention. Employing superconducting cables can tackle this problem, making the transmission loss-free. Superconductors are well used in cleaning water, via removing the magnetic particles, etc. This will help to make available good-quality water, an essential requirement for the sustainability of human life.

The zero resistance and the associated zero loss of superconductors make them very valuable in creating sustainable energy resources. All the clean energy harvested from hydrothermal plants, solar panels, and wind-based plants contribute a little part of the whole energy. To increase the efficiency of this clean energy's usage, we need superconducting technology, e.g., for a local energy storage. In future, DC electric energy from solar plants in Africa (365 sunny days) could be transported worldwide, especially to cold countries, using superconducting cables. Synthesis and contraction of new applications of superconductors require materials scientists and engineers. Based on new education opportunities, this field has a great capability to create job opportunities. Superconducting technology is an efficient way to create sustainability as well as reduce greenhouse gases. Superconductivity is an interdisciplinary field where materials scientist, physicists, and industrialists play important roles. This volume is designed to summarize the recent advances in materials science of high-T_c superconductors and their development to the SDGs. The volume consists of 16 chapters, each defining the subject and surveying the recent development in the field. The main objective of this volume is to summarize the recent advances in materials science of high-T_c superconductors, including their properties, processing, and applications. The new topics in this book cover experts' opinion on high-T_c superconductors and their relevance for SDG goals, and superconductivity in biomedicine, the powder technology of MgB_2, a cost-effective way to develop bulk MgB_2 superconducting supermagnets, Fe-based superconductors, and quasi one-dimensional layered superconductors. To justify the broad picture of the currently ongoing research on high-T_c superconducting materials, the book covers large-scale applications starting from HTS cable applications up to ship magnetic propulsion, bearing design for high-performance machines, and also applications in sensitive devices, superconducting electric motors, hybrid-type superconducting magnetic bearings for rotating machinery, compact magnetic field generators, and recent developments in the application of superconducting supermagnets for a variety of industrial applications and their development with respect to the SDGs.

We would like to take this opportunity to express our sincere gratitude to all chapter contributors for their great endeavor in completing this book in time. We also wish to acknowledge Jenny

Rompas from Jenny Stanford Publishing for offering invaluable advice at every stage of the editing of this book.

The authors and the editor sincerely hope that the presented ideas and information in this book will be helpful for a broad range of readers, scientists, new-coming young researchers, and bachelor's and master's students and will encourage further development in the field, especially work towards the Sustainable Development Goals.

Muralidhar Miryala
Editor
Shibaura Institute of Technology
Graduate School of Science and Engineering &
College of Engineering
Materials for Energy and Environmental Laboratory
Superconducting Materials
Tokyo, Japan
miryala1@shibaura-it.ac.jpp

Chapter 1

Expert Opinion: Relevance of High-T_c Superconductors for SDG Goals

1.1 Superconductivity and Sustainable Development Goals

Masato Murakami
Shibaura Institute of Technology, Japan

The most important contribution of superconductivity to the global sustainable development is its clean energy-saving feature. Superconducting materials exhibit zero resistivity, which enables the storage of the energy in the form of electric persistent currents known as SMES (superconducting magnetic energy storage).

Today, however, almost all the applications of superconductors are based on superconducting magnets with high field generation. Magnetic resonance imaging (MRI) uses 5 T superconducting magnet to observe human body for medical diagnosis without dissection. It can detect small diseased parts like early cancers. So far, MRI has been the most successful commercial application of superconductors.

High-T_c Superconducting Technology: Towards Sustainable Development Goals
Edited by Muralidhar Miryala

ISBN 978-981-4877-65-7 (Hardcover), 978-1-003-16468-5 (eBook)
www.jennystanford.com

Another interesting application area is the power machines. The power of all the electric machines is given by a simple equation:

$$F = I \times B$$

where I is the electric current and B is magnetic induction. Thus, if one can have a high magnetic field density B, the current I is minimized, which is effective in saving a great amount of energy.

The magnetic induction of a normal conducting magnet is 1 T at maximum. One can achieve 5 T even in a normal conducting magnet. However, it must be cooled with tremendous amount of water, which requires a huge energy. In contrast, one can generate 5–15 T with a superconducting magnet with no energy consumption, since superconducting currents are persistent. Thus, the superconducting magnets have advantages in power applications over normal magnets.

For magnetic field generation, superconducting solenoids are commonly used, where superconducting wires are wound in the form of coils. One drawback of such superconducting magnet is a small energy density. Recently, new type of superconducting magnets were developed in a bulk form based on RE-Ba-Cu-O (RE: rare earth) system. They are simple blocks and magnetized using electromagnetic induction. They are attractive for various applications since their sizes are 2–15 cm in diameter with extremely high energy density.

Using bulk RE-Ba-Cu-O, one can levitate a permanent magnet in midair. One can rotate a huge floating ring magnets levitated over the bulk superconductors, which can be used as another type of energy storage system called a flywheel.

The employment of superconducting materials for transmission cables has attracted interests of many people. However, the critical temperatures stayed extremely low until 1986. The cost for cooling with liquid helium made it difficult to bring the superconducting cables into a real market. With the discovery of high-temperature superconductors (HTS) (Y-Ba-Cu-O, Bi-Sr-Ca-Cu-O) with T_C above the boiling point of liquid nitrogen, electric power companies showed strong interests in the development of practical superconducting cables worldwide. In Japan, the United States, Korea, and China, superconducting cable projects have been launched, where the test cables are already installed in power transmission lines.

In many continents, desertification has become a serious problem. From a different perspective, however, the desert is a great place for solar power generation. If one quarter of the Sahara desert is covered with photovoltaic cells, the electricity needed for the entire world can be supplied. The question is how to transmit the power, since losses always occur in long-distance transmission. This problem is countered by using superconducting cables with zero electrical resistance. In other words, the energy problem of our planet can be solved at once by using superconducting power transmission to send electricity generated by solar power to the world.

The geomagnetism protects us from the solar wind with radioactive particles. Without the earth magnetic field, human beings cannot survive. There is a report that the field is decreasing and may disappear in 780 years. This problem can be solved by winding a superconducting cable around the equator, which can create artificial geomagnetism. It will be costly, but it is possible if we work together.

We are also facing a serious problem related to the solar wind. It is known that massive solar storms often hit the earth. In 2012, NASA warned the world that the most powerful solar storm in the history might hit the earth. Fortunately, it was just a near miss, and therefore the earth was saved. Had it hit the earth, it might have caused worldwide damage to modern power grids and electric devices including computers and artificial intelligence. Superconductors are known to be the only materials that can shield magnetic fields completely. Hence, we need to install superconducting magnetic shield devices to protect our modern lives.

Superconductivity provides the key technologies for global sustainability. However, one should bear in mind the fact that international collaboration is also the key to the achievements of sustainable development goals (SDGs).

1.2 High-T_c Superconducting Technology: Towards Sustainable Development Goals

David Cardwell
University of Cambridge, UK

The discovery of HTS by Bednorz and Muller in 1986 was accompanied by a rise in expectation for a new generation of sustainable high field, high current, and low-loss engineering applications for which there is simply no competing technology. Although the interim 35 years have not delivered new applications on the scale envisaged, there have been some remarkable developments in all forms of these materials, which, like all embryonic technologies, have taken time to mature. Coupled with advances in cryogenic coolers and fuelled by the drive to global sustainability, these technologically important materials are re-emerging as key to achieving a low carbon, low energy consumption future across a wide variety of industries. Significantly, HTS technology is relevant to many of the 17 sustainability goals identified by the United Nations required to transform the world, and to good health and well-being, affordable and clean energy, industry, innovation and infrastructure, sustainable cities and communities and climate action, in particular.

HTS exhibit a range of unique properties, which will potentially enable a wide range of novel engineering devices. These include the ability to superconduct electrical current at technically and economically achievable temperatures (typically between 50 and 77 K), carry current densities that exceed those achievable in copper by a factor of 100, support DC transport current up to a critical value without any dissipation of energy, to generate some of the highest, stable magnetic fields in stand-alone samples observed in the history of the planet (greater than 20 T) and to respond to extremely small levels of magnetic flux on a quantum scale. This gives rise to a myriad of potential sustainable applications, including high current transmission cables, magnetic levitation devices, magnetic separators, energy storage, motors and generators, MRI and nuclear magnetic resonance (NMR), high-frequency mixers and generators, sensitive magnetic field detectors, and microwave filters.

HTS materials are fabricated generally in three principle forms: wires and tapes, bulk materials, and thick and thin films. Each of these forms of material has advantages and disadvantages for a myriad of applications, but each has its own unique processing challenges. A fundamental problem in fabricating wires and tapes and bulk forms of high temperature superconductor, for example, is that the presence of grain boundaries within the material microstructure almost inevitably limits severely its current carrying capability. As a result, sophisticated, state of the art processing techniques have been developed to generate long, crystallographically aligned wires and tapes in various geometrical configurations and bulk materials in the form of large single grains, primarily for the generation of large magnetic fields. So-called second-generation coated conductors are deposited on technical substrates over long lengths usually via vacuum technology, although a variety of emerging processing technologies, such as liquid phase epitaxy, are being developed to enable more cost-effective and reliable processing of HTS conductors. Thick films of HTS also, therefore, suffer from the effects of granularity and are generally deposited instead in thin film form on flat, single crystal substrates over relatively small surfaces areas (typically 10 cm in diameter). All known forms of superconductor require cooling to cryogenic temperatures for their technological operation, often with liquid nitrogen or bespoke cryocoolers, prior to operation, which generates significant thermal stress during thermal cycling. As a result, all material forms need to be sufficiently robust to withstand these forces.

There are currently a number of high temperature superconductor compositions, of which the (RE)-Ba-Cu-O [(RE)BCO] family is the best known and arguably the most useful from an applications perspective. Other HTS, including the Bi-Sr-Ca-Cu-O (BSCCO) system, tend to be easier to process but generally offer lower levels of performance compared to (RE)BCO, and hence tend to find application in relatively niche devices. It is the complex oxide nature of the systems that has led inevitably to 4 decades of development and optimization of the materials properties for different applications. In addition, bulk forms of these materials tend to be brittle and have required the development of processes to reinforce their mechanical properties in order to withstand the forces associated with very large magnetic fields.

The ultimate goal of the development of HTS for large-scale, high power applications is to fabricate a long wire of many kilometers length, economically, and reliably with high current carrying capability (at least of the order of 10^6 A/m^2 at 77 K). Such a wire would enable the generation of extremely large magnetic fields that could be controlled easily by simple adjustment of the current for almost any application. In addition, such wires could be used to support the transfer of large amounts of energy in a very cost-effective and efficient way, often without the need for major infrastructure development. Given the complexity of the process required to fabricate such conductors, however, bulk materials melt processed in the form of discs of diameter up to 10 cm can be used as an intermediate, affordable technology in devices that enable more efficient transportation and lower loss operation in the short term.

The potential of HTS to contribute to a sustainable global future is considerable. From removing contact surfaces in transportation systems, and therefore reducing energy loss associated with friction in Maglev devices, to enabling essentially lossless DC transmission in smart cities and solenoids for in wind turbines in the generation of renewable energy, the potential impact of these remarkable materials is profound. Like all technological developments, the development of state-of-the-art materials is never complete and overall device performance can always be improved by further optimization of their properties and through improved design. HTS have posed many processing challenges that have been overcome systematically in all forms of material and are now at the stage where their potential for incorporation in sustainable devices is imminent. As a result, we can look forward to technological advances in the fields of transportation, healthcare, and the pharmaceutical industry.

Finally, I would offer my strongest possible encouragement to students and young scientists who are considering a technical career to pursue options within the general area of superconductivity. As the United Nations has so clearly articulated, now is the time for the best young minds on the planet to focus on global sustainability, and there is no other technology in the world that offers greater opportunity, potential impact, and scope than superconductivity to do exactly that.

1.3 The Potential of Superconductor Technology: Towards Sustainable Development Goals

Nick Long
Robinson Research Institute, Victoria University of Wellington

The world seems to be in an increasing state of crisis, with destructive storms, wildfires, droughts, and pandemics becoming more common. In this context, the UN SDGs seem far away. The goals are about ending poverty, improving health and education, reducing inequality, creating economic growth, mitigating climate change, and conserving the environment. These are the goals we would all like to see achieved in our lifetimes. The great privilege of being a scientist or engineer is the opportunity to work towards these goals through the generation of new knowledge and the application of this knowledge in useful technologies.

Since its discovery in the 1980s, high-temperature superconductivity has had immense promise for the betterment of society. The most striking characteristic of superconductivity is the ability to carry DC current without resistance. This immediately suggests the possibility of reducing electrical losses, which can result in more efficient production, transmission, and use of electricity. A less obvious characteristic of superconductivity is the high current density in the material. This is a key property which allows high field magnets and producing magnetic devices which are more compact and lightweight. But superconductivity also has other stranger properties which can lead to completely new devices.

Progress has been slow towards these applications and devices due to the immense difficulties of producing high-temperature superconductor materials in forms with the performance and cost needed for applications. HTS are ceramic materials which are brittle and require a high degree of phase purity and crystalline alignment between grains to effectively transport current. Producing either long lengths of wire or large bulk materials was a significant scientific and engineering challenge. Although many improvements and refinements are still to be made wires and bulks are now commercially available or near commercialization which can meet application needs.

Wires based on the bismuth oxide HTS family of materials have been commercially available since the 1990s. With the performance steadily improving these wires are now routinely used in specialist magnets. Wires based on the yttrium-barium-copper-oxide family have been slower to develop but there are now multiple companies providing this wire at reasonable cost and excellent current capacity, and the wire has been used in numerous technology demonstrations. The newer magnesium diboride superconductor is also readily available in wire and bulk forms. Bismuth oxide and Y-Ba-Cu-O based bulks are also commercially available and have been demonstrated to create fields approaching 18 T. Working on new materials is always a good strategy for those who want to make a difference in the world and new superconductor materials continue to be discovered. Advances in some of these materials are presented in this volume.

There have been many demonstrations of HTS cables throughout the world. These have demonstrated that power can be delivered not just efficiently but with a lower use of scarce space and disruption to cities. The HTS cables have demonstrated the potential to reduce greenhouse gas emissions and have more resilient power grids. These cables can carry MW of power or be scaled down to deliver power within facilities such as data centers. In this book, you will find a discussion of the use of cables on ships which shows the creative uses to which superconductors can be put.

Superconductivity gives rise to the phenomenon of persistent currents. Once a current is set up in the material it can flow almost indefinitely if a low temperature is maintained. This property is linked to another startling property, that of magnetic flux trapping. A magnetic field configuration once established in a material is "locked in" and maintained. These properties of persistent currents and flux trapping are well exploited in bulk materials, where the flowing currents create large fields and the flux trapping can transfer large forces. This leads to using bulk superconductors in innovative ways to create trapped field magnets which can be used within motors or be used in applications such as NMR. The trapped field magnets can also be used as bearings within motors or flywheels allowing for the support of ultra-high-speed frictionless rotation. Very-high-speed motors have higher power/weight ratios enabling lighter motors for applications such as electric aircraft.

Using superconducting wire in magnets is well established. But with HTS wire the possibility is to create much higher fields and operate the magnets using mechanical cryocoolers, allowing for easier installation and operation. Already HTS systems have been demonstrated at record fields and magnets for NMR and MRI have demonstrated they can meet the demanding stability requirements for these applications. For future accelerators the possibility is for dipole and other specialist magnets at fields beyond 20 T, allowing access to energies previously unobtainable.

Incorporation of renewable energy sources into power grids has increased the need to stabilize grids and protect electrical assets from current surges. The unique ability of superconductors to carry enormous currents but then to become resistive above a critical current means they can act as natural reusable fuses in a power grid. Recent years have seen the uptake of this exciting new technology as part of creating a power grid for the future.

To make progress towards the SDGs, we need scientists and engineers who think creatively about the problems of the field—how to improve the performance of materials and devices and reduce costs. We also need those who will form new companies and join existing companies to make usable products. These products will save energy, create wealth for their communities, lead to breakthroughs in health care, and mitigate climate change. Although the technical and commercial problems are hard, we have the tools we need in our hands and in our heads, so let us get to work!

1.4 Superconducting Technology: A Step to the United Nation's Sustainable Development Goals

Frank N. Werfel
Adelwitz Technologiezentrum GmbH (ATZ), Torgau, Germany

In the shadow of dominant long-length HTS, textured bulk materials show a considerable effort on magnetic engineering application. The present volume on superconductivity is designed to cover recent achievements in material processing and application of high-T_c superconductors. Scientists and engineers are balancing science and technology towards applications with the specific opportunity to contribute to improvement and protection our life and our shared home, Earth. The scientific community as a reasonable call of nature has to decide in the present time: Saving energy and stopping the climate change, improving the essential health care, and taking the responsibility of basic education for all.

In 2015, in a global action the United Nations has provided a 2030 blueprint for shared prosperity in a sustainable world—a world where all people can live peacefully, vibrantly, and productively on a healthy planet. To achieve these SDGs in just less than a decade is a challenge for society. The general goal targets span from rising sea level due to climate change, global hunger, missing education, lack of essential health service, up to violent conflicts within and between nations, and facing disadvantages and discrimination (woman).

Where and how can scientists and engineers, including the young generation taken their responsibility to protect our home, our world. In the past, every generation has believed to live in an exceptional time. The begin of industrialization started 200 years ago, the times before and after world wars ended with a new world order, and the digital www generation growth after 2000 will change our life fundamentally.

Among these discussions a sharp break, never seen before or expected, met our Earth at the beginning of 2020. A new highly infectious virus with millions of infections and hundred thousands of died people worldwide has brought a sharp thinking about our sensible relation to nature. The COVID-19 pandemic has affected many science, space, and technology institutions. I know the reaction of most of our colleagues who take and care for their profession passionately. Our greatest responsibility is the safety of people, and I know all of us in the science community understand why this is necessary.

When I started in the 1960s and 1970s, the electronic structure investigation of materials with jumps in the electrical resistivity (vanadium-, titanium-oxides, A15-phases and Chevrel chalcogenides M_xMoX_8), weekly, my Professor ask me about news of materials with unexpected large resistivity jumps. He was euphoric about all what have been done that time in superconductivity. Transition metal oxides, alloys, and superconducting A15 compounds were in our focus. I should study materials with resistivity jumps up to Mott mechanism by writing a PhD on that exciting field.

Then came in 1986, completely unexpected, in "Zeitschrift für Physik," in which Klaus Bednorz, Alex Müller, and 1987 Paul Chu in *Physical Review Letters* reported about the discovery of high-temperature (high-T_c) cuprate superconductors (HTS) switching T_c with 35 K, later 92 K well above the longstanding record of 23.3 K of Nb_3Ge. Superconductivity had been on the front page of the world's largest newspapers. With that scientific news and the basic discovery of superconductivity in 1911 by Kamerlingh Onnes, superconductivity was entering a golden era and fired the hope of new applications to solve the world's longstanding magnetic problems: copying powerful electromagnets above 10 T, aiming Sun's fusion mechanism up to NASA's Astronaut protection program from Wernher von Braun (1969).

The HTS discovery generated a rush in the development, the scientific protagonists were basking in the light of fame. They promised applications at much higher operating temperatures. The easy use of liquid nitrogen (77 K) transferred superconductors from a few low-temperature laboratories to the daylight brightness of universities, high schools up to the public life. Levitation pictures of free suspending permanent magnets (PM) several centimeters above a new superconductor cooled by liquid nitrogen had a run in journals and digital media screens. Public explanations of the levitation phenomenon given by the Meissner effect were wrong... It's pinning plus a few per mil Meissner.

Many of the key points made in that time remain valid today. The driving power behind the superconductivity development was and is the application. The alternating biennial two large conferences, ASC (Applied Superconductivity Conference) held in US and the EUCAS (European Conference on Applied Superconductivity) except other large meetings and workshops (ISS, MT) demonstrate the continuously successful way of superconductivity.

Today we observe transition temperatures up to 138 K (in between hydrides under pressure coming close to room temperature) but learned that T_c is not the only important parameter for practical HTS devices. Thin films, wires, and bulks must carry large currents and are limited by the critical current I_c or current density J_c. Unfortunately, significant large magnetic fields $\mathbf{H}_c$ also have their critical limitations.

In addition, the Cuprate material structure is uncomfortable, being anisotropic (planar) and having a short coherence length, which brings factors as microstructure alignment achieving low-angle grain boundaries into the practical wire implementation. In that sense, MgB_2 is much simpler in structure and processing.

The critical parameters mentioned so far are not the only important ones. For high-field applications mechanical properties like Lorentz stress, brittleness, hardness, strain tolerance must be well accounted. And finally, not directly correlated with HTS application, we are still fishing in a dark sea when trying to understand the mechanism of Cooper pairing in most of the HTS: Either the legendary BCS theory of Bardeen, Cooper and Schrieffer or, on the contrary, electron correlation effects with $d_{x^2-y^2}$ symmetry and an order parameter in both hole- and electron-doped Cuprate superconductors are favored for the high-T_c mechanism! In this sense, we are (still) blind but producing thousands of kilograms HTS for applications.

Basically, three critical parameters determine practical applications: temperature T_c, current I_c, and magnetic field $\mathbf{H}_c$, whereby the parameters are functionally strictly coupled. Having the basic parameters, the engineering limitations will determine the practical use to establish safety margins for functional reliability and lifetime. In addition, geometrical parameters, as long wires and cables dominate the application, local devices as thin films are specific in electronics and sensors, bulk superconductors can trap magnetic fields of almost 18 T, generating magnetic forces more than three hundred times larger than the corresponding PM values.

Magnetic resonance imaging (MRI) is the only commercial success based on superconductivity with some ten thousand devices in hospitals and clinics worldwide achieving a global sales of USD 3.5 billion per year. Compared to MRI none other device have been successfully commercialized. The limiting factor is the cost of superconducting devices versus other non-superconducting solutions. Nevertheless, a number of prototypes and demonstrator devices in the power area have been built: Motors and generators in MW scales, underground power cables in km length, transformers, ton-load rotational magnetic bearings and levitated trains above 10 ton mass. All of these applications will save energy and contribute towards the engaged SDG goals.

A prediction of the future of superconductors is uncertain. In this book, more than 14 internal experts present their updated HTS research contributions, ideas, and prospects for the next decade. Applications covering cable, cryomagnets, bearings and high-efficiency machines, reports about trapped field values, and proposals to overcome application limitations are carefully documented. Materials research will continue, but more important is the consequent improvement of the surrounding system technology including industrial-like manufacturing process, cryogenics, perfect thermal insulation, and cost reduction. The high-T_c community has to accept that the conditions for continuous application success will always be in competition with other, often non-superconducting methods.

And here is my proposal and experience. We should follow our goals instantly, not to switch from one material to another. When improving the efficiency of our HTS machines, saving energy or contributing to digital or medical networks with new HTS components, we are on the right way to the sustainable goals. Thereby, after more than 30 years' HTS experience, our research knowledge and scientific responsibility is most imposed, absolutely essential, and should drive the public debate to save our climate and nature. Together, we can change the world.

1.5 Superconductors as Friends of Our Environment

Milos Jirsa
Institute of Physics ASCR, Na Slovance 2, CZ-182 21 Praha 8, Czech Republic

This book has been written during the hard days of the COVID-19 viral pandemic. This catastrophe has affected our everyday life. It forced us to stay home and when going out, then only with the covered face. The disease has taught us, how even in the modern age with an advanced medicine and the enormous progress in technique, humanity is still vulnerable. Not only the human health but also the health of economy are fragile. Nobody could imagine such a dramatic, fast economic recession. Many voices say the society after this pandemic will never return to the original level. I believe it will.

However, we should grasp this opportunity as a challenge to change our mind and start a new era with reasonable, SDGs. The pandemic killed hundreds of thousands of people and caused an enormous social devastation. On the other hand, it showed us how effectively our environment recovers after its crazy devastation is stopped.

Superconductors are exceptional materials in many aspects. They are front products in developing environment-friendly devices. Already the classical superconductors have been used for 70 years in a number of equipment that are used in our everyday life; they are principal and today ordinary devices. Superconducting magnets with conventional wiring are nowadays a laboratory standard on which physics of elementary particles, studies of thermal, magnetic, and electric properties of solids, medical screening by a magnetic tomography or encephalography, or the idea of just now constructed magnetic levitated train line between Tokyo and Nagoya (MAGLEV) are based. All these applications of superconductors employ the unique property of a loss-less passing of huge electric currents through the superconducting wire. On the other hand, Josephson junctions used in the superconducting quantum interference devices (SQUID) enable for detection of extremely small electric currents, like those of heart or brain. The basic problem of all these applications is that the superconducting state exists only at low temperatures, to which the superconductor has to be cooled by relatively expensive and rare liquid helium.

Since 1986, a new category of superconductors has been studied and developed for use, so-called high-temperature ones. In contrast to the metallic conventional superconductors, the new ones are of ceramic, polycrystalline nature and their critical parameters, such as critical temperature and critical fields, largely prevail those of metallic counterparts. Their discovery was a big surprise for physicists, bringing into the field of seemingly well-understood phenomenon completely new challenges. Even more peculiar was the development of new products employing these novel materials for practical use. The idea of current transport at temperatures as high as some 50–90 K, mainly above the boiling point of liquid nitrogen, was so exciting that wire fabrication prevailed in the field of high-T_c materials development. In the beginning, BiSrCaCuO powders were filled into silver tubes, rolled and pressed by various techniques, then annealed at high temperatures to prepare wires of polycrystalline BiSrCaCuO for current transport. This technique matured and hundreds kilometers of these tapes were fabricated. However, the method could not be scaled up and this technique was finally abandoned. The reason was that the grain boundaries of this type of superconductors were weak links for current flow and no way was found, how to overcome this limitation. Anyway, these tapes were further developed; they are still fabricated, especially in Japan, and used for special purposes. One of such special applications was successfully tested coils wound of BiSrCaCuO tapes for one of MAGLEV trains tested on Jamanashi test line.

The problem of grain-boundary weak links was addressed and successfully solved by a huge effort devoted to development of the second generation of tapes, called coated conductors, based on REBaCuO thin films. They combine the ultrahigh critical current density and big magnetic field safety margins with the possibility to operate the devices at a relatively high temperature, being cooled by the cheap liquid nitrogen or a cryocooler.

Another story are bulk high-T_c super-magnets. In principle, REBaCuO single-crystals of 1–2 cm size can be grown by a classical Bridgman method. However, such a product is very expensive and its fabrication takes a lot of time. Much easier is re-crystallize a polycrystalline pellet by one of the seeded-growth techniques into a single-grain form. In this way, discs up to 15 cm diameter can be prepared. Why speaking about super-magnets? It was experimentally manifested that nearly 3 cm in diameter single-grain YBaCuO superconductor can trap at 29 K magnetic field as high as 17–18 T. It is order of magnitude more than the best today known NdFeB ferromagnets. In other words, bulk superconductors can be a source of extremely high magnetic field suitable for constructions of a wide range of new compact designs, like mobile medical diagnostic devices, effective water purification units. They can replace the classical superconducting coils in MAGLEV trains or be used in levitation, serving in loss-less bearings, both axial and radial, in superconducting motors, in flywheel energy storage units like UPS or stabilizators of electric current in the public network (up to ten features stabilized simultaneously) or as position stabilizators in space units, boats etc. In all these cases the solution is always much smaller and by far more effective than the conventional designs.

In electronics, the microwave frequency filters are needed to define separate frequency ranges. These filters are metallic cavities where resistivity of walls is the issue. The lower the resistance, the higher is the filter quality. Resistance of superconductors in zero. The best filters are covered by a layer of superconductor and represent basic devices for mobile telephony systems.

In 2001, superconductivity was discovered in MgB_2. This material has critical temperature of only 39 K, but it is very attractive by its chemical simplicity, thus low cost and relatively simple, fast, and cheap fabrication. It also possesses a fascinating property that being polycrystalline, its grain boundaries are not weak links for current flow like BiSrCaCuO. It also does not need the rather lengthy oxygenation process. Thus, it represents an effective alternative to the previously mentioned solutions. Recently, in 2006, superconductivity was found in iron-based materials, know now as iron-pnictides, compounds, where ferromagnetism and superconductivity coexist, which was before considered impossible. These superconductors have T_c up to 50 K, have a relatively simple chemical composition and layered structure like cuprates (REBaCuO). Their electromagnetic properties and optimal composition are still under study.

All these high- and middle-temperature superconductors are not only revolutionary new technical opportunities, but also a huge material for theory. They show that the principles of superconductivity, believed until 1986 to be well understood, is in fact much broader and universal than considered before. In any case, for environment-friendly applications, these materials are a big help and potential. Middle-temperature superconductors represent some transition state, part of which can be described by classical BCS theory, while the rest requires a completely new approach.

Chapter 2

Dense and Robust (RE)BCO Bulk Superconductors for Sustainable Applications: Current Status and Future Perspectives

Devendra K. Namburi and David A. Cardwell
Department of Engineering, University of Cambridge, Cambridge CB2 1PZ, UK
dkn23@cam.ac.uk, ndevendra@gmail.com

Rare earth (RE)-based Ba-Cu-O bulk, high-temperature superconductors (HTS) have considerable potential for number of sustainable engineering and technological applications due to their ability to carry large currents and thereby to trap very large magnetic fields. Some existing and envisaged applications employing these materials include high-performance trapped field magnets, magnetic bearings, flywheels, portable magnetic resonance imaging (MRI), magnetic separators, superconducting motors, and levitated transportation systems. In this chapter, the primary focus will be on the processing and properties of (RE)BCO-based HTS cuprate

High-T_c Superconducting Technology: Towards Sustainable Development Goals
Edited by Muralidhar Miryala

ISBN 978-981-4877-65-7 (Hardcover), 978-1-003-16468-5 (eBook)
www.jennystanford.com

materials. We will elucidate the importance of obtaining single grains of (RE)BCO: how potential seed crystals are employed to enable heterogeneous nucleation, procedures to prepare precursor powders, and ways to tune the heat treatments to transform the compacted preform into a robust single grain. Top-seeded melt growth (TSMG) has emerged as an effective technique to fabricate single grains via a so-called cold-seeding approach, which will be described and the recent advancements made to this processing methodology will be outlined. The first part of the chapter will describe an alternative fabrication approach based on a top-seeded infiltration and growth (TSIG) process. The near-net shaping TSIG technique will be described in detail, including recent advances in processing that have opened pathways for producing dense and robust high-performance (RE)BCO bulk superconductors required for practical applications. The challenging task of controlling the formation and distribution of the presence of non-superconducting phase in samples processed by IG, which has been addressed recently, will be presented. The development of a novel, two-step buffer-aided TSIG (2-step BA-TSIG) advanced fabrication route will be described in detail. The high field performance of YBCO bulk superconductors, fabricated by a 2-step BA-TSIG approach, both at liquid nitrogen and lower temperatures will be summarised and correlated with their microstructural and mechanical properties. The performance of the IG-processed materials will be compared to those fabricated via traditional melt growth processing, where appropriate.

The second part of the chapter will focus on the GdBCO single grain bulk superconductors fabricated by the recently developed, novel 2-step BA-TSIG process. The ability of these samples to trap magnetic field at both 77 K and 40 K is described. An as-processed 25 mm diameter GdBCO sample (i.e. without internal or external reinforcement) is tested for robustness by applying a large magnetic field of 8 T at 40 K. The sample was found to trap a magnetic field of 5.6 T at 40 K and also exhibited a levitation force of 55 N at 77 K. Superconducting properties including critical temperature and critical current density both within the *a*-growth and *c*-growth regions of the sample are measured and results reported. The sample exhibits almost uniform superconducting properties with zero-field current density of 50 kA/cm^2 at 77 K with corresponding irreversibility fields in the range 4–6 T at 77 K. Mechanical properties measured for the GdBCO samples indicate a consistent flexural strength in excess of 60 MPa. Detailed macro- and microstructural studies carried out on the sample showed a good correlation between the measured superconducting and mechanical properties.

In the final section, a short summary of novel experiments performed recently on (RE)BCO bulk superconductors is presented. Bulk materials containing artificial holes produced both via TSMG and TSIG fabrication techniques, developed primarily to reduce oxygenation time and to achieve more rapid cooling in liquid cryogen will be outlined. Magnetic shielding performance obtained in both bulk and thick film configurations of (RE)BCO cavities, in general and YBCO, in particular, will be highlighted. Attempts to obtain large (RE)BCO bulk samples via multi-seeding approaches with recent processing advancements enabling the fabrication of quasi single grains will be summarised in this section. Finally, recent reinforcement studies carried out employing metallic and SiC fibres in YBCO, in an attempt to improve the mechanical properties of the bulk single grains whilst retaining their superconducting properties, are presented. The section will conclude with a summary of the range of potential engineering and technological applications that may utilise these advanced and robust (RE)BCO bulk superconductors.

2.1 Introduction

Superconductors have been long well-known for their spectacular physical properties and true quantum behaviour, and since their discovery in mercury in 1911 by Kamerlingh-Onnes [1, 2]. These materials offer no resistance to the flow of dc electrical current in their superconducting state at a temperature below a critical value, T_c. Resistance to the flow of dc current has been established in some superconducting materials to be less than 1 in 10^{15} of their normal state resistance (i.e. the resistance of the material at an operating temperature above the T_c of the material). Magnetic flux expulsion from the interior of the superconductor at relatively low levels of applied field, known as the 'Meissner effect' (as discovered by Meissner and Ochsenfeld in 1933 [3]) is yet another signature of these extraordinary materials. Persistent currents generated in bulk superconductors have been observed to flow in annular-shaped samples without dissipation for more than a year, which underlines the vanishing-small levels of resistance present in this class of materials [4]. There are other critical parameters that, in turn, characterise the superconducting state, including critical magnetic field H_c, critical current density J_c, and critical frequency ω_c. The limits of the superconducting state within these critical parameters for both low temperature superconductors (LTS) and high temperature superconductors (HTS) are shown in Fig. 2.1(a). These critical parameters are related frequently to each other and their dependences are influenced largely by the superconducting energy gap '2Δ'. For example, $\omega_c = 2\Delta/\hbar$, where $\hbar = h/2\pi$ and h is Planck's constant. Superconductors are classified based on their response to applied magnetic field as either type-I or type-II. When magnetic field is applied to type-I superconducting materials, the superconductor undergoes an abrupt transition from the superconducting state to the normal state at the critical field H_c, as shown in Fig. 2.1(b). This change-over in behaviour is evident in the case of a type-II superconductor as the appearance of the so-called mixed state, as shown schematically in Fig. 2.1(b). Typical critical fields are in the range of 10–100 Oe and 100–300 T for type-I and type-II superconductors, respectively (note that 1 T = 10,000 Oe). The corresponding temperature dependences of critical field (i.e. the *H-T* phase diagram) for type-I and type-II superconducting materials are shown in Fig. 2.1(c). In the superconducting state, the conduction electrons form Cooper pairs with net integral spin and behave consequently as bosons, condensing to a low energy, ordered state. The origin and the character of this ordering were developed originally by Bardeen, Cooper and Schrieffer and this theory is generally referred to as the highly successful BCS theory [5].

Single grain bulk (RE)-Ba-Cu-O [(RE)BCO] HTS materials (where RE represents a rare-earth element such as Gd, Nd, Sm, Dy or yttrium) are all type-II superconducting materials and hence can be attractive from an applications perspective given they are known to trap large magnetic fields even at liquid nitrogen temperatures [6–8]. The principal performance figure of merit of a single-grain, bulk HTS is its ability to trap magnetic field. The trapped magnetic field (B_t) is known generally to depend on the product of two critical parameters: The diameter of the single grain '*d*' and the critical current density 'J_c', i.e. $B_t \propto J_c{*}d$ [9]. As a result, larger the size of a

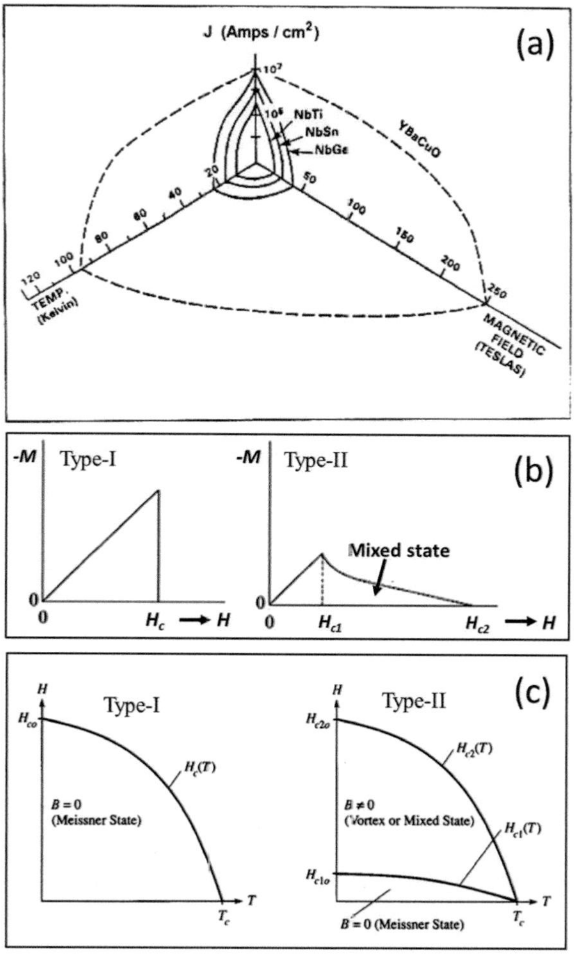

Figure 2.1 (a) Critical parameters (critical temperature T_c, critical magnetic field H_c and critical current density J_c) below which the superconducting phase exists is shown schematically for both low (LTS) and high temperature superconductors (HTS). (b) Field dependence of magnetisation (M Vs H) and (c) temperature dependence of critical field (H_c Vs T) observed in type-I and type-II superconducting materials.

single grain and higher the current carrying ability, the greater is its ability to trap magnetic field. The ability of these materials to trap magnetic fields increases significantly at lower temperatures due to their enhanced magnetic flux pinning strength. Direct evidence of this is the generation of trapped magnetic fields in the range of 8–10 T at the surface of bulk superconductors [10, 11] and in excess of 17 T in the temperature range 25–30 K in stacks of YBCO and GdBCO+Ag single grain samples [12–14]. Some of the practical and possible applications employing these materials are magnetic bearings, separators, flywheels, portable magnetic resonance imaging (MRI), motors, levitated transportation systems, shielding devices, and others [6–8, 15–17]. These applications are possible due to the higher critical temperature of these materials and further due to the fact that they possess enhanced superconducting properties associated with strong magnetic flux pinning in the presence of high fields.

Yttrium-based (RE)BCO bulk superconductors have been studied extensively in the literature [18–24] due to two prime factors. The first is that the peritectic temperature T_p of this system is relatively low compared to many other potential (RE)BCO materials such as Sm, Gd, and Nd [25, 26]. For example, the T_p of YBCO in air is ~1005 °C, whereas the corresponding values for the Gd, Sm and Nd-systems are ~1030 °C, 1054 °C, and 1068 °C, respectively. The second is that no RE/Ba solid solution formation occurs in the YBCO system due partially to the relatively different ionic radii of Y and Ba, which is unlike the other RE-based systems for which the RE and Ba radii are similar [27]. For the last three decades, several fabrication methodologies such as melt growth (MG), quench melt growth (QMG), melt powder melt growth (MPMG), and infiltration and growth (IG) have been developed in order to fabricate single grains of (RE)BCO bulk superconductors for practical applications [28–33]. Of these, top-seeded melt growth (TSMG) has emerged as the most successful and has enabled the fabrication of single grain (RE)BCO materials consisting of a superconducting $(RE)Ba_2Cu_3O_{7-x}$ (RE-123) phase matrix with embedded $(RE)_2BaCuO_5$ (RE-211) particle inclusions to aid flux pinning strength and thereby enabling the bulk material to exhibit extra-ordinary field-trapping properties [34–36].

2.1.1 Top-Seeded Melt Growth Technique

A melt growth process developed initially by Jin et al. [30] was further tuned and optimised by several researchers that enabled the development of the TSMG process. There are three important steps involved in fabricating (RE)BCO single grains via TSMG: (i) Preparation of the appropriate precursor powders with appropriate composition (usually 75 wt.% RE-123 + 25 wt.% RE-211), mixing and homogenisation; (ii) employing a suitable seed crystal (with higher melting temperature, possessing a similar crystal structure and having phase stability with the melt) for heterogeneous nucleation and; (iii) employing appropriate heat treatments that enable successful single grain growth. Details of T_p of various (RE)BCO materials are presented in Table 2.1.

Table 2.1 Peritectic temperature (T_p) in air, for various $(RE)Ba_2Cu_3O_{7-x}$ compounds [26]

RE in ($REBa_2Cu_3O_{7-x}$)	Peritectic temperature, T_p (°C) (Error: ±5 °C)
La	1080
Nd	1068
Sm	1054
Eu	1046
Gd	1030
Dy	1010
Y	1005
Ho	995
Er	990
Yb	960

A schematic illustration of a single grain, including the *a*-growth and *c*-growth regions of the sample, is shown in Fig. 2.2(a). These growth regions and the growth sector boundaries observed in a YBCO sample fabricated by TSMG are shown in Fig. 2.2(b). It has been observed that the addition of grain-refining agents, such as Pt or CeO_2, to the precursor powders reduces significantly the RE-211 particle size in the single grain microstructure, thereby improving the magnetic flux-pinning strength in these materials. Several growth-related studies have been carried out to understand the nucleation and temperature window for single-grain growth, which are crucial not only to fabricate (RE)BCO single grains but to enable their microstructures to be engineered to support large J_c (i.e. in excess of 10,000 A/cm^2 at 77 K) as required for practical applications.

Recently, two important improvements have been made that have enhanced the reliability of single grain growth process of all (RE)BCO superconductors, including YBCO. The first is the development of buffer strategy and its integration within the TSMG approach. The second is the development of generic seed crystal with higher T_p for seeding any (RE)BCO superconducting composition. These aspects are discussed in Sections 2.1.2 and 2.1.3, respectively.

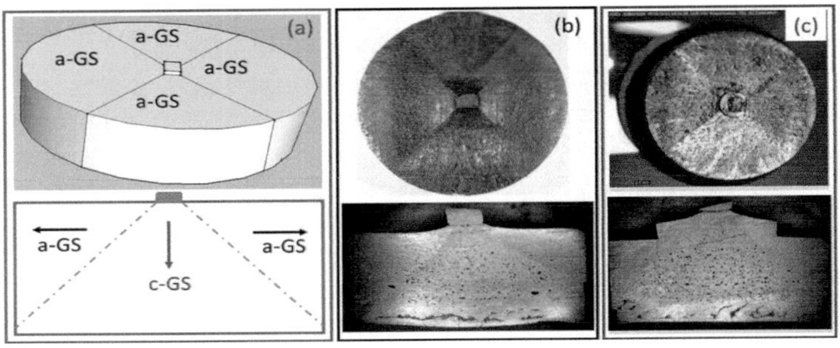

Figure 2.2 (a) A schematic illustration showing the top- and cross-sectional views of a single-grain (RE)BCO bulk superconductor. The *a*-growth and *c*-growth sectors emanating from the seed crystal can be seen in the figure. The growth sectors and the growth sector boundaries as observed in YBCO samples fabricated employing the TSMG and buffer-assisted TSMG (BA-TSMG) processes are shown in (b) and (c), respectively.

2.1.2 Buffer Strategy in TSMG

Although the TSMG approach has enabled the fabrication of high-quality (RE)BCO bulk superconductors, the relatively poor success rate (of ~30–40%) had hindered the routine manufacture of these materials in the form of single grains. To overcome this potentially serious problem, the so-called buffer-technique has been developed and integrated within the TSMG approach. This advanced fabrication method is now referred to as the buffer-aided top seeded melt growth (BA-TSMG) process [37–40]. This integrated fabrication process has aided the growth of single grains appreciably by preventing failure of the seed crystal, thereby improving the reliability of the (RE)BCO single-grain fabrication process. The top and cross-sectional views indicating the growth regions observed in a BA-TSMG processed YBCO sample are shown in Fig. 2.2(c). It can be seen that small-sized seed crystal is sufficient to produce a large, single grain [40]. Comparison of the growth regions for TSMG and BA-TSMG processes are shown schematically in Fig. 2.3(a). Additionally, it was found that the buffer pellet shields the seed crystal from the main (RE)BCO sample and therefore limits the diffusion of seed crystal elements into the growing sample, as illustrated in Fig. 2.3(b).

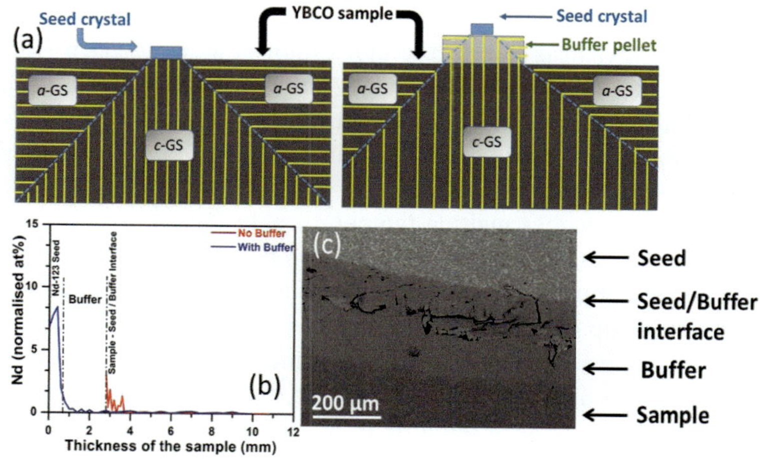

Figure 2.3 (a) Schematic representations of the *a*-growth and *c*-growth regions as observed in the TSMG and buffer-assisted TSMG processes. The confinement of the seed crystal element diffusion within the buffer pellet is illustrated in (b). (c) A scanning electron micrograph obtained through the cross-section from the seed crystal to the sample through the buffer pellet. Cracks originating due to seed/sample lattice mismatch were found to be confined to the buffer pellet thereby reducing significantly the number of defects present in the bulk single grain.

A scanning electron micrograph obtained for the cross-section capturing the seed crystal, single grain, and the buffer pellet is shown in Fig. 2.3(c). This demonstrates clearly that macro-defects, such as substantive cracks, occurring due to the lattice mismatch stress are confined to the buffer pellet, which effectively protects the main single grain from mismatch effects. In summary, the buffer pellet acts as a gateway channel that allows the crystal texture to propagate but prevents both the seed crystal elements diffusing into the single grain or liquid phase infiltration into the seed to aid greatly the reliability of the single grain growth process. In (RE)BCO+Ag systems, where

Ag is known to deteriorate the seed crystal properties, the newly developed BA-TSIG approach has been particularly beneficial as it prevents Ag from reaching the seed crystal, and therefore improves the reliability of the single grain growth of both (RE)BCO and (RE) BCO+Ag systems [39, 40].

2.1.3 Generic Seed Crystals and NdBCO Film Seeds

Seeding is a critical step in the fabrication of a single grain (RE) BCO. In the complicated so-called hot-seeding approach [41], the parent composition of the target single grain can be used as seed crystal. In the cold-seeding technique, however, the seed has to be from a different system [and usually within the (RE)BCO family] with a higher melting temperature. In the melt process, it is important to subject the sample assembly to a high maximum processing temperature ($T_{\max}$) that is at least 20–30 °C above the peritectic temperature of the (RE)BCO material being fabricated, since this helps to enlarge the growth window and simultaneously eliminates impurities that can otherwise cause unwanted satellite nucleation(s) leading to the growth of multiple grains. A material needs to possess the following three important characteristics in order to work effectively as a seed: (i) A higher melting temperature (at least by 20–30 °C than the (RE)BCO material being grown; (ii) the same crystal structure with similar lattice parameters as the sample being grown and; (iii) phase stability with the melt that occurs at elevated processing temperature. A cleaved crystal of NdBCO, for example, works as an effective seed for the YBCO system in that it aids heterogeneous nucleation and enables single-grain growth. Until fairly recently, however, it was not possible to grow other (RE) BCO systems, such as GdBCO, SmBCO and NdBCO, in the form of large single grains due to the lack of availability of an appropriate seed crystal. In this context, Shi et al. [42, 43] developed a generic seed crystal based on Mg-doped NdBCO that can potentially and successfully seed any single grain in the (RE)BCO family of bulk superconductors. This material exhibits an increase in T_p by ~20 °C (bringing the T_p value to 1088 °C) when compared to undoped NdBCO (with T_p of 1068 °C) enabling its use as a generic seed crystal, which can be used successfully to seed a broad range of (RE)BCO single grains.

A thin NdBCO film deposited on MgO has also emerged fairly recently as an alternative potential seed crystal for the growth of (RE)BCO and (RE)BCO+Ag bulk materials. Superheating properties of NdBCO films result in an elevated melting temperature to the order of 1100 °C and have enabled improved processing reliability of (RE)BCO single grains [44, 45]. The thermal stability that the film seed can offer is usually associated with the orientation, microstructure, and quality of the film. The superheating properties of the film seeds were enhanced subsequently by the introduction of a YBCO buffer layer of ~25 nm in thickness to the NdBCO film seed configuration (of thickness ~500–600 nm). This integration of (RE)BCO materials enabled an increase in peritectic temperature by about 20 °C, yielding an increased tolerance of the NdBCO/YBCO/MgO film seed to processing temperatures of up to ~1120 °C in air. A combination of these advanced seeding methodologies has enabled potential pathways for the batch production of single-grain (RE)BCO superconductors, which is required for commercial applications.

2.2 Infiltration and Growth Process

Extensive research to tune and optimise the parameters in the TSMG process has enabled the successful production of single-grain (RE)BCO bulk superconductors that exhibit significantly improved superconducting properties. However, there are certain limitations to the TSMG process, such as: (i) sample shrinkage during processing; (ii) the formation of macro-cracks in the single grain; (iii) the formation of large pores/voids; (iv) the presence of large-sized RE-211 inclusions in the final microstructure and; (v) inhomogeneity in the distribution of RE-211 inclusions causing non-uniform superconducting properties throughout the sample volume. These limitations, which are associated directly with the processing technique itself and occur during heat treatment, are inherent to the process and are unavoidable. They affect both the superconducting and mechanical properties of the as-processed single-grain samples. As a result, an alternative fabrication approach based on the so-called infiltration and growth (IG) technique has emerged to overcome these limitations. A modified IG fabrication approach, which is a near-net shaping technique [46, 47], is discussed in this section.

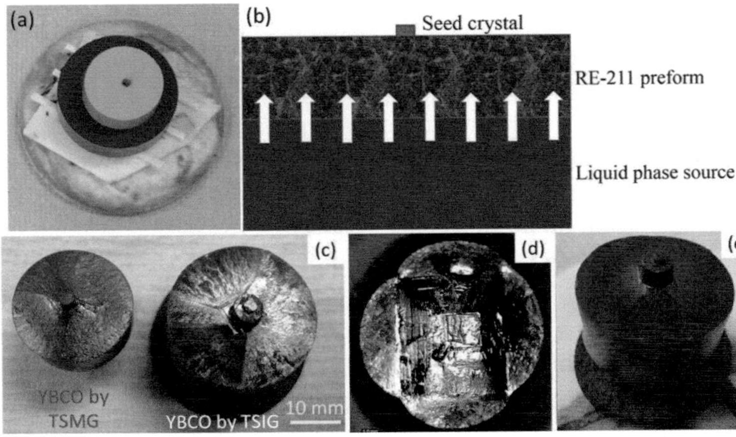

Figure 2.4 (a) Sample arrangement in TSIG processing. A RE-211 preform, capped with an NdBCO seed crystal, is placed in contact with the liquid phase reservoir pellet. The whole assembly is supported on yttria-stabilised zirconia rods and an alumina plate. (b) A schematic illustration showing the liquid phase infiltration into the RE-211 preform during the heat treatment step. (c) Shrinkage observed in YBCO samples fabricated via melt growth (~20%) and infiltration growth (~0.5%) processes. (d) Use of an unsintered RE-211 preform (with poor mechanical rigidity) causing steps in the as-grown sample along with small deformations created by growth fronts. (e) Sintering the RE-211 preforms at about 1050 °C prior to processing enabled YBCO bulk superconductors to retain their shape.

In the IG process, a RE-211 preform is placed in contact with a liquid-phase reservoir source pellet, as shown in Fig. 2.4(a), and subjected to either a top-seeded IG (TSIG) [48–50] or a bottom-seeded IG (BSIG) [51, 52] process. Seeding involves placing a seed crystal in contact with a RE-211 preform prior to heat treatment so that heterogeneous nucleation occurs as the sample assembly cools back though the peritectic temperature of the (RE)BCO material being grown. The liquid phase from the reservoir pellet infiltrates into the RE-211 preform at elevated temperature and subsequently reacts with it to form the RE-123 phase during the IG processing step, as illustrated schematically in Fig. 2.4(b). An excess of the RE-211 phase following the peritectic reaction is retained within the sample microstructure to constitute an effective source of flux pinning [53, 54]. One of the primary merits of the IG processing is near-net shaping, and this is illustrated in Fig. 2.4(c).

It can be seen in this figure that the IG-processed sample shows no shrinkage, whereas the MG-processed sample shrinks considerably by ~18–20% [55]. Although the IG process has been known for last two decades, it has not gained sufficient traction due primarily to the fact that: (i) The success rate for growing single-grains via this approach has been poor and; (ii) the superconducting properties of IG-processed materials tend to be significantly inferior compared to those associated with samples fabricated via conventional melt-growth.

Recently, the IG process has been studied in a greater detail and significant modifications have been made to improve the processing methodology considerably. An initial development was the appropriate choice of liquid phase reservoir component by Namburi et al. [56], who carried out a detailed study examining the potential of various liquid phase components including Y-123, $Ba_3Cu_5O_8$, a mixture of Y-123 and $Ba_3Cu_5O_8$, Yb-based liquid phase and Y-based liquid phase on the efficacy of IG processing and single-grain growth. This detailed study enabled a Yb-based liquid phase mixture [57] (comprising of $BaCuO_2$, CuO and Yb_2O_3) to be identified as the best composition for the liquid-phase reservoir pellet. The appropriate choice of liquid phase enabled effective infiltration and also supported single-grain growth in the subsequent reaction step, thereby eliminating unwanted sub-grain formation in the material during grain growth. Additionally, the fact that the IG processing enables near-net shaping makes it attractive from a processing perspective. Studies also revealed that sintering of the preforms prior to IG processing also aids retention of the shape of the preform and provides mechanical rigidity during the relatively aggressive infiltration stage of the processing step [56]. These aspects of the process are illustrated in Figs. 2.4(d) and (e). Although this improved version of IG processing enabled reliable production of dense (RE)BCO bulk superconductors, the properties exhibited by these samples were still found to be inferior compared to those obtained in bulk sample fabricated via the traditional TSMG approach.

IG processing enables the fabrication of (RE)BCO single grains with dense microstructures that contain fewer pores and voids compared to those obtained via melt processing. This is illustrated in Fig. 2.5(a), which shows the complete cross-section of YBCO single grains processed by TSMG and TSIG. Optical micrographs obtained under a magnification of 50× for these samples are also shown in Fig. 2.5(b) for purposes of comparison.

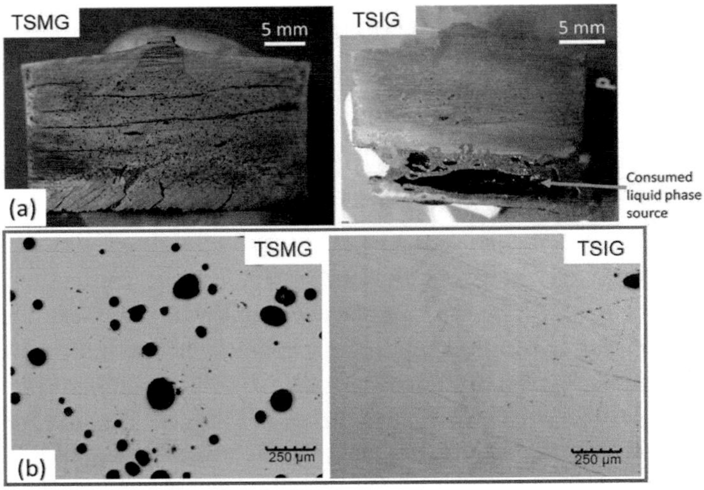

Figure 2.5 (a) Micrographs of the cross-sections of YBCO samples fabricated by TSMG and BA-TSIG techniques. (b) Microstructural images obtained for these samples at a magnification of 50×.

2.2.1 Development of 2-Step BA-TSIG Process

Recently, a detailed study [58] was carried out to understand the underlying reasons for the inferior performance of single grain samples fabricated by IG compared to those processed by TSMG. A typical microstructure observed for an IG-processed YBCO sample is shown in Fig. 2.6, which illustrates the presence of a uniform distribution of Y-211 inclusions in the Y-123 phase matrix. Image analysis carried out in several locations in the sample indicated that the conventional IG processing leads to a non-superconducting RE-211 phase content in the range 40–42%, which is too large for optimum supercurrent conduction (the optimum value is ~28–30% [59, 60]). Such a large concentration of non-superconducting phase clearly affects significantly the microscopic superconducting properties of the single grain sample.

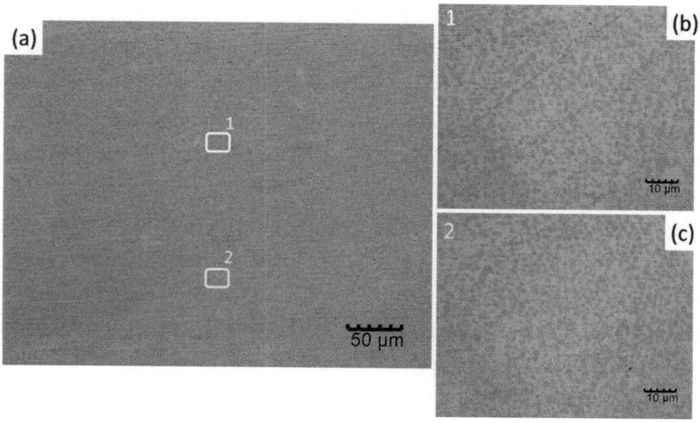

Figure 2.6 (a) Optical micrograph obtained under a magnification of 200× of a bulk YBCO sample processed by TSIG. A uniform distribution of Y-211 inclusions can be seen in the Y-123 phase matrix. Magnified images obtained under a magnification of 1000×, from the regions indicated as (b) '1' and (c) '2'.

Controlling and optimising the RE-211 content in the single grain microstructure is critical for improving the magnetic flux pinning strength and thereby achieving enhanced superconducting properties bulk, melt processed samples. It is relatively easy to control the amount of residual RE-211 in the single grain microstructure in the TSMG approach since this depends greatly on the initial quantity of RE-211 added to the precursor powder mixture. In IG processing, on the other hand, the tuning and optimisation of RE-211 content is not straightforward since there is no control on amount of liquid phase infiltrating the RE-211 preform and hence its subsequent reaction to form RE-123. Within this context, control of the RE-211 content in the IG process was studied recently via two different approaches: (i) Through tuning, adjusting and optimising the infiltration temperature and infiltration time and; (ii) by adding additional liquid phase component(s) to the RE-211 preform prior to melt processing [58]. The details of these two approaches are illustrated in Fig. 2.7. The critical parameters of infiltration temperature (T_i) and the infiltration time (t_i) were tuned and optimised in Approach-1. This involved melt processing nominally identical samples at three different infiltration temperatures (950 °C, 1050 °C, and 1150 °C) but with the same infiltration time (1 h). It can be seen that infiltration temperature of 950 °C was found to fail to provide sufficient kinetic activation to support infiltration, whereas the infiltration temperature of 1150 °C, on the other hand, resulted in the coarsening of Y-211 inclusions in the final sample microstructure. As a result, 1050 °C was selected as a close-to-optimum infiltration temperature. Optical micrographs obtained both under low (50×) and high (1000×) magnification in the optimised sample are shown in Fig. 2.7(a). In Approach-2, an attempt was made to consume the additional RE-211 phase during the reaction stage by adding extra (10 wt.% and 20 wt.%) liquid phase (LP) to the Y-211 preform at the powder preparation stage. Optical micrographs demonstrating a clear and systematic variation of Y-211 content in the Y-123 phase matrix are shown in Fig. 2.7(b). Although the RE-211 content could be adjusted in the second approach, the macrostructure was found to contain large-sized pores/voids due to the consumption of the liquid phase component in the preform.

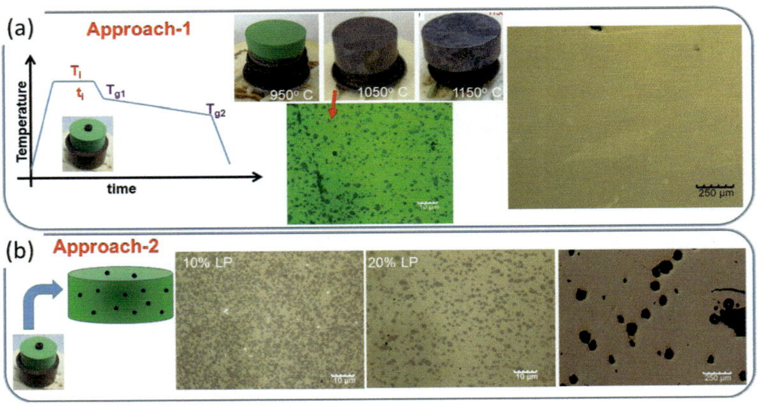

Figure 2.7 Two approaches (labelled as Approach-1 and Approach-2) used to control the RE-211 content in the final single grain microstructure in IG processing are shown in (a) and (b), respectively.

In fact, both Approaches-1 and -2 facilitated tuning and optimisation of the RE-211 content in the sample microstructure. However, the first approach enabled the formation of dense microstructures whilst achieving simultaneously control of the RE-211 content in the as-processed single grain, as desired. Approach-1 was further extended subsequently to incorporate an additional processing step to form the so-called 2-step BA TSIG technique [61] in which the infiltration and the growth steps were carried out separately and sequentially to improve significantly the success rate of fabrication of single grain (RE)BCO bulk superconductors. A schematic representation, indicating the critical steps of the 2-step BA-TSIG process is shown in Fig. 2.8(a).

An YBCO bulk superconductor of diameter ~32 mm fabricated by the novel 2-step buffer-assisted TSIG approach, as shown in Fig. 2.8(b), trapped a magnetic field of ~1 T at 77 K, with a measured value of 0.86 T at a height of 1.5 mm above the sample surface is shown in Fig. 2.8(c). The magnetic properties of the specimens extracted from the *c*-growth region of the YBCO sample, with locations shown schematically in Fig. 2.8(d), generally exhibited uniform superconducting properties. T_c of ~90 K with a transition width of ~ 1 K is observed typically in these samples, as shown in Fig. 2.8(e), which indicates their quality. The sample also exhibited a uniform field dependence of J_c with self-field J_c [$J_c(0)$] values exceeding 50–60 kA/cm^2 at 77 K, as shown in Fig. 2.8(f).

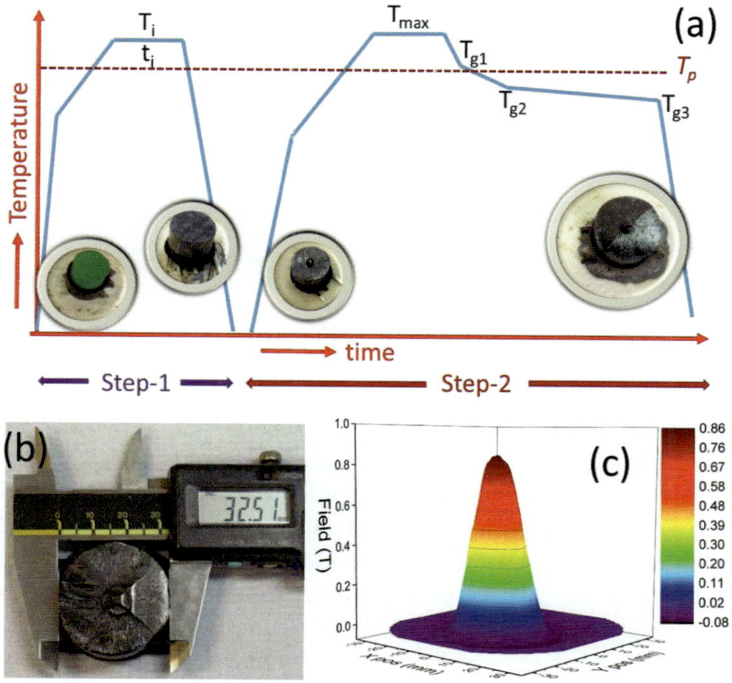

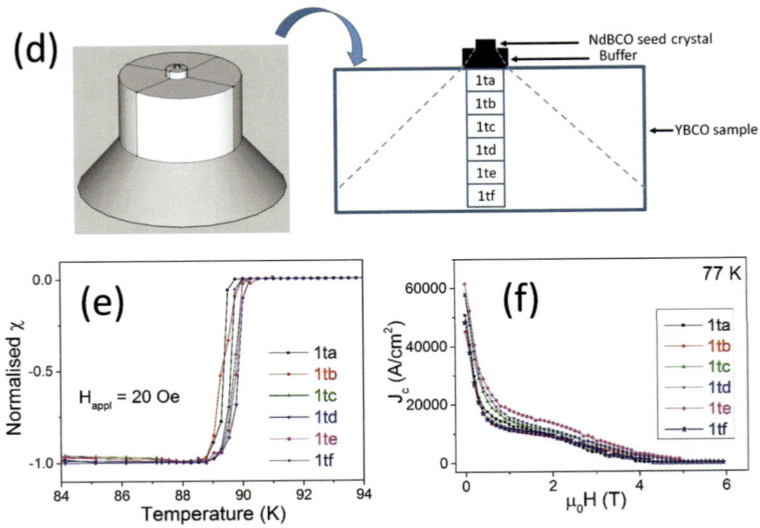

Figure 2.8 (a) A schematic illustration of the critical steps (step-1 and step-2 referring to 1. infiltration and 2. growth) associated with the novel 2-step buffer-aided top-seeded infiltration and growth (2-step BA-TSIG) technique. (b) YBCO single grain sample grown by this advanced fabrication approach and (c) the measured trapped field of the sample at 77 K. (d) A schematic illustration indicating the locations from where the specimens were extracted from the parent single grain (along *c*-growth region) for the measurements of magnetic properties. (e) The temperature dependence of normalised susceptibility (χ) of the specimens. (f) Field dependence of critical current density $J_c(H)$ at 77 K obtained in the specimens extracted along the *c*-growth region of the sample indicated in (d).

2.2.2 High-Field Studies of TSIG-Processed Samples

Two standard YBCO bulk superconductors (of diameter ~25 mm; sample codes #621 and #673) with engineered microstructures were fabricated via the 2-step BA-TSIG approach, as shown in Fig. 2.9(a). Each of these samples exhibited a trapped field of ~0.85 T at 77 K. These were machined and subsequently shrink-fitted with stainless steel rings, as shown in Fig. 2.9(b). Five Hall sensors were mounted on the surface of one of the samples. Both shrink-fitted samples were then arranged in a two-sample stack, as shown in Fig. 2.9(c), and the trapped field ability of the assembly measured in high applied field at low temperature. For this purpose, the sample stack was field cooled to 28 K in an applied magnetic field of 18 T. It is encouraging to observe that the sample stack trapped a field of 14.3 T [62] at 28 K at the centre of the stack, as shown in Fig. 2.9(d). The fact that a relatively simple, 2-sample YBCO stack (with no Ag or resin impregnation) exhibits such a large trapped field is an extraordinary advancement in terms of a practical fabrication process. The ability of dense (RE)BCO samples fabricated via TSIG to withstand large tensile forces without the addition of Ag is attributed to the reduced presence of intrinsic cracks/pores in the single grain microstructure, which is characteristic of this advanced processing technique.

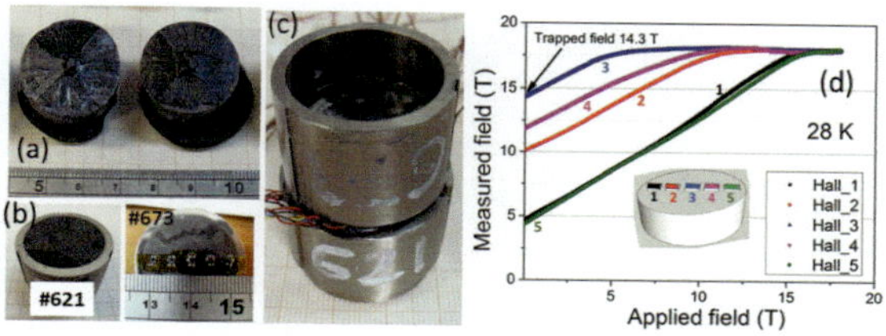

Figure 2.9 (a) Two standard YBCO samples (with sample codes #621 and #673) fabricated by the novel 2-step BA-TSIG process. (b) Machined samples reinforced with shrink-fit steel with five Hall sensors mounted on the top surface of one of the samples. (c) A two-sample stack with Hall sensors between the two reinforced samples. (d) Trapped field of ~14.3 T for the two-sample stack achieved at 28 K obtained after removing the applied magnetic field.

Trapped magnetic field has been measured in a variety of stacked sample arrangements, including: (a) 7–8 T in a Ag-free YBCO stack [63]; (b) 11.5 T in a YBCO+Ag sample stack encapsulated and reinforced by stainless steel tubes [64]; (c) 12.6 T in a YBCO stack encapsulated in epoxy, impregnated in resin and further reinforced by steel rings [65] and; (d) 13.3 T in a more complicated stacked configuration in which an irradiated sample of YBCO containing Zn and reinforced with epoxy filling and resin impregnation [66]. It is clear, therefore, that the recent measurement of a trapped field of 14.3 T in simple, Ag-free YBCO sample stack reinforced with shrink-fit steel represents significant field trapping performance [62]. It has been observed that enhancement in trapped field in IG-processed single grains can be attributed to the dense microstructure of the YBCO material achieved by the novel 2-step BA-TSIG approach [61], as shown in Fig. 2.10(a). This processing development has enabled a significant improvement in the mechanical properties of individual single grains, measured via the Brazilian technique [67–69], as shown in Fig. 2.10(b). The indirect tensile strength σ is determined from the relation: $\sigma = 2P/(\pi Dt)$, where P is the applied load on the sample when it fractures, and D and t are the diameter and thickness of the sample, respectively. The tensile strength measured for YBCO fabricated by a typical melt growth and infiltration growth process is shown in Fig. 2.10(b).

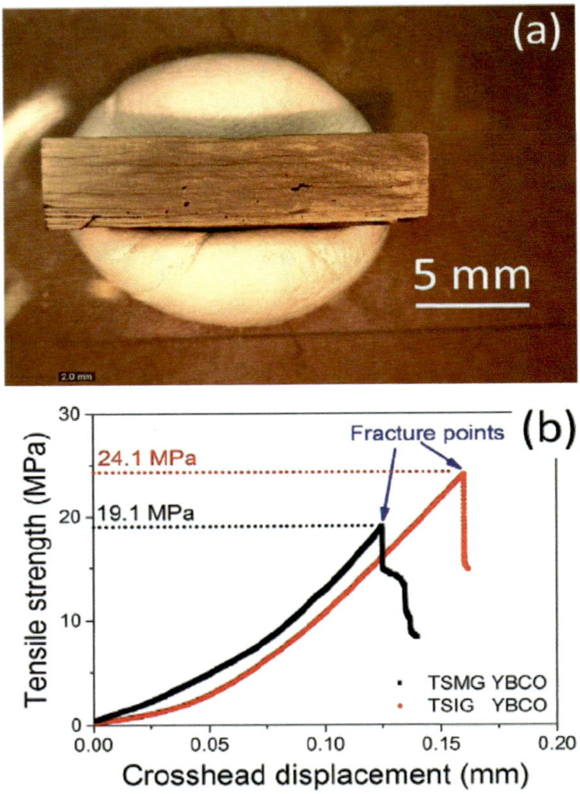

Figure 2.10 (a) Typical cross-section of a YBCO sample fabricated via the 2-step BA-TSIG technique. A dense microstructure with reduced porosity and cracking is evident from the optical micrograph. (b) A comparison of the tensile strength of YBCO bulk single grains fabricated by TSMG and TSIG measured by the Brazilian test.

2.2.3 Generic Seeds Fabricated by the TSIG Approach

Mg-doped NdBCO seed crystals with higher T_p, [42], as described in Section 2.1.3, are potentially suitable to seed all bulk (RE)BCO HTS materials. In practice, the success rate of producing single grain

(RE)BCO by using these seeds has turned out to be moderate. A selection of failed (RE)BCO and (RE)BCO+Ag samples in which the seeds failed either partially or completely in TSMG, TSIG or buffer-aided fabrication processes are shown in Fig. 2.11(a). A recent study has revealed that the presence of a large porosity within the seed crystal is the main reason for this poor yield [70]. This can be seen in Fig. 2.11(b), where liquid phase tracks were found in the seed crystal after processing. Since the IG process enables the fabrication of dense single grains, attempts were made to fabricate pore-free, Mg-doped generic seed crystals using this approach [70]. This yielded dense generic seed crystals, which addressed one of the major challenges associated with melt processing and enabled the reliable growth of single grains. A comparison of the microstructures of a conventional NdBCO seed and a generic seed fabricated by the TSMG and TSIG processes is made in Fig. 2.11(c), from which the reduced porosity in the seeds fabricated using the TSIG route is clear. The fact that robust and dense generic seeds are now available has provided renewed motivation to explore higher T_p systems. In this context, GdBCO bulk superconductors have been fabricated via the 2-step buffer-assisted TSIG processing and these are discussed in Section 2.3.

Figure 2.11 (a) Either partial or complete failure in single-grain growth as observed in several (RE)BCO and (RE)BCO+Ag systems. The reason for the failure was found to be associated with the porosity of the seed crystal. (b) Segregation of liquid phase in the seed crystal was found to be the primary cause of failure for heterogeneous nucleation and single grain growth. (c) Comparison of optical micrographs of a conventional NdBCO and generic seed as obtained via TSMG and IG processing.

2.3 High Performance Bulk Superconductors

GdBCO bulk superconductors are characterised by higher T_c and hence higher J_c and trapped field compared to YBCO. Relatively fewer attempts have been made to fabricate and measure the properties of GdBCO single grains via melt growth approaches [71–74], and an even smaller number that discuss the fabrication of GdBCO single grains via infiltration and growth based approaches [75–78]. Some of this research is summarised in Section 2.3.1, along with an outline of the motivation for the work presented in subsequent sections.

2.3.1 Existing Literature on GdBCO Bulk Superconductors

Detailed work on the growth-related studies of GdBCO fabricated via the melt growth approach (using Mg-doped generic seeds [42]) has been reported by Shi et al. [71]. This work discusses the spatial variation of T_c both along a/b and c-axis of the individual single grains. The effect of BaO_2 content both on the growth rate and T_c was reported as part of this investigation. It was observed further that the Gd_2BaCuO_5 (Gd-211) content in the sample varied between 10% and 35% by area fraction along a/b growth region for the vicinity of the seed crystal to a location towards the edge of the sample. Cardwell et al. [72] further studied the effect of Gd-211 addition (in different sample sets with 20–30 wt.% of Gd-211 added to Gd-123 precursor powders) on the J_c of GdBCO samples and found that the best performance was achieved in a sample containing 25 wt.% Gd-211 with current densities reaching 30–40 kA/cm^2 at 77 K. Attempts were also made to multi-seed and to recycle both GdBCO and GdBCO+Ag bulk superconductors by melt processing [73, 74].

A GdBCO bulk superconductor was grown under a reduced oxygen atmosphere to overcome the effects of Gd/Ba substitution by employing the infiltration growth route and adjusting the maximum melting temperature, which enabled the fabrication of a GdBCO bulk sample of diameter 20 mm that trapped a magnetic field of 0.32 T at 77 K [75]. Research by Li et al. [76] reported that a mixture of Gd-211 and $Ba_3Cu_5O_8$ can be used as a single component mixture for obtaining a GdBCO single grain by employing the TSIG approach, with the resulting sample exhibiting a levitation force of ~23 N at 77 K. Yuan and Yang [77] investigated the effect of sintering the preform pellet (comprising of Gd_2O_3 + $BaCuO_2$) in the temperature range 950–1200 °C, before subjecting it to TSIG, which subsequently improved the levitation force from 24 N to 38 N at 77 K, due partly to an improved microstructure made possible by the densification of the preform. More recently, Yang et al. [78] showed that a GdBCO bulk superconductor of diameter 20 mm trapped a field of 0.26 T at 77 K, with another of diameter 31 mm containing nano-inclusions trapping 1.1 T at 77 K [76]. Although the fabrication of GdBCO bulk single grains has been reported in literature for both TSMG and IG approaches, no single study has been performed where all the relevant properties, such as mechanical strength, crystal growth and correlations between the microstructures and the superconducting properties, are discussed in a coherent way. The present section describes the fabrication of GdBCO bulk superconductors by the recently developed, novel 2-step BA-TSIG process [61]. The resulting sample is assessed for its ability to trap magnetic field both at 77 K and 40 K. Furthermore, the levitation performance of the sample, its T_c and $J_c(H)$ both along the *a*-growth and *c*-growth regions of the sample, is discussed. The flexural strength was measured on bars extracted from the parent single grain and the results presented. A detailed microstructural study carried out on the sample showed a good correlation between the observed single grain microstructure and its measured superconducting and mechanical properties. Finally, the potential of the 2-step BA-TSIG process is investigated by attempting to batch processing of single grain GdBCO bulk samples.

2.3.2 GdBCO Bulk Superconductors Fabricated via 2-Step BA-TSIG

GdBCO single-grain bulk samples described in the present work were fabricated by employing the recently developed 2-step BA TSIG process [61]. Powders of Gd-211 (99.9% Toshima), $Ba_3Cu_5O_8$ (99.9% Toshima), Yb_2O_3 (American elements, 95%), BaO_2 (Sigma Aldrich, 95%), and CeO_2 (Sigma Aldrich, 99.9%) were employed in this study. BaO_2 was used to suppress Gd/Ba substitution in air [79, 80] and two main constituent pellets were used, the preform pellet and the liquid phase reservoir pellet, to fabricate samples by the infiltration and growth process. The preform pellet was made from a mixture of powders with the following composition: Gd-211 + 1 wt.% CeO_2 + 2 wt.% BaO_2. CeO_2 was used as a Gd-211 grain-refiner and BaO_2 was added to minimise the effect Gd/Ba solid solution formation. A Yb-based phase was used as the source of liquid phase during the infiltration and growth process. For this purpose, a mixture of Yb_2O_3 and $Ba_3Cu_5O_8$, referred as the Yb-based liquid phase (composed of Yb_2O_3: $Ba_3Cu_5O_8$ = 1:1.313, as discussed elsewhere [81]) was employed.

Powders for each of the configurations (preform and the liquid phase) were each mixed separately for about 3 h employing an auto-mixer equipped with a mortar and pestle. Both the preform and the liquid phase reservoir pellets were compacted separately and sequentially in steel dies using a uniaxial press. Details of the weight of powders used and the dimensions of the compacted pellets are given in Table 2.2. Finally, the liquid phase reservoir pellet was supported on an Yb_2O_3 substrate.

Table 2.2 Details of powders used for preparing pellets to fabricate the preform and the liquid phase reservoir pellets

	Set-A: 25 mm dia		Set-B: 20 mm dia	
	Diameter (mm)	**Weight of powder used (g)**	**Diameter (mm)**	**Weight of powder used (g)**
Gd-211 preform pellet	25	20	20	10
Yb-based liquid phase reservoir pellet	31	35	25	15
Yb_2O_3 substrate	31	3	25	1.5

In the first processing step, the sample assembly comprising of the Gd-211 preform pellet supported by the Yb-based liquid phase reservoir pellet was subjected to infiltration at 1050 °C for 1 h, as shown as step-1 of Fig. 2.12. During this step, the liquid phase (comprising mainly of BaCuO2 and CuO) is formed in the reservoir pellet and infiltrates into the Gd-211 preform, with which it is in contact. Once the infiltration process is complete, the sample assembly was quenched to room temperature and supported subsequently with a mini-sized buffer pellet [39, 40] (comprising of 75 wt.% Gd-123 + 25 wt.% Gd-211) capped with a generic seed crystal [70], as shown as step-2 of Fig. 2.12. This assembly was then subjected to a standard melt growth heat treatment involving slow cooling through the peritectic temperature, which enabled heterogeneous nucleation to occur at the position of the seed crystal and subsequent single grain growth into a fully melt processed sample, as can be seen in the final part of Fig. 2.12. The 2-step, BA-TSIG GdBCO sample fabrication process produced single-grain bulk samples in two different sizes: (i) 25 mm and (ii) 20 mm in diameter with near-net shaped geometries (the 25 mm diameter GdBCO sample is shown in Fig. 2.13). The characteristic four-fold facet lines from the position of the seed crystal to the edge of the sample, as can be seen clearly in Fig. 2.13, confirm that the sample is indeed a single grain. The second set of samples (20 mm in diameter) are discussed in the Section 2.6. All the samples were oxygenated at 400 °C, in flowing oxygen for 200 h following single grain growth to transform the crystal structure from the non-superconducting tetragonal to the superconducting orthorhombic phase, as is standard practice for (RE)BCO bulk materials.

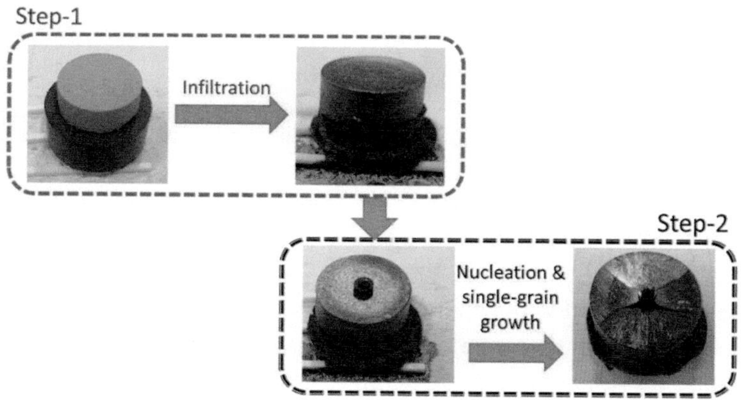

Figure 2.12 The fabrication of GdBCO samples via the 2-step buffer-aided top-seeded infiltration and growth (2-step BA-TSIG) process. The two distinct steps involved in the fabrication (infiltration and single grain growth) are shown in the figure.

Figure 2.13 A GdBCO sample fabricated via the 2-step buffer-aided top-seeded infiltration and growth (2-step BA-TSIG) process. Both the top and side views of the sample are shown in (a) and (b), respectively.

2.3.3 Trapped Field Performance

The GdBCO sample was field cooled in an applied magnetic field of 1.4 T to liquid nitrogen temperature. Once the sample reached thermal equilibrium, the applied field was withdrawn and the trapped magnetic field at 77 K measured at a height of 1.5 mm above the sample surface using a rotating array of Hall sensors. The 3-dimensional trapped field profile obtained at 1.5 mm above the sample surface is shown in Fig. 2.14. The single peak observed confirms that the sample is indeed a single grain with a field trapping ability in excess of 0.7 T at 77 K. The 2-dimensional profile obtained (showing a single set of concentric circular profile pattern) is shown in the inset of the figure.

The trapped field ability of the sample was measured further at lower temperature (40 K) and in a larger applied magnetic field of 8 T. For this purpose, the sample was cooled in an applied magnetic field of 8 T down to 40 K using a cryogen-free superconducting magnet (ICE Oxford, DRY-ICE-12T-LEMON model). Once the temperature was stable at 40 K, the applied field was removed and the magnetic field trapped in the sample measured using six Hall sensors placed on the surface of the sample (as shown schematically in the inset of Fig. 2.15). The single basic GdBCO sample processed by TSIG with no internal or external reinforcement (i.e. without Ag addition, steel shrink-fitting or resin impregnation) trapped a magnetic field of 5.6 T at 40 K, as can be seen from Fig. 2.15(a).

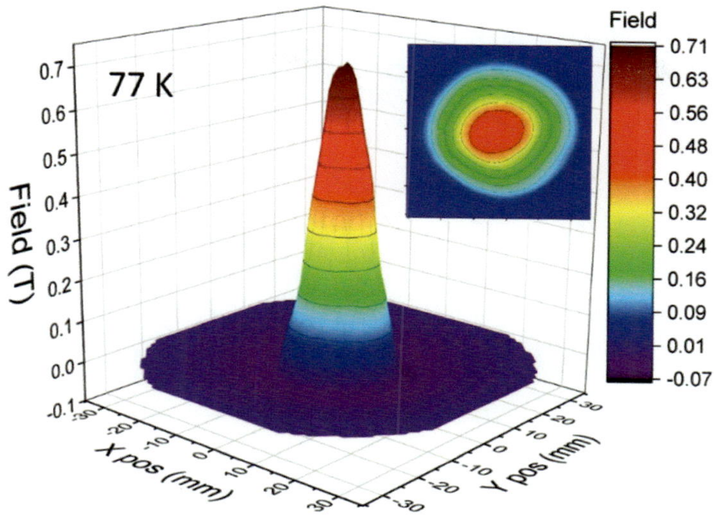

Figure 2.14 Trapped field measured at a distance of 1.5 mm above the surface of a 25 mm diameter GdBCO sample, at 77 K. The corresponding 2-dimensional field profile is shown in the inset of the figure.

It is encouraging to see that the sample successfully survived an applied field of 8 T without cracking. Usually the brittle nature of these ceramic-like materials forces them to fracture when exposed to large magnetic fields (typically above 7 T) [82] due to the generation of large Lorentz forces. The present sample, however, was able to resist these forces, despite its basic construction, providing an initial indication that GdBCO bulk samples fabricated by the 2-step, BA-TSIG technique exhibit improved mechanical strength. This is consistent with the earlier observation that TSIG fabrication methodology enables the fabrication of materials with dense microstructures (i.e. reduced porosity and cracks) and hence improved mechanical properties. It can be observed from Fig. 2.15 that the first Hall sensor was not functional and hence the corresponding data are not shown. The sample was stabilised at the magnetising temperature for about 15 min in order to allow flux creep to occur (i.e. the redistribution of magnetic flux), before being warmed up whilst its trapped field was monitored, as shown in Fig. 2.15 (b).

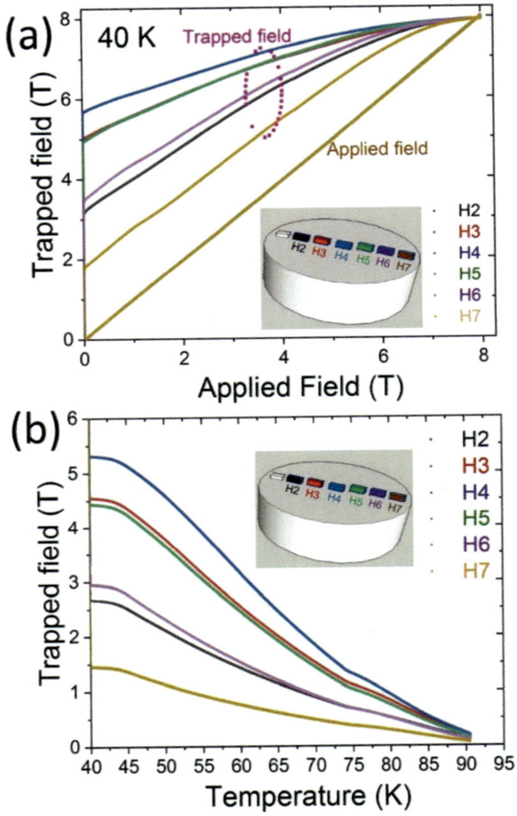

Figure 2.15 (a) Trapped field measured at the surface of a GdBCO sample of diameter 25 mm at 40 K. The sample was field cooled in an applied magnetic field of 8 T to 40 K. The position of the Hall sensors placed on the surface of the sample is shown schematically in the inset. (b) Temperature dependence of the remnant trapped field over the range 40–90 K.

2.3.4 Levitation Force Measurements

Levitation force was measured for the 25 mm diameter GdBCO sample at 77 K. For this purpose, a magnet stack was employed comprising three NdFeB permanent magnets of the same thickness as the sample being measured. The net field strength of the magnet assembly was about 0.5 T at the sample surface and this was brought into close proximity to the sample surface (which had been zero-field cooled to 77 K). The experimental arrangement of the measurement system is shown in Fig. 2.16(a). A stepper motor was used to control the position of the magnet stack and the distance between the magnet and the sample measured with an accuracy of microns. The force, referred as the 'levitation force', experienced by the magnet due to repulsion by the superconductor was measured using a load cell (Omega Engineering Inc., model LCM703). All measurement units were synchronised using a LabVIEW program to enable systematic measurements to be performed. Levitation force as a function of separation between the sample and the magnet for displacement to and from the sample is shown in Fig. 2.16(b). The 25 mm diameter, single-grain GdBCO sample exhibited a levitation force of 55 N at 77 K, which is promising when compared to the levitation force performance of other similar samples reported to date.

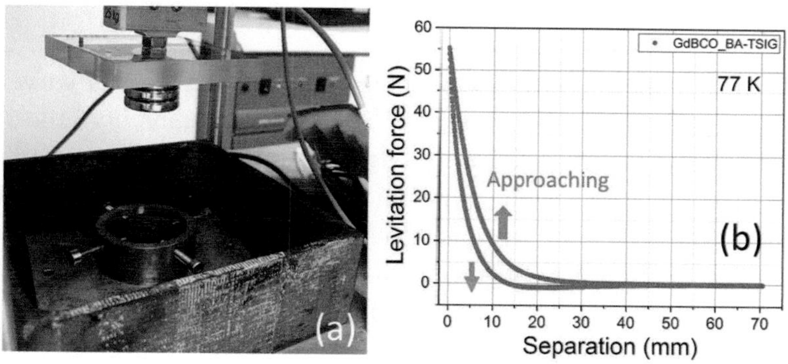

Figure 2.16 (a) Experimental arrangement showing a NdFeB magnet stack and superconductor aligned for measurement of levitation force. (b) Levitation force measured as a function of separation between the magnet and sample for both approaching and receding paths. The upper curve (indicated by the green arrow, online) indicates the levitation force when the magnet is approaching the superconductor and the lower curve (indicated by the blue arrow, online) when the magnet is receding.

2.3.5 Critical Temperature and Critical Current Density

A slice of the single grain sample was cut from the centre of the sample (i.e. along the *c*-growth region) was sliced into strips using a diamond saw. Specimens of typical size 2 mm × 1.5 mm × 1 mm ($a \times b \times c$)

were then cut from these strips to enable measurement of T_c and J_c as a function of sample position using the labelling scheme shown schematically in Fig. 2.17. A Superconducting Quantum Interference Device (SQUID) magnetometer was employed to obtain the volume magnetisation M as a function of both temperature and field.

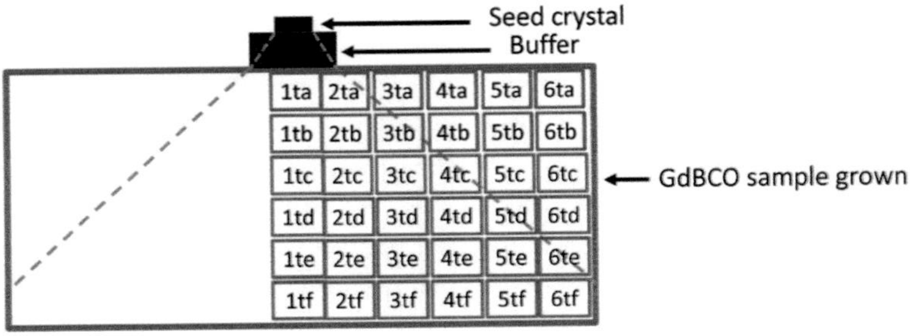

Figure 2.17 Schematic illustration of the locations from where the specimens were extracted from the GdBCO single grain sample to evaluate the various superconducting properties.

Magnetisation as a function of temperature (i.e. the 'M-T curve') was measured in applied field of 20 Oe to determine both the T_c and the width of superconducting transition ΔT_c for each of the specimens. It is well-known from previous studies that T_c of the bulk single grain in the vicinity of the seed crystal (position '1ta' in Fig. 2.17) is particularly sensitive to the diffusion of the seed crystal elements into the main sample [39, 40]. As a result, the initial measurements of T_c were performed on the specimen '1ta', as shown in Fig. 2.18(a).

It can be seen that this sample exhibits a superconducting transition with an onset, T_c^{onest}, of 92.8 K and a transition width of ΔT_c of 1 K, which is a clear indication that the specimen is free from any contamination by the seed crystal elements. This is a direct consequence of integrating the buffer technique into TSIG process. The magnetisation hysteresis (M-H) curves were obtained subsequently for this specimen at different temperatures in the range 85 K–30 K. The critical current densities were determined from the width of the hysteresis loops following extended Bean's critical state model [83], and the resulting $J_c(H)$ graphs for different temperatures are shown in Fig. 2.18(b).

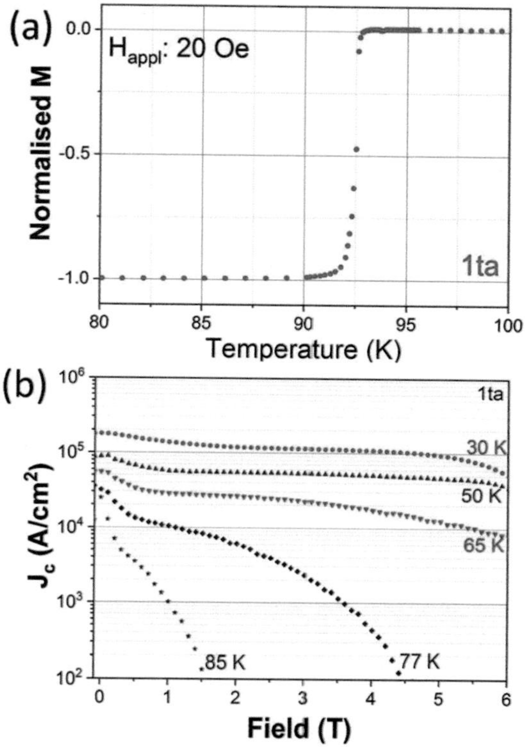

Figure 2.18 (a) Normalised magnetisation (*M*) as a function of temperature, showing the normal to superconducting phase transition. A sharp superconducting transition with a ΔT_c of 1 K observed in the specimen indicates that no Gd/Ba substitution is present in the sample. (b) Critical current density J_c as a function of applied magnetic field at different temperatures in the range 85 K to 30 K.

The field dependences of J_c at 77 K for different locations both along the *a*-growth and *c*-growth regions of the bulk single grain are shown in Fig. 2.19.

It can be seen from Fig. 2.19 that T_c and $J_c(H)$ are similar for all specimens and across in the *a*- and *c*-growth regions of the sample. The sample exhibits a uniform T_c of about 92.5 K with a transition width <1 K, indicating its relatively high quality. In addition, zero-field current densities $J_c(0)$ exceeding 20 KA/cm^2 and reaching in excess of 50 kA/cm^2, at 77 K are observed. These results are greatly encouraging and suggest that the 2-step BA-TSIG is effective for the fabrication of bulk samples that exhibit superior superconducting properties. An irreversibility field (H_{irr}) in excess of 4.5 T and reaching 6 T at 77 K is encouraging from the perspective of applications that require a large applied magnetic field. The Gd-211 content was found to be slightly higher at the edge/bottom region of the sample due to the growth-related features, in-line with particle pushing/trapping theory [84], as will be discussed further in the detailed microstructural study presented in Section 2.5. It can be observed further that $J_c(H)$ is more uniform along the a-growth region [Fig. 2.19(b)] in comparison to that obtained along the c-growth region [Fig. 2.19(d)]. The reason for this small variation in $J_c(H)$ can be attributed to the strong dependence on the fraction of Gd-211 present in the sample microstructure, as discussed in Section 2.5.

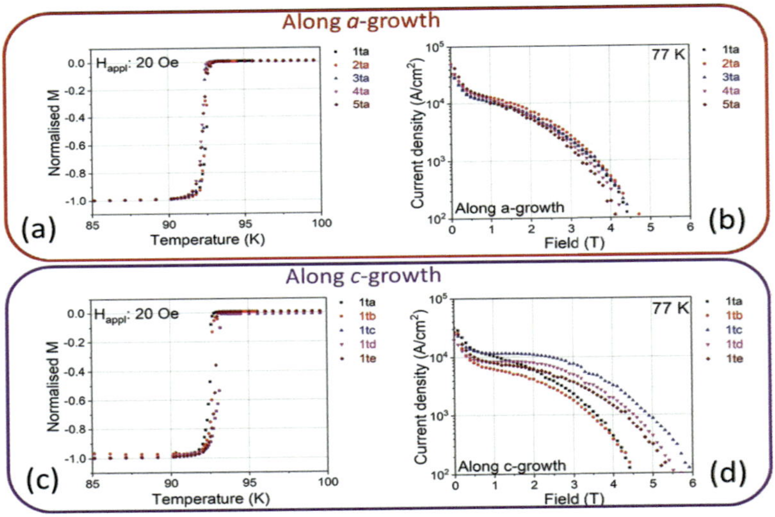

Figure 2.19 (a) Superconducting properties [T_c and $J_c(H)$] along the a- and c-growth regions of the GdBCO sample. Normalised magnetisation as a function of temperature for specimens extracted along the a- and c-growth regions are presented in (a) and (c), respectively. (b) and (d) J_c as a function of applied magnetic field, at 77 K for specimens both along a- and c-growth regions. The labels indicate the position from where the specimens were extracted for the measurements (see Fig. 2.17).

2.3.6 Flux Pinning Force

The flux pinning force 'F_p' ($\propto J_c{}^*B$) was determined as a function of applied field at different positions in the single grain. Irreversibility field H_{irr} (defined as the field at which the critical current density J_c decreases to a magnitude of 100 A/cm^2) was estimated from each of the individual $J_c(H)$ curves at 77 K and then re-plotted as a function of reduced field, defined as $h = H_{appl}/H_{irr}$. The variation of normalised flux pinning force ($F_p/F_{p,max}$) as a function of reduced field 'h' is shown in Fig. 2.20 for the various sub-specimens for both growth regions (a-growth and c-growth) of the sample. It can be seen that the mechanism for flux pinning is very similar for the entire sample studied in the present work.

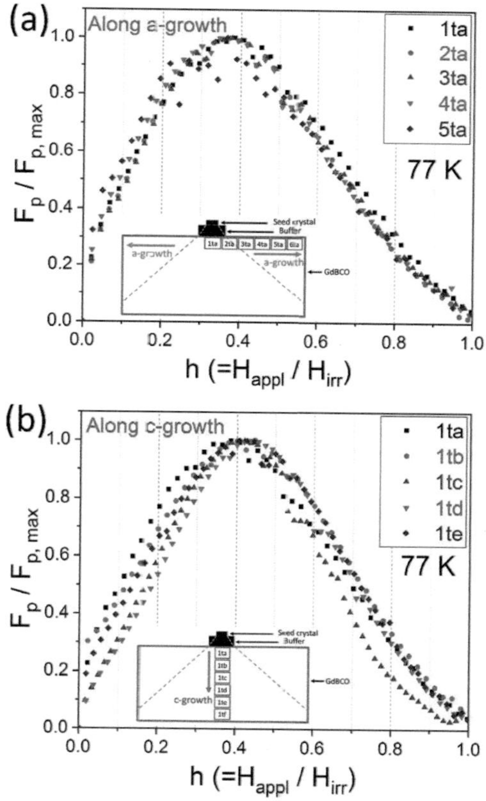

Figure 2.20 (a) and (b) Normalised flux pinning force ($F_p/F_{p,\,max}$) is shown as a function of normalised field h (= $\mathbf{H}_{appl}/\mathbf{H}_{irr}$) for both a-growth and c-growth regions of the sample. Growth regions are shown schematically in the insets of the respective figures. It can be seen that the positions of the peak are almost coincident indicating, that the mechanism for flux pinning is similar at each point in the sample.

2.4 Mechanical Property Measurement

The mechanical strength of the sample was estimated using the 3-point bend test method following the method reported by Huang et al. and others [69, 85]]. The sample was sliced through its centre (location '1t') using a diamond saw for the purpose of these measurements and four bars (labelled as 1A, 1B, 1C and 1D) of dimensions ~25 mm × 2 mm × 1.5 mm (length '*L*' × width '*w*' × thickness '*t*') extracted from the *c*-growth region, as shown both schematically and for the actual sample in Fig. 2.21. The flexural strength of each bar was measured for these samples following the 3-point bend test method. An Instron testing machine (model 5584) was employed and a load '*P*' applied to the test bar-shaped specimen with a feed rate of 0.015 mm/s until the sample fractured (i.e. either by cracking of breaking completely). The flexural strength was estimated by taking the magnitude of the load '*P*' into account and using the following relation:

$$\text{Flexural strength } \sigma \text{ (Pa)} = 3PL/(2wt^2) \qquad (2.1)$$

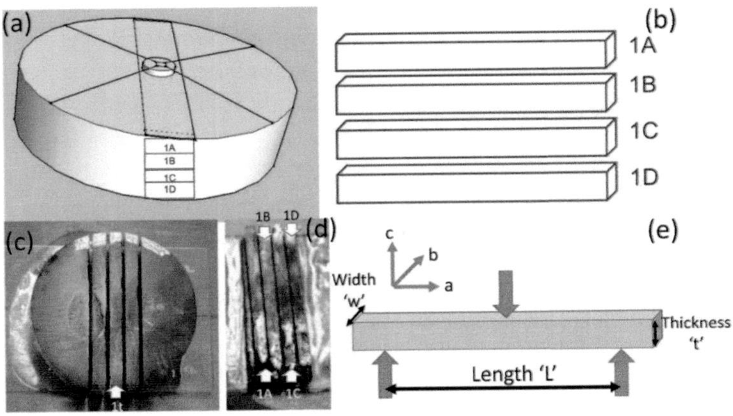

Figure 2.21 (a) and (b) Schematic illustration of the extraction of bars from the centre of the sample for mechanical strength measurement. (c) and (d) the sliced bars employed in actual experiments (e) Schematic illustration of the crystallographic orientation of each bar during application of the load during the 3-point bend test measurement.

A typical stress-strain curve represented by the calculated flexural strength as a function of extension measured for one of the bar samples (1B) at room temperature is shown in Fig. 2.22(a). It can be seen that the sample exhibited a flexural strength of 63.5 MPa above which it cracked prior to brittle failure. The magnitude of flexural strength is shown in Fig. 2.22(b) as a function of the position along the thickness of the sample, i.e. along the *c*-growth region of the sample (1A to 1D).

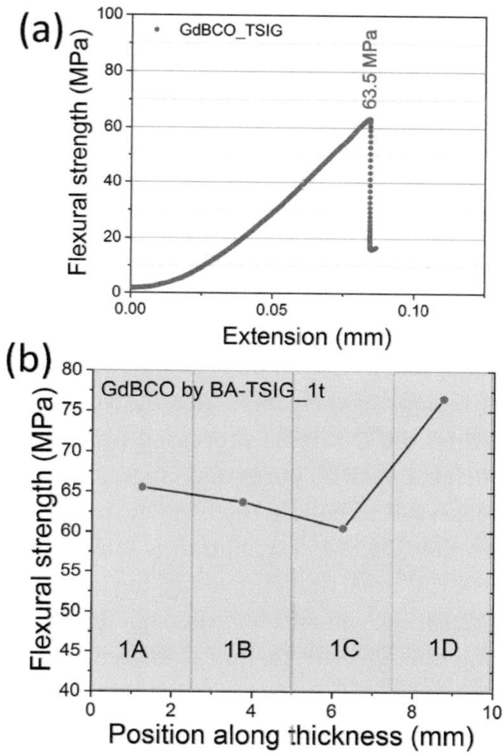

Figure 2.22 (a) Flexural strength as a function of extension for a GdBCO bar extracted from location 1B (see Fig 2.21). (b) The magnitude of the flexural strength as a function of position along the thickness of the sample (along the *c*-growth region).

The flexural strength of the sample varied in the range 60–75 MPa. Three bars '1A', '1B' and '1C' showed almost similar values of strength due to rather uniform macrostructure and reduced porosity associated with the TSIG fabrication methodology. However, the bottom-most bar, labelled '1D', exhibited improved strength of about 75 MPa. This may be attributed to (i) reduced porosity in the specimen microstructure and (ii) the presence of a relatively large fraction of Gd-211 inclusions in the matrix of Gd-123, as observed at location '1D' of the sample microstructure. Further optimisation of the preform fabrication method and the processing parameters, including the addition of silver (Ag), are anticipated to improve the mechanical properties further, which would be important for practical applications.

2.5 Microstructural Studies

An auto-polisher/grinder was employed to polish the surface of the sliced sample down to the 1 μm level using successive applications of SiC paper and diamond paste, using ethanol as a lubricant. The polished surfaces were observed using an optical microscope equipped with a polariser. The polarised optical micrographs obtained under a magnification of 1000×, both along the *a*-growth and *c*-growth regions of the sample are shown in Fig. 2.23. It can be seen that the GdBCO sample processed by TSIG is almost free from macro-defects, such as pores and cracks. The porosity in this sample was observed to be less than 1% by area fraction. Further, it can be seen that Gd-211 distribution is fairly uniform between 25% and 30% (by area fraction) along the *a*-growth region (as shown in Fig. 2.23(b)), whereas it varies significantly between 15% and 35% along the *c*-growth region, as shown in Fig. 2.23(c). The presence of RE-211 in the matrix of RE-123 is desirable to aid the current density performance through enhanced flux pinning [86]. The highly homogeneous microstructure characterised by a uniform Gd-211 distribution along the *a*-growth region explains the uniform superconducting properties observed along this growth region, as shown in Fig. 2.19, Fig. 2.20, and Fig. 2.23. The variation in Gd-211 distribution observed along the *c*-growth region correlates well to the small variation observed both in terms of the superconducting and mechanical properties, as illustrated in Fig. 2.19(d) and Fig. 2.22(b).

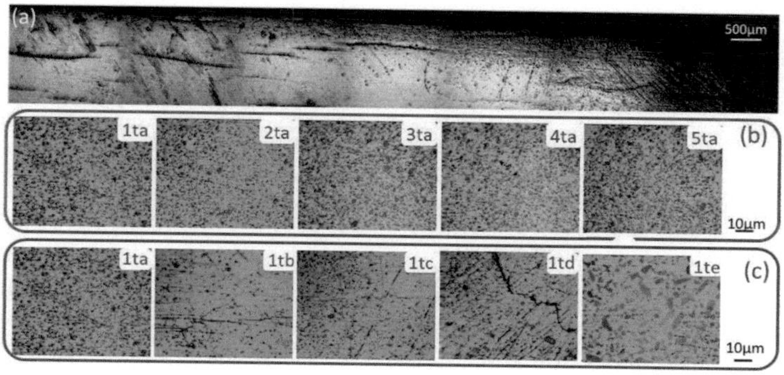

Figure 2.23 Microstructural study of the GdBCO sample fabricated by TSIG. (a) Low magnification optical micrographs obtained from the sample along the *a*-growth region (b) and (c) polarised optical micrographs obtained from the sample under higher magnification of 1000×, both along the *a*-growth and *c*-growth regions.

2.6 Reliability of Fabrication

In order to confirm if sample production by the 2-step-BA-TSIG process is reliable, a set of five GdBCO samples (20 mm in diameter) was batch processed into single grains in a box furnace. The batch process incorporated small improvements in sample fabrication and heat treatment conditions identified in the development of the single sample process. As-grown single-grain GdBCO samples after processing are shown in Fig. 2.24(a).

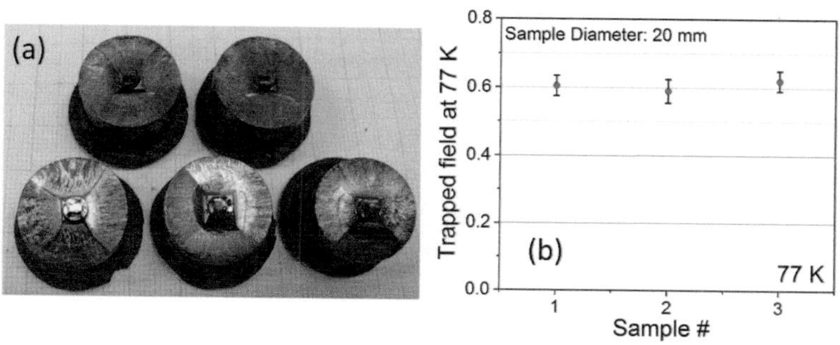

Figure 2.24 (a) A batch of five GdBCO samples of 20 mm in diameter, fabricated via the 2-step BA-TSIG process. The potential of generic seeds and the incorporation of a buffer enable the successful seeding and hence batch processing of GdBCO single grain bulk samples. (b) Trapped field measured at 77 K in three of the batch-processed samples.

It is encouraging to see that the generic seed crystals employed with the buffer technology are reliable in successfully seeding (RE) BCO single grains, even in the complex fabrication methodology such as infiltration and growth where the excess liquid phase can easily compromise the integrity of the seed crystal. The trapped field obtained via field-cooled magnetisation was measured at 77 K in three of these samples, as shown in Fig. 2.24(b). It is encouraging to see that the trapped field is ~0.6 T at 77 K for all the measured samples (each of 20 mm in diameter).

2.7 Novel Experiments Investigated

2.7.1 (RE)BCO Bulk Superconductors with Artificial Holes

YBCO bulk superconductors containing artificial holes have been investigated recently in order to expedite the oxygenation process. This approach has been found to be particularly advantageous for the IG processing methodology, where oxygen permeation through the dense microstructure is a challenge. Additionally, this configuration was found to aid more rapid cooling of the (RE)BCO bulk sample by immersion in liquid cryogen due to the increased effective surface area of the single grain. YBCO single grain bulk samples containing perforations of ~1 mm in diameter were fabricated by TSMG [87, 88] and TSIG [89] processes. Significantly, the presence of perforations in the precursor preform pellet does not inhibit the propagation of the growth front during peritectic solidification. In some studies, these perforations were back-filled with BiPbSnCd alloy post heat-treatment to improve the mechanical properties of the bulk sample, which also simultaneously improved the thermal conductivity of the composite material. The BiPbSnCd alloy typically distributes homogeneously within the holes of the sample. It is also possible to fill these artificial holes with certain metals and tough fibres, also to improve the mechanical strength of these brittle ceramic-like materials primarily to facilitate their use in high-field engineering applications.

2.7.2 YBCO Cavities for Magnetic Shielding

A fundamental signature of superconductivity is the expulsion of magnetic flux from the interior of the sample on cooling through the superconducting transition temperature, which is known as the Meissner effect. As a result, superconductors are often described as perfect diamagnets, with magnetic susceptibility, $\chi = M/H = -1$, as long as the applied field is maintained below a certain critical value. Electromagnetic shielding is an important phenomenon to safeguard both sensitive devices and the environment [90]. Conventionally, metals, and mu-metals, in particular, have been applied traditionally as magnetic shields. When the source field is >1 kHz, metals can attenuate the electromagnetic field due to the so-called skin effect [91]. Conductors arranged in the form of Faraday cage are able to shield electrical signals at lower frequencies, although they fail to shield magnetic fields. In this context, high-permeability ferromagnetic materials are potential candidates for shielding magnetic field in the low frequency domain (i.e., < 1 kHz) [92]. Superconducting HTS cavities, however, have emerged as the best option for providing frequency independent shielding for applications that operate at cryogenic temperatures, such as high-field magnets. Superconducting cavities can be effectively and efficiently used as shielding devices for two different scenarios: (i) To protect field-sensitive sensors/measurements such as SQUID magnetometers from stray electromagnetic noise and; (ii) to protect the outside environment from typically large remnant magnetic fields, such as those produced by high-field magnets.

Two characteristic parameters that determine the quality of superconducting shields are the 'limiting field B_{lim}' (the maximum applied field at which there is no induced field inside the superconducting cavity) and the 'shielding factor *SF*' (defined as the ratio of applied magnetic field B_{appl} to the induced field B_{in}, i.e. $SF = B_{appl}/B_{in}$) [93]. The higher the values of B_{lim} and *SF*, the better the performance of the superconducting cavity as a magnetic shield. For example, a YBCO single grain of tube geometry with dimensions ~30 mm in height, ~15 mm in inner diameter, and ~4.5 mm wall thickness, fabricated employing the buffer-aided top-seeded melt-growth process, successfully shielded a magnetic field of 1 T at 20 K [94]. A photograph of the YBCO tube with one end sealed along with its shielding performance at different temperatures in the range 77 K–20 K are shown in Fig. 2.25(a). It is challenging to produce longer tubes of single-grain (RE)BCO bulk material, however, and efforts have therefore been made to produce thick films of (RE)BCO on Ag or Ni tubes via electrophoretic deposition [95, 96] or solution spray techniques [97, 98]. A YBCO thick film of thickness ~100 μm and length 80 mm deposited on Ag tube via electrophoretic deposition technique is shown in Fig. 2.25(b) [99]. The macro-structure, which shows the fusion of grains, X-ray diffractogram, and the shielding performance of the thick film cavity in the temperature range 77 K–20 K are shown in Figs 2.25(c)–(e).

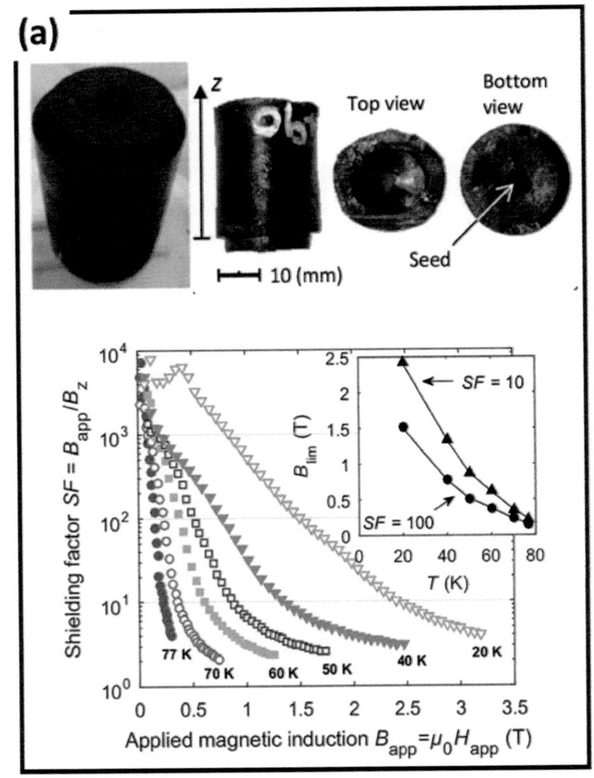

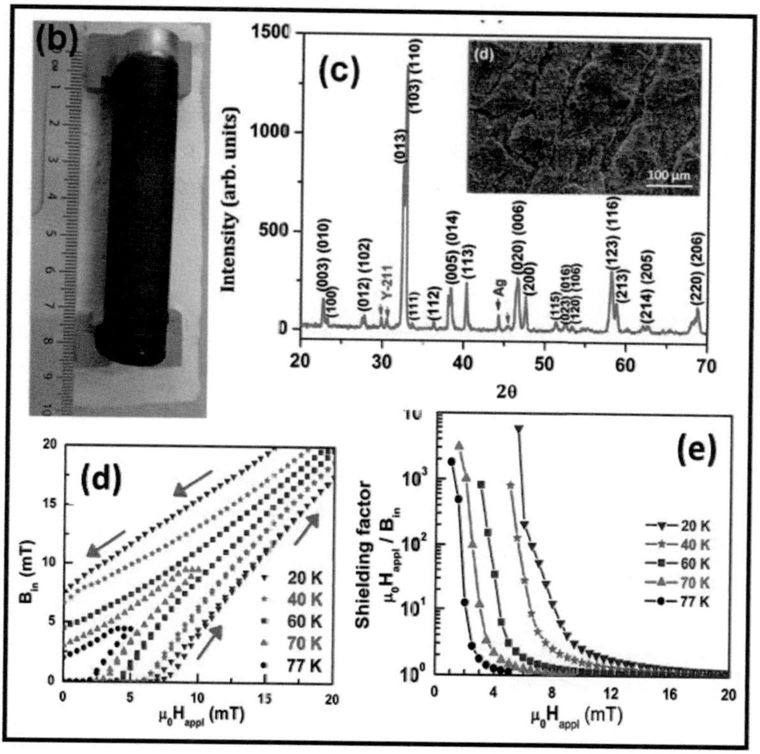

Figure 2.25 (a) YBCO cavity (both as-pressed and the heat-treated single grain) produced via buffer-assisted top-seeded melt growth process. The shielding factor as a function of applied magnetic field in the temperature range 77 K–20 K are also shown. (b) YBCO deposited as a thick film (~100 µm) on an Ag tube via electrophoretic deposit on technique. (c) X-ray diffractogram obtained on melt processed YBCO thick film and macrostructure. (d) and (e) Limiting field B_{lim} and shielding factor *SF* as a function of applied magnetic field of the YBCO thick film cavity.

2.7.3 Multi-Seeding Experiments

The growth of bulk (RE)BCO in the form of a single grain is fundamental to obtaining larger current loops in the material and hence the generation of a large magnetic field. However, there is a practical limitation to the size of the single grain that can be successfully produced in single-grain samples. Even with today's extraordinary advancements in this area of research, it is safe to state that achieving single grains in the size range of 25–30 mm in diameter is practical and that obtaining samples up to 60 mm diameter is challenging but achievable if the processing parameters are carefully tuned, well-optimised, and detailed attention is paid both to the processing step and heat treatments involved. However, a single grain of diameter greater than 60 mm is a real challenge to process successfully and the success rate for this size range could be <10% [100, 101]. Even if a single grain of diameter 60–80 mm is achieved, the field trapping properties of the sample are not particularly good given that trapped field does not generally scale with the diameter of the single grain in this size range [102]. Further, microstructural defects such as extensive porosity, cracks, misorientations, and large-angle grain boundaries determine predominantly the macroscopic superconducting properties of these large diameter single grains [73, 103].

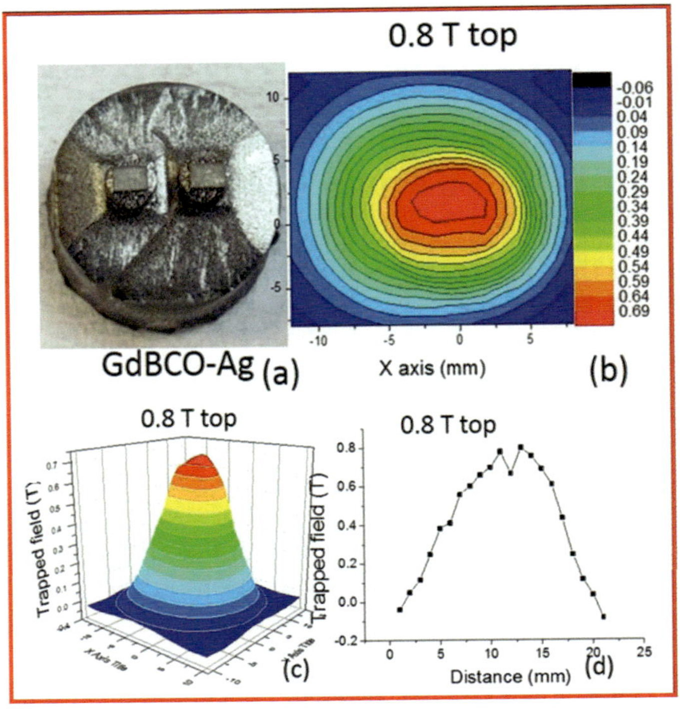

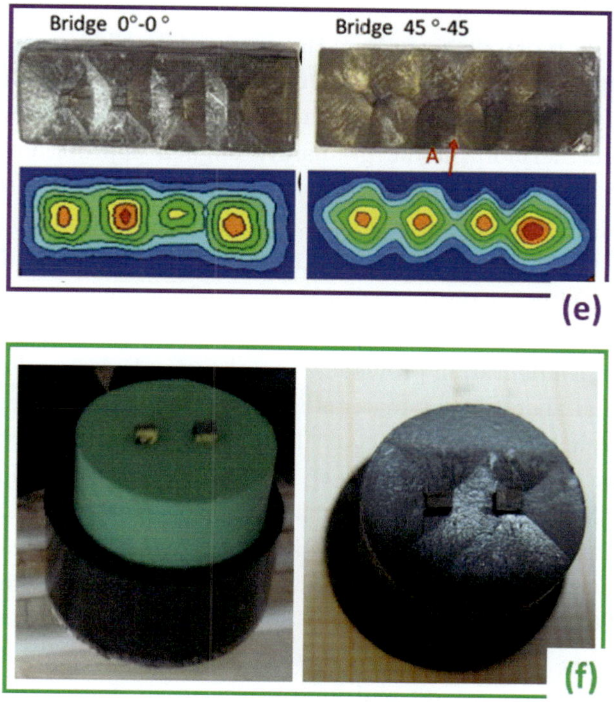

Figure 2.26 (a–d) Appropriate alignment of two-seeds using the buffer technique was found to yield a quasi-single grain with enhanced superconducting properties. A photograph of the top view of the sample and the trapped field measured at 77 K are shown. (e) Multi-seeded samples fabricated from seeds oriented in a 0°–0° and 45°–45° configuration in bar-shaped (RE)BCO sample and the associated 2-dimensional trapped profiles. (f) A similar 0°–0° configuration attempted in IG processing.

The seed acts as a heterogeneous nucleation centre in the seeding process and the optimised heat treatment enables the RE-123 growth front to propagate and transform the entire bulk precursor into a single grain. The smaller growth rate of the RE-123 phase, which is of the order of 0.2–0.3 mm/h [104], has motivated researchers to explore and develop a multi-seeding approach [105, 106], where more than one seed crystal is employed to obtain a quasi-single grain. Through this approach, it is possible to reduce the growth times significantly and simultaneously to obtain high performance, large-sized (RE)BCO quasi-single grains [107]. Additionally, the misorientation problems that are known to increase in severity with increasing radial distance from the seed can be addressed directly by employing a multi-seeding approach. The number of grains in the parent sample formed is determined by the number of seed crystals used. Furthermore, if the grains are aligned appropriately after growth, then a quasi-single grain with good performance in terms of trapped field and levitation force can be achieved. In this context, Shi et al. [108] investigated two multi-seeding approaches, with seeds arranged either in the 0°–0° or 45°–45° configurations placed on the upper surfaces of compacted preform pellets. It was observed that 45°–45° configuration provided a better fusion between the grains and simultaneously eliminated RE-211 particle accumulation at the inter-grain interface. Deng et al. [109] and Shi et al. [110] observed that the multi-seeded (RE)BCO samples exhibit a characteristic higher levitation force in comparison to single-seeded samples. These results illustrate the relative importance of multi-seeding, and particularly in the context of Maglev-based applications [111]. (RE)BCO bulk superconductors both in cylindrical and cuboidal geometries have been fabricated by a multi-seeding technique employing melt-growth and infiltration-growth based fabrication approaches [112]. More recent efforts have been made to grow a quasi-single grains via multi-seeding from just one seed crystal but by employing different buffer pellets designed and arranged appropriately on the parent precursor sample [73]. This technique has emerged as a novel approach for eliminating variations in misalignment between the pre-determined orientation of the seed crystals employed in the multi-seeding process. A selection of results from multi-seeding using both TSMG and TSIG-based approaches are shown in Fig. 2.26.

2.7.4 Reinforcement Studies

Single-grain (RE)BCO superconductors with engineered microstructures have the potential to trap large magnetic fields. In reality, however, these materials undergo brittle failure at applied fields in excess of 8–10 T, due to their ceramic nature and associated poor mechanical properties. For example, when a (RE)BCO superconductor is exposed to an applied magnetic field of 15 T, the magnetic pressure that the sample experiences is ~100 MPa, which is typically 4–5 times larger than its tensile strength. The failure mechanism is accentuated further by the unavoidable presence of macro- and micro-cracks in the single grain microstructure. Very often, majority of these cracks do appear during the melt processing and oxygenation steps. Cracks and pores present in the single grain microstructure further promote cracking in the material, thereby compromising the integrity of the (RE)BCO samples in practical applications. It has been observed that addition of Ag to the (RE)BCO microstructure either partially or completely fuses the macro- and micro-cracks and hence improves the mechanical strength of the bulk sample, at least moderately. Further reinforcement of the sample is necessary, however, if these materials are to be employed routinely and reliably in high-field engineering applications. In this context, metallic and SiC fibres have recently been introduced successfully into the single grain microstructure through a modification of the melt process and employing both buffer-assisted TSMG and TSIG fabrication approaches [113].

The tensile strength of the YBCO bulk superconductors reinforced with W-cored SiC fibres produced an improvement by ~40% whilst simultaneously retaining their superconducting properties. Aspects of these developments, which are shown in Fig. 2.27, are particularly attractive from the perspective of developing practical applications.

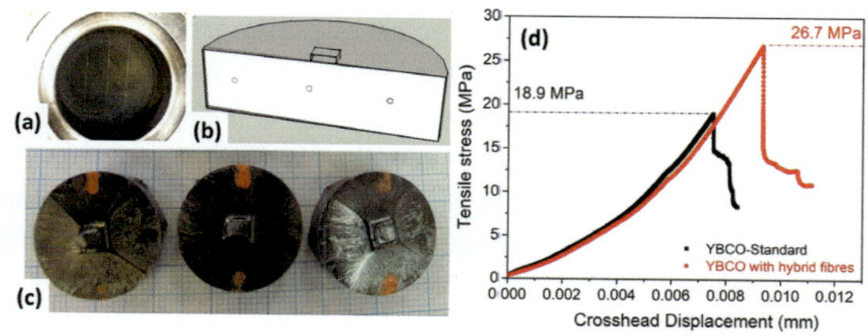

Figure 2.27 (a) and (b) SiC fibres were introduced into preform YBCO compacts. (c) Single grain YBCO samples containing SiC fibres produced via the buffer-assisted TSMG process. (d) Tensile stress as a function of crosshead displacement as measured employing Brazilian technique in both a standard YBCO sample and the sample reinforced with SiC fibres. An enhancement of tensile strength by ~40% has been achieved.

2.8 Summary and Conclusions

When fabricated in the form of a single grain, the (RE)BCO material can support large current loops and hence can trap large magnetic fields. The current state of the art in the processing of these materials along with recent advancements made to the TSMG and the TSIG fabrication methodologies have been presented. In particular, the integration of a buffer to both these processes has improved considerably the reliability of the single grain growth process, with a success rate exceeding 90%. In addition, the buffer was found to overcome problems of seed crystal element diffusion into the main (RE)BCO single grain, thereby improving the uniformity of superconducting properties across the volume of the sample. Recent developments in the TSIG approach, which have enabled the challenging task of controlling the RE-211 content in the superconducting matrix of RE-123 to be addressed successfully, have been summarised. The potential of the recently developed novel, 2-step BA TSIG process in producing dense, near-net shaped single grain (RE)BCO bulk superconductors has been described. YBCO single grains of diameter ~25 mm fabricated via the 2-step BA-TSIG process exhibited superior superconducting and mechanical properties than those grown by TSMG. YBCO single grains fabricated by TSIG generated a trapped field of ~1 T at 77 K on the surface of an individual sample and ~14.3 T at 28 K in a two-sample stack configuration. The mechanical properties of TSIG-processed YBCO samples were found to be higher by at least 25% when compared the conventional melt-grown samples. These improved mechanical and superconducting properties are particularly attractive from an applications perspective. The development of Mg-doped 'generic' seeds both via TSMG and, more recently, via TSIG approaches have opened pathways for the fabrication of enhanced-performance (LRE)BCO and Ag-added (RE)BCO bulk superconductors for high-field engineering devices.

GdBCO single-grain bulk superconductors have been fabricated successfully and reliably via a recently developed 2-step buffer-aided top-seeded infiltration and growth process. A GdBCO sample of diameter 25 mm trapped a magnetic field of 0.82 T at 77 K and 5.8 T at 40 K at its surface. Furthermore, it is encouraging to observe that an as-processed GdBCO sample (i.e. without reinforcement, such as Ag-alloying, resin impregnation or steel shrink fitting) did not crack or break when exposed to an applied magnetic field as high as 8 T. This is due partly to the dense microstructure and improved mechanical properties of these single grain ceramic-like materials due, in turn, directly to an improved fabrication methodology. This sample exhibited a levitation force of greater than 55 N at 77 K and a zero-field current density $J_c(0)$ in excess of 40 kA/cm^2 (and as high as 50 kA/cm^2). Furthermore, the material exhibited almost uniform superconducting, microstructural and mechanical properties, with a flexural strength measured by a 3-point bend test at room temperature of greater than 60 MPa.

Novel experiments carried out recently on bulk superconducting materials fabricated by the TSIG-based process have been summarised. Bulk samples processed with artificial holes, cavities for magnetic shielding applications, long bar/cylindrical and disc-shaped samples fabricated via multi-seeding and, finally, reinforced (RE)BCO bulk single grains with metal and SiC-fibres have enhanced significantly the mechanical properties of these ceramic-like materials. With these advancements, robust, single grain (RE)BCO bulk superconductors are being developed rapidly towards high-field technological applications that include portable MRI systems, high-torque rotating machines, magnetic separation, levitation vehicles (Maglev), energy storage flywheels, and non-contact mixers.

Acknowledgements

The Henry Royce Institute (Equipment grant ref. EP/P024947/1) is acknowledged for financial support that enabled the procurement of a 12 T superconducting magnet used for some of the measurements reported in the present work.

Acronyms/Symbols

Acronyms/symbols in the order of their appearance in the chapter:

T_c:	Critical temperature
H_c:	Critical magnetic field
J_c:	Critical current density
ω_c:	Critical frequency
LTS:	Low temperature superconductors

HTS: High temperature superconductors
2Δ: Superconducting energy gap
H: Planck's constant
BCS theory: Bardeen Cooper Schrieffer theory
(RE)BCO: (RE)-Ba-Cu-O
B_t: Trapped field
d: Diameter of the single grain
MRI: Magnetic resonance imaging
T_p: Peritectic temperature
MG: Melt growth
QMG: Quench melt growth
MPMG: Melt powder melt growth
IG: Infiltration and growth
TSMG: Top-seeded melt growth
RE-123: $(RE)Ba_2Cu_3O_{7-x}$
RE-211: $(RE)_2BaCuO_5$
BA-TSMG: Buffer-assisted top-seeded melt growth
T_{max}: Maximum processing temperature
TSIG: Top-seeded infiltration and growth
BSIG: Bottom-seeded infiltration and growth
T_i: Infiltration temperature
t_i: Infiltration time
LP: Liquid phase
SQUID: Superconducting quantum interference device
M: Magnetisation
ΔT_c: Width of superconducting transition
M–H curves: Magnetisation hysteresis curves
H_{irr}: Irreversibility fields
$J_c(0)$: Zero-field current density
F_p: Flux pinning force
h: Reduced field (defined as H_{appl}/H_{irr})
$(F_p/F_{p,max})$: Normalised flux pinning force
χ: Magnetic susceptibility
B_{lim}: Limiting field or threshold field
SF: Shielding factor
μ: Magnetic permeability
B_{appl}: Applied magnetic field
B_{in}: Induced magnetic field

References

1. Onnes, H. K. (1911). The resistance of pure mercury at liquid helium temperatures, *Leiden Commun.*, **120b**, 1226.
2. Onnes, H. K. (1913). The sudden disappearance of the ordinary resistance of tin, and the super-conductive state of lead, *Commun. Phys. Lab. Univ. Leiden*, **d**, 133.
3. Meissner, W., and Ochsenfeld, R. (1933). Ein neuer effekt bei eintritt der supraleitfähigkeit, *Naturwissenschaften*, **21**, 787–788.
4. Kittel, C. (2005). *Introduction to Solid State Physics*, Chapter-10: Superconductivity 8th Ed. (John Wiles & Sons Inc., USA), ISBN: 978-0-471-41526-8, WIE ISBN 978-0-471-68057-4.
5. Bardeen, J., Cooper, L. N., and Schrieffer, J. R. (1957). Microscopic theory of superconductivity, *Phys. Rev.*, **106**, 162–164.
6. Cardwell, D. A. (1998). Processing and properties of large grain (RE) BCO, *Mater. Sci. Eng. B*, **53**, 1–10.
7. Murakami, M. (1992). Melt Processed High Temperature Superconductors (World Scientific Publishing Co. Pte. Ltd., Singapore).
8. Bray J. W. (2009). Superconductors in applications: some practical aspects, *IEEE Trans. Appl. Supercond.*, **19**, 2533–2539.
9. Ainslie, M. D., and Fujishiro, H. (2015). Modelling of bulk superconductor magnetization, *Supercond. Sci. Technol.*, **28**, 053002.
10. Ren, Y., Weinstein, R., Liu, J., Sawh, R. P., and Foster, C. (1995). Damage caused by magnetic pressure at high trapped field in quasi-permanent magnets composed of melt-textured Y-Ba-Cu-O superconductor, *Physica C*, **251**, 15–26.
11. Ikuta, H., Mase, A., Yanagi, Y., Yoshikawa, M., Itoh, Y., Oka, T., and Mizutani, U. (1998). Melt-processed Sm-Ba-Cu-O superconductors trapping strong magnetic field, *Supercond. Sci. Technol.*, **11**, 1345–1347.
12. Tomita, M., and Murakami, M. (2003). High-temperature superconductor bulk magnets that can trap magnetic fields of over 17 tesla at 29 K, *Nature*, **421**, 517–520.
13. Durrell, J. H., Dennis, A. R., Jaroszynski, J., Ainslie, M. D., Palmer, K. G. B., Shi, Y. H., Campbell, A.M., Hull, J., Strasik, M., Hellstrom, E. E., and Cardwell, D. A. (2014). A trapped field of 17.6 T in melt-processed, bulk Gd-Ba-Cu-O reinforced with shrink-fit steel, *Supercond. Sci. Technol.*, **27**, 082001.
14. Huang, K-Y., Shi, Y-H., Srpčič, J., Ainslie, M. D., Namburi, D. K., Dennis, A. R., Zhou, D., Boll, M., Filipenko, M., Jaroszynski, J., Hellstrom, E.,

Cardwell, D. A., and Durrell, J. H. (2020). Composite stacks for reliable >17 T trapped fields in bulk superconductor magnets, *Supercond. Sci. Technol.*, **33**, 02LT01.

15. Murakami, M. (2007). Processing and applications of bulk RE–Ba–Cu–O superconductors, *Int. J. Appl. Ceram. Technol.*, **4**, 225–241.
16. Claassen, J. H., Osofsky, M. S., Namburi, D. K., Vanderbemden, Ph. (2014). Thin-film superconducting shields, *IEEE Trans. Appl. Supercond.*, **25**, 7500204.
17. Durrell, J. H., Ainslie, M. D., Zhou, D., Vanderbemden, Ph., Bradshaw, T., Speller, S., Filipenko, M., and Cardwell, D. A. (2018). Bulk superconductors: a roadmap to applications, *Supercond. Sci. Technol.*, **31**, 103501.
18. Ogawa, N., Yoshida, M., Hirabayashi, I., and Tanaka, S. (1992). Preparation of YBCO bulk superconductor by platinum doped melt growth method, *Supercond. Sci. Technol.*, **5**, S89-S92.
19. Cloots, R., Koutzarova, T., Mathieu, J-P., and Ausloos, M. (2005). From RE-211 to RE-123. How to control the final microstructure of superconducting single-domains, *Supercond. Sci. Technol.*, **18**, R9-R23.
20. Noudem, J. G., Meslin, S., Horvath, D., Harnois, C., Chateigner, D., Ouladdiaf, B., Eve, S., Gomina, M., Chaud, X., and Murakami, M. (2007). Infiltration and top seed growth-textured YBCO bulks with multiple holes, *J. Am. Ceram. Soc.*, **90**, 2784–2790.
21. Devendra Kumar, N., Rajasekharan, T., Gundakaram, R. C., and Seshubai, V. (2011). Extensive nanotwinning: origin of high current density to high fields in preform-optimized infiltration-growth-processed $YBa_2Cu_3O_{7-\delta}$ superconductor, *IEEE Trans. Appl. Supercond.*, **21**, 3612–3620.
22. Devendra Kumar, N., Rajasekharan, T., and Seshubai, V. (2013). Preform optimization in infiltration growth process: an efficient method to improve the superconducting properties of $YBa_2Cu_3O_{7-\delta}$, *Physica C*, **495**, 55–65.
23. Muralidhar, M., Jirsa, M., Nariki, S., and Murakami, M. (2001). Influence of combined Pt and CeO_2 additions on microstructure and magnetic properties in (Nd, Eu, Gd)–Ba–Cu–O, *Supercond. Sci. Technol.*, **14**, 832–838.
24. Salama, K., and Lee, D. F. (1994). Progress in melt texturing of $YBa_2Cu_3O_x$ superconductor, *Supercond. Sci. Technol.*, **7**, 177–193.
25. Cardwell, D. A., and Hari Babu, N. (2006). Processing and properties of single grain (RE)–Ba–Cu–O bulk superconductors, *Physica C*, **445–448**, 1–7.

26. Devendra Kumar, N., Shi, Y., and Cardwell, D. A. (2016). *Superconductivity: Applications Today and Tomorrow*. Chapter-1: Fabrication of bulk (RE)BCO superconductors by the Infiltration and Growth process: Past, Present and Future, p 1–35, Ed. Muralidhar, M. (Nova Science Publishers, New York), ISBN: 978-1-63483-756-9.

27. Kambara, M., Umeda, T., Tagami, M., Yao, X., Goodilin, E. A., and Shiohara, Y. (1998). Construction of the quasi-ternary phase diagram in the $NdO_{1.5}$–BaO–CuO_x system in an air atmosphere, Part I: 'Equilibrium tie lines in the $Nd_{1+x}Ba_{2-x}Cu_3O_{6+d}$ solid solution and liquid region, *J. Am. Ceram. Soc.*, **81**, 2116–2124.

28. Jin, S., Tiefel, T. H., Sherwood, R. C., Davis, M. E., van Dover, R. B., Kammlott, G. B., Fastnacht, R. A., and Keith, H. D. (1998). High critical currents in Y-Ba-Cu-O superconductors, *Appl. Phys. Lett.*, **52**, 2074–2076.

29. Endo, A., Chauhan, H. S., Nakamura, Y., and Shiohara, Y. (1996). Relationship between growth rate and undercooling in Pt-added $Y_1Ba_2Cu_3O_{7-x}$, *J. Mater. Res.*, **11**, 1114–1119.

30. Jin, S., Tiefel, T. H., Sherwood, R. C., van Dover, R. B., Davis, M. E., Kammlott, G. W., and Fastnacht, R. A. (1989). Melt-textured growth of polycrystalline $YBa_2Cu_3O_{7-\delta}$ with high transport J_c at 77 K, *Phys. Rev. B*, **37**, 7850–7853.

31. Fujimoto, H., Murakami, M., Gotoh, S., Koshizuka, N., and Tanaka, S. (1990). Melt processing of YBaCuO oxide superconductors, *Adv. Supercond.*, **2**, 285–288.

32. Muralidhar, M., Koishikawa, S., Koblischka, M. R., and Murakami, M. (1999). Study of superconducting properties of OCMG processed (Nd, Eu, Gd)–Ba–Cu–O with Pr doping, *Physica C*, **314**, 277–284.

33. Viswanath, N. V. N., Rajasekharan, T., Harish Kumar, N., Menon, L., and Malik, S. K. (1998). Infiltration-growth processing of $SmBa_2Cu_3O_y$ superconductor, *Supercond. Sci. Technol.*, **11**, 420–425.

34. Hari Babu, N., Iida, K., and Cardwell, D. A. (2006). Enhanced magnetic flux pinning in nano-composite Y-Ba-Cu-O Superconductors, *Physica C*, **445–448**, 353–356.

35. Muralidhar, M., Jirsa, M., Sakai, N., and Murakami, M. (2003). Progress in melt-processed (Nd-Sm-Gd)$Ba_2Cu_3O_y$ superconductors, *Supercond. Sci. Technol.*, **16**, R1–R16.

36. Hari Babu, N., Iida, K., Shi, Y., Withnell, T. D., and Cardwell, D. A. (2006). $YBa_2Cu_3O_{7-\delta}/Y_2Ba_4CuMO_y$ single grain nano-composite superconductors with high critical current densities, *Supercond. Sci. Technol.*, **19**, S461-S465.

37. Li, T. Y., Cheng, L., Yan, S. B., Sun, L. J., Yao, X., Yoshida, Y., and Ikuta, H. (2010). Growth and superconductivity of REBCO bulk processed by a seed/buffer layer/precursor construction, *Supercond. Sci. Technol.*, **23**, 125002.

38. Zhou, D., Xu, K., Hara, S., Li, B., Deng, Z., Tsuzuki, K., and Izumi, M. (2012). MgO buffer-layer-induced texture growth of RE–Ba–Cu–O bulk, *Supercond. Sci. Technol.*, **25**, 025022.

39. Shi, Y-H, Namburi, D. K., Zhao, W., Durrell, J. H., Dennis, A. R., and Cardwell, D. A. (2016). The use of buffer pellets to pseudo hot seed (RE)–Ba–Cu–O–(Ag) single grain bulk superconductors, *Supercond. Sci. Technol.*, **29**, 015010.

40. Devendra Kumar, N., Shi, Y., Zhai, W., Dennis, A. R., Durrell, H. H., and Cardwell, D. A. (2015). Buffer pellets for high-yield, top-seeded melt growth of large grain Y-Ba-Cu-O Superconductors, *Cryst. Growth Des.*, **15**, 1472–1480.

41. Chauhan, H. S., and Murakami, M. (2000). Hot seeding for the growth of c-axis-oriented Nd-Ba-Cu-O, *Supercond. Sci. Technol.*, **13**, 672–675.

42. Shi, Y-H., Hari Babu, N., and Cardwell, D. A. (2005). Development of a generic seed crystal for the fabrication of large grain (RE)–Ba–Cu–O bulk superconductors, *Supercond. Sci. Technol.*, **18**, L13-L16.

43. Hari Babu, N., Shi, Y-H., Iida, K., and Cardwell, D. A. (2005). A practical route for the fabrication of large single-crystal (RE)–Ba–Cu–O superconductors, *Nature Mater.*, **4**, 476–480.

44. Tang, C., Yao, X., Hu, J., Rao, Q., Li, Y., and Tao, B. (2005). YBCO melt-textured growth seeded by superheating YBCO/MgO thin film, *Supercond. Sci. Technol.*, **18**, L31-L34.

45. Volochová, D., Kavečanský, V., Antal, V., Diko, P., and Yao, X. (2016). Thermal stability of NdBCO/YBCO/MgO thin film seeds, *Supercond. Sci. Technol.*, **29**, 044004.

46. Chen, Y. L., Chan, H. M., Harmer, M. P., Todt, V. R., Sengupta, S., and Shi, D. (1994). A new method for net-shape forming of large, single-domain $YBa_2Cu_3O_{6+x}$, *Physica C*, **234**, 232–236.

47. Sudhakar Reddy, E., and Rajasekharan, T. (1998). Fabrication of textured $REBa_2Cu_3O_{7-\delta}/RE_2BaCuO_5$ (RE = Y, Gd) composites by infiltration and growth of RE2BaCuO5 preforms by liquid phases, *Supercond. Sci. Technol.*, **11**, 523–534.

48. Kim, C. J., Jee, Y. A., Lee, K. W., Sung, T. H., Han, S. C., Kuk, I. H., Hong, G. W. (1998). Low temperature melt process of $SmBa_2Cu_3O_{7-y}$ using a liquid infiltration technique, *Appl. Supercond.*, **6**, 149–156.

49. Devendra Kumar, N., Rajasekharan, T., Muraleedharan, K., Banerjee, A., and Seshubai, V. (2010). Unprecedented current density to high fields in $YBa_2Cu_3O_{7-\delta}$ superconductor through nano-defects generated by preform optimization in infiltration growth process, *Supercond. Sci. Technol.*, **23**, 105020.

50. Li, G-Z., Li, D-J., Deng, X-Y., Deng, J-H., and Yang, W-M. (2013). Infiltration growth and crystallization characterization of single-grain Y–Ba–Cu–O bulk superconductors, *Cryst. Growth. Des.*, **13**, 1246–1251.

51. Yang, P., Yang, W., Zhang, L., Chen, L., Wang, M. (2019). Bottom-seeded infiltration and growth for fabrication of single-grain GdBCO superconducting ring, *IEEE Trans. Appl. Supercond.*, **29**, 6802705.

52. Devendra Kumar, N., Rajasekharan, T., and Seshubai, V. (2011). YBCO/Ag composites through a preform optimized infiltration and growth process yield high current densities, *Supercond. Sci. Technol.*, **24**, 085005.

53. Matsushita, T. (2000). Flux pinning in superconducting 123 materials, *Supercond. Sci. Technol.*, **13**, 730–737.

54. Murakami, M., Gotoh, S., Fujimoto, H., Yamaguchi, K., Koshizuka, N., and Tanaka, S. (1991). Flux pinning and critical currents in melt processed YBaCuO superconductors, *Supercond. Sci. Technol.*, **4**, S43-S50.

55. Hari Babu, N., Kambara, M., Smith, P. J., and Cardwell, D. A. (2000). Fabrication of Large Single-grain Y–Ba–Cu–O Through Infiltration and Seeded Growth Processing, *J. Mater. Res.*, **15**, 1235–1238.

56. Namburi, D. K., Shi, Y., Palmer, K. G., Dennis, A. R., Durrell, J. H., and Cardwell, D. A. (2016). An improved top seeded infiltration growth method for the fabrication of Y–Ba–Cu–O bulk superconductors, *J. European Ceram. Soc.*, **36**, 615–624.

57. Li, G. Z., and Yang, W. M. (2011). Improvements of liquid source for infiltration and growth of $GdBa_2Cu_3O_{7-x}$ superconducting single grains, *Mat. Chem. Phys.*, **129**, 288–291.

58. Namburi, D. K., Shi, Y., Palmer, K. G., Dennis, A. R., Durrell, J. H., and Cardwell, D. A. (2016). Control of Y-211 content in bulk YBCO superconductors fabricated by a buffer-aided, top seeded infiltration and growth melt process, *Supercond. Sci. Technol.*, **29**, 034007.

59. Murakami, M., Fujimoto, H., Gotoh, S., Yamaguchi, K., Koshizuka, N., and Tanaka, S. (1991). Flux pinning due to nonsuperconducting particles in melt processed YBaCuO superconductors, *Physica C: Superconductivity*, **185–189**, 321–326.

60. Shiohara, Y., and Endo, A. (1997). Crystal growth of bulk high-T_c superconducting oxide materials, *Mater. Sci. Eng.*, **19**, 1–86.

61. Namburi, D. K., Shi, Y., Palmer, K. G., Dennis, A. R., Durrell, J. H., and Cardwell, D. A. (2016). A novel, two-step top seeded infiltration and growth process for the fabrication of single grain, bulk (RE)BCO superconductors, *Supercond. Sci. Technol.*, **29**, 095010.

62. Namburi, D. K., Durrell, J. H., Jaroszynski, J., Shi, Y-H., Ainslie, M. D., Huang, K., Dennis, A. R., Hellstrom, E., and Cardwell, D. A. (2018). A trapped field of 14.3 T in Y–Ba–Cu–O bulk superconductors fabricated by buffer-assisted seeded infiltration and growth, *Supercond. Sci. Technol.*, **31**, 125004.

63. Fuchs, G., Krabbes, G., Schätzle, P., Gruβ, S., Stoye, P., Staiger, T., Müller, K-H., Fink, J., and Schultz, L. (1997). High trapped fields in bulk $YBa_2Cu_3O_{7-\delta}$ / $YBa_2Cu_3O_{7-\delta}$ samples at temperatures around 50 K, *Appl. Phys. Lett.*, **70**, 117–119.

64. Gruss, S., Fuchs, G., Krabbes, G., Verges, P., Stöver, G., Müller, K-H., Fink, J., and Schultz, L. (2001). Superconducting bulk magnets: very high trapped fields and cracking, *Appl. Phys. Lett.*, **79**, 3131–3133.

65. Fuchs, G., Schätzle, P., Krabbes, G., Gruß, S., Verges, P., Müller, K-H., Fink, J., and Schultz, L. (2000). Trapped magnetic fields larger than 14 T in bulk $YBa_2Cu_3O_{7-x}$, *Appl. Phys. Lett.*, **76**, 2107–2109.

66. Krabbes, G., Fuchs, G., Canders, W-R., May, H., and Palka, R. (2006). *High Temperature Superconductor Bulk Materials: Fundamentals, Processing, Properties, Control and Application Aspects* (Wiley-VCH Verlag GmbH& Co., Weinheim) pp 179–89.

67. Fairhurst, C. (1964). On the validity of the 'Brazilian' test for brittle materials, *Int. J. Rock Mech. Mining Sci. Geomech.*, **1**, 535–546.

68. Andreev, G. E. (1991). A review of the Brazilian test for rock tensile strength determination, *Min. Sci. Technol.*, **13**, 445–456.

69. Huang, K-Y., Zhou, D., Srpčič, J., Shi, Y-H., Namburi, D. K., Dennis, A. R., Ainslie, M. D., Boll, M., Bause, R., Filipenko, M., Durrell, J. H., and Cardwell, D. A. (2018). Spatial distribution of flexural strength in Y–Ba–Cu–O bulk superconductors, *IEEE Trans. Appl. Supercond.*, **28**, 6801505.

70. Namburi, D. K., Shi, Y-H., Dennis, A. R., Durrell, J. H., and Cardwell, D. A. (2018). A robust seeding technique for the growth of single grain (RE) BCO and (RE)BCO–Ag bulk superconductors, *Supercond. Sci. Technol.*, **31**, 044003.

71. Shi, Y., Hari Babu, N., Iida, K., and Cardwell, D. A. (2007). Growth rate and superconducting properties of Gd-Ba-Cu-O bulk superconductors melt processed in air, *IEEE Trans. Appl. Supercond.*, **17**, 2984–2987.

72. Cardwell, D. A., Shi, Y-H., Hari Babu, N., Pathak, S. K., Dennis, A. R., and Iida, K. (2010). Top seeded melt growth of Gd-Ba-Cu-O single grain superconductors, *Supercond. Sci. Technol.*, **23**, 034008.

73. Shi, Y., Durrell, J. H., Dennis, A. R., Huang, K., Namburi, D. K., Zhou, D., and Cardwell, D. A. (2017). Multiple seeding for the growth of bulk GdBCO–Ag superconductors with single grain behaviour, *Supercond. Sci. Technol.*, **30**, 015003.

74. Shi, Y., Namburi, D. K., Wang, M., Durrell, J. H., Dennis, A. R., and Cardwell, D. A. (2015). A reliable method for recycling (RE)-Ba-Cu-O (RE: Sm, Gd, Y) bulk superconductors, *J. Amer. Ceram. Soc.*, **98**, 2760–2766.

75. Iida, K., Babu, N. H., Shi, Y-H., and Cardwell, D. A. (2006). Seeded infiltration and growth of single-domain Gd-Ba-Cu-O bulk superconductors using a generic seed crystal, *Supercond. Sci. Technol.*, **19**, S478-S485.

76. Li, G-Z., Yang, W-M., Cheng, X-F., Fan, J., and Guo, X-D. (2010). Fabrication of single-grain $GdBa_2Cu_3O_{7-x}$ bulk superconductors with a new kind of liquid source by the top seeded infiltration and growth technique, *Pramana J. Phys.*, **74**, 827–832.

77. Yuan, X., and Yang, W. (2018). Effects of Gd211 particles characteristic in the precursor pellets on the levitation force of single-domain GdBCO bulks prepared by the Gd+011 TSIG method, *Adv. Mater. Sci. Eng.*, **2018**, 5481625.

78. Yang, P., Yang, W., Zhang, L., and Chen, L. (2018). Novel configurations for the fabrication of high quality REBCO bulk superconductors by a modified RE + 011 top-seeded infiltration and growth process, *Supercond. Sci. Technol.*, **31**, 085005.

79. Caixuan, X., Anming, H., Naomichi, S., Mitsuru, I., and Izumi, H. (2005). Effect of BaO_2 and fine $Gd_2BaCuO_{7-\delta}$ addition on the superconducting properties of air-processed $GdBa_2Cu_3O_{7-\delta}$, *Supercond. Sci. Technol.*, **18**, 229–233.

80. Shi, Y-H., Hari Babu, N., Iida, K., and Cardwell, D. A. (2009). The effect of very high Barium content in the precursor on the properties of GdBCO single grain bulk superconductors, *J. Mater. Res.*, **24**, 10–18.

81. Li, G-Z., Yang, W. M., Cheng, X. F., Fan, J., and Guo, X. (2009). A modified TSIG technique for simplifying the fabrication process of single-domain GdBCO bulks with a new kind of liquid source, *J. Mater. Sci.*, **44**, 6423–6426.

82. Gonzalez-Arrabal, R., Eisterer, M., Weber, H. W., Fuchs, G., Verges, P., and Krabbes, G. (2002). Very high trapped fields in neutron irradiated and reinforced $YBa_2Cu_3O_{7-\delta}$ melt-textured superconductors, *Appl. Phys. Lett.*, **81**, 868–870.

83. Gyorgy, E. M., van Dover, R. B., Jackson, K. A., Schneemeyer, L. F., and Waszczak, J. V. (1989). Anisotropic critical currents in $Ba_2YCu_3O_7$ analyzed using an extended Bean model, *Appl. Phys. Lett.*, **55**, 283–285.

84. Endo, S., Chauhan, H. S., Egi, T., and Shiohara, Y. (1996). Macrosegregation of $Y_2Ba_1Cu_1O_5$ particles in $Y_1Ba_2Cu_3O_{7-\delta}$ crystals grown by an undercooling method, *J. Mater. Res.*, **11**, 795–803.

85. Nariki, S., Sakai, N., Murakami, M., and Hirabayashi, I. (2004). Effects of Dy_2BaCuO_5 contents on microstructure and mechanical strength of Ag-added Dy–Ba–Cu–O bulk superconductors, *Physica C*, **412–414**, 651–656.

86. Murakami, M., Yamaguchi, K., Fujimoto, H., Nakamura, N., Taguchi, T., Koshizuka, N., and Tanaka, S. (1992). Flux pinning by non-superconducting inclusions in melt-processed YBaCuO superconductors, *Cryogenics*, **32**, 930–935.

87. Lee, N., Jang, G. E., Kim, C., Sung, T., Han, Y., and Jung, S. (2007). The effects of artificial holes with different diameters on the surface magnetic properties in YBCO superconductor, *Physica C*, **463–465**, 320–324.

88. Dias, D. H. N., Sotelo, G. G., Moysés, L. A., Telles, L. G. T., Bernstein, P., Kenfaui, D., Aburas, M., Chaud, X., and Noudem, J. G. (2015). Application of textured YBCO bulks with artificial holes for superconducting magnetic bearing, *Supercond. Sci. Technol.*, **28**, 075005.

89. Chaud, X., Kenfaui, D., Louradour, E., and Noudem, J. G. (2012). Thin-wall bulk high temperature superconductor as a permanent cryomagnet, *IEEE Trans. Appl. Supercond.*, **22**, 6800304.

90. Panagopoulos, D. J., and Chrousos, G. P. (2019). Shielding methods and products against man-made electromagnetic fields: protection versus risk, *Sci. Total Environ.*, **667**, 255–262.

91. Denis, S., Dusoulier, L., Dirickx, M., Vanderbemden, Ph., Cloots, R., Ausloos, M., and Vanderheyden, B. (2007). Magnetic shielding properties of high-temperature superconducting tubes subjected to axial fields, *Supercond. Sci. Technol.*, **20**, 192–201.

92. Paul, C. R. (1992). Introduction to Electromagnetic Compatibility (Wiley, New York), ISBN: 978-0-471-75500-5.

93. Pavese, F. (1998). *Handbook of Applied Superconductivity*, G10 Magnetic shielding (Institute of Physics Publishing, CRC Press, Bristol) ed. B. Seeber, pp 1461–1483.

94. Wera, L., Fagnard, J-F., Namburi, D. K., Shi, Y-H., Vanderheyden, B., and Vanderbemden, Ph. (2016). Magnetic shielding above 1 T at 20 K with bulk, large grain YBCO tubes made by buffer-aided top seeded melt growth, *IEEE Trans. Appl. Supercond.*, **27**, 6800305.

95. Denis, S., Grenci, G., Dusoulier, L., Cloots, R., Vanderbemden, P., Vanderheyden, B., Dirickx, M., and Ausloos, M. (2006). Characterisation of the magnetic shielding properties of YBaCuO thick films prepared by electrophoretic deposition on silver substrates, *J. Phys.*, **43**, 509–512.

96. Closset, R., Kumar, N. D., Wera, L., Dellicour, A., Henrist, C., Boschini, F., Cloots, R., Vanderbemden, P., and Vertruyen, B. (2014). $YBa_2Cu_3O_{7-\delta}$ thick films for magnetic shielding: Electrophoretic deposition from butanol-based suspension, *Mat. Lett.*, **119**, 154–156.

97. Pavese, F., Bianco, M., Andreone, D., Cresta, R., and Rellecati, P. (1992). Magnetic shielding properties of $YBa_2Cu_3O_{7-x}$ thick films deposited on silver cylinders with the continuous detonation spray technique, *Physica C*, **204**, 1–7.

98. Vanolo, M., Pavese, F., Giraudi, D., and Bianco, M. (1994). Preliminary results for YBCO thick films sprayed on stainless steel with the HVOF technique, and for multilayers using Ag or CuO, *Nuovo Cimento D*, **16**, 2119–2126.

99. Devendra Kumar, N., Closset, R., Wera, L., Cloots, R., Vanderbemden, Ph., and Vertruyen, B. (2015). Magnetic shielding performances of $YBa_2Cu_3O_{7-\delta}$ coated silver tubes obtained by electrophoretic deposition, *Supercond. Sci. Technol.*, **28**, 015007.

100. Teshima, H., and Morita, M. (2012). Recent progress in HTS bulk technology and performance at NSC, *Physics Procedia*, **36**, 572–575.

101. Werfel, F. N., Floegel-Delor, U., Rothfeld, R., Riedel, T., Goebel, B., Wippich, D., and Schirrmeister, P. (2012). Superconductor bearings, flywheels and transportation, *Supercond. Sci. Technol.*, **25**, 014007.

102. Sakai, N., Nariki, S., Nagashima, K., Miryala, M., Murakami, M., and Hirabayashi, I. (2007). Magnetic properties of melt-processed large single domain Gd–Ba–Cu–O bulk superconductor 140 mm in diameter, *Physica C*, **460–462**, 305–309.

103. Ogasawara, K., Sakai, N., and Murakami, M. (2002). Subgrains in large-grain RE123 bulk superconductors, *Supercond. Sci. Technol.*, **15**, 791–795.

104. Shi, Y-H., Yoeh, W., Dennis, A. R., Hari Babu, N., Pathak, S., Xu, Z., and Cardwell, D. A. (2010). Growth rate of YBCO single grains containing Y-2411(M), *J. Phys. Conf. Series.*, **234**, 012039.

105. Sawamura, M., Morita, M., and Hirano, H. (2004). A new method for multiseeding RE–Ba–Cu–O superconductor, *Supercond. Sci. Technol.*, 17, S418-S421.

106. Shi, Y-H., Dennis, A. R., Zhou, D., Namburi, D. K., Huang, K., Durrell, J. H., and Cardwell, D. A. (2016). Factors affecting the growth of multiseeded superconducting single grains', *Cryst. Growth Des.* 16, 5110–5117.

107. Kim, C-J., Hong, G-W. Oh, H-W. (2001). Multi-seeded melt growth processed YBCO superconductors, *Physica C*, **357–360**, 635–641.

108. Shi, Y-H., Durrell, J. H., Dennis, A. R., Zhang, Z., Zhai, W., Hari Babu, N., and Cardwell, D. A. (2013). A comparison of 0°–0° and 45°–45° bridge-seeded, YBCO single grains, *J. Am. Ceram. Soc.*, **96**, 1757–1762.

109. Deng, M. Z., Izumi, M., Miki, M., Tsuzuki, K., and Felder, B. (2012). Trapped flux and levitation properties of multiseeded YBCO bulks for HTS magnetic device applications—part II: practical and achievable performance, *IEEE Trans. Appl. Supercond.*, **22**, 6800210.

110. Shi, Y-H., Dennis, A. R., Huang, K., Zhou, D., Durrell, J. H., and Cardwell, D. A. (2018). Advantages of multi-seeded (RE)–Ba–Cu–O superconductors for magnetic levitation applications, *Supercond. Sci. Technol.*, **31**, 095008.

111. Bernstein, P., and Noudem, J. (2020). Superconducting magnetic levitation: principle, materials, physics and models, *Supercond. Sci. Technol.*, 33, 033001.

112. Namburi, D. K., Shi, Y-H., Ainslie, M., Dennis, A. R., Durrell, J. H., and Cardwell, D. A. (2020). Buffer-assisted top-seeded infiltration and growth for fabricating dense, single-grain (RE)-Ba-Cu-O bulk superconductors, *IEEJ Trans. Power Energy*, **140**, 148–153.

113. Namburi, D. K., Huang, K., Lau, W., Shi, Y-H., Palmer, K. G., Dennis, A. R., Cardwell, D. A., and Durrell, J. H. (2020). A simple, reliable and robust reinforcement method for the fabrication of (RE)–Ba–Cu–O bulk superconductors, *Supercond. Sci. Technol.*, **33**, 054005.

Chapter 3

Growth, Microstructure, and Superconducting Properties of Ce Alloyed YBCO Bulk Single-Grain Superconductors

P. Diko,[a] K. Zmorayova,[a] L. Vojtkova,[a] V. Antal,[a] V. Kucharova,[a] R. Pagacova,[a] V. Kavečanský,[a] M. Radusovska,[a] M. Rajnak,[b] T. Hlasek,[c] and J. Plechacek[c]

[a]*Institute of Experimental Physics, Slovak Academy of Sciences, Watsonova 47, 04001 Košice, Slovakia*
[b]*Faculty of Electrical Engineering and Informatics, Technical University of Košice, Slovakia*
[c]*CAN Superconductors, Ringhofferova 66, 251 68 Kamenice, Czech Republic*
dikos@saske.sk

3.1 Introduction

The trapped magnetic field in YBCO bulk single-grain superconductors (BSS) is an important feature for their practical applications

High-T_c Superconducting Technology: Towards Sustainable Development Goals
Edited by Muralidhar Miryala

ISBN 978-981-4877-65-7 (Hardcover), 978-1-003-16468-5 (eBook)
www.jennystanford.com

as levitators or superconducting permanent magnets [1]. Since the trapped magnetic field in YBCO BSS is directly proportional to the critical current density, J_c, refining Y211 particles, which are efficient pinning centres of magnetic flux lines in these bulks, is an important technological challenge [2–4]. The Y211 particles increase in size by diffusion of Y through the melt when the sample is in a partially molten state and the Y211 particles and the xBaO·yCuO melt are in equilibrium (Ostwald ripening process [5, 6]). It has been shown experimentally that this ripening process can be inhibited by the addition of a small amount of Pt or Ce [7–9]. Pt is added in metallic form or as a compound of Pt. Cerium is mostly added as CeO_2 [10, 11]. Both additives also increase the melt viscosity and thus suppress melt leakage during YBCO BSS growth at slow cooling [12]. Since Ce is much cheaper than Pt, there is a tendency to replace the addition of Pt by the addition of Ce. Interestingly, the mechanism of Ce (as well as Pt) influence on the Ostwald ripening process is not yet clear [13, 14]. Therefore, more studies on this aspect of YBCO BSS production are useful to understand the effect of Ce addition on the growth, structure, and superconducting properties of YBCO bulks. In this paper we present studies on YBCO BSS alloyed with cerium, also in combination with rare earth elements (RE) for inducing chemical pining, which were prepared in our research laboratories or in a commercial company CAN superconductor using top-seeded melt-growth (TSMG) or top-seeded infiltration-growth (TSIG) processes. It has been shown that when Ce is added in the form of nanosized $BaCeO_3$, the refinement of Y211 particles is more effective in TSMG YBCO BSS, which is reflected in the high-trapped magnetic field. The addition of CeO_2 to the Y211 preform for the TSIG process has a positive effect on the Y211 refinement, but the YBCO BSS growth rate is slower. In addition, the YBCO BSS with addition of CeO_2 have much longer oxidation cracks than the CeO_2-free samples, which can negatively affect mechanical properties. The contamination of TSIG YBCO samples with Yb from the substrate is lower in the samples with CeO_2 addition, suggesting that Ce addition may affect solubility of RE in xBaO·yCuO melt or diffusivity of RE in that melt. Ytterbium doping or a combination of samarium and ytterbium doping results in a secondary peak effect. The relationship between the local microstructure and the superconducting properties of the commercially produced TSMG YBCO BSS with the addition of CeO_2

was investigated. It is shown how the size and volume fraction of Y211 as well as the subgrain structure affect the local critical current density.

3.2 Influence of Addition of Nanosize Barium Cerate on the Microstructure and Properties of TSMG YBCO Bulk Superconductors

It is important to have cerium distributed homogeneously in the starting pellet of the pressed powders to grow single-grain YBCO to ensure cerium activity on the micrometer scale. Cerium can be added in two forms, either CeO_2 or $BaCeO_3$. In both cases, the addition is at least partially dissolved in the peritectically melted Y123 ($YBa_2Cu_3O_7$) compound and support our hypothesis that Ce in the melt segregates to the surface of Y211 (Y_2BaCuO_5) particles and thereby prevents their growth by the Ostwald Ripening process [15]. The disadvantage of using CeO_2 is that it produces copper(II) oxide by the equation: $2YBa_2Cu_3O_7 + 3CeO_2 \rightarrow 3BaCeO_3 + Y_2BaCuO_5 + 5CuO$, which at some critical concentration prevents Y123 crystal growth [4, 5]. In addition, $BaCeO_3$ was shown to suppress Y211 particles more efficiently than CeO_2 [16].

Barium cerate ($BaCeO_3$) is a ceramic material with perovskite structure. It has a wide range of uses, for example, in the preparation of hydrogen gas sensors, hydrogen membranes, fuel cells and the like [17, 18], but also in the fabrication of YBCO BSS. There are many ways to prepare barium cerate which could be divided into two main groups. The first group is solid-state reactions which are mostly calcination of mixtures of CeO_2 and $BaCO_3$ [19, 20] or mixtures of CeO_2 and BaO_2 [21]. The second group is "wet" chemical methods of preparation, that is, chemical synthesis from solutions. Co-precipitation methods are most commonly used in which a solution containing barium and cerium salts is mixed with a solution of a particular co-precipitant, which may be, for example, hydroxides, oxalate, carbonate or acrylate salts and the like [22–24]. In addition to co-precipitation methods, $BaCeO_3$ can be prepared by many other chemical methods, such as solvothermal and hydrothermal processes [25–28] or sol-gel methods [29–32].

In this work we prepared $BaCeO_3$ powders in three different ways: By solid-state reaction, by co-precipitation of oxalate, and by sol-gel method. The final calcination was held in air or in vacuum. The homogeneity of the distribution of added $BaCeO_3$ powder in the prepared YBCO single-grains was tested. $BaCeO_3$ prepared by a sol-gel process with vacuum calcination which exhibits the best properties was used to grow YBCO single-grain bulks. The optimal $BaCeO_3$ addition was determined from the measurement of Y211 particle size and critical current density.

$BaCeO_3$ was prepared by different ways using $BaCO_3$, 98.5%, CeO_2, 99%, both from ACROS Organics, $Ba(NO_3)_2$, 99%, $Ce(NO_3)_3$, 99.5% from Alfa Aesar, ethanol, 99% from Sigma-Aldrich, diammonium oxalate monohydrate, p.a., citric acid, p.a. and ethylene glycol, p.a. from Centralchem. YBCO samples were prepared from Y123 (Solvay GmbH, 99.9% purity, 30 μm) and Y211 (Toshima, 1–2 μm). All chemicals were used as received.

$BaCeO_3$ by solid-state reaction was prepared by mechanical mixing of $BaCO_3$ and CeO_2 in molar ratio 1:1 (n = 30 mmol) followed by calcination 4 h at 1100°C in air or 6 h at 1000°C in vacuum. In preparation of $BaCeO_3$ by oxalate co-precipitation, 0.5 mmol of $Ba(NO_3)_2$ and 0.5 mmol $Ce(NO_3)_3$ were dissolved in 5 mL of water. The solution was added dropwise to aqueous-ethanol solution (4 mL of water and 1 mL of ethanol) of diammonium oxalate monohydrate (n = 2 mmol) at 80°C. White precipitate was formed, filtered off, dried on air and calcined 2 h at 1100°C in air or 4 h at 1000°C in vacuum. For preparation of $BaCeO_3$ by sol-gel reaction the appropriate amount of $Ba(NO_3)_2$ and $Ce(NO_3)_3$ (n = 7.7 mmol) was dissolved in 40 mL of water. While continuously stirring, citric acid (77 mmol), that was previously dissolved in 20 mL of water was added. The solution was stirred 2.5 h at room temperature. After that, 154 mmol of ethylene glycol was added and stirred for another 10 min followed by evaporation and gelation phase. Water was first evaporated by continuous stirring during heating at 100°C on a hot plate. The resulting gel-type precursors were dried 4 h at 450°C in the air atmosphere. The dried precursors were calcined 2 h at 1100°C in air or 4 h at 1000°C in vacuum.

The studied bulk YBCO samples were produced by a TSMG process from a mixture of Y123, Y211, and $BaCeO_3$ powders. The Sm-123 seed was placed in the centre of the top surface of the pellet. The samples were processed in a chamber furnace in the air using the following time-temperature profile (Fig. 3.1): Heating up to 940°C/dwell 24 h, heating up to maximum temperature T_{max} = 1040°C/dwell 1 h, fast cooling to 1002°C then slow cooling to isothermal dwell temperature T_{is} = 995°C/dwell 80 h and finally cooling to room temperature by furnace rate cooling.

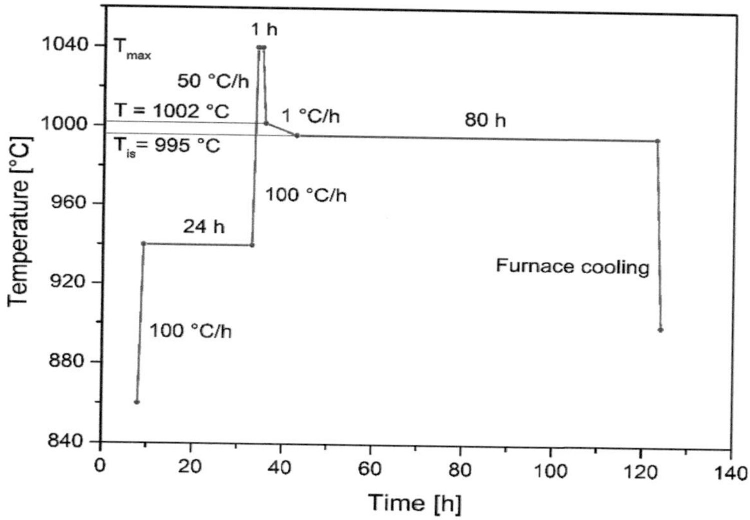

Figure 3.1 Schematic illustration of time temperature profile for YBCO samples produced by TSMG process.

Grown bulks were annealed at 420°C for 150 h in oxygen atmosphere. The microstructure of prepared samples was studied using an optical polarising microscope in normal and polarised light (NIKON UDMLV 100 D-U) and by scanning electron microscope (SEM) MIRA3 TESCAN on the polished halves of the bulks cut along the *a*/*c*-plane. Quantitative microstructural analysis was performed using image processing software. Prepared $BaCeO_3$ samples were examined by X-ray powder diffraction (XRD) (Rigaku Ultima IV, CuK_{α} radiation λ = 0.15417 nm) in air and vacuum at T = 25–950°C, and the XRD patterns were taken at selected temperatures.

The trapped magnetic field profile was measured at 77 K. The samples were magnetised in 1.4 T field. The surface of the samples was scanned by Hall probe with 0.3 mm step at a distance of about 130 μm from the surface of each sample. The field dependence of the magnetic moment was measured for the samples at 77 K using commercial Magnetic Property Measurement System MPMS3

(Quantum Design), with magnetic field of up to 7 T at a constant sweep rate of 100 A/m · min^{-1}. Magnetic moment was calibrated to an accuracy of less than 1% using palladium sample. The samples had the shape of a slab with dimensions 1.50 × 1.50 × 0.50 mm^3, where the smallest dimension was parallel to the *c*-axis of the crystal. During the magnetisation measurements, the applied external magnetic field was parallel to the *c*-axis of the crystal. The critical current densities, J_c, were calculated from magnetisation hysteresis loops using the Bean model for rectangular samples, $J_c(B) = \{\Delta m(B)/V\}\{2/[b(1-b/3a)]\}$, where Δm is the difference in magnetic moments between the ascending and descending branches of the magnetic hysteresis loops, and V is the sample volume, $a \times b \times c$. The superconducting transition temperatures were determined from the magnetic moment transition curves taken after zero-field cooling in an applied external magnetic field of 10 A/m.

We added 1.90 wt % $BaCeO_3$ prepared by three different ways to the testing charges for growth of YBCO single-grains. This $BaCeO_3$ concentration is equivalent to 1.0 wt % CeO_2 which is usually used for preparation of YBCO bulks. The top surfaces of grown YBCO bulks are shown in Fig. 3.2. The growth of single-grains was hindered probably by high concentration of $BaCeO_3$, therefore samples are not fully overgrown by single-grain. Nevertheless, we cut the samples along the *a*/*c*-plane to see the distribution of $BaCeO_3$ in the grain (Fig. 3.3). Besides typical porosity (dark spots), bright agglomerates of $BaCeO_3$ are present in the bulks grown with the addition of solid state prepared (annealed in air or vacuum) $BaCeO_3$ (Fig. 3.3a,b). These agglomerates were not observed in the samples with addition of $BaCeO_3$ prepared by chemical methods (Fig. 3.3c–f).

At higher magnification the details of distribution of small $BaCeO_3$ particles can be seen (Fig. 3.4). Huge agglomerates of $BaCeO_3$ crystals are seen in the sample with powder prepared by solid-state method annealed in air (Fig. 3.4a). The highest homogeneity in the distribution of small $BaCeO_3$ particles is observed for $BaCeO_3$ prepared by chemical processes with final annealing in vacuum (Fig. 3.4d,f). While in the sample with $BaCeO_3$ prepared by solid-state method annealed in air we counted only about 1,000 $BaCeO_3$ particles mm^2 on the polished surface (agglomerates were counted as one particle), for $BaCeO_3$ prepared by solid-state method annealed in vacuum it was 5,100, oxalate-coprecipitation annealed in air 5,200, sol-gel method annealed in air 14,000, oxalate-coprecipitation annealed in vacuum 19,000 and for $BaCeO_3$ prepared by sol-gel method annealed in vacuum 21,000 particles per mm^2.

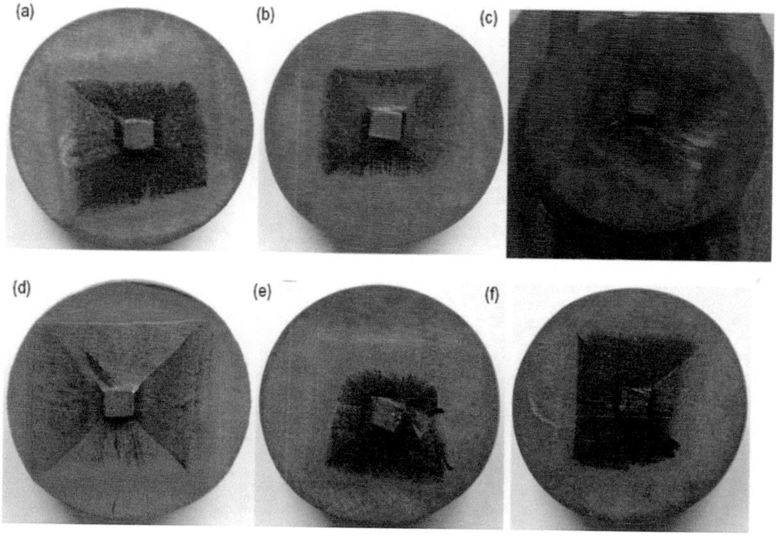

Figure 3.2 Macrographs of the top surfaces of samples after the growth process. Addition of 1.90 wt % $BaCeO_3$ prepared by solid-state (annealed in air (a) or vacuum (b)), oxalate-coprecipitation (annealed in air (c) or vacuum (d)), and sol-gel (annealed in air (e) or vacuum (f)) method.

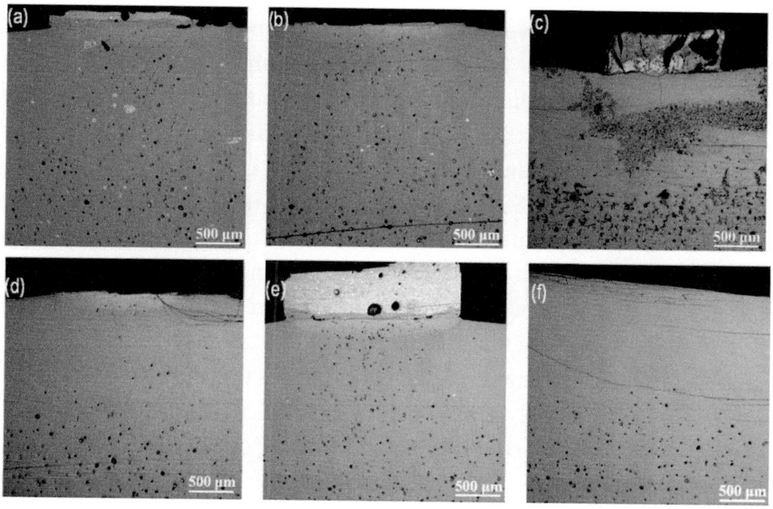

Figure 3.3 SEM micrographs of *a*/*c*-sections of YBCO bulk single-grain samples at low magnification. Addition of $BaCeO_3$ prepared by solid-state (annealed in air (a) or vacuum (b)), oxalate-coprecipitation (annealed in air (c) or vacuum (d)) and sol-gel (annealed in air (e) or vacuum (f)) method. $BaCeO_3$ particles are white.

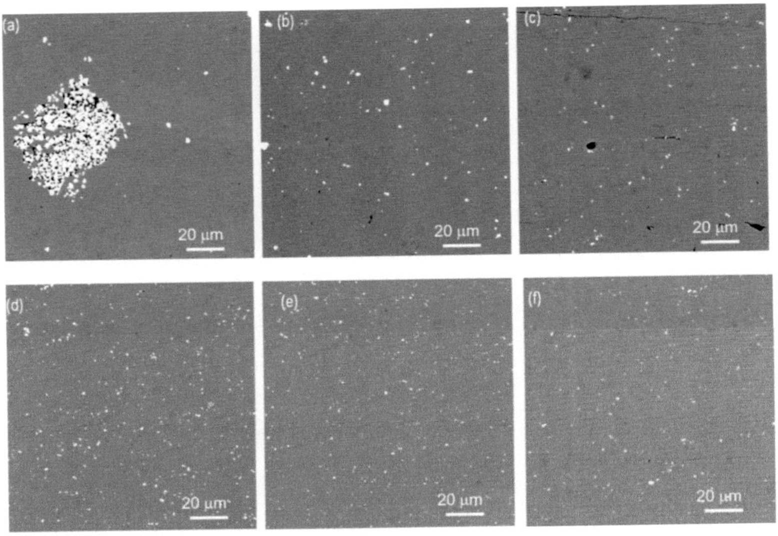

Figure 3.4 SEM micrographs of *a/c*-sections at the beginning of *a*-growth sector of the single-grain samples with addition of $BaCeO_3$ prepared by solid-state (annealed in air (a) or vacuum (b)), oxalate-coprecipitation (annealed in air (c) or vacuum (d)) and sol-gel (annealed in air (e) or vacuum (f)) method. $BaCeO_3$ particles are white.

From the results we can conclude, the samples with the addition of powders prepared by chemical routes and finally annealed in vacuum exhibited higher density of $BaCeO_3$ particles. Higher density and homogenous distribution of $BaCeO_3$ particles in the samples leads to more efficient refinement of Y211 particle size during fabrication of YBCO BSS this has a significant impact on the resulting superconducting properties. The best results were obtained with $BaCeO_3$ prepared by a sol-gel process. The morphology of the final sol-gel product is shown in Fig. 3.5 after annealing in air (Fig. 3.5a) and annealing in vacuum (Fig. 3.5b). The nanosize crystals are smaller after annealing in vacuum, and the crystalline agglomerates are porous, apparently with weak connectivity. These agglomerates are easily crushed during mixing and milling of powder charge, and $BaCeO_3$ nanoparticles can be homogeneously distributed in the powder sample. XRD analysis of vacuum-annealed sol-gel powder presented in Fig. 3.6 confirms that the prepared $BaCeO_3$ is single-phase without detectable impurities.

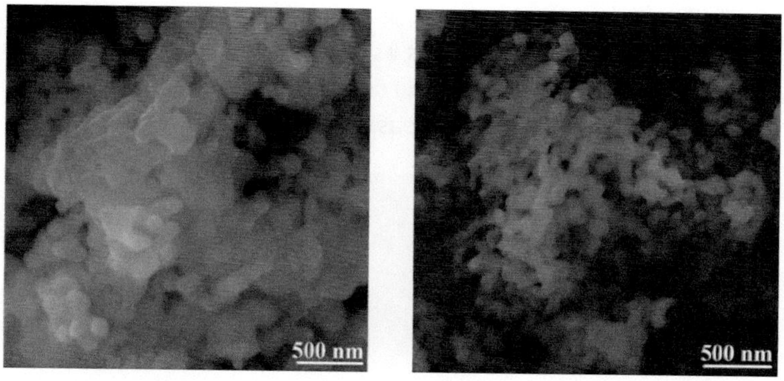

Figure 3.5 Nanosize $BaCeO_3$ particles observed by SEM in $BaCeO_3$ prepared by sol-gel method annealed in air (left) and vacuum (right).

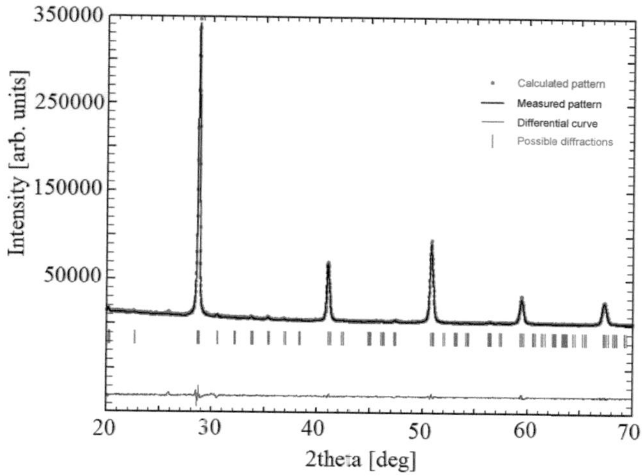

Figure 3.6 XRD pattern of $BaCeO_3$ prepared by sol-gel method and annealed in vacuum at 1000°C for 4 h (the pattern refined by LeBail method).

$BaCeO_3$ prepared by sol-gel method annealed in vacuum with the best properties was tested with the aim to look for optimum concentration of added powder. Thus, a total of six Y123 + Y211 (weight ratio of Y123/Y211 is 7/3) samples with addition 0.38, 0.76, 0.95, 1.14, 1.52, and 1.90 wt % $BaCeO_3$ were grown (Fig. 3.7).

For the study of microstructural and physical properties, three samples of each concentration were prepared. The microstructure of all samples was analysed on the *a*/*c*-central cut of the samples

(Fig. 3.8) and quantitative data on the volume fraction (V_{211}) and the size (d_{211}) of Y211 particles as a function of the distance from the seed were evaluated, then the critical current density, J_c, and the critical temperature, T_c, were measured.

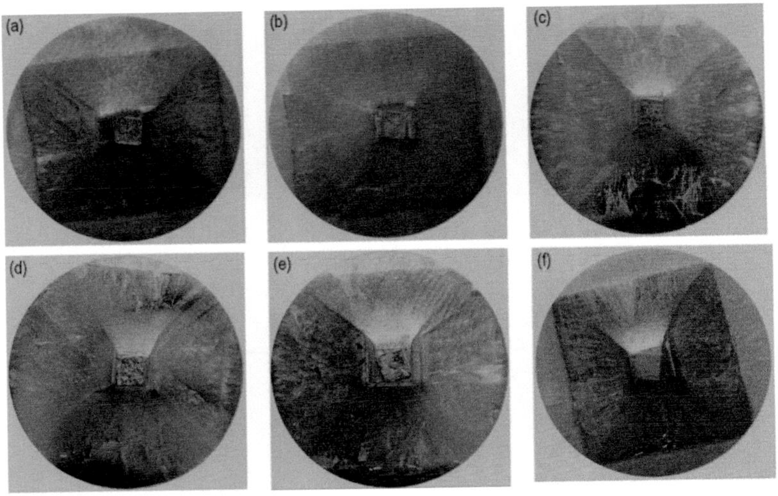

Figure 3.7 Macrographs of the top surfaces of samples with different additions of $BaCeO_3$, 0.38 wt % (a), 0.76 wt % (b), 0.95 wt % (c), 1.14 wt % (d), 1.52 wt % (e) and 1.90 wt % (f).

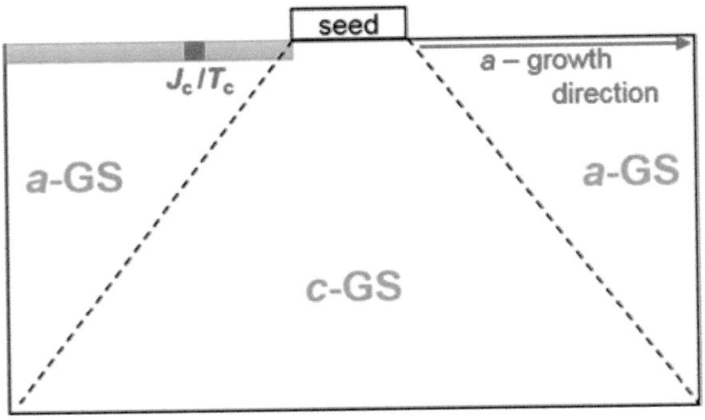

Figure 3.8 Schematic illustration of the growth sectors (GS), the line along which the size and volume fraction of Y211 particles was measured and the position of sample separated for J_c and T_c measurements.

The representative microstructure at the beginning and the end of *a*-sector of prepared samples are presented in Fig. 3.9. As mentioned above, for these samples we focused on assessing the size and volume fraction of Y211 particles. The measurement results of V_{211} and d_{211} are shown in Fig. 3.10. The refinement of Y211 particles occurred over the entire interval of $BaCeO_3$ addition since without the addition of Ce the mean size of Y211 particles would be about 10 μm. It is clear from the dependences measured that higher concentration of $BaCeO_3$ leads to refinement of Y211 particles. Their size increases with the distance from the seed due to the Ostwald ripening process. Consequently, more Y211 particles have subcritical dimension at the beginning of the crystal grows and will be pushed. Increasing the cooling rate of the undercooled Y123 crystal leads to an increase in the volume fraction of Y211 particles to values in the range of 30 to 40 vol % at the end of crystallisation. At the highest concentrations of $BaCeO_3$, the volume fraction of Y211 particles at the beginning of the *a*-growth sector decreases to about 12 vol %, while at the lowest concentration of $BaCeO_3$, the volume fraction of Y211 is twice that [33–35].

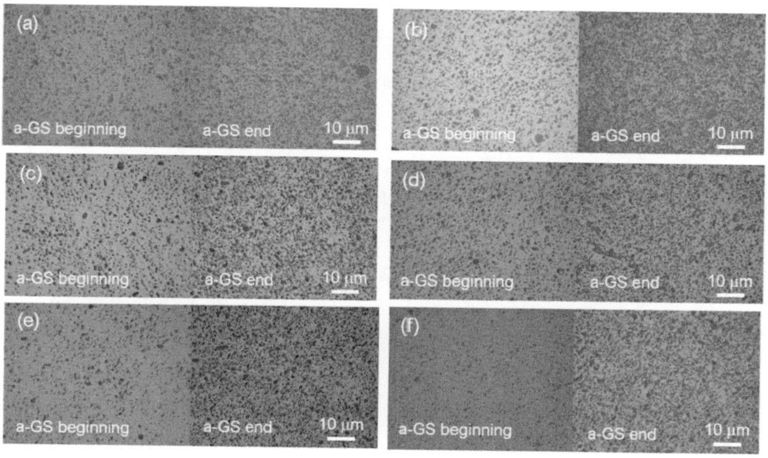

Figure 3.9 Polarised light microscopy micrographs from the YBCO samples with addition of 0.38 wt % (a), 0.76 wt % (b), 0.95 wt % (c), 1.14 wt % (d), 1.52 wt % (e) and 1.90 wt % (f) wt % of $BaCeO_3$ prepared by sol-gel method annealed in vacuum taken at the beginnings (close to the seed) and ends (end of Y123 crystal) of *a*-growth sector. Micrographs were taken on the central *a*/*c*-cut of the pellet along the central pellet axis (*c*-direction) and along the line parallel to the *a*-direction close to the sample top surface.

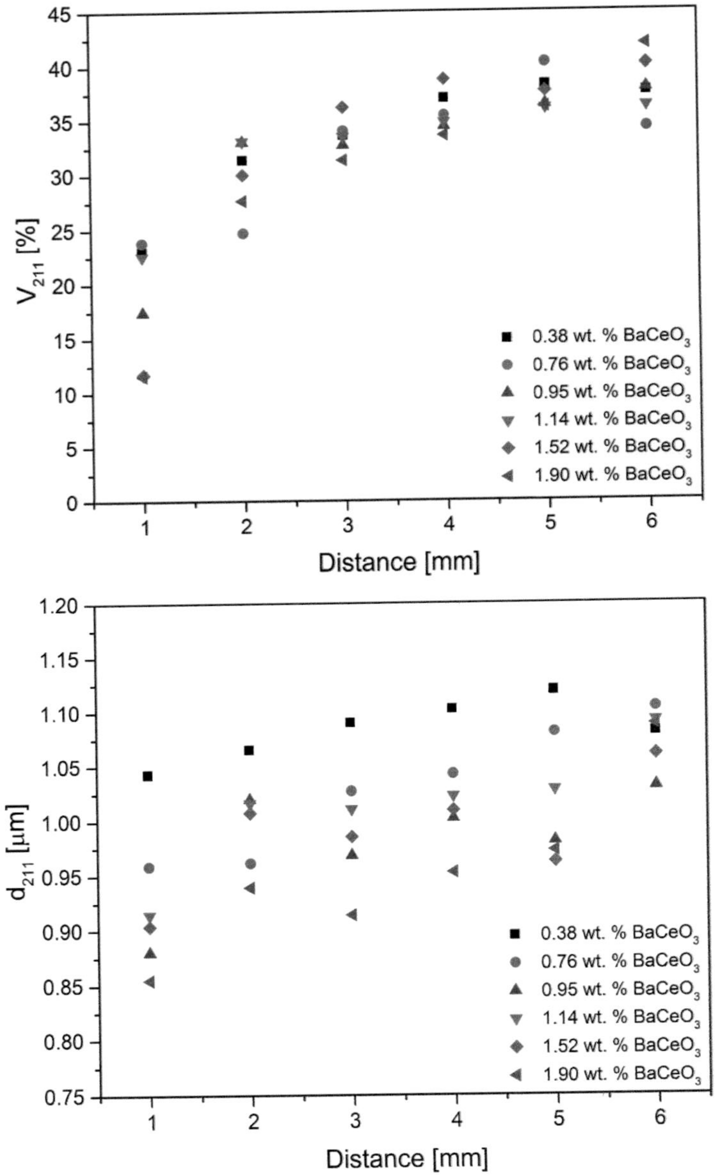

Figure 3.10 Dependences of the Y211 volume fraction (V_{211}) and the size of Y211 (d_{211}) on the distance from the seed measured for in the *a*-growth sector for different additions of $BaCeO_3$ prepared by sol-gel method annealed in vacuum.

Transition temperature and critical current density were measured on the 1.5 × 1.5 × 0.5 mm^3 samples cut from the upper surface of grown single-grain YBCO samples 2 mm from the seed (a-growth sector).

The superconducting transition temperatures are shown in Fig. 3.11. All onset temperatures $T_{c,onset}$ were above 91 K, and the superconducting transition widths were about 1 K or less. From these results, it is clear that addition of $BaCeO_3$ does not affect the superconducting transition of Y123. Some differences in the transition can be ascribed to the contamination of YBCO crystal from the Sm123 seed [36].

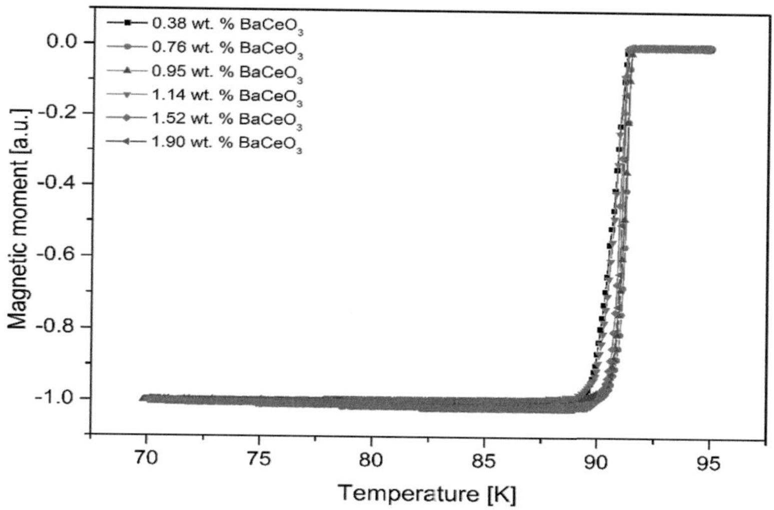

Figure 3.11 Transition temperature for the samples with different amounts of $BaCeO_3$.

The field dependence of the critical current density, J_c, at 77 K for samples with all $BaCeO_3$ additions is presented in Fig. 3.12. The appearance of weakly pronounced secondary peak effect is probably caused by some sample contamination from the Sm123 seed as it was observed in the superconducting transition (Fig. 3.11). As the size and volume fraction of Y211 particles have the strongest influence on pinning in low fields [4, 37] we compared the values of critical current in zero field, J_{c0} (Fig. 3.13). The J_{c0} increases slightly with the amount of $BaCeO_3$ added. In further experiments, we decided to use an addition of 0.95 wt % $BaCeO_3$, since at this concentration the Y211 particles are effectively refined and the pushing effect does not cause as much macroscopic inhomogeneity of the Y211 concentration as in the samples with a higher $BaCeO_3$ addition. Addition of 0.95 wt % $BaCeO_3$ is equivalent to addition of 0.5 wt % CeO_2 which is 2 times lower than the standard amount of 1.0 wt % CeO_2 [38–41].

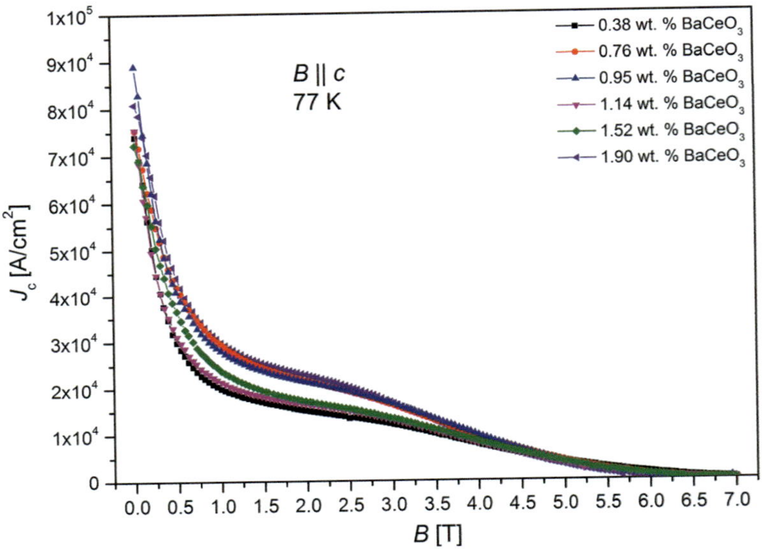

Figure 3.12 Dependences of critical current density at 77 K on magnetic field for the samples with different amounts of $BaCeO_3$.

The pinning force induced by non-superconducting Y211 particles is proportional to the Y123/Y211 interface [37]. If Y211 particles are effectively refined, then lower Y211 content may be optimal to obtain the highest trapped field as the effective cross-sectional area for the supercurrent increases. Therefore, we prepared YBCO single-grain samples with lower Y211 addition than standard 30 wt %. The top surfaces of the samples with the addition 20, 25, and 30 wt % Y211 and 0.95 wt % $BaCeO_3$ prepared by sol-gel method annealed in vacuum are presented in Fig. 3.14.

Microstructure micrographs from the *a*-growth sector presented in Fig. 3.15 clearly show that the Y211 particle size distribution has a bimodal character which is most pronounced in the sample with the lowest Y211 amount.

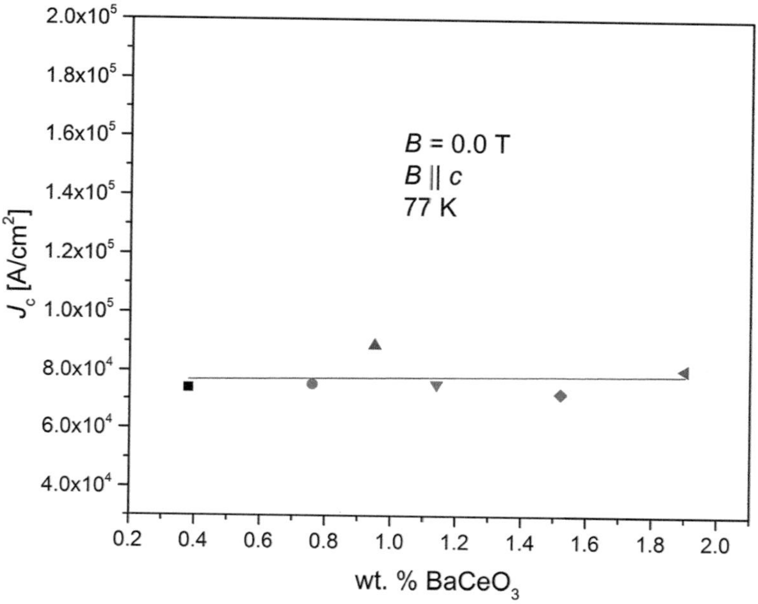

Figure 3.13 Dependence of the critical current density in zero magnetic field, J_{c0}, on the concentration of added $BaCeO_3$.

Figure 3.14 Macrographs of the top surfaces of samples with 20 (a), 25 (b), and 30 (c) wt % Y211 and addition of 0.95 wt % $BaCeO_3$ prepared by sol-gel method annealed in vacuum.

The measured Y211 particle size and their volume fraction depending on the distance from the seed in the *a*-growth sectors are shown in Fig. 3.16. The mean Y211 particle size was effectively reduced to less than one micrometer by the addition of an optimal amount of $BaCeO_3$ which is low compared to the published TSMG YBCO BSS [42–45]. The amount of Y211 powder added affects the mean Y211 particle size. The smaller mean Y211 particle size at the higher Y211 addition can be related to the ratio of Y211 particles added to the $Y211_P$ particles produced by the peritectic melting of the Y123 phase. It was observed that the $Y211_P$ particles formed during peritectic melting are larger in size than Y211 particles added [46, 13]. If we suppose that added Y211 particles represent the nucleation points for melting of Y123 phase through reaction: $Y123 = Y211_P + L$ and peritectic $Y211_P$ grow on the added Y211, then the number of nucleation points is lower at lower amount of added Y211 and consequently $Y211_P$ grow bigger. Due to the growth of bulk crystals in the slow cooling mode, an increase in the volume fraction of Y211 particles with the distance from the nucleus was observed in all three samples (Fig. 3.16).

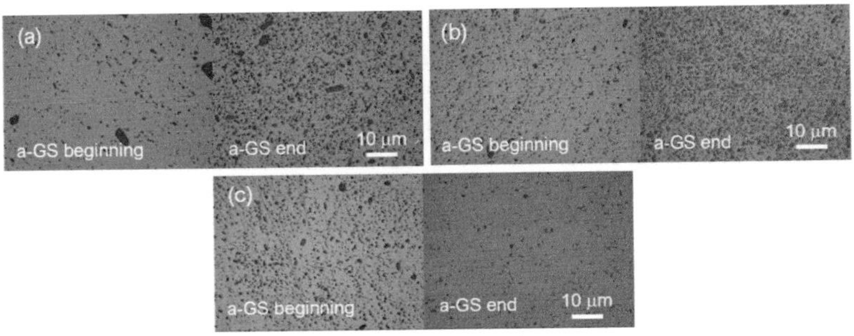

Figure 3.15 Polarised light microscopy micrographs from the YBCO samples with 20 (a), 25 (b) and 30 (c) wt % Y211 and addition of 0.95 wt % $BaCeO_3$ prepared by sol-gel method annealed in vacuum taken at the beginning and end of *a*-growth sector.

Measurement of trapped magnetic field at liquid nitrogen temperature showed that the addition $BaCeO_3$ prepared by sol-gel method annealed in vacuum improved magnetic flux pinning resulting in higher values of the trapped magnetic field (Fig. 3.17), compared to the report on YBCO samples of the same mm diameter prepared by the TSMG method [2]. As the amount of Y211 added decreases, the values of the trapped magnetic field are lower as the Y123/Y211 phase interface area decreases. This is because the positive effect of $BaCeO_3$ prepared by sol-gel method annealed in vacuum is partially suppressed by more pronounced bimodal Y211 size distribution in the samples with lower addition of Y211 (Fig. 3.15).

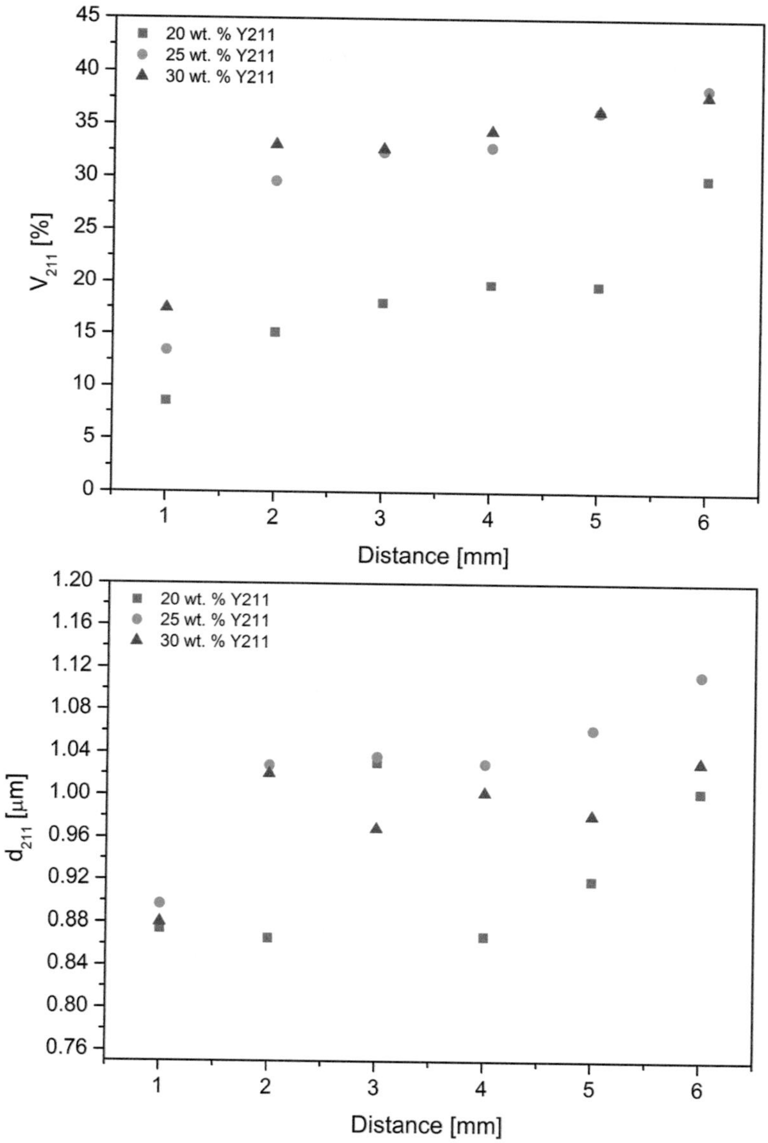

Figure 3.16 Dependences of the Y211 volume fraction (V_{211}) and the size of Y211 (d_{211}) on the distance from the seed measured in the *a*-growth sector for the samples with 20, 25 and 30 wt % Y211 and 0.95 wt % $BaCeO_3$ prepared by sol-gel method annealed in vacuum.

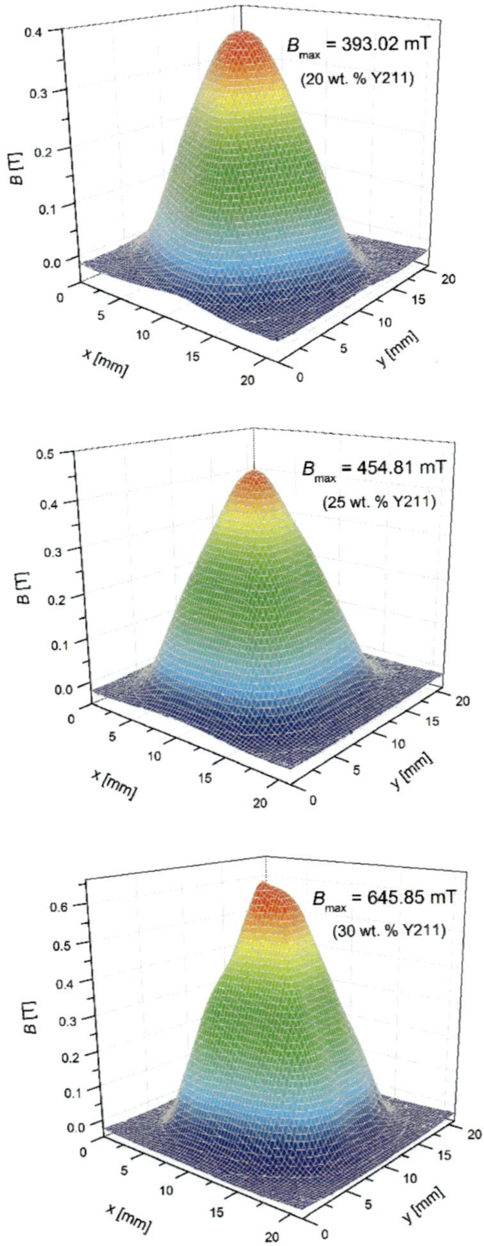

Figure 3.17 Profiles of trapped field at 77 K for the samples with 20 (a), 25 (b) and 30 (c) wt % Y211 and 0.95 wt % $BaCeO_3$ prepared by sol-gel method annealed in vacuum.

In summary, crystalline barium cerate ($BaCeO_3$) was synthesized by solid-state reaction, oxalate co-precipitation, and sol-gel methods. Final calcination was held in air or in vacuum. Solid-state synthesis was shown to produce barium cerate with the lowest efficiency in Y211 particle refining. Oxalate co-precipitation and sol-gel methods as well as final annealing under vacuum are much more efficient. It is shown that the refinement of Y211 particles occurred over the entire interval of nanocrystalline $BaCeO_3$ addition from 0.38 to 1.90 wt %. Based on microstructural analysis and characterisation of superconducting properties, the optimal concentration of nanosize barium cerate was chosen to be 0.95 wt %. The effect of Y211 content on trapped field in YBCO bulks with addition of nanocrystalline barium cerate is shown. Positive effect of nanocrystalline $BaCeO_3$ is partially suppressed in the samples with lower addition of Y211 by more pronounced bimodal Y211 size distribution.

3.3 Influence of CeO_2 on Microstructure, Cracking, and Trapped Field of TSIG YBCO Single-Grain Superconductors

Another important property, especially at lower temperatures, is the good mechanical stability of the samples. The YBCO superconductors are inherently brittle, therefore they are sensitive to cracking. Its fracture toughness, K_{IC}, is low, as is typical for most ceramic materials. The *a*/*b*-plane ((001) plane) is the first cleavage plane, and K_{IC} = 0.32 MPa $m^{1/2}$ is the lowest there. K_{IC} in the second cleavage plane which is the *a*/*c*-plane ({100} type plane) is a bit higher (0.8 MPa $m^{1/2}$). The strength of 123/211 composites increases with the Y211 particle volume fraction, and for standard samples K_{IC} = 1.5 MPa $m^{1/2}$ [47] which is still low. Since as grown YBCO bulk single-grain superconductors are in the tetragonal state, they need to be transformed into the orthorhombic phase $YBa_2Cu_3O_7$ by oxygenation at approximately 400°C [48]. During the oxygenation process so-called oxygenation cracks are formed [47, 49]. They may be parallel to the *a*/*b*-plane and are called *a*/*b*-oxygenation cracks (*a*/*b*-OC) or parallel to the *a*/*c*-plane called *a*/*c*-oxygenation cracks (*a*/*c*-OC). Oxygen cracks negatively affect mechanical properties. Careful control of the microstructure will lead to improved superconducting and mechanical properties of YBCO bulk materials [50–54]. In this work, we focus on YBCO bulk single-grain samples fabricated by the TSIG process which overcomes many limitations of the TSMG technique [55–62]. It is claimed that the TSIG process is an efficient process for preparing bulk single crystals with finely dispersed particles of Y211 and can be used to produce even large specimens of nearly perfect shape. The main subject of this paper, therefore, focuses on the characterisation of the microstructure of YBCO bulk single-grain superconductors fabricated by TSIG process. The influence of CeO_2 addition on Y211 particles and crack formation in the single-grain YBCO is shown. Measured trapped field at 77 K is related to the structure of prepared samples.

Two different YBCO bulk samples without (IG) and with CeO_2 addition (IG + CeO_2) were prepared. All the precursor powders were manufactured at CAN SUPERCONDUCTORS from raw oxides and barium carbonate which are used in standard commercial production. The green phase (Y211, Y_2BaCuO_5) powder was pressed into disks of 24 mm in diameter (each pellet mass 20 g). For the IG + CeO_2 sample, 1 wt % of CeO_2 was mixed into the Y211 powder prior to pressing. Liquid source pellets were composed of 85 wt % $Ba_3Cu_5O_8$ + 12.5 wt % Yb_2O_3 + 2.5 wt % BaO_2 (32 mm in diameter, 32 g).

The thin film seeds 2 × 2 mm, Nd123 (500 nm) on 15 nm YBCO buffer layer was fixed in the centre of top green pellets, and the samples were grown on MgO substrate. The samples were fabricated by a two-step TSIG process according to earlier work by Namburi [63]. The diameter of samples after the TSIG process was 24 mm. After the fabrication the samples were annealed in a flowing oxygen atmosphere at 400°C during 250 h. Top surface morphology of the YBCO bulk samples is presented in Fig. 3.18.

For microstructural study of *a*/*c*-plane the YBCO bulk samples were cut in half parallel to their *c*-axis and then embedded with a conductive material for further grinding and polishing. The grinding of the samples was carried out on the SiC paper and then polishing with diamond spray with decreasing particle size from 9 μm to 0.25 μm. Before investigation of cracks the surface of samples was etched in 0.5 wt % HCl solution in ethyl alcohol. The microstructure of YBCO bulk samples was analysed by polarised light or normal light microscopy (Nikon UDMLV 100 D-U) on polished *a*/*c*-plane and *a*/*b*-plane. For maximum magnification, a 100× magnification objective with a numerical aperture of NA = 0.9 was used, giving a spatial resolution of 350 nm in green light.

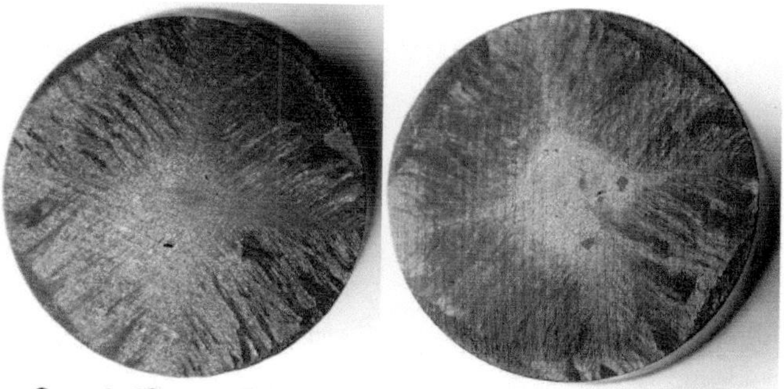

Figure 3.18 Top surface morphology of the YBCO bulk samples prepared by TSIG process.

Quantitative measurement

Quantitative microstructural data of the Y211 particle size, d_{211}, and Y211 volume fraction, V_{211}, in YBCO bulks were obtained using Image–Pro Plus image processing software. The dependences of the d_{211} and V_{211} on the distance from the seeds were done on the micrographs taken at the maximum resolution of the optical microscope.

Trapped magnetic field profile

The samples were magnetised by an electromagnet in the magnetic field 1.4 T for 3 min at 77 K. The growth direction "c" was parallel to the magnetic field. Trapped magnetic field measurement was performed after 15 min of flux relaxation using the Hall probe with a 0.3 mm step, and the Hall probe scanned the *a*/*b*-plane at a fixed distance of 50 μm.

Macrographs of the sample cuts along *a*/*c*-plane show that the single crystal does not cover the full sample area mainly in the case of the sample with CeO_2 addition (Fig. 3.19). In particular, the solidified melt and the Y123 spheroids [64] are present at the rims and at the bottom of the samples (Fig. 3.20). As both samples grew at the same time/temperature profile, the slower crystal growth rate of the sample with CeO_2 addition is unexpected because it was supposed that the growth rate should be higher for the sample with smaller Y211 particles [65].

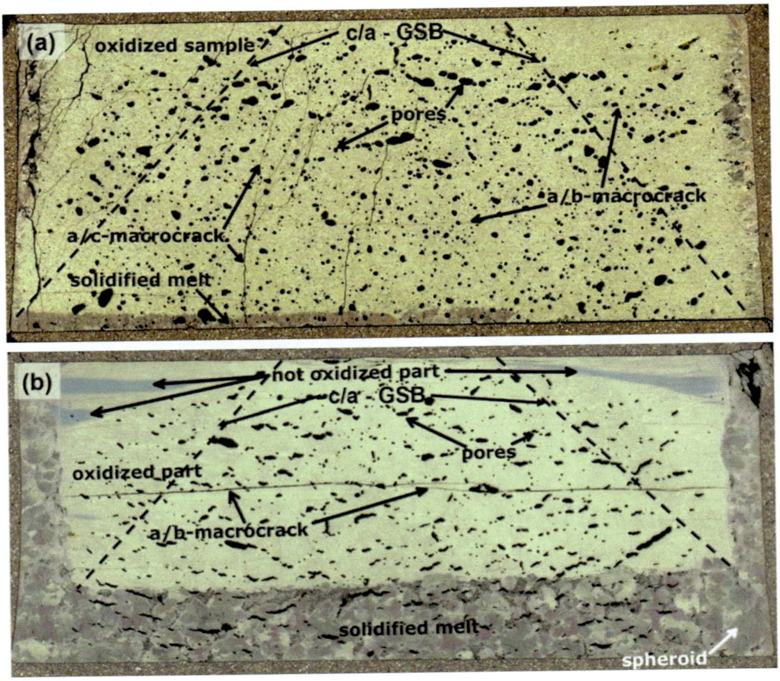

Figure 3.19 Optical macrographs of the *a*/*c*-section: (a) sample IG and (b) sample IG + CeO_2.

Figure 3.20 Polarised light optical micrograph of Y123 spheroids at the bottom of the sample IG + CeO_2. The pores (black) are present in the form of elongated bubbles and small dots in the solidified melt between the spheroids.

Another important observed feature is that the sample with CeO_2 addition has elongated regions of tetragonal phase in the *a*-GS close to the top surface. This insufficient oxygenation can be caused by differences in porosity but also by the intensity of cracking in these regions during the oxygenation process caused by the size and distribution of Y211 particles [47].

One of the advantages of the samples prepared by TSIG process is the fact that porosity is much lower than in the conventional TSMG method [66]. The source of the porosity in the bulk TSMG Y123 superconductors are gases evolving from both the original porosity of the sintered samples and peritectic melting of Y123 phase [67]. Residual carbonates present in the sample easily decompose in the melt and therefore carbon dioxide (CO_2) formed during this decomposition may also contribute to porosity. In the case of samples prepared by the TSIG process, one source of porosity is eliminated by the absence of melting of the Y211 precursor. The porosity of both samples was analysed on the *a*/*c*-sections. The microstructure observation of the samples revealed not only different volume fraction of pores but also their different shape (see Fig. 3.19a,b).

In both samples, the porosity free region at the sample surface is formed by the diffusion of gases through the melt. The volume fraction of pores (Table 3.1) counted by image processing in the central area of the samples is higher for the IG sample (5 vol %) than for the IG + CeO_2 sample (3.7 vol %). While the pores in the sample IG have mostly globular shape, in the sample with CeO_2 addition the shape is more elliptical.

Table 3.1 Volume fraction and ellipticity of pores in the IG and IG + CeO_2 samples

Pores	IG	ID + CeO_2
Volume fraction	$V_p = 5\%$	$V_p = 3.7\%$
Ellipticity	$R_E = 1.6$	$R_E = 3.5$

The reason for more elliptical pores in the sample with CeO_2 is unclear but we suppose that it is caused by changes in viscosity and/or surface tension of the peritectic melt containing dissolved cerium that may influence the interaction of gases in the Y211 precursor with the soaking melt [68]. In principle, the sample porosity in the form of gas bubbles can be regarded as a kind of defect that affects the flowing supercurrent and decreases the levitation force.

Another important microstructural feature which was influenced by CeO_2 addition was the dependence of the size and the volume fraction of Y211 particles on the distance from the seed [35]. Such inhomogeneous Y211 distribution could be explained by both the pushing/trapping of Y211 particles by the growth front and by different crystal growth rate in the *a*- and *c*-growth directions [34, 69] and by growth of Y211 particles by Ostwald ripening process [6].

This inhomogeneity can be very strong when the mean Y211 particle size in the Y123 bulk matrix is small (i.e., below or around 1 μm) under conditions of a relatively low Y123 crystal growth rate, mainly at the beginning of crystallisation [34].

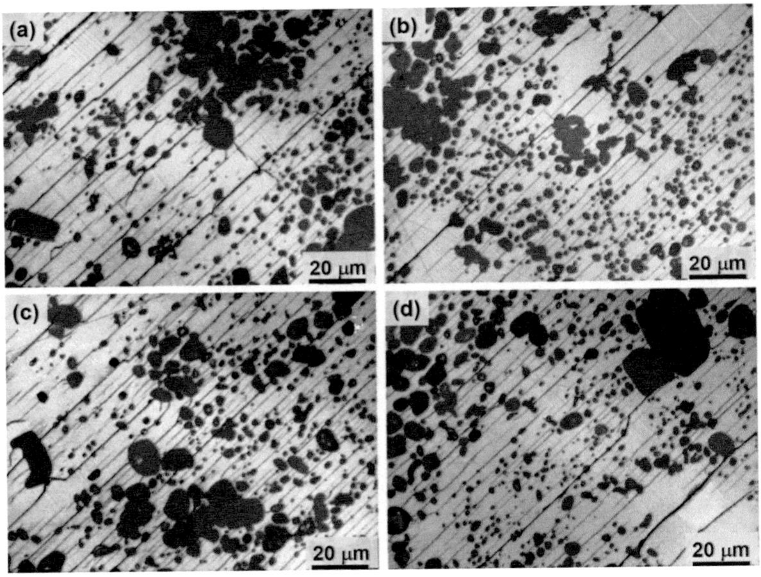

Figure 3.21 (a–d) Polarised light optical micrograph of Y211 particle concentration in the growth sector; beginning of the *a*- (a) and *c*-growth sector (c); end of the *a*- (b) and *c*-growth sector (d) in IG single-domain sample.

The microstructure at the beginning and the end of the growth sectors is shown in Fig. 3.21a–d for the IG sample and Fig. 3.22a–d for the IG + CeO_2 sample. It should be noted that the first measured value of the Y211 particle concentration and Y211 particle diameter in the *a*- and *c*-growth sectors was done approximately 2 millimetres away from the seed because the top surface of the samples was flattened for the measurements of the trapped magnetic field.

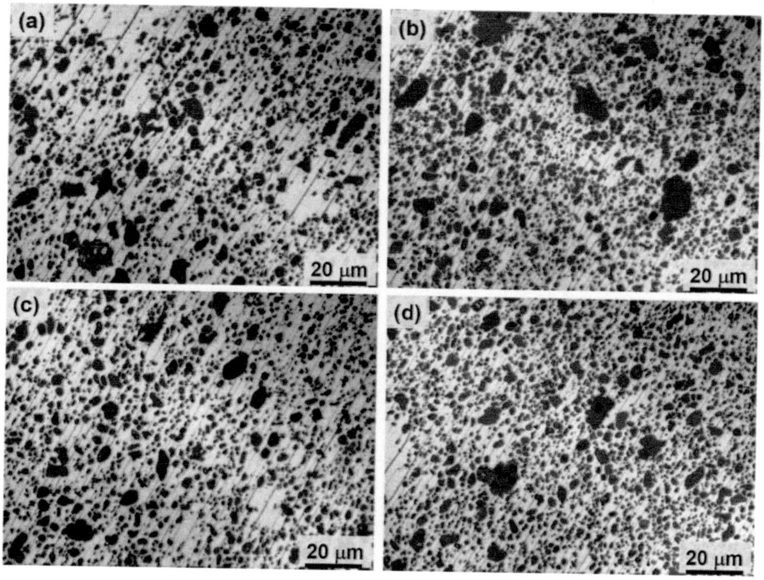

Figure 3.22 (a–d) Polarised light optical micrograph of Y211 particle concentration in the growth sector; beginning of the *a*- (a) and *c*-growth sector (c); end of the *a*- (b) and *c*-growth sector (d) in IG + CeO_2 single-domain sample.

The results of image processing measurements of Y211 volume fraction are presented in Fig. 3.23a,b. We divided the Y211 measurements into three parts: All Y211 particles, Y211 particles smaller or larger than 3 μm. Agglomerates of Y211 particles, which were not automatically divided into individual particles by Image-Pro software itself, were considered as one particle.

The behaviour typical of Ostwald-ripening mechanism is confirmed in the IG sample: Larger Y211 particles grow and smaller Y211 particles are dissolved, thus the volume fraction of the Y211 particles smaller than 3 μm decreases with the distance from the seed (Fig. 3.24b), while the volume fraction of the Y211 particles above 3 μm increases (Fig. 3.24a). In the case of IG + CeO_2 sample we observed an increasing volume fraction of the Y211 particles both smaller (Fig. 3.24b) and larger than 3 μm (Fig. 3.24a). This could be explained by the fact that the size of some Y211 particles was smaller than critical, and these were pushed and trapped later at higher crystal growth rate. The samples with CeO_2 admixture have more Y211 particles below 3 μm (Fig. 3.24b). Their average volume fraction is approximately $V_{211 < 3\ \mu m}$ = 10.5% in the *a*-GS and $V_{211 < 3\ \mu m}$ = 9.8% in the *c*-GS, whereas in the sample IG the Y211 volume fraction in *a*-GS is five times lower ($V_{211 < 3\ \mu m}$ = 2%) and in the *c*-GS is up to seven times lower ($V_{211 < 3\ \mu m}$ = 1.4%).

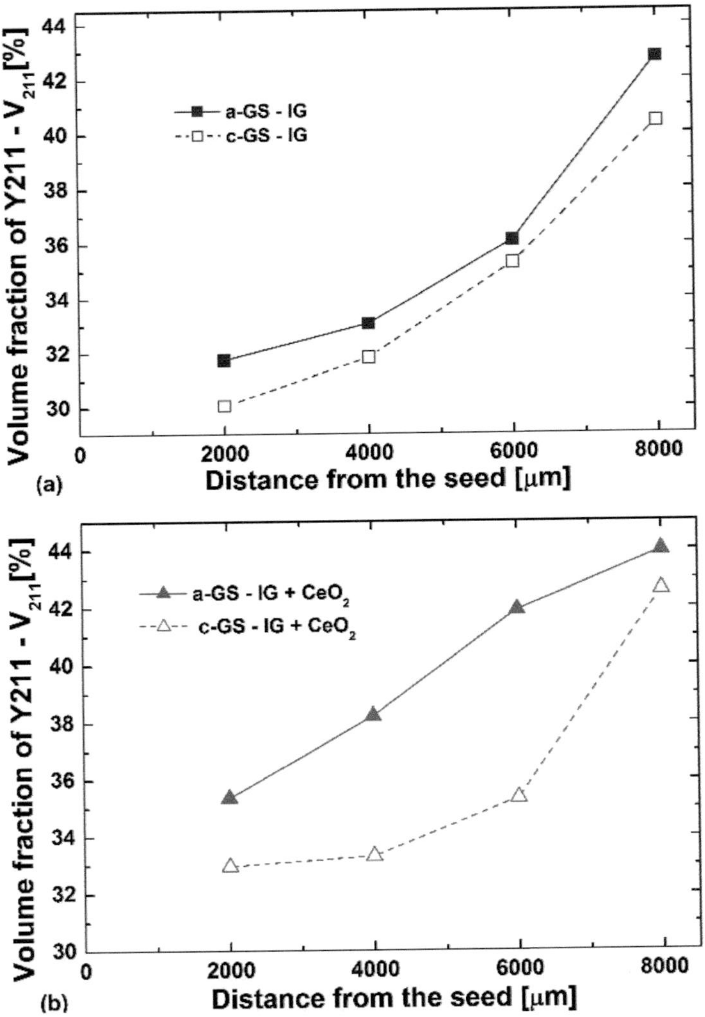

Figure 3.23 Dependences of the Y211 volume fraction (V_{211}) on the distance from the seed in the *a*-growth sector (*a*-GS) and in the *c*-growth sector (*c*-GS) for the samples: (a) IG and (b) IG + CeO_2.

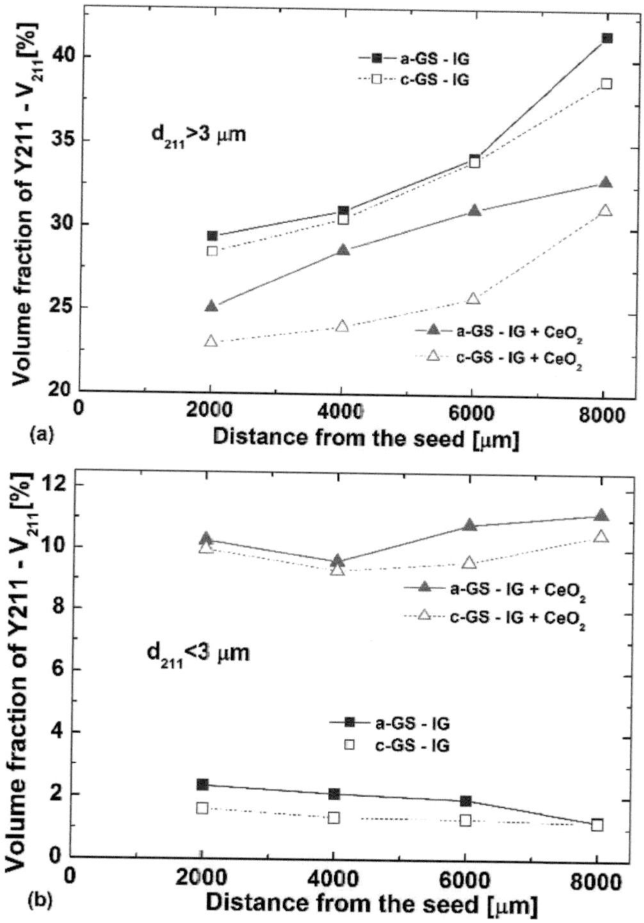

Figure 3.24 Dependences of the Y211 volume fraction (V_{211}) on the distance from the seed in the *a*-GS and in the *c*-GS for the sample IG and sample IG + CeO_2 (a) d_{211} above 3 μm and (b) d_{211} below 3 μm.

The measured values of Y211 particle size in the growth sectors (GS) for both samples are illustrated in Fig. 3.25. The mean Y211 particle size in the *a*-GS and in the *c*-GS is increasing with the distance from the seed and is much higher for the IG sample. From the Y211 particle size distribution in the samples (Fig. 3.26) it is apparent that the addition of CeO_2 effectively reduces the amount of large Y211 particles and increases the number of Y211 particles smaller than 3 μm.

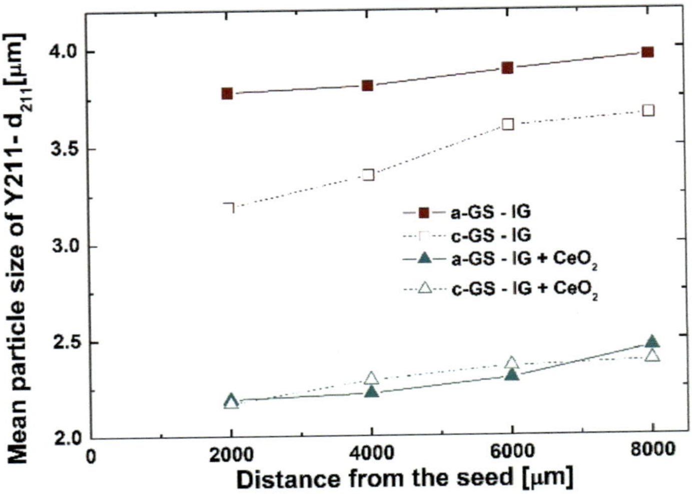

Figure 3.25 Dependences of mean particle size of Y211 (d_{211}) on the distance from the seed in the growth sectors.

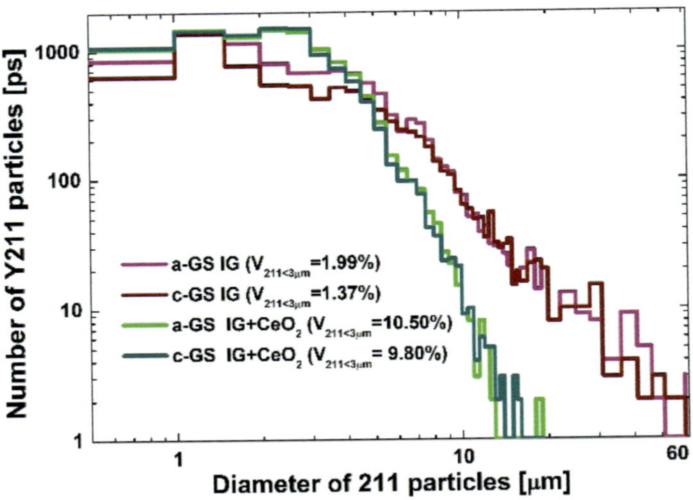

Figure 3.26 Distribution of the Y211 particle size in the growth sectors.

To summarise, the addition of CeO_2 to the studied IG sample refines Y211 particles that should improve critical current density in low magnetic fields. This refinement is not as effective as in the best Y211 sample where mean Y211 particle size is close to 1 μm [34]. The IG sample without CeO_2 has high local inhomogeneity in the

Y211 volume fraction and in the size of the Y211 particles leading to clusters of large Y211 particles and to Y211-free regions.

Observation on the *a*/*c*- and *a*/*b*-sections at low magnification revealed *a*/*b*- and *a*/*c*-oxygenation cracks (Figs. 3.27 and 3.28). These cracks are formed due to the shortening of lattice parameters at higher oxygen concentration in Y123 phase [47]. The *a*/*c*-oxygenation cracks (*a*/*c*-OC) start from the surface of *a*/*b*-oxygenation cracks (*a*/*b*-OC). They are also called macrocracks because their length is much higher than the length of *a*/*b*-microcracks induced by individual Y211 particles which length is not more than some Y211 interparticle distances [47].

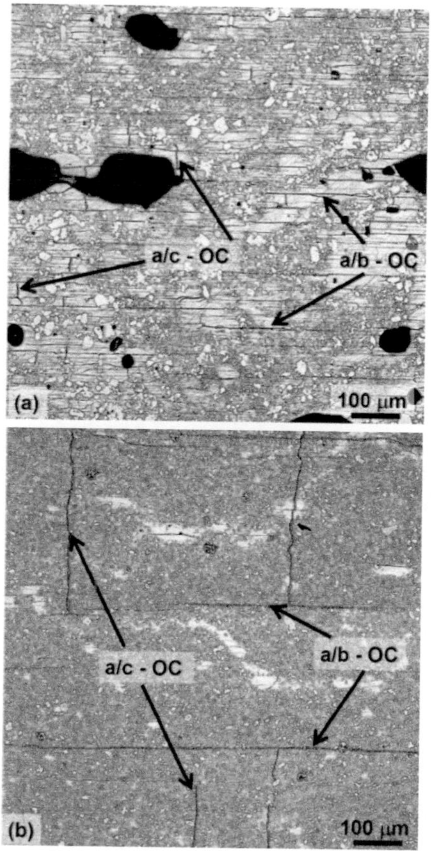

Figure 3.27 Optical micrographs of the oxygenation cracks (OC) in the *a*/*c*-plane by normal light after etching for the sample IG (a) and IG + CeO_2 (b). The *a*/*c*-oxygenation cracks (*a*/*c*-OC) start at the surface of *a*/*b*-oxygenation cracks (*a*/*b*-OC).

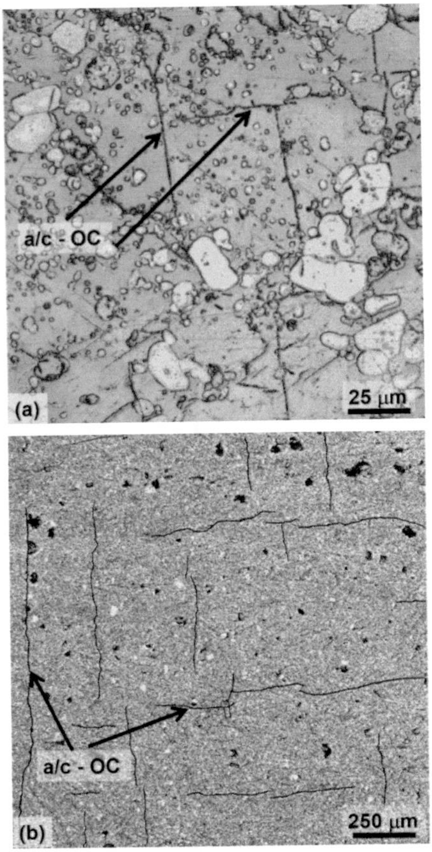

Figure 3.28 Optical micrographs of the oxygenation cracks (OC) in the *a/b*-plane by normal light after etching for the sample IG (a) and IG + CeO_2 (b). The large Y211 particles stop the *a/c*-OC propagation in the IG sample.

The oxygenation cracks were counted on *a/c*-section along the sample axis and on the top *a/b*-surface of the sample. The counted mean length and spacing of oxygenation cracks for the *a*- and *c*-growth sectors are presented in Table 3.2.

The *a/c*-OCs appear in both growth sectors (*a*-GS and *c*-GS), and their length and spacing is shorter in the *c*-GS than in the *a*-GS. The other even more important fact is that the length and spacing of *a/c*-OCs for the IG sample is much lower than for the IG + CeO_2 sample in both growth sectors (Table 3.2 and Figs. 3.29 and 3.30).

Table 3.2 Average values of *a/c*-OCs spacing and length in both sectors on the *a/b*- and *a/c*-plane for the IG and IG + CeO_2 samples

	a/b-Plane				*a/c*-Plane			
	IG		IG + CeO_2		IG		IG + CeO_2	
	Length	Spacing	Length	Spacing	Length	Spacing	Length	Spacing
a-GS	50 μm	314	504 μm	775	20 μm	437	90 μm	623
c-GS	34 μm	157	188 μm	364	17 μm	338	40 μm	422

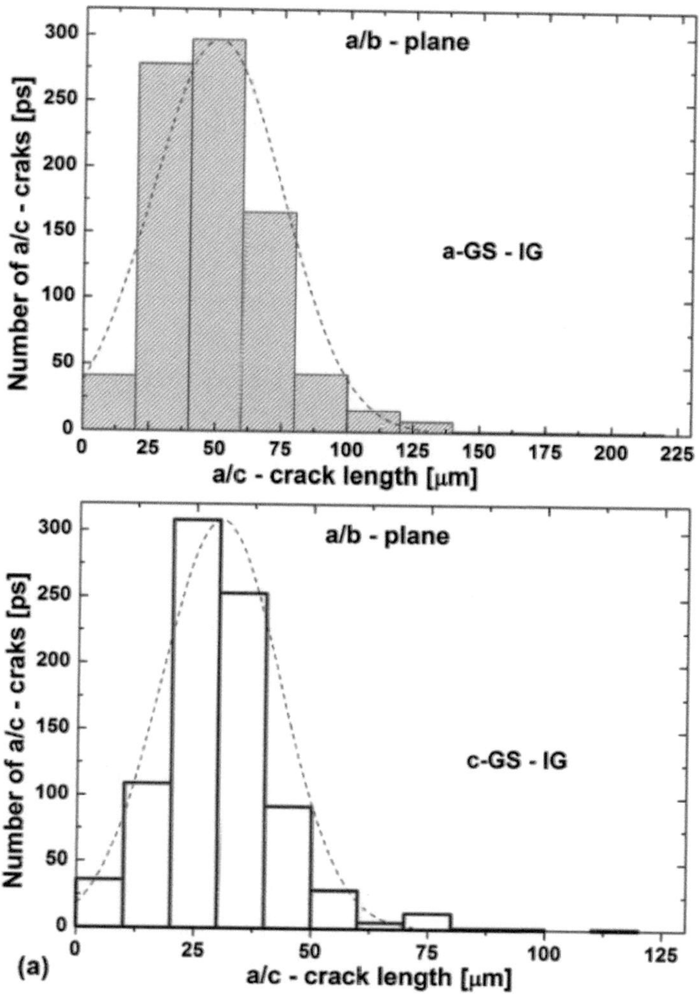

(a)

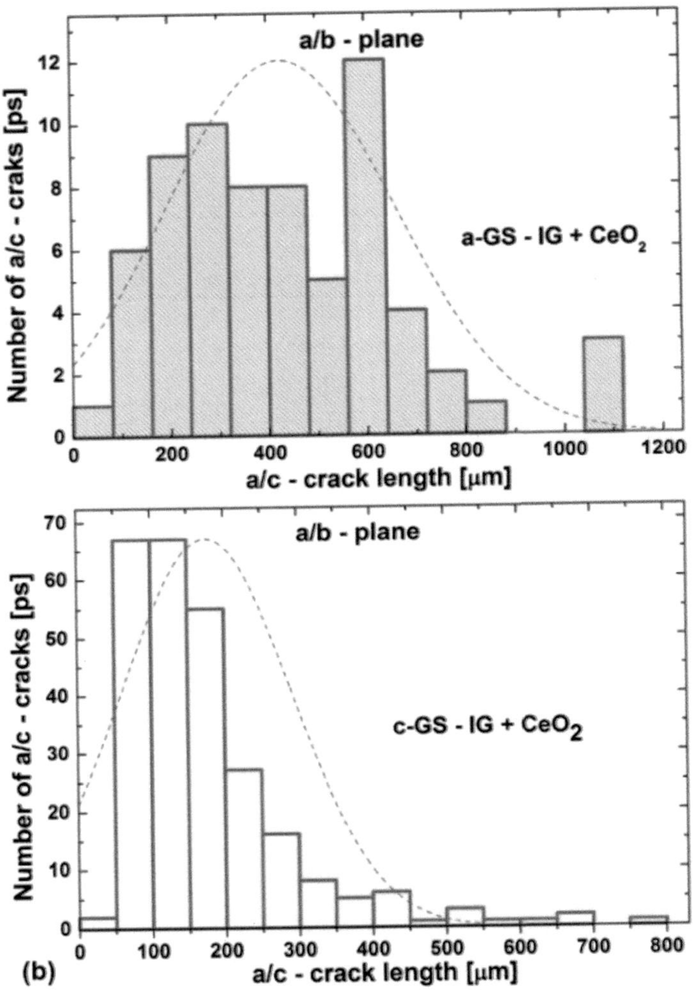

Figure 3.29 Distribution of *a/c*-crack length on the *a/b*-plane in the *a*-GS and *c*-GS for the samples: (a) IG, and (b) IG + CeO_2.

The spacing of oxygenation cracks is mainly influenced by fracture toughness of the Y123 + Y211 composite [47], therefore we may expect differences of spacing in the *a*- and *c*-GSs as higher volume fraction of the Y211 particles in the *a*-GS leads to higher fracture toughness. The length of *a/c*-OCs observed on the *a/c*-section is limited by the spacing of *a/b*-OCs as they start at the surface of *a/b*-OCs. The spacing of *a/b*-OCs is again a function of fracture strength and consequently of the Y211 volume fraction that reflects macroscopic inhomogeneity in Y211 concentration as well as macroscopic stresses induced by this sample inhomogeneity. Very short length and low spacing of *a/c*-OCs in the IG sample is apparently caused by the presence of the clusters of large Y211 particles. The *a/c*-OCs evidently stop at such a cluster (Fig. 3.27a).

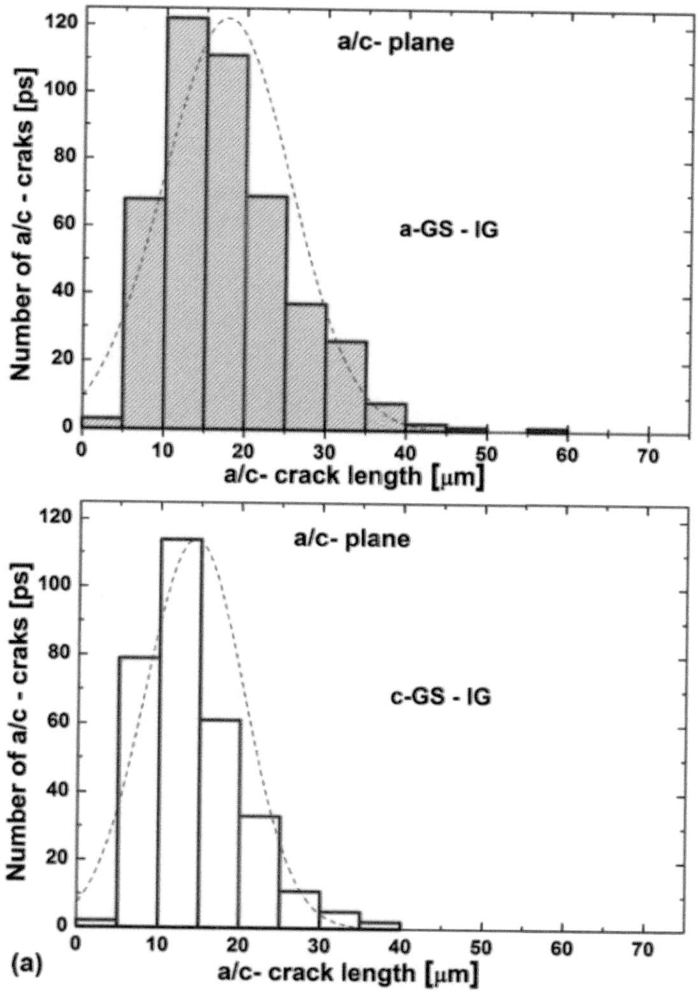

(a)

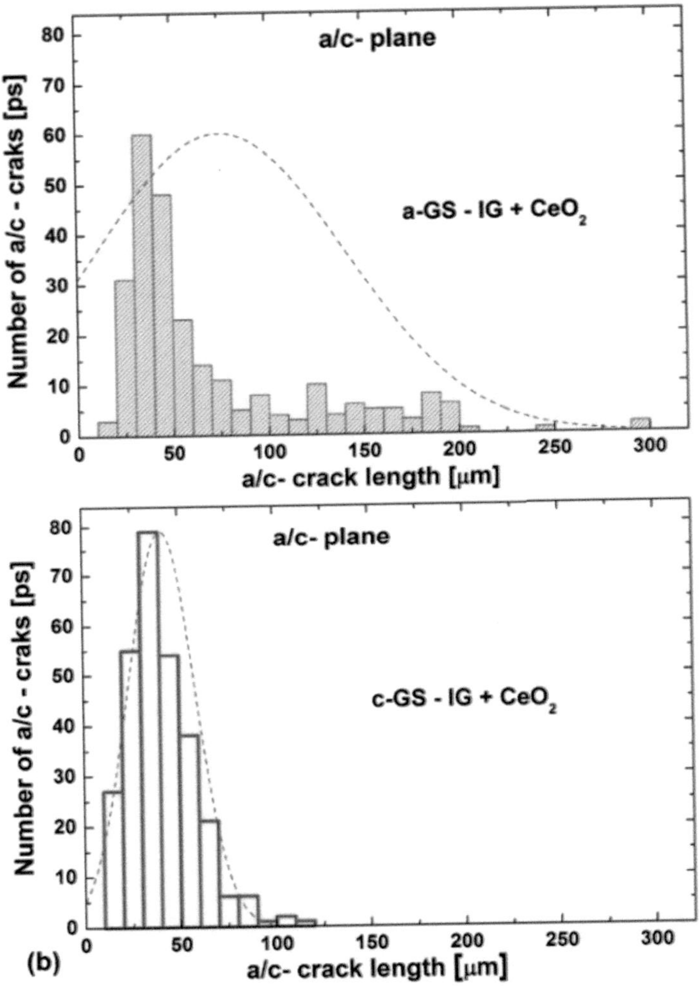

Figure 3.30 Distribution of *a*/*c*-crack length on the *a*/*c*-plane in the *a*-GS and *c*-GS for the samples: (a) IG, and (b) IG + CeO_2.

Significant difference is also observed in the distribution of a/c-crack length measured in both studied samples (Figs. 3.29 and 3.30). While the distribution of length of a/c-OCs in the IG sample is quite narrow, in the IG + CeO_2 sample the distribution is very broad, and the maximum length of a/c-OCs exceeds one millimetre. The defect of this size has then the main role in sample strength degradation.

The oxygenation cracks parallel to the *a*/*b*-plane do not significantly affect the superconducting properties of the sample, since they are parallel to the direction of the super-current flow in field trapping applications. The formation of cracks parallel to the *a*/*c*-plane (*a*/*c*-OCs), however, limits the superconducting properties of the bulk grain since they reduce the effective cross-section of the sample. The *a*/*c*-cracks can significantly influence the mechanical stability of the bulk because of the tensile stress in the *a*/*b*-plane induced during sample magnetisation.

Figure 3.31 shows the results of the measured trapped magnetic field profile (3D and 2D maps) for the IG and IG + CeO_2 samples at 77 K. Both samples show only one peak in their field profile which signals a well-grown single domain. Magnetic field profile reflects macro- and microstructure of the studied samples. As mentioned earlier, the size of the Y211 particles is much larger in the IG sample than in the IG + CeO_2 sample. We suppose that Y211 size was the most significant parameter affecting the trapped field of studied samples that was higher in the IG + CeO_2 sample (597 mT) than in IG sample (370 mT). On the other hand, an irregular shape is visible at the bottom of the profile of the IG + CeO_2 sample. We suppose that this is caused by imperfect oxygenation of the sample (Fig. 3.32).

In summary: The effect of CeO_2 addition to the YBCO bulk samples produced by TSIG process on the growth and microstructure formation was investigated. Quantitative analysis of porosity, the size and volume fraction of the Y211 particles as well as their impact on the formation and propagation of *a*/*b* and *a*/*c*-cracks was performed in the samples with (IG + CeO_2) and without CeO_2 addition (IG). The slower crystal growth rate and elongated regions of tetragonal phase in the *a*-GS close to the top surface of the sample with CeO_2 were observed.

The volume fraction of pores is higher in the IG sample (5 vol %) than the IG + CeO_2 sample (3.7 vol %). While the pores in the IG sample have mostly globular shape, in the sample with CeO_2 the pore shape is more elliptical which was related to possible changes in viscosity and/or surface tension of the peritectic melt containing dissolved cerium.

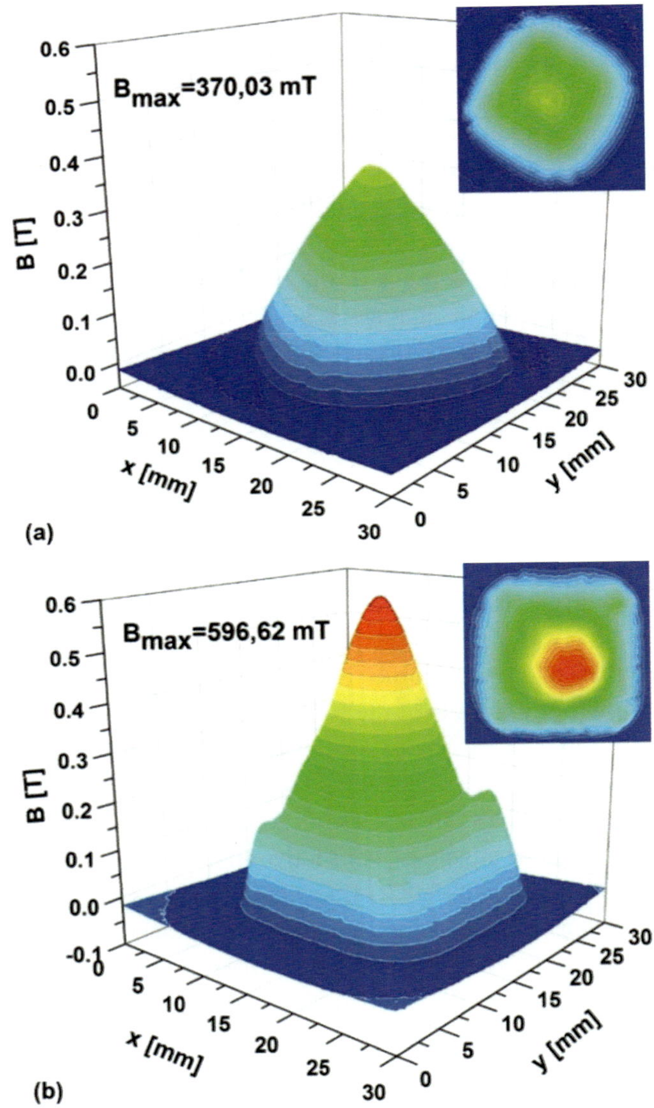

Figure 3.31 Trapped magnetic field profile at 77 K of the sample IG (a) and sample IG + CeO_2 addition (b).

The microstructure study showed smaller Y211 particles in the sample with CeO_2 where the volume fraction of particles below 3 μm was approximately 11%. In the case of the sample without admixture the microstructure was inhomogeneous and contained only 2% of the Y211 particles below 3 μm and clusters of large Y211 particles.

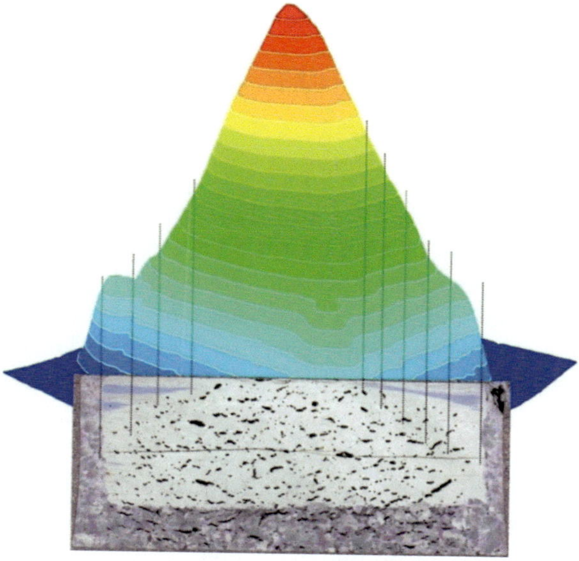

Figure 3.32 Trapped magnetic field profile reflects the non-oxidised part of the sample.

While the *a*/*c*-OC length distribution in the IG sample is quite narrow around the mean length below 50 μm, in the case of the IG + CeO_2 sample the distribution is very broad, and the maximum length of *a*/*c*-OCs is over one millimetre. The defects of this size may be critical to the sample strength degradation.

The size of the Y211 particles was the most significant parameter affecting measured trapped field in the studied samples. Trapped magnetic field in the IG + CeO_2 sample was higher (597 mT) than in IG sample (370 mT). An irregular shape of the field profile of the IG + CeO_2 sample was caused by imperfect oxygenation of the sample.

3.4 Microstructural Aspects of Infiltration Growth YBCO Bulks with Chemical Pinning

The main advantage of infiltration growth method is almost the same dimensions of resultant sample in compared to the starting precursors. This property is very important in the preparation of samples for real applications. Study of these materials is concerned with improving the superconducting properties and understanding the mechanisms of creation of their inherent structure [55, 57, 63, 70–74].

The samples prepared by synthesis of Y211 (Y_2BaCuO_5) solid phase by reaction (3.1):

$$Y_2O_3 + BaCuO_2 \rightarrow Y_2BaCuO_5 \tag{3.1}$$

have sufficient porosity [75] and they are suitable for standard oxygenation process. In addition, the Y211 particles in Y123 matrix are very small even without CeO_2 addition [63, 71].

It is well known that Y211 particle size has significant influence on pinning and consequently on critical current density in zero magnetic field [35, 76–78]. The substitutions in Y123 crystal lattice (dopants) induce so-called chemical pinning associated with the appearance of a secondary peak at higher fields in the dependence of critical current density, J_c, on applied magnetic field, B [79–82]. It was already shown that the Yb_2O_3 substrate which is used to retard grain nucleation at the bottom of the melt grown sample can be the source of chemical pinning by the contamination of the sample by Yb [83].

In this work, the single-grain YBCO bulk superconductors prepared by infiltration growth process were analysed. The influence of the composition of the source of melt for infiltration and the CeO_2 addition on Yb contamination, microstructure, transition temperature, and critical current density at 77 K was studied.

The samples were prepared by infiltration growth method. All precursor pellets were prepared by mixing in a grinding mill for 20 min and subsequently uniaxially pressed under pressure 6 MPa to form cylindrical pellets. For the sample S_1 the solid phase pellet was prepared by the synthesis of 1 mol Y_2O_3 and 1.2 mol $BaCuO_2$ [75] and the composition of the liquid source pellet was Y123 + $3 \cdot BaCuO_2 + 2 \cdot CuO$ (16.5 g). The sample S_2 was prepared in the same way but with the addition of 0.5 wt % CeO_2 to the solid phase pellet. The sample S_3 was without CeO_2 addition, but the composition of the liquid source pellet was $Y_2O_3 + 10 \cdot BaCuO_2 + 6 \cdot CuO$ (16.5 g). Solid phase pellets were synthesised in the form of pellets with 20 mm diameter (12 g) at 910°C for 12 h. During this annealing Y211 particles were formed according to Eq. 3.1. The liquid source pellets were prepared by pressing to diameter 32 mm. The composition of layers in the crystal growth experiment consisted of liquid source pellet, on which the synthesised solid phase pellet was placed. A thin layer Nd123 seed was placed on top in the centre of the solid phase pellet.

The composition consisting of liquid source pellet, solid source pellet, and seed was laid on the base. The base had three layers composed of Al_2O_3 plate, MgO single crystal, and Yb_2O_3 base which was in contact with liquid source pellet. This composition was used for the single crystal growth in a chamber furnace.

The time-temperature regime with slow cooling through peritectic reaction temperature was used for obtaining one single crystal by epitaxial growth from the seed. First, in order to ensure infiltration of the melt from liquid source pellet to solid Y211 pellet, the sample was heated to T_{max} = 1040°C with holding time one h. Then the sample was cooled near the temperature of peritectic reaction T_p = 1015°C, and subsequently slowly cooled in three steps: To 1003°C at the rate of 0.3°C/h, continuing to 995°C at 0.2°C/h, and finally to 985°C at 0.1°C/h. After the single crystal growth by slow cooling of the sample, all system was cooled with the furnace to ambient temperature.

The arrangement of individual pellets and time-temperature regime used for the single crystal growth are presented in Fig. 3.33.

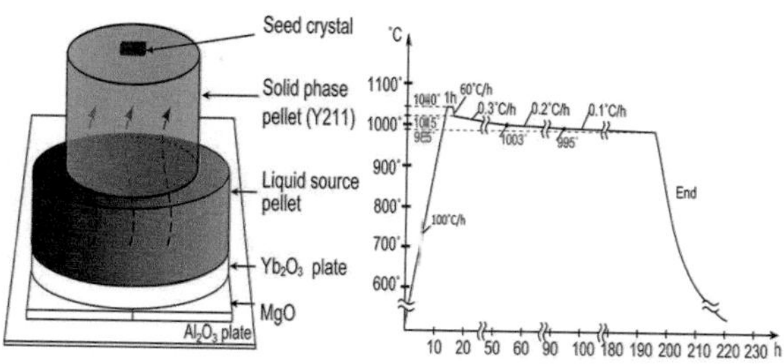

Figure 3.33 Schematic arrangement of pellet layers and time-temperature regime for preparations of YBCO single crystal by infiltration growth process.

Samples for the measurement of superconducting properties were cut from the upper part of single crystals at a distance of 1 mm from the seed. The dimensions of samples were 0.5 mm in "c" direction, and 1.5 × 1.5 mm in "a" and "b" directions. Samples were oxygenated in flowing oxygen at 410°C for 400 h. The measurements of superconducting properties were performed at liquid nitrogen temperature. The critical temperature of the transition from normal to superconducting state was estimated from the magnetic transition curves. Microstructure of the samples was observed with an optical microscope in polarised light. The size distribution of Y211 particles was evaluated from the obtained images using image analysis software. The composition of the samples was analysed by energy dispersive spectroscopy (EDS) at a SEM.

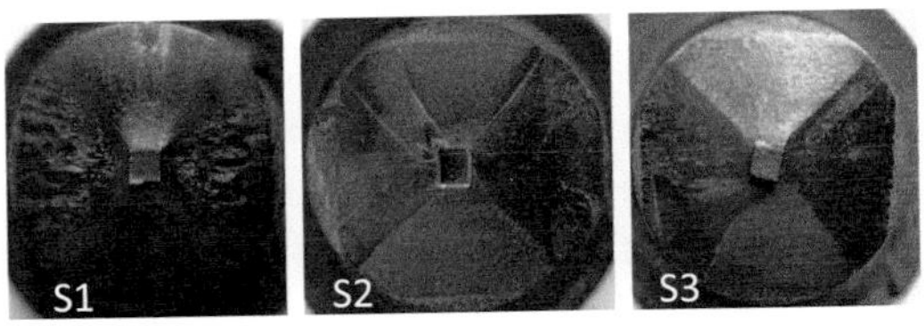

Figure 3.34a Macrographs of grown samples.

Macrostructure and microstructure of prepared samples are presented in Fig. 3.34. Presented micrographs were taken at the beginning of the *a*-growth sector close to the sample top surface. Sample S_1 has bimodal Y211 particle size distribution (Fig. 3.34a). Besides small Y211 particles, larger Y211 particles can also be seen in the microstructure micrograph. Mean particle size of Y211 particles in this sample is d_{mean} = 1.3 μm.

Volume fraction of the Y211 particles with the size larger than 4.5 μm is $V_{211>4.5}$ = 3.1 vol % (Fig. 3.35). The *a*/*b*- and *a*/*c*-cracks are clearly seen in the orthorhombic Y123 phase after oxygenation. Sample S_2 prepared by the same procedure but with the addition of 0.5 wt % CeO_2 exhibits much different microstructure (Fig. 3.34b). The Y211 particles are smaller and do not have bimodal character. The mean Y211 particle size is d_{mean} = 1.0 μm and the particles larger than 4.5 μm are present in the sample in a very small amount ($V_{211>4.5}$ = 0.1 vol %). The *a*/*b*-cracks are not observed in the structure of this sample because its microstructure was analysed before oxygenation in the tetragonal state. Sample S_3 with different composition of the source of melt and without CeO_2 addition exhibits similar microstructure as sample S_1 but with a higher fraction of large Y211 particles. The mean Y211 particle size is d_{mean} =

1.4 µm. Volume fraction of the Y211 particles with the size larger than 4.5 µm is $V_{211>4.5} = 6.8$ vol %.

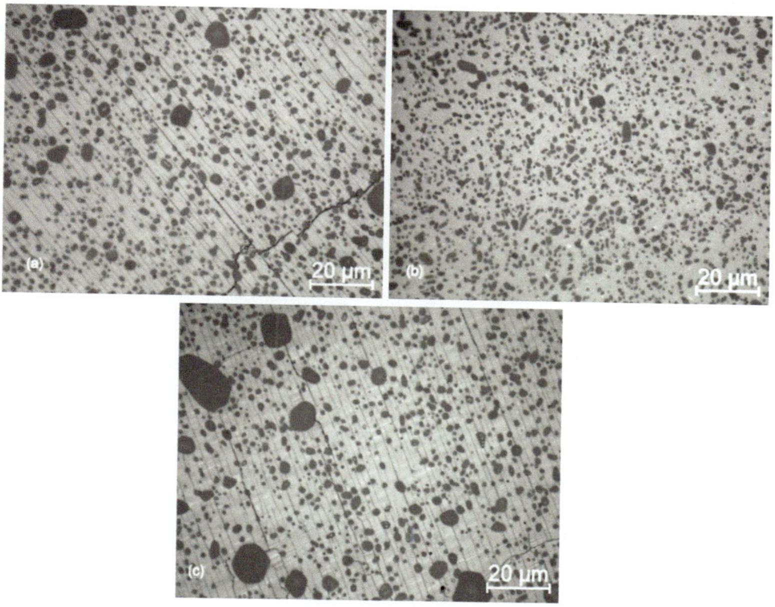

Figure 3.34b Microstructures of the YBCO single crystals with Y211 particles of samples S_1, S_2 and S_3, taken at the beginning of *a*-growth sector close to the sample top surface.

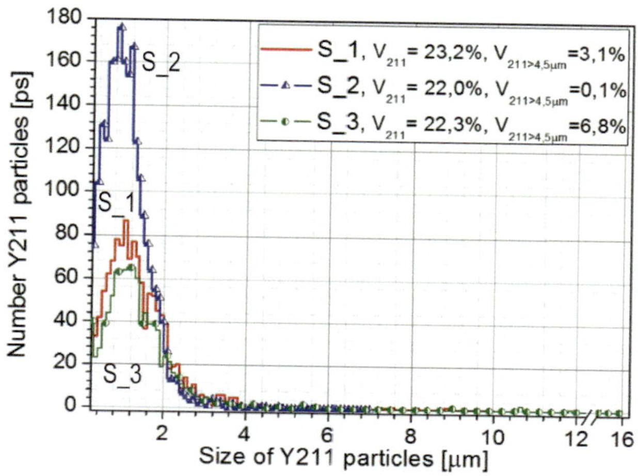

Figure 3.35 Y211 particle size distribution measured in samples S_1, S_2, and S_3.

As the studied samples were grown on the Yb_2O_3 substrate we checked the contamination of the samples with ytterbium. The concentration of Yb in the sample was measured by EDS microanalysis at SEM as a function of the distance from the bottom of the grown YBCO single-grain sample. The results of these measurements are presented in Fig. 3.36.

All three dependences have a typical diffusion profile and prove that Yb from Yb_2O_3 plate was dissolved in the melt and diffused in the melted stage to the sample.

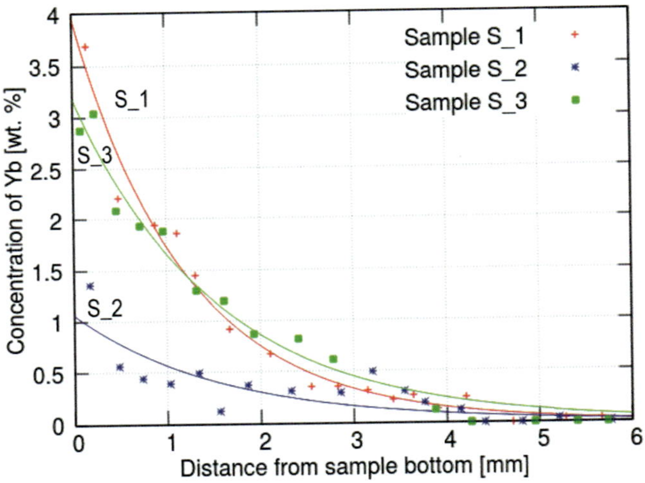

Figure 3.36 Yb concentration measured by EDS in samples S_1, S_2, and S_3.

The concentration of Yb is below the resolution limit of EDS (0.1 wt %) at the distance of about 6 mm from the seed, but we may suppose that some low concentration of Yb will be present also at the top surface of the samples. The maximum measured Yb concentration is obviously at the sample bottom surface. The lowest Yb concentration at the bottom surface was measured in the sample S_2 with CeO_2 addition and reached 1 wt %. For samples S_1 and S_3 this value was 3.5 and 3.0 wt %, respectively. Our measured Yb concentrations are the sum of Yb content in the Y123 phase and Yb content in the Y211 phase. Therefore, we measured Yb concentration in the Y211 and Y123 phases.

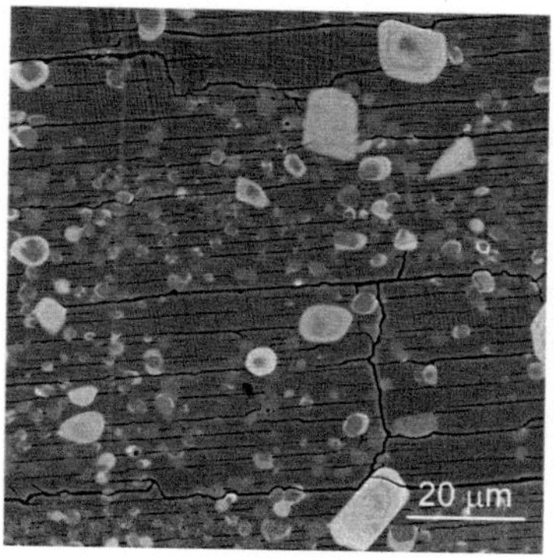

Figure 3.37 Microstructure of sample S_1 close to the sample bottom observed by SEM in backscattered electrons. Brighter parts of the 211 particles are enriched in ytterbium.

The microstructure of sample S_1 close to the bottom surface is presented in Fig. 3.37. In a backscattered electron SEM image, a higher concentration of Yb is reflected by higher brightness because the atomic mass of Yb is higher than that of Y (atomic mass: Y = 88.91, Yb = 173.0). One can see that more Yb is present in the 211 particles than in the 123 phase. In both phases, Yb substitutes for Y in the crystal lattice. Apparently, Yb diffused through the melt to the Y211 particles, and the surface parts of Y211 particles were converted into $(Y,Yb)_2BaCuO_5$ solid solution. Microstructure in Fig. 3.36 shows that bigger 211 particles have more Yb than smaller one. The middle parts of 211 particles are dark and EDS analyses done on large 211 particles confirmed that in the central part it does not contain Yb at all. The highest Yb concentration was measured in the bright part of large 211 particles at 1.8 wt %. The measured concentration of Yb in the 123 phase was 0.8 wt %.

During the formation of YBCO bulk crystal by back peritectic reaction:

$$((Y,Yb)_2BaCuO_5)_S + (3BaO{\cdot}5CuOy((Y,Yb)O_{1.5}))_L = ((Y, Yb)Ba_2Cu_3O_x)_S \tag{3.2}$$

where symbol, S, means solid and symbol, L, means liquid, Y and Yb from the (Y,Yb)211 particles are participating in formation of (Y,Yb)123 crystal. The local concentration of Yb in the diffusion layer ahead of growing 123 crystal, influenced by local 211 particle size distribution, will influence Yb concentration in the 123 phase. We may also suppose that Yb concentration will be higher than average around large 211 particles, i.e., inhomogeneous Yb distribution in the 123 phase can be expected.

Measured values of transition temperature T_c and critical current density J_c are presented in Figs. 3.38 and 3.39, respectively. Values of critical current density were evaluated from magnetic loop measurements by means of extended Bean model [84].

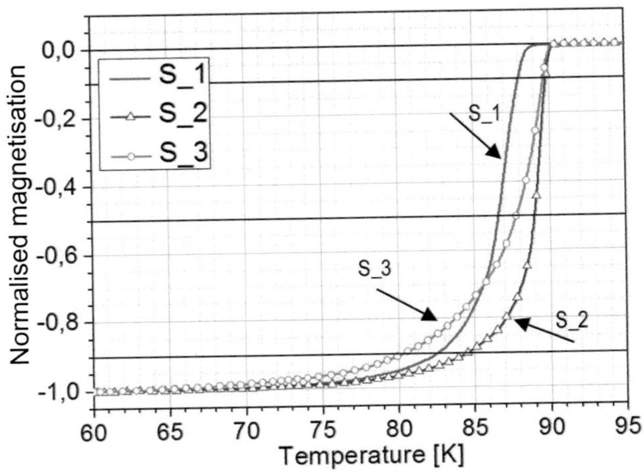

Figure 3.38 The temperature dependence of magnetisation of samples S_1, S_2, and S_3.

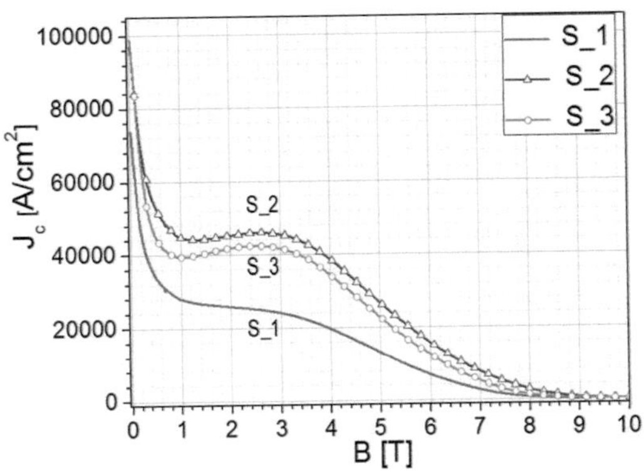

Figure 3.39 J_c-B properties at 77 K for samples S_1, S_2 (CeO_2 addition), and S_3 (different source of melt) prepared by infiltration growth method with the synthesis of Y211 solid phase.

Analysed samples exhibit differences in the transition to superconducting state but also in the field dependences of critical current density. The highest transition temperature was measured for the sample S_2 with CeO_2 addition. Sample S_1 with the highest Yb concentration at the bottom surface has the lowest T_c while sample S_3 with the highest volume fraction of large 211 particles has the broadest transition. A characteristic feature in J_c-B dependences is the appearance of a secondary peak that suggests the presence of chemical pinning in the studied samples caused by Yb substitution. Besides that, high values of critical current density at zero field reaching 10^5 A/cm^2 for sample S_2 are typical of samples prepared by the infiltration process [55, 70, 73].

Microstructure analysis and measured superconducting properties confirmed that the studied samples contain ytterbium impurities from the Yb_2O_3 substrate. This admixture occurs at the melting stage because the diffusion of Yb in the melt is several orders of magnitude higher than in the solid state. First, Yb is diffused to the melted batch. Then it infiltrated to the solid-state Y211 tablet with the melt, and finally, it diffused from the Yb_2O_3 substrate to the melted sample during the maximum temperature step as well as to the melted part of the sample during YBCO crystal growth. Apparently, this process is influenced by the composition of the source of melt but also by CeO_2 addition. The composition of the melt and CeO_2 addition may affect the viscosity of the melt and consequently the diffusion rate of Yb in the melt. Measured diffusion profiles of Yb concentration versus distance from the bottom of the sample may reflect the concentration of Yb at the top surface of the YBCO crystal where superconducting properties were measured.

Another reason for different behaviour of studied samples are the differences in the 211 particle size distribution. Yb concentrates in the 211 particles, and during the peritectic reaction, Yb from these particles participates in the formation of (Y,Yb)123 crystal. The concentration of Yb in the (Y,Yb)123 phase depends on the kinetics of 211 particle dissolution, and consequently the 211 particle size distribution influences Yb concentration in the 123 phase. We also suppose that Yb concentration will be higher than average around large 211 particles the, i.e., inhomogeneous Yb distribution in the 123 phase can be expected. This may be the reason for the broader transition to the superconducting state of sample S_3 which exhibits largest 211 particles.

The appearance of a secondary peak in J_c-B dependences is typical for chemical pinning induced by substitutions in Y123 crystal lattice. Based on measured diffusion profiles of Yb concentration, we may suppose that observed behaviour is caused by Yb substitutions in Y123 crystal. The concentration of dopant for optimum pinning is related to the state when $\lambda = 2\ \xi$ (λ—mean distance between chemical pinning centres, ξ—coherence length) [80] and is only one hundredth of weight percent, which is below the resolution limit of EDS analysis.

In summary: Microstructure and superconducting properties of single-grain YBCO bulk superconductors prepared by infiltration growth process were studied. It was shown that:

- The composition of the source of melt for infiltration and CeO_2 addition influence the transition to superconducting state and field dependence of critical current density at 77 K.
- Studied samples are doped by Yb from the Yb_2O_3 substrate and exhibit chemical pinning effect.
- Yb is concentrated in 211 particles.
- Concentration of Yb in measured samples can be influenced by the diffusion rate of Yb in the melt and by the kinetics of 211 particle dissolution during the growth of YBCO crystal.
- Diffusion rate of Yb in the melt can be influenced by the composition of the melt.

3.5 Influence of Sm_2O_3 Microalloying and Yb Contamination on Y211 Particles Coarsening and Superconducting Properties of IG YBCO Bulk Superconductors

The latest results show that the samples prepared by the synthesis of Y211 (Y_2BaCuO_5) solid phase by reaction (3.3):

$$Y_2O_3 + BaCuO_2 \rightarrow Y_2BaCuO_5 \tag{3.3}$$

exhibit small Y211 particles in Y123 matrix even without CeO_2 or Pt addition [63, 71]. Besides that, they have sufficient porosity [75] and are suitable for the standard oxygenation process. Small Y211 particles in Y123 matrix have positive influence on critical current density in zero magnetic field [35, 76–78]. It was already shown that the size of Y211 particles can also be influenced by microalloying with RE elements (doping) in the concentrations below 1 wt %. Doping with RE elements may lead to substitutions in Y123 crystal lattice that induce so-called chemical pinning associated with the appearance of a secondary peak at higher fields in the dependence of critical current density, J_c, on applied magnetic field, B [79–82, 85]. The Yb_2O_3 substrate which is used to retard grain nucleation at the bottom of the sample can also be the source of chemical pinning by the contamination of the sample by Yb [83].

In our previous work [86], we have shown that RE dopants into YBCO single-grain bulks can influence both coarsening of Y211 particles and chemical pinning. In this work, the microstructure, contamination from the substrate and superconducting properties of single-grain YBCO bulk superconductors prepared by infiltration growth process were analysed. The influence of Sm microalloying and the composition of layers in the crystal growth experiments on Yb contamination, coarsening of Y211 particles, transition temperature, and field dependence of critical current density at 77 K were studied.

The samples were produced by infiltration growth method. The solid phase pellet (Y211 pellet) was synthesised from the mixture of $Y_2O_3 + 1.2 \cdot BaCuO_2$ powders. This ratio was chosen according to work [75]. Sm dopant in the form of Sm_2O_3 powder in the quantity 0.21 wt % was added into the precursor of solid phase before synthesis [82]. The powders were mixed for 20 min and then uniaxially pressed under pressure 6 MPa into a cylindrical form of diameter 20 mm and weight 12 g.

The precursor of liquid-phase pellet consisted of the mixture of powders in the ratio $Y_1Ba_2Cu_3O_{7-x} : 3 \cdot BaCuO_2 : 2 \cdot CuO$. The powders were mixed for the same time as solid phase precursor. 16.5 g of mixed powder was pressed into a cylindrical form with 32 mm diameter at analogous conditions as in the previous case.

The thinner solid phase pellet precursor was put onto the thicker liquid-phase pellet. Precursor pellets of sample S1 without dopant and sample S2 with Sm_2O_3 addition were placed on the Yb_2O_3 substrate plates. In the case of sample S3, ZrO_2 bars were inserted between the liquid source pellet and Yb_2O_3 layer. The solid- and liquid-phase pellet with Yb_2O_3 substrate plate were put on single-crystalline MgO laying on Al_2O_3 plate. In order to ensure epitaxial growth of the crystals, Sm123 single crystal seeds were placed in the centre on the top of the whole assembly. Schematic arrangement of layers for YBCO bulk superconductors production is shown in Fig. 3.40.

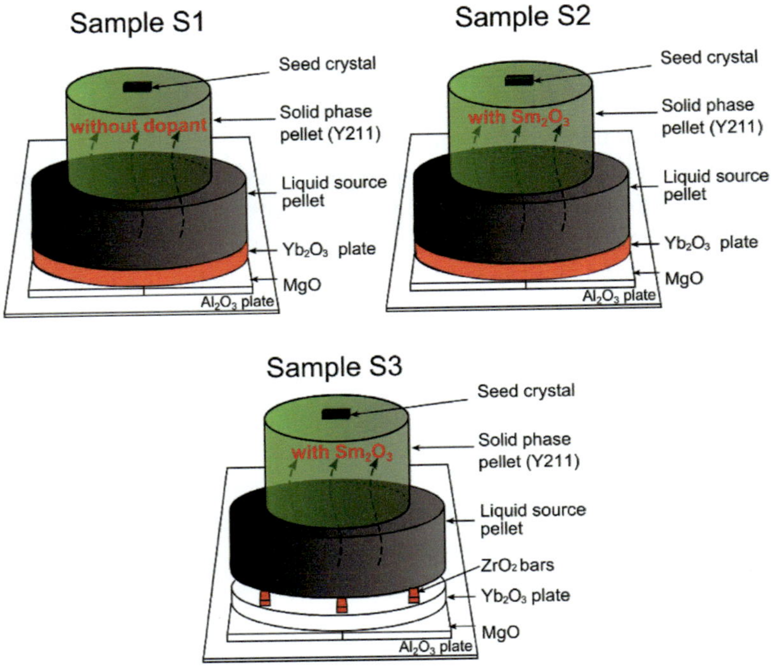

Figure 3.40 Schematic arrangement of YBCO bulk superconductors production by infiltration growth method without/with Sm_2O_3 (S1/S2, S3, respectively) addition and with used Yb_2O_3 substrate plate for samples S1, S2, and ZrO_2 bars for sample S3.

The precursors of Y211 solid phase pellet were sintered for 20 h at 910°C according to Eq. 3.3. Time-temperature diagram of heat treatment of solid phase pellet is shown in Fig. 3.41a.

The growth of single crystals was realised in accordance with the time-temperature schedule presented in Fig. 3.41b. The process started with heating the samples to the maximum temperature T_m = 1040°C, heating rate 100°C/h, at which point the liquid phase is formed and infiltrated into the solid phase pellet during 1 h. Then the system was cooled to near peritectic temperature 1015°C at the rate of 60°C/h, and the samples were slowly cooled through the peritectic temperature over 180 h. The cooling rate was decreased in three steps (0.3°C/h; 0.2°C/h; 0.1°C/h) and subsequently the samples were cooled with furnace to room temperature (Fig. 3.41b).

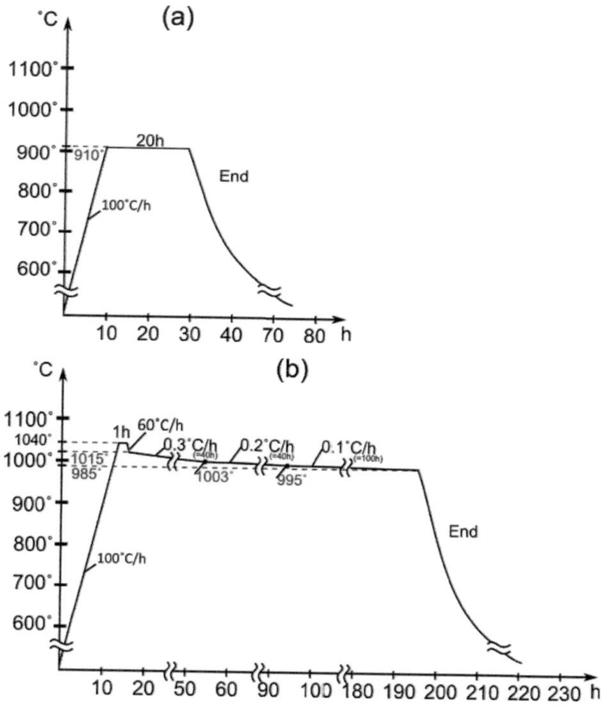

Figure 3.41 Time-temperature schedule of single crystal preparation by infiltration growth method: (a) for synthesising Y211 solid phase pellet; (b) for crystal growth process from the seed.

To measure superconductive properties of samples with and without Sm_2O_3 addition and to investigate the effect of Yb_2O_3 substrate on the system, the samples were cut out in two places: First 1A—at the end of the *a*-growth sector near the top surface (Fig. 3.42 (1A-position)); second 2A—at the end of the *a*-growth sector but at its middle (Fig. 3.42 (2A-position)). Samples were cut out with dimensions approximately 1.5 mm × 1.5 mm × 0.5 mm (*a* ×

$a \times c$-directions, respectively). Subsequently small specimens were oxygenated in flowing oxygen at 410°C for 200 h.

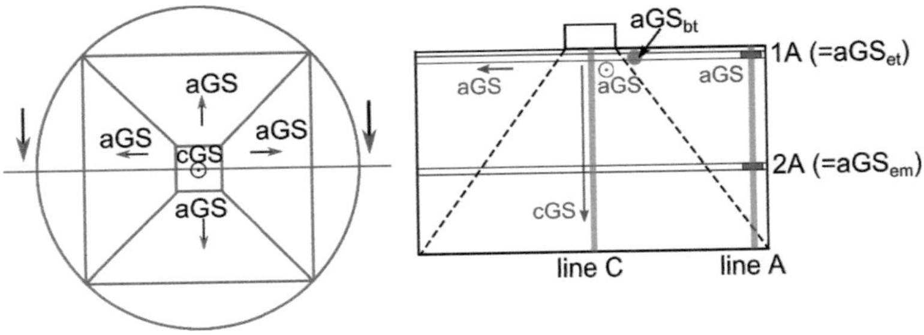

Figure 3.42 Scheme of crystal growth directions and sectors (left) and cutting out of the samples (right) for the measurements of superconducting properties (1A, 2A). Line A and C represent the measurement of Yb concentration in the sample. Microstructure photos were made at the positions indicated by red marks in *a*-growth sector (*a*GS): $a\text{GS}_{bt}$—at the beginning of *a*-GS near the seed; $a\text{GS}_{et}$—at the end top in the area 1A; $a\text{GS}_{em}$—at the end of *a*GS and in the middle of the sample rim (area 2A).

The measurements of transition temperature T_c and critical current density J_c were performed by VSM magnetometer at liquid nitrogen temperature. Critical transition temperature to superconducting state was estimated from the curves of magnetic transition (1 mT). Critical current density values were calculated from the measurements of magnetic loop by means of extended Bean mode [84].

The composition of the samples and quantitative analysis of ytterbium penetration from substrate into crystal at line A and C from bottom to top (Fig. 3.42) was analysed by EDS at a SEM.

The microstructure of the samples was observed with an optical microscope in polarised light. The size distribution of Y211 particles was evaluated from the obtained images using image analysis software.

As the studied samples were grown on the Yb_2O_3 substrate, we checked the contamination of the samples with ytterbium. Quantitative analysis of ytterbium penetration from the substrate into crystal was performed along line A and C (Fig. 3.42), and the dependences of Yb concentration as a function of the distance from the bottom of the grown YBCO single-grain sample are presented in Fig. 3.43. All three dependences have a typical diffusion profile and prove that Yb diffused through the melt in melted stage to the sample. The concentration of Yb is below the resolution limit of EDS (0.1 wt%) at the distance of about 6 mm from the seed for the measurement along the C line and about 7 mm for measurement along the A line; this is apparently caused by the fact that the parts at the rim of the sample were in the melted state for a longer time. We suppose that corresponding but low concentration of Yb will be present also at the top surface of the samples. The maximum measured Yb concentration is obviously at the sample bottom surface. The lowest Yb concentration at the bottom surface was measured in sample S3 with ZrO_2 bars placed between the liquid source pellet and Yb_2O_3 layer (Fig. 3.40). For samples S1 and S2 this value is higher.

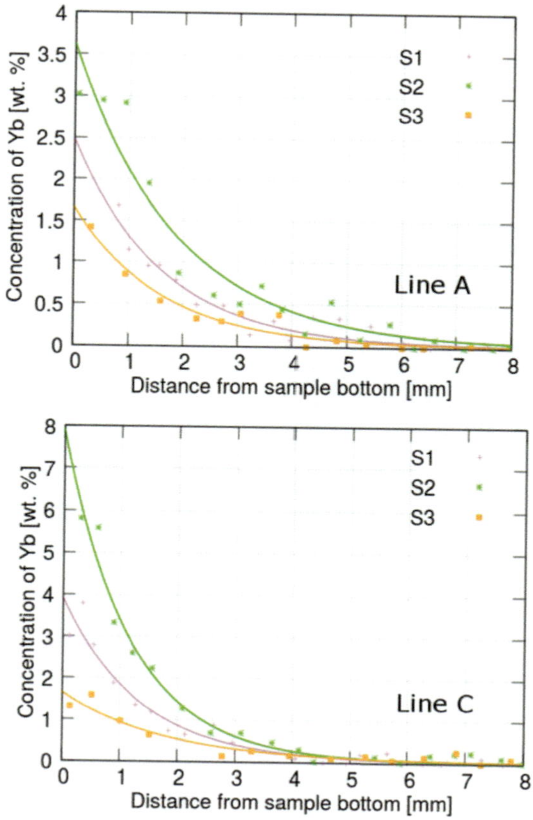

Figure 3.43 Dependence of Yb concentration in the Y123 + Y211 mixture on the distance from the sample bottom.

Our measured Yb concentrations are the sum of Yb content in the Y123 phase and Yb content in the Y211 phase. Therefore, we measured Yb concentration in the Y211 and Y123 phases close to the sample bottom. The microstructure of sample S1 close to the bottom surface is presented in Fig. 3.44. In a backscattered electron mode of SEM image, a higher concentration of Yb is reflected by higher brightness because the atomic mass of Yb is higher than that of Y (atomic mass: Y = 88.91 amu, Yb = 173.0 amu). One can see that more Yb is present in the 211 particles than in the 123 phase. In both phases, Yb substitutes for Y in the crystal lattice. Apparently, Yb diffused through the melt to the Y211 particles, and the surface parts of Y211 particles were converted into $(Y,Yb)_2BaCuO_5$ solid solution. Microstructure in Fig. 3.44 shows that larger 211 particles contain more Yb than smaller ones. The middle parts of the 211 particles are dark, and EDS analysis of a large 211 particle confirmed that its central part does not contain Yb at all. The highest Yb concentration was measured in the bright part of large 211 particle at 1.8 wt %. The measured concentration of Yb in the 123 phase was 0.8 wt %.

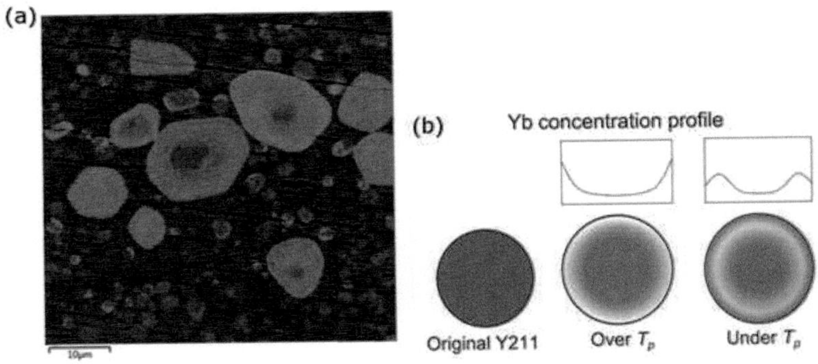

Figure 3.44 Local EDS analysis of sample S1 at the sample bottom (a) and schematic representation of enrichment processes of Y211 particle with ytterbium from the Yb_2O_3 substrate which take place above peritectic temperature, T_p, and depletion of the 211 particle surface at the crystal growth below peritectic temperature (b).

During the formation of YBCO bulk crystal by the peritectic reaction:

$$((Y,Yb)_2BaCuO_5)_S + (3BaO \cdot 5CuO \cdot y((Y,Yb)O_{1.5}))_L \leftrightarrow ((Y,Yb)Ba_2Cu_3O_x)_S \quad (3.4)$$

where symbol S means solid and symbol L means liquid, Y and Yb from the (Y,Yb)211 particles are participating in the formation of the (Y,Yb)123 crystal.

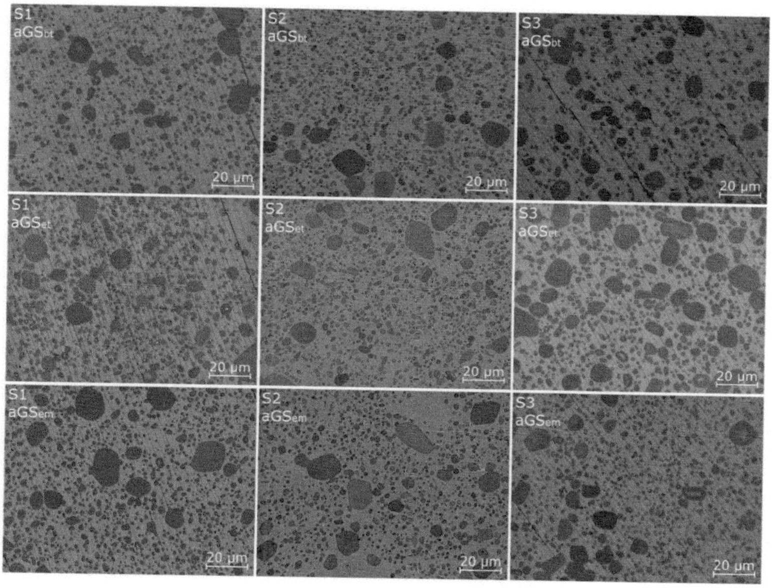

Figure 3.45 Polarised light optical micrographs from the polished halve sample section parallel to the *a*/*c*-plane. Positions on the section: $a\mathrm{GS}_{bt}$ – beginning of *a*-growth sector at the top sample surface, $a\mathrm{GS}_{et}$ – end of *a*-growth sector at the top sample surface (position 1A), $a\mathrm{GS}_{em}$ – end of *a*-growth sector in the middle of the sample rim (position 2A).

Sm and Yb may influence coarsening of Y211 particles by the Ostwald ripening process and trapping/pushing of the 211 particles by the growing 123 crystal. Therefore, we analysed the microstructure of the samples in three different positions in the *a*-growth sector (Fig. 3.45) with supposed different concentration of Yb and constant concentration of Sm (samples S2 and S3). In all positions, some amount of large Y211 particles is present but the microstructure in those positions differs mainly in the size of smaller Y211 particles. The quantitative data of the Y211 particle size, d_{211}, and the 211 volume percentage, V_{211}, for all analysed positions are presented in Table 3.3. The dependence of the size of smaller Y211 particles on the position in the *a*-GS is shown in Fig. 3.46. It can be seen that the position in the sample has significant influence on the size of small 211 particles mainly in the samples with Sm addition (S2 and S3). The observed size of the 211 particles can be related to competition of two processes. The first process is a growth of the 211 particles with the time in melted state by Ostwald ripening mechanism. The second process is a decrease of the critical 211 particle size for pushing with increasing undercooling, i.e., higher growth rate.

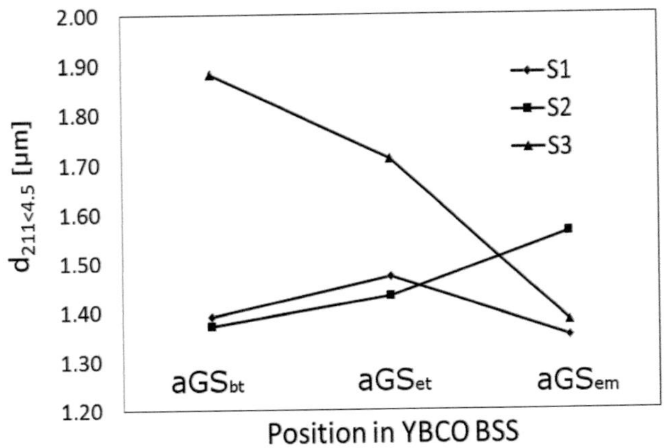

Figure 3.46 Dependence of the size of the 211 particles larger than 4.5 mm, $d_{211>4.5}$, on the position in the *a*-growth sector (*a*GS). Positions: aGS_{bt} – beginning of *a*-growth sector at the top sample surface, aGS_{et} – end of *a*-growth sector at the top sample surface (position 1A), aGS_{em} – end of *a*-growth sector in the middle of the sample rim (position 2A).

The inclusion trapping/pushing theory [69] proposed that when a sample solidifies at a certain growth rate, R_1, inactive inclusion particles smaller than a certain critical radius, r^* are not trapped in the solid but are instead excluded. At a higher growth rate, R_2, inclusion of particles with the radius of r^* are then trapped. The relation between the growth rate, R^*, and the critical radius, r^*, is predicted by

$$R^* = \Delta\sigma_0/\eta r^{*n} \tag{3.5}$$

Table 3.3 Data of the measurement of mean size, d_{211}, and volume percentage, V_{211}, of Y211 particles in the *a*-growth sector.

Sample	S1						S2						S3					
Measured value	d_{211} [μm]			V_{211} [%]			d_{211}[μm]			V_{211} [%]			d_{211} [μm]			V_{211} [%]		
Position in aGS	bt	et	em	bt	et	em	bt	et	em	bt	et	em	bt	et	em	bt	et	em
All Y211	1.54	1.63	1.48	26.55	32.24	35.65	1.46	1.54	1.70	35.15	36.24	35.97	2.17	2.17	1.55	32.43	43.16	32.54
Y211<4.5 μm	1.39	1.47	1.35	16.31	19.76	21.84	1.37	1.43	1.56	25.41	25.23	22.99	1.88	1.71	1.38	18.51	17.18	19.56
Y211>4.5 μm	7.22	7.39	7.92	10.24	12.48	13.80	8.35	7.77	8.37	9.74	11.02	12.98	6.7	8.16	7.43	13.92	25.98	12.99

Note: Positions in the sample: bt—beginning of *a*-growth sector at the top sample surface, et—end of *a*-growth sector at the top sample surface (position A1), em—end of *a*-growth sector in the middle of the sample rim (position A2).

where $\Delta\sigma_0 = \Delta\sigma_{sp} - \Delta\sigma_{sl} - \Delta\sigma_{pl}$. The interface energy between the solid phase and the inclusion particle is represented by $\Delta\sigma_{sp}$, $\Delta\sigma_{sl}$ represents the interface energy between the solid phase and the liquid phase, and $\Delta\sigma_{pl}$ represents the interface energy between the inclusion particle and the liquid phase; η is the viscosity coefficient of the liquid phase and n is a constant between 1 and 2. This theory has been successfully applied to the trapping/pushing phenomena in the solidification of the 123–211 system [34]. The growth rate R is proportional to the undercooling ΔT. The samples presented were grown at a slow cooling rate, and thus the undercooling and the growth rate increased with the distance from the seed. The result should be an increase of V211 with a distance from the seed and a decrease of d_{211}.

Ostwald ripening process also depends on the interface energy between a 211 particle and the peritectic melt, L_p, [6] by

$$d^3_t - d^3_0 \approx DC\sigma t, \tag{3.6}$$

where d_t is the final and d_0 is the starting 211 particle size, D is the diffusion coefficient of Y in the peritectic melt L_p, C is the Y concentration in the melt, σ is the $211/L_p$ interface energy, t is the holding time.

The presence of Sm and Yb in the studied samples can influence $211/L_p$ interface energy as well as the viscosity of the peritectic melt. The measured size of the 211 particles, $d_{211>4.5}$, in different positions is then the result of Yb and Sm concentration at different positions of the growing 123 crystal. In the sample without Sm doping (S1) and the highest Yb contamination the size of the 211 particles is influenced only by Yb from the substrate, and observed dependence of the 211 size on the position in the sample suggests that the concentration of Yb is close to optimum. In the sample S3 with Sm doping, the contamination by Yb at the beginning of crystallisation is too low (Fig. 3.43) and the 211 particles are larger (Fig. 3.46, position $a\text{GS}_{bt}$). Increasing contamination with longer time in the melted state (Fig. 3.46 position $a\text{GS}_{et}$) and the position closer to Yb_2O_3 substrate (Fig. 3.46, position $a\text{GS}_{em}$) in this sample leads to lower size of the 211 particles suggesting some optimum Sm + Yb concentration. In the sample S2 with higher Yb contamination (Fig. 3.43), the Sm + Yb concentration is apparently increasing above optimal, and that causes coarsening of the 211 particle.

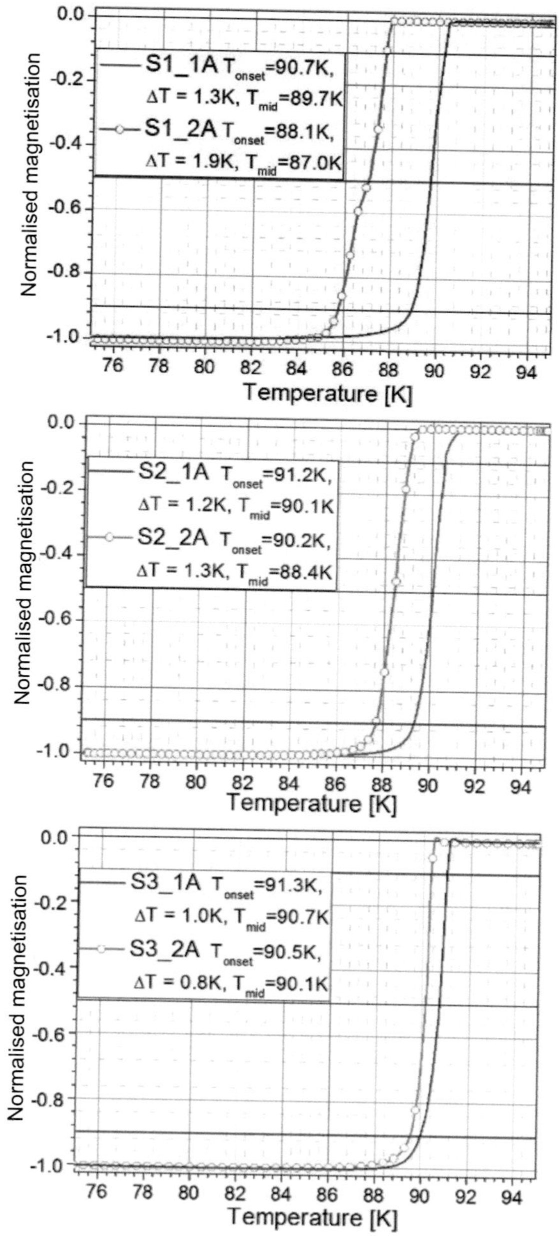

Figure 3.47 The temperature dependence of magnetisation for samples S1 (without Sm doping), S2 (Sm doping), and S3 (Sm doping, ZrO_2 bars).

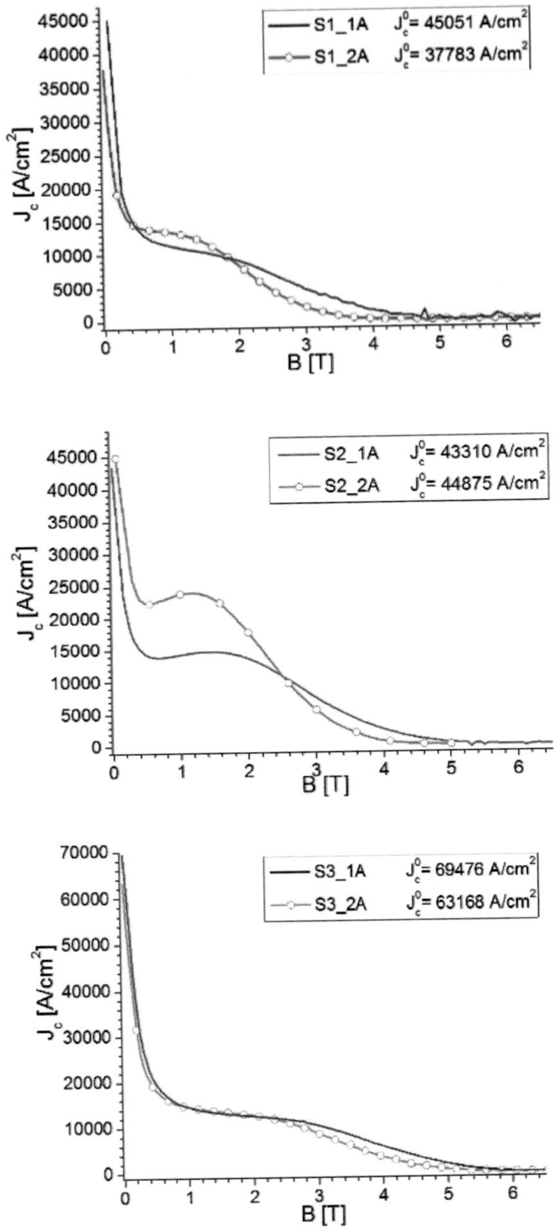

Figure 3.48 J_c-B dependences at 77 K for samples S1 (without Sm doping), S2 (Sm doping), and S3 (Sm doping, ZrO_2 bars).

Measured values of critical current density J_c at 77 K and transition temperature T_c for the positions 1A ($a\text{GS}_{\text{et}}$) and 2A ($a\text{GS}_{\text{em}}$) are presented in Figs. 3.47 and 3.48, respectively. Values of critical current density were evaluated from magnetic loop measurements by means of extended Bean model. Analysed samples exhibit differences in the transition to superconducting state and in the field dependences of critical current density. The sample S1 without Sm doping has higher T_c at the $a\text{GS}_{\text{et}}$ position ($T_c = 89.7$ K, $\Delta T_c = 1.3$ K) where we may expect lower Yb contamination than at the $a\text{GS}_{\text{em}}$ position ($T_c = 87.0$ K, $\Delta T_c = 1.9$ K). Higher Yb contamination even leads to double T_c like behaviour suggesting the presence of two 123 phases with different Yb concentration (Fig. 3.47, 2A).

Higher Yb contamination is also reflected in a more pronounced peak effect in the field dependence of critical current density at 77 K and lower irreversibility field (Fig. 3.48, S1) for the $a\text{GS}_{\text{em}}$ position. The sample S2 with Sm doping and Yb contamination exhibits nicely pronounced secondary peak effects for both measured positions (Fig. 3.48). The highest J_c at 1.5 T reaching $2.5 \cdot 10^4$ A/cm^2 was measured for the $a\text{GS}_{\text{em}}$ position in this sample. Similarly, to the sample S1, higher Yb contamination leads to lower irreversibility field. It is interesting that T_c values at the measured positions are higher in this sample than in the sample S1; this may be caused by some interaction of Sm and Yb dopants. The sample S3 with Sm doping and lower Yb contamination has the highest values of T_c and the highest J_c values at zero field reaching $6.9 \cdot 10^4$ A/cm^2 (Fig. 3.48). T_c and irreversibility field values in this sample are the highest for the position with supposed lowest Yb contamination, and both of them only slightly decrease with higher Yb contamination.

Observed behaviour of the superconducting properties suggest that some optimum Sm and Yb concentration and Sm/Yb ratio may exist for obtaining required superconducting properties.

In summary: Quantitative characterisation of microstructure, Yb impurities and superconducting properties of single-grain YBCO bulk superconductors microalloyed with Sm prepared by infiltration growth process were studied. It was shown that:

- The contamination of samples from the Yb_2O_3 substrate exhibits diffusion like dependence of Yb concentration on the distance from the sample bottom. The contamination is influenced by Sm doping and by insertion of ZrO_2 interlayer placed between liquid source pellet and Yb_2O_3 substrate.

- At the melting stage, Yb is concentrated in the 211 particles.
- The size of the 211 particles in the 123 bulk single crystal is influenced by both Sm dopant and Yb contamination, through the influence of trapping and growth of 211 particles.
- Optimum combination of added Sm and Yb contamination leads to the most pronounced secondary peak effect and to the highest critical current density in higher magnetic fields.
- Observed double T_c like behaviour suggests inhomogeneity of Yb distribution in the 123 phase.

3.6 Relationship between Local Microstructure and Superconducting Properties of Commercial $YBa_2Cu_3O_{7-\delta}$ Bulk

It is well known that $YBa_2Cu_3O_{7-\delta}$ (Y123) single-grain bulk superconductor is highly anisotropic material. This anisotropy is manifested not only in physical properties [87] but also in crystallographic anisotropy that is directly reflected in the microstructure of the superconductor. Formation of different inhomogeneities in Y123 high-temperature bulk superconductors, such as micro- and macro-cracks, twinning structure, subgrain boundaries, dislocations, chemical inhomogeneities, oxygen deficient regions, and presence of the growth boundaries between the growth sectors [88] as well as formation of Y_2BaCuO_5 (Y211) particles can substantially influence the quality of the superconductor [89, 90] and consequently limit its practical application.

In this work we have investigated both physical properties and microstructure of commercially prepared Y123 single-grain bulk superconductors that are used in practice [91]. Microstructures of different parts of the bulk were supplemented with measurements of the superconducting properties. We showed that the superconducting transition temperature did not depend on the arrangement of the non-superconducting Y211 particles within the sample and the subgrain structure whereas the local critical current densities did. Knowledge of the relationship of local microstructure and superconducting properties in a broad temperature range will be useful in practical applications of Y123 bulk superconductors.

Commercial Y123 single-grain bulk superconductor (Fig. 3.49) was fabricated by CAN Superconductors company using the top-seeded melt growth (TSMG) process with $NdBa_2Cu_3O_{7-\delta}$ thin film seed (500 nm $NdBa_2Cu_3O_{7-\delta}$ on MgO) as a nucleation centre [92, 93] for the crystal growth in an air atmosphere. As a nominal composition for the crystal growth, the powder mixture of Y123, Y211 and 0.5 wt % CeO_2 was used. After the fabrication process the sample was oxygenated at 400–430°C during 240 h, and after that six small samples for magnetisation measurements were cut from different parts of the bulk as shown in Fig. 3.50a,b. Samples 1*a* and 1*aa* were cut from the *a*-growth sector (GS) and the *a*/*a*-growth sector boundary at a distance of 2 mm from the seed, respectively. Samples 2*a* and 2*aa* were cut from the same position of the bulk but with a distance of 8 mm from the seed. Samples 1*c* and 2*c* were cut from the *c*-growth sector with a distance of 2 mm and 8 mm below the seed, respectively. All samples had the shape of a slab with dimensions 1.5 × 1.5 × 0.5 mm^3, where the smallest dimension was parallel to the *c*-axis of the crystal.

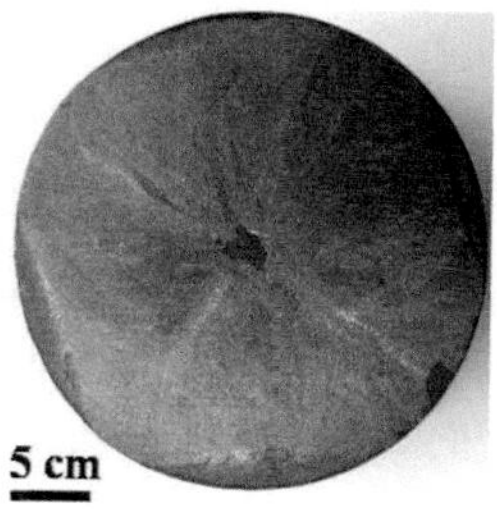

Figure 3.49 Top view of Y123 single-grain bulk.

The field dependence of the magnetic moment was measured at 77 K and 65 K using the commercial Magnetic Property Measurement System with magnetic field of up to 7 T (Quantum Design) at a constant sweep rate of 100 Oe sec^{-1} and at 4.2 K with magnetic field of up to 18 T (Cryogenic Ltd.) at a constant sweep rate of 35 Oe sec^{-1}. During the magnetisation measurements, the applied external magnetic field was parallel to the *c*-axis of the crystal. The critical current densities, J_c, were calculated from the magnetic hysteresis loops using the Bean model for rectangular samples,

$J_c(B) = \{\Delta m(B)/V\}\{2/[b(1-b/3a)]\}$, where Δm is the difference in magnetic moments between the ascending and descending branches of the magnetic hysteresis loops, and V is the sample volume, $a \times b \times c$. The superconducting transition temperatures were determined from the magnetic moment transition curves taken after zero-field cooling in the applied external magnetic field of 10 Oe. Microstructure of the sample was analysed by scanning electron microscopy (Tescan Mira3) and optical microscopy (Nikon Eclipse LV100) as well as an image processing software package used for the quantitative microstructural data analysis.

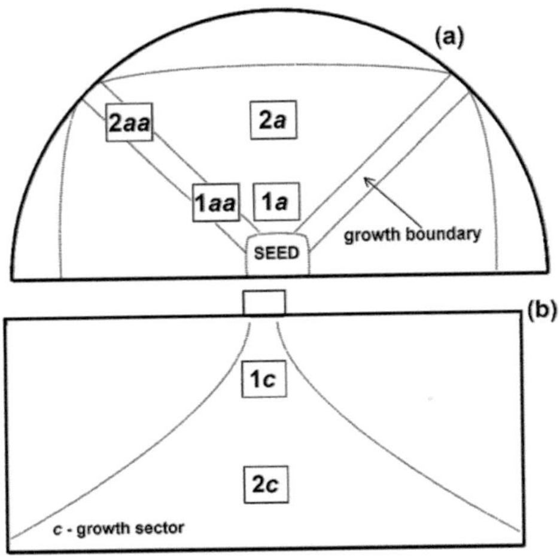

Figure 3.50 Schematic illustration of the bulk, where small samples were cut. (a) Top view of the bulk and (b) the *a*/*c*-growth section. Squares 1*a*, 1*aa*, 2*a*, 2*aa*, 1*c* and 2*c* indicate sample position.

Both optical microscopy and scanning electron microscopy (SEM) analysis confirmed that all parts of the sample consisted of Y123 phase which is superconducting if Y123 has orthorhombic structure [94], and Y211 phase which is embedded into the Y123 matrix in the form of Y211 particles [95]. Micrographs of the sample in polarised light taken from the same sides as small samples for magnetisation measurements were cut are presented in Fig. 3.51 for the *a*-GS, and in Fig. 3.52 for the *c*-GS. It is clearly visible that for both growth sectors the size (d_{211}) and mainly volume distribution (V_{211}) of Y211 particles vary with a distance from the seed. Such inhomogeneity in the Y211 arrangement could be explained by both the pushing/

trapping of Y211 particles by the growth front and different crystal growth rate in both the *a*- and *c*-growth directions [35, 96] and by growth of Y211 particles by Ostwald-ripening process [97] as it is described in section 3.5 (Eqs. 3.5 and 3.6).

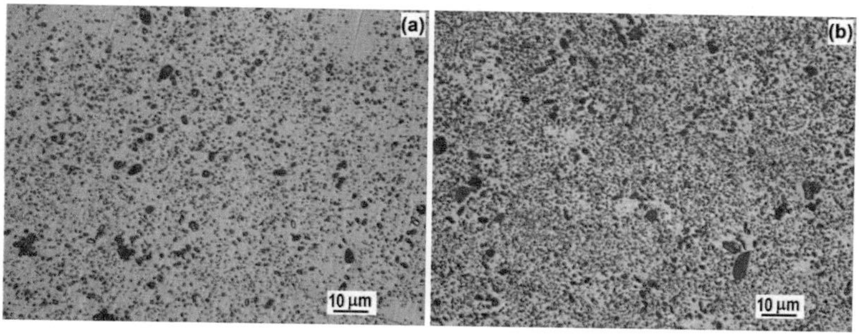

Figure 3.51 Polarised optical micrograph taken from the *a*-GS at a distance of 2 mm (position 1*a* in Fig. 3.1a) (a) and 8 mm (position 2*a* in Fig. 3.1a) from the seed (b). The Y211 particles are visible as darker parts of the micrograph.

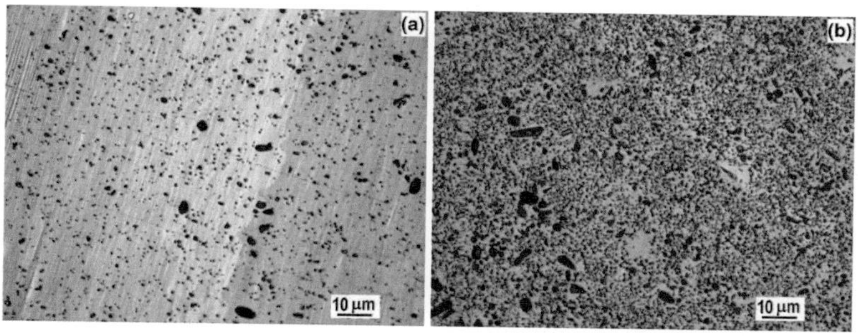

Figure 3.52 Polarised optical micrograph taken from the *c*-GS at a distance of 2 mm (position 1*c* in Fig. 3.1b) (a) and 8 mm (position 2*c* in Fig. 3.1b) below the seed (b). The Y211 particles are visible as darker parts of the micrograph.

The change of d_{211} and V_{211} of Y211 particles with distance from the seed in the *a*-GS and the *c*-GS are presented in Fig. 3.53a and 3.53b, respectively. These changes were investigated on the bulk from the seed to the rim in the same direction as small samples for magnetisation measurements were cut. The value of V_{211} in the vicinity of the seed is much higher in the *a*-GS in comparison to the *c*-GS. Such deficiency of Y211 particles in the *c*-GS was caused by stronger pushing of Y211 particles by the *c*-growth front than by the *a*-growth front at the beginning of the solidification process due to higher $\Delta\sigma_0$ (Eq. 3.5) for the *c*-GS.

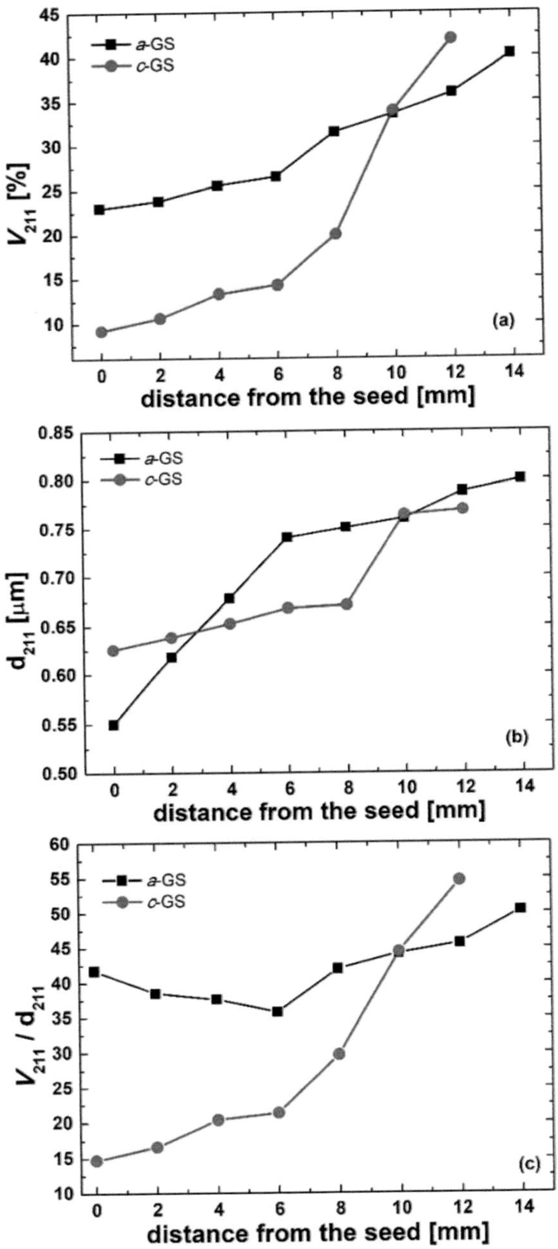

Figure 3.53 Volume distribution of Y211 particles (a) and the size of Y211 particles (b) along the *a*-GS and the *c*-GS. (c) Calculated V_{211}/d_{211} ratio for the *a*-GS and the *c*-GS.

For longer distances from the seed both processes affect measured V_{211} and d_{211}. Increasing undercooling leads to higher growth rate (more Y211 are trapped) and longer time leads to the growth of Y211 particles and consequently a higher number of larger particles is trapped in Y123 matrix (Fig. 3.53b). It should be mentioned that microstructural analysis showed that the size and volume distribution of Y211 particles is constant along the growth front in the same growth sector, e.g., in the centre of GS or at the growth sector boundary. If we compare the V_{211}/d_{211} ratio between the *a*-GS and the *c*-GS (Fig. 3.53c), it is clear that the ratio between Y211 arrangement and its size [35] is higher in the *a*-GS at a distance less than 10 mm from the seed. Higher V_{211}/d_{211} values reflect increasing critical current density at self-field that we directly observed in investigated samples, and the results are presented below.

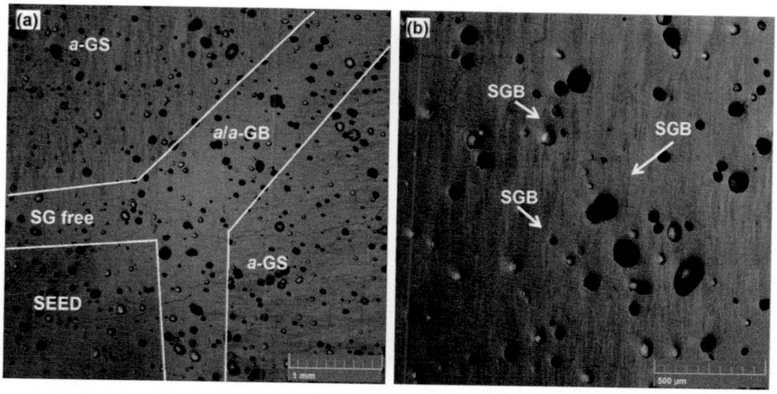

Figure 3.54 SEM image of the *a*-GS (a) with subgrains free (SG free) regions in the vicinity of the seed (SEED), where *a*/*a*-GB means the *a*/*a*-growth sector boundary and *a*-GS means the *a*-growth sector. (b) Backscattered electron image taken from the *a*-GS where subgrain boundaries (SGB) are visualised.

The subgrain structure of the sample [35, 88] was investigated using SEM. This showed that the sample has subgrain-free regions in the vicinity of the seed and along the *a*/*a*-growth boundaries (Fig. 3.54a). In other parts of the sample, subgrains were observed and subgrain boundaries were well visualised in backscattered electron image (Fig. 3.54b). The subgrains were parallel to the *a*-axis of the crystal. Formation of subgrains could have significant influence on the superconducting properties of Y123 due to the formation of weak links [35, 98]. It was evident from the SEM investigation that samples taken from the *a*/*a*-growth sector boundary (1*aa* and 2*aa* in Fig. 3.50a) have regions that are free of subgrains.

The superconducting transition temperatures for all samples are presented in Fig. 3.55. The curves have sharp transition without any secondary steps, and all samples have zero transition temperatures, $T_{c,0}$, above 91 K, as well as the superconducting transition widths, ΔT_c, in the range of less than 0.8 K (Table 3.4) that also indicated the orthorhombic structure and high quality of prepared Y123 single-grain sample.

Table 3.4 Dependence of onset ($T_{c,\text{onset}}$), 50% curve ($T_{c,50}$) and zero ($T_{c,0}$) superconducting transition temperatures, and transition width, ΔT_c, on the sample position

Sample position	$T_{c,\text{onset}}$ [K]	$T_{c,50}$ [K]	$T_{c,0}$ [K]	ΔT_c [K]
1*a*	90.88	91.06	91.25	0.35
1*aa*	90.80	90.98	91.20	0.38
1*c*	91.01	91.13	91.28	0.34
2*a*	90.50	90.79	90.95	0.37
2*aa*	90.74	90.92	91.08	0.30
2*c*	90.19	90.71	91.55	0.80

The field dependence of the critical current densities, J_c, at 77 K and 65 K are presented in Fig. 3.56a and 3.56b, respectively. The highest J_c at 0 T at both 77 K and 65 K, where mainly Y211 particles are effective pinning centres [99], was observed in the sample 2*aa* (Table 3.5), whereas samples 1*a* and 1*aa* had almost the same value of J_c that was also observed for samples 1*c* and 2*c* (Table 3.5). While J_c for samples 1*a* and 1*aa* was almost identical in the whole range of the applied magnetic field in both 77 K and 65 K, in the case of samples 1*c* and 2*c* J_c differs with increasing applied magnetic field. Also, the critical current in the sample 1*c*, unlike other samples, starts to increase with the magnetic field above 2 T, due to the appearance of the peak effect (Fig. 3.56b).

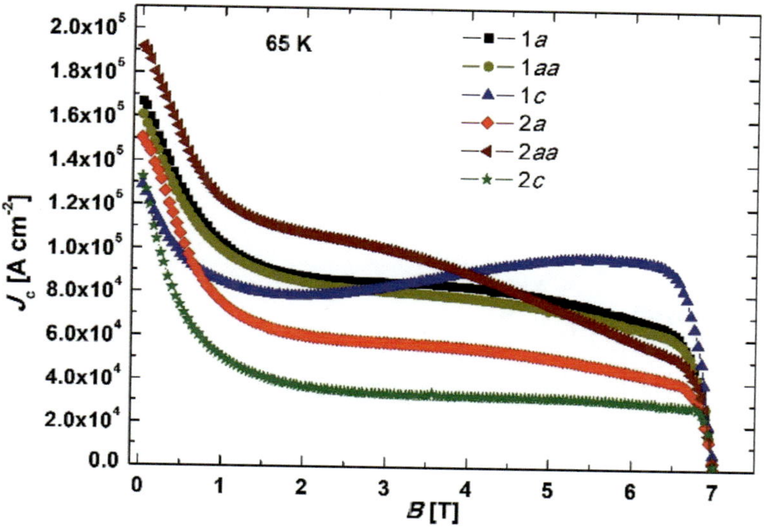

Figure 3.55 Superconducting transition temperature curves.

Similarity in J_c for 1*a* and 1*aa* samples could be explained by the same V_{211}/d_{211} ratio (Fig. 3.53c) as well as the absence or very low amount of subgrains in the sample 1*aa* and small angle between subgrain boundaries in the sample 1*a* that do not have effect on the critical current density. It is well known that with the distance from the seed angle between subgrain boundaries become larger that strengthens their action as weak links [35, 98]. If we compare J_c at 0 T for samples 2*a* and 2*aa* (Table 3.5) we see that J_c for the sample 2*a* is lower due to the presence of subgrains with larger angle between subgrain boundaries in comparison to the sample 2*aa* with the subgrain-free regions.

Table 3.5 Dependence of the critical current density, J_c (A cm^{-2}), at 0 T on temperature and sample position

T [K]	1*a*	1*aa*	1*c*	2*a*	2*aa*	2*c*
77	9.28×10^4	8.98×10^4	6.93×10^4	8.28×10^4	1.07×10^5	7.09×10^4
65	1.66×10^5	1.60×10^5	1.28×10^5	1.50×10^5	1.91×10^5	1.32×10^5
4.2	1.81×10^6	1.84×10^6	2.32×10^6	1.39×10^6	2.01×10^6	1.21×10^6

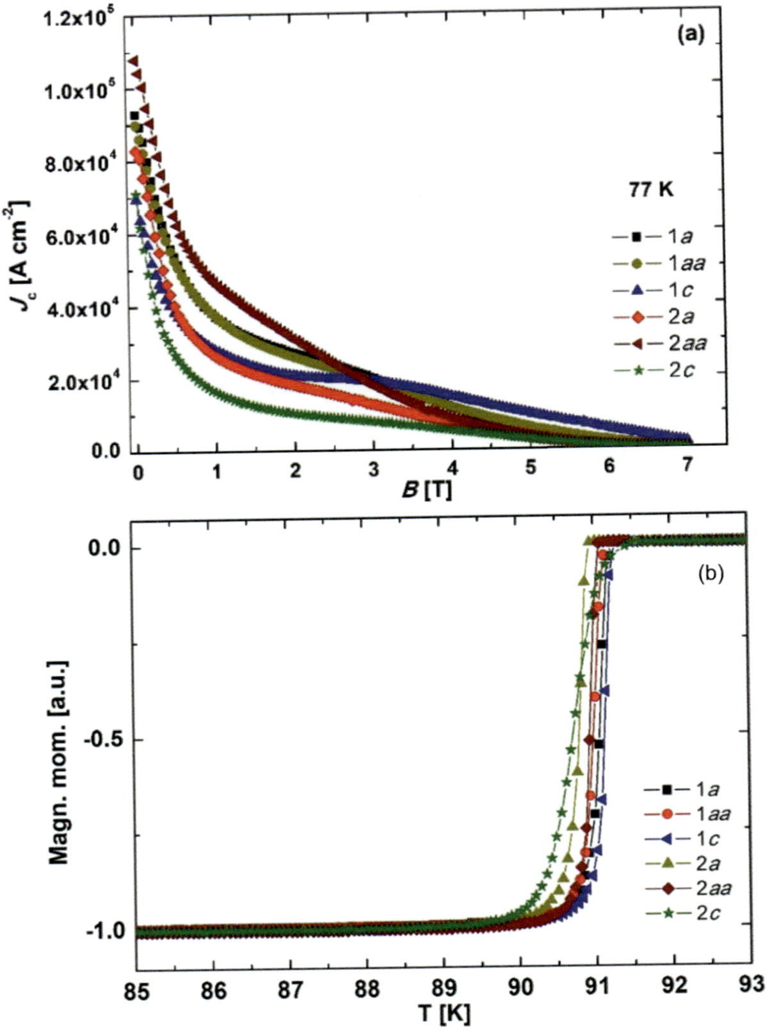

Figure 3.56 Dependence of the critical current density, J_c, on the external magnetic field, B, at 77 K (a) and 65 K (b).

In the case of 1*c* and 2*c* samples J_c substantially increased for the sample 1*c* at higher magnetic fields (Fig. 3.56a) due to the presence of inhomogeneities in the crystal structure which improve the flux pinning process. We suppose that these inhomogeneities in the sample 1*c* were created by possible neodymium contamination of Y123 from the seed during the solidification process [83, 100]. However, chemical elemental mapping of the bulk using SEM in the vicinity of the seed did not detect neodymium that most probably could be caused by its concentration below energy dispersive X-ray spectroscopy detection limit which is 0.1 wt %. The inhomogeneities in the sample 1*c* interact with the flux lines as effective pinning centres similarly to Y123-doped superconductors [101–103] and display the appearance of clear peak effect at temperatures below 77 K (Fig. 3.57a). The pronounced peak effect in the sample 1*c* indicated that the size of created pinning centres or the mean distance between them should be close to two coherence lengths, $\sim\xi_{ab}$ [80, 104], that could be obtained, for example, in the samples with low chemical doping [101, 103]. Moreover, the sample 1*c* had higher J_c at 4.2 K in the whole range of the applied magnetic field in comparison to the other samples (Fig. 3.57b). Such increase in J_c value could be explained by a change in dominant pinning mechanism [78, 105] in the sample 1*c* with decreasing temperature that manifested in higher efficiency of flux pinning.

In summary: Microstructural analysis showed that Y211 particles were inhomogeneously arranged within the commercially prepared Y123 single-grain bulk. The size and volume fraction of Y211 increase further from the seed. It was shown that subgrains were formed in the sample and grow larger away from the seed in both the *a*- and *c*-GS that eventually had significant influence on the formation of weak links in these parts of the bulk. The subgrain-free regions were observed in the *a*/*a*-growth boundaries and in the vicinity of the seed. Magnetisation measurements confirmed that the critical current density strongly depends on the microstructure of the sample. Variation of the critical current density observed to be in agreement with microstructural analysis and showed that J_c was higher in the samples that contained subgrain-free regions. It was also shown on sample 1*c* that the contamination from the seed enhances local critical current density due to effective flux pinning, especially at temperatures below 77 K.

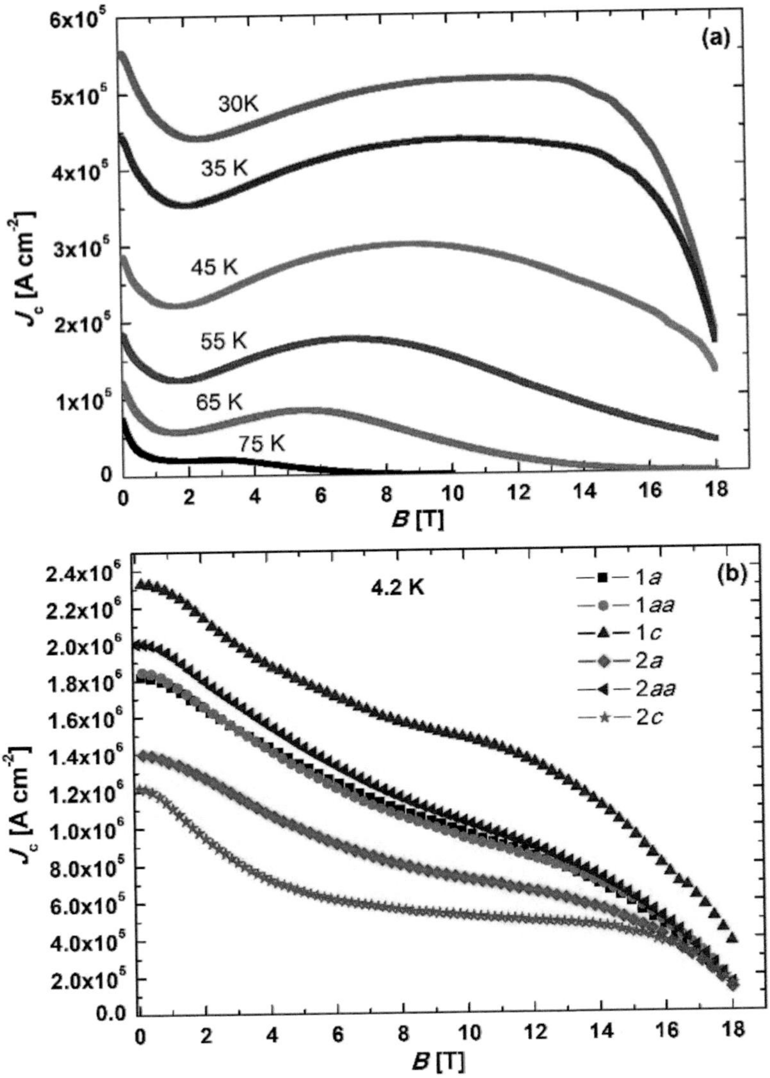

Figure 3.57 Dependence of the critical current density, J_c, on the external magnetic field, B, at temperature range from 30 K to 75 K for the sample 1*c* (a) and J_c at 4.2 K for all samples (b).

Acknowledgements

This work was realised within the framework of the projects: Research Centre of Advanced Materials and Technologies for Recent and Future Applications "PROMATECH" (ITMS 26220220186), APVV-17-0625, VEGA No. 2/0044/19. This work was partly financially supported by Czech Ministry of Industry and Trade, project Aplikace (CZ.01.1.02/0.0/0.0/15_018/0001791).

References

1. Durrell J H, Ainslie M D, Zhou D, Vanderbemden P, Bradshaw T, Speller S, Filipenko M and Cardwell D A 2018 *Supercond. Sci. Technol.* **31** 103501.
2. Campbell A M, Cardwell D A 1997 *Cryogenics* **37** 567.
3. Fujimoto H, Murakami M, Nakamura N, Gotoh S, Kondoh A, Koshizuka N and Tanaka S 1991 *Advances in Superconductivity IV, Proc. 4th Int. Symp. on Superconductivity (ISS '91)* (Tokyo) ed H Hayakawa and N Koshizuka (Berlin: Springer) p 339.
4. Martinez B, Obradors X, Gou A, Gomis V, Pinol S, Fonkuberta J, Van Tol H 1996 *Phys. Rev. B* **53** 2795.
5. Ostwald W 1901 *Z. Phys. Chem.* **37** 385.
6. Lifshitz I M, Slyozov V V 1961 **19** 35.
7. Varanashi C, Black M A, McGinn P J 1994 *Supercond. Sci. Technol.* **4** 10.
8. Ogava N, Hirabayashi I, Tanaka S 1991 *Physica C***177** 101.
9. Kim C J, Lai S H and McGinn P J 1994 *Mater. Lett.* **19** 185.
10. Litzkendorf D, Habisreuther T, Muller R, Kracunovska S, Surzhenko O, Zeisberger M, Riches J, Gawalek W 2002 *Physica C* **372** 1163.
11. Shlyk L, Nenkov K, Krabbes G, Fuchs G 2005 *Physica C* **423** 22.
12. Monot I, Verbist K, Hervieu M, Lafez P, Delamare M P, Wang J, Desgardin G, Tendeloo van G 1997 *Physica C* **274** 253.
13. Diko P, Šefčiková M, Zmorayová K, Antal V 2011 *Int. J. Mater. Prod. Technol.* **40** 215.
14. Diko P, Šefčíková M, Kanuchová M, Zmorayová K 2008 *Mat. Sci. Eng. B* **151** 7/10.
15. Diko P, Volochová D, Radušovská M, Zmorayová K, Šefčíková M, Antal V, Jurek K, Jirsa M, Kováč J 2013 *Physica C* **494** 31.
16. Volochová D, Diko P, Antal V, Radušovská M, Piovarči S 2012 *J. Cryst. Growth* **356** 75.
17. Shimada T, Wen C, Taniguchi H, 2004 *J. Pow. Source* **131** 289.
18. Munch W, Kreuer K D, Seifert G, Maier J 2000 *Solid State Ion* **136** 183.

19. Kim C J, Quadrin N, Mahmood A, Han Y H, Sung T 2007 *Physica C* **463** 344.
20. Mahmood A, Park S D, Jun B H, Sung T H, Lee S H, Kim C J 2008 *Physica C* **468** 1355.
21. Nariky S, Sakai N, Murakami M, Hirabayashi I, 2005 *Physica C* **426** 602.
22. Shao Z, Zhou W, Zhu Z 2012 *Prog. Mater. Sci.* **57** 804..
23. Moure A, Castro A, Martinez I, Moure C, Tartaj J 2012 *Ceram. Int.* **38** 5907.
24. Orlov A V, Shlyakhtin O A, Vinokurov A L, Knotko A V, Ttretyakov Y D 2005 *Inorg. Mater.* **41** 1194.
25. Zhu X, Zhu J, Zhou S, Liu Z, Ming N, Hesse D 2005 *J. Cryst. Growth* **283** 553.
26. Lu S W, Lee B I, Wang Z L, Samuels W D 2000 *J. Cryst. Growth* **219** 269.
27. Xu C, Zhu J, Yang X, Lu L, Wang X 2008 *J. Rare Earths* **26** 51.
28. Liu J, Li Y, Wang W, Wang H, Zhang F, Ma G 2010 *J. Mater. Sci.* **45** 5860.
29. Magdaluyo E R Jr, Gapasin J E 2015 *Adv. Mater. Res.* **1098** 69.
30. Princivalle A, Martin G, Viazzi C, Guizard C, Grunbaum N, Dessemond L 2011 *J. Power Sources* **196** 9238.
31. Lin H L, Chiang R K, Kuo C L, Chang C W 2007 *J. Non-Cryst. Solids* **353** 1188.
32. Subramaniyan A, Tong J, O´Hayre R P, Sammes N M 2011 *J. Am. Ceram. Soc.* **94** 1800.
33. Diko P, Pelerin N, Odier P 1995 *Physica C* **247** 169.
34. Endo A, Chauhan H S, Egi T and Shiohara Y 1996 *J. Mater. Res.* **11** 795.
35. Diko P 2000 *Supercond. Sci. Technol.* **13** 1202.
36. Volochová D, Antal V, Piovarči S, Kavečanský V, Diko P 2018 High-Temperature Superconductors: Occurence, Synthesis and Applications 1 Book chapter.
37. Murakami M, Yamaguchi K, Fujimoto H, Nakamura N, Taguchi T, Koshizukaand N, Tanaka S, 1992 *Cryogenics* **32** 930.
38. Muralidhar M, Jirsa M, Iida K, Murakami M 2011 *Physica C* **357** 665.
39. Radušovská M, Volochová D, Antal V, Diko V 2016 *IEEE T. Appl. Supercond.* **26** 7201604.
40. Yeoh W K, Pathak S K, Shi Y H, Dennis A R, Cardwell D A, Hari Babu N, Iida K, Strasik M 2009 *Supercond. Sci. Technol.* **22** 065011.
41. Diko P, Todt V R, Miller D J, Goretta K C 1997 *Physica C* **278** 192.

42. Congreve J V J, Shi Y, Dennis A R, Durrell J H, Cardwell D A 2018 *Supercond. Sci. Technol.* **31** 035008.

43. Dias D H N, Sotelo G G, Moysés L A, Telles L G T, Bernstein P, Kenfaui D, Aburas M, Chaud X, Noudem J G 2015 *Supercond. Sci. Technol.* **28** 075005.

44. Naik S P K, Muralidhar M, Koblischka M R, Koblischka-Veneva A, Oka T, Murakami M 2019 *Appl. Phys. Express* **12** 063002.

45. Kracunovska S, Diko P, Litzkendorf D, Habisreuthera T, Gawalek W 2003 *Physica C* **397** 123.

46. Diko P, Kračunovská S, Gawalek W 2007 *Int. J. Mater. Prod. Tecnol.* **28** 270.

47. Diko P 2004 *Supercond. Sci. Technol.* **17** R45.

48. Cardwell DA, Ginley DS, 2003 *Handbook of superconducting Materials* I (Bristol and Philadelphia: IOP Publishing).

49. Zmorayova K, Diko P, Krabbes G 2006 *Physica C.* **445** 436.

50. Sakai N, Seo SJ, Inoue K, Miyamoto T, Murakami M 2000 *Physica C* **335** 107.

51. Pathak SK, Babu NH, Shi YH, Dennis AR, Strasik M, Cardwell DA 2010 *J. Mater. Res.* **25** 1243.

52. Kenfaui D, Sibeud PF, Louradour E, Chaud X, Noudem JG 2014 *Adv. Funct. Mater.* **24** 3996.

53. Congreve JVJ, Shi YH, Huang KY, Dennis AR, Durrell JH, Cardwell DA, 2019 *IEEE Trans. Appl. Supercond.* **29** 6802305.

54. Hlasek T, Huang KY, Esnoz-Larraya J, Plechacek V, Durrell, J, Valiente-Blanco, I, Cardwell, DA 2019 *IEEE Trans Appl. Supercond.* **29** 8400704.

55. Iida K, Hari Babu N, Shi Y and Cardwell DA 2005 *Supercond. Sci. Technol.* **18** 1421.

56. Wang M, Yang W M, Wang M Z and Wang X J 2013 *J. Supercond. Nov. Magn.* **26** 3221.

57. Namburi DK, Raju PMS, Naik, SPK, Rajasekharan T, Seshubai V 2014 *Physica C* **496** 18.

58. Namburi DK, Shi Y, Palmer KG, Dennis AR, Durrell JH, Cardwell DA 2016 *J. Eur. Ceram. Soc.* **36** 615.

59. Das Dhruba, Muralidhar M, Ramachandra RMS, Murakami M 2017 *Supercond. Sci. Technol.* **30** 105015.

60. Ouerghi A, Moutalbi N, Noudem J G, M'chirgui A 2017 *Physica C* **534** 37.

61. Vojtkova L, Diko P, Kovac J, Vojtko M 2018 *Supercond. Sci. Technol.* **31** 065003.
62. Naik SPK, Muralidhar M, Murakami M 2019 *IEEE Trans. Appl. Supercond.* **29** 6800305.
63. Namburi DK, Shi Y, Palmer KG, Dennis AR, Durrell JH, Cardwell DA 2016 *Supercond. Sci. Technol.* **29** 095010.
64. Diko P, Sefcikova M, Zmorayova K, Kalmanova M, Volochova D, Piovarci S 2012 *J. Cryst. Growth* **338** 239.
65. Shiohara Y, Endo A 1997 *Mater. Sci. Eng., R* **R19** 1.
66. Namburi DK, Rajasekharan T, Seshubai V 2013 *Physica C* **495** 55.
67. Lo W, Cardwell DA, Dewhurst CD, Dung SL 1996 *J. Mater. Res.* **11** 786.
68. Diko P, Wende CH, Litzkendorf D, Klupsch T, Gawalek W 1998 *Supercond. Sci. Technol.* **11** 49.
69. Uhlmann DR, Chalmers B, Jackson KA 1964 *J. Appl. Phys.* **35** 2986.
70. Chen P-W, Chen I-G, Chen S-Y, Wu M-K 2011 *Supercond. Sci. Technol.* **24** 0850211.
71. Liu Y, Pan B, Xiang H, Qian J, Du G, Huang S, Yao X, Izumi M, Wang Y 2017 *J. Am. Ceram. Soc.* **100** 3858.
72. Yang W M, Wang M 2013 *Physica C* **493** 128.
73. Muralidhar M, Ide N, Koblischka M R, Diko P, Inoue K, Murakami M 2016 *Supercond. Sci. Technol.* **29** 0540031.
74. Muralidhar M, Kenta N, Zeng X L, Koblischka M R, Diko P, Murakami M 2016 *Phys. Status Solidi A* **213** 443.
75. Vojtkova L, Diko P 2017 *Mater. Sci. Forum* **891** 489.
76. Fujimoto H, Murakami M, Nakamura N, Gotoh S, Kondoh A, Koshizuka N, Tanaka S 1991 *Advances in Superconductivity IV, Proc. 4th Int. Symp. On Superconductivity (ISS '91)* p 339.
77. Mahmood A, Park S D, Jun B H, Youn J S, Han Y H, Sung T H, Kim C J 2009 *Physica C* **469** 1165.
78. Koblischka M R, Murakami M 2000 *Supercond. Sci. Technol.* **13** 738.
79. Shlyk L, Krabbes G, Fuchs G, Nenkov K, Verges P 2003 *Physica C* **392** 540.
80. Ishii Y, Yamazaki Y, Nakashima T, Ogino H, Shimoyama J, Horii S, Kishio K 2008 *Mater. Sci. Eng. B* **151** 69.
81. Pavan Kumar Naik S, Muralidhar M, Jirsa M, Murakami M 2017 *J. Appl. Phys.* **122** 193902.
82. Volochova D, Antal V, Piovarci S, Kováč J, Jirsa M, Noudem J G, Diko P 2016 *IEEE Trans. Appl. Supercond.* **26** 7200604.

83. Volochova D, Jurek K, Radusovska M, Piovarci S, Antal V, Kovac J, Jirsa M, Diko P 2014 *Physica C* **496** 14.

84. Bean C P 1962 *Phys. Rev. Lett.* **8** 250.

85. Volochova D, Antal V, Kovac J, Diko P 2017 *Mater. Sci. Forum* **891** 483.

86. Vojtkova L, Diko P, Rajňák M 2018 *IEEE Trans. Appl. Supercond.* **28** 6801804.

87. Dinger T R, Worthington T K, Gallagher W J, Sandstorm R L 1987 *Phys. Rev. Lett.* **58** 2687.

88. Diko P, Gawalek W, Habisreuther T, Klupsch Th, Görnert P 1995 *Phys. Rev.* B **52** 13658.

89. Eisterer M et al 2006 *Supercond. Sci. Technol.* **19** S530–6.

90. Eisterer M, Haindl S, Wojcik T, Weber H W 2003 *Supercond. Sci. Technol.* **16** 1282–5.

91. Plechacek V, Jirsa M, Rames M, Muralidhar M 2012 *Phys. Proc.* **36** 538.

92. Hlásek T, Shi Y, Durrell J H, Dennis A R, Namburi D K, Plecháček V, Rubešová K, Cardwell D A, Jankovský O 2019 *Solid State Sci.* **88** 74

93. Volochová D, Kavečanský V, Antal V, Diko P, Yao X 2016 *Supercond. Sci. Technol.* **29** 044004

94. Jorgensen J D, Veal B W, Paulikas A P, Nowicki L J, Crabtree G W, Claus H, Kwok W K 1990 *Phys. Rev.* B **41** 1863.

95. Kim C J, Hong G W 1999 *Supercond. Sci. Technol.* **12** R27.

96. Vandewalle N, Ausloos M, Mineur N, Cloots R, Hong G W, Kim C J 1996 *Supercond. Sci. Technol.* **9** 665.

97. Diko P 2006 *Physica C* **445–448** 323.

98. Diko P, Gawalek W, Habisreuther T, Th K, Görnert P 1996 *J. Microsc.* **184** 46.

99. Murakami M, Fujimoto H, Gotoh S, Yamaguchi K, Koshizuka N, Tanaka S 1991 *Physica C* **185** 321.

100. Volochová D, Diko P, Piovarči S, Antal V, Kováč J, Jirsa M 2017 *Acta Phys. Pol. A* **131** 1009.

101. Shlyk L, Krabbes G, Fuchs G, Stöver G, Gruss S, Nenkov K 2002 *Physica C* **377** 437.

102. Antal V, Volochová D, Kavečanský V, Kováč J, Diko P 2017 *Physica C* **541** 22.

103. Antal V et al 2009 *Supercond. Sci. Technol.* **22** 105001.

104. Takezawa N, Fukushima K 1997 *Physica C* **290** 31.

105. Matsushita T 2000 *Supercond. Sci. Technol.* **13** 730.

Chapter 4

Superconductivity in Biomedicine: Enabling Next Generation's Medical Tools for SDGs

Santosh Miryala
Nuffield Department of Surgical Sciences, University of Oxford, Headington, Oxford, OX3 9DU, UK
santoshjuly6@gmail.com

As new technologies and scientific innovations emerge, superconducting materials are playing a phenomenal role in applications associated with healthcare and biomedicine. The magnificent properties of ideal magnetism, zero electrical resistance, and presence of quantized magnetic flux lines clearly indicate that superconductors are an attractive option for primary care and biomedical applications. The current superconducting research had proved its world-recognized strength in innovative magnet manufacturing and design towards low- and high-temperature superconducting magnetic applications in healthcare systems; an example is the significant role of a superconducting magnet in diagnostic imaging where a successful clinical implementation of

High-T_c Superconducting Technology: Towards Sustainable Development Goals
Edited by Muralidhar Miryala

ISBN 978-981-4877-65-7 (Hardcover), 978-1-003-16468-5 (eBook)
www.jennystanford.com

nuclear magnetic resonance and magnetic resonance imaging was conducted. On the other hand, compact superconducting cyclotrons are also playing an increasingly important role in medicine due to its emission of external beam therapy with carbon ions and protons. In the first section of this chapter, an extensive overview of fundamental properties of superconductivity is discussed. Moreover, the second section of this chapter primarily focuses on the essential role of superconducting materials in biomedical applications and its significance towards attaining United Nations sustainable development goals (SDGs).

4.1 Introduction

The United Nations member states have pledged to accomplish Agenda 2030 for 17 SDGs in order to tackle unprecedented global challenges across three characteristic dimensions of sustainable development: Economic, social, and environmental [1]. In comparison to previous development agendas that prioritized financial prosperity, the 17 SDGs and 169 targets provide a framework for multidimensional view for societal development through the cooperation of local, regional, national, and international efforts across all segments of society [2]. In fact, the achievement of this new universal agenda would crucially depend on a strong focus on inclusiveness and equity and coping with all needs of disadvantaged citizens within a short time frame of 15 years.

To shed light on this significant issue, health has been universally acknowledged as a central component towards global development for the past 30 years and extensive efforts have been made towards reducing mortality and morbidity rate, facilitating access to reproductive health care, reversing the spread of malaria, HIV/AIDS, and tuberculosis [3]. Without a doubt, the attainment of the following tasks was instrumental towards stabilizing global resources in low- and middle-income nations. Pursuing this topic further, it is evident to note that health has been specifically addressed in one of the 17 SDGs to "ensure healthy lives and promote well-being for all ages"; with this purpose in mind, health is also indirectly interwoven with nine other SDGs and underpinned by thirteen targets suggesting that a wide spectrum of determinants of health are holistically embraced due to interconnected nature of SDGs [4]. In short, health is a cross cutting issue within SDGs that requires an integrated implementation of scalable and innovative approaches along with novel partnerships that are guided by principles of social justice and ecology in a global context.

Since the discovery of superconductivity by a notable Dutch physicist Heike Kamerlingh Onnes in 1911, there has been a vast array of new materials discovered and remarkable progress has been made towards the development of promising superconducting technologies in the biomedical field that addresses health in SDGs by achieving a broader range of goals. In particular, numerous large-scale research initiatives had effectively utilized large magnetic fields of superconductors for successful healthcare applications of magnetic resonance imaging and magneto-encephalography to obtain a stronger foundation of human physiology. Thus, this chapter outlines the basic phenomenon and properties of bulk superconducting materials, examines the extensive applications and instrumental role of superconducting magnets within prominent biomedical engineering applications, and reflects on the approaches and target fulfillment of SDGs.

4.2 The Basic Phenomenon of Superconductivity

4.2.1 Zero Resistance

A notable Dutch physicist—Professor Heike Kammerlingh Onnes—had discovered a miraculous phenomenon called superconductivity: certain substances possess the following physical properties at low absolute temperatures where electrical resistance will immediately vanish along with magnetic flux fields being expelled [5]. After Onnes had attained to liquefy helium, his research team was able to investigate electrical resistance of mercury at low temperature as mercury is a chemical element that was easy to be produced without any impurities and had maintained liquid state in room temperature. During this experimentation, Onnes and his team aimed to reduce the effect of impurities towards the resistance of mercury. Furthermore, the team had lowered the temperature of mercury as they measured its electric resistance; as a result, the resistance of mercury had disappeared at a certain temperature and this indicated the existence of zero resistance superconductivity occurrence [6]. In recognition, Professor Heike Kammerlingh Onnes was acknowledged with the Nobel Peace Prize in Physics during 1913 "for his investigations on the properties of matter at low temperatures which led, inter alia, to the production of liquid helium" [7].

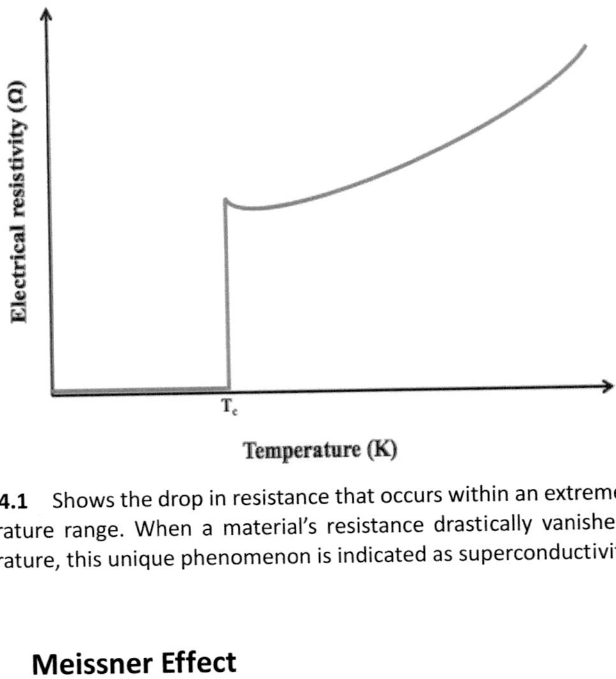

Figure 4.1 Shows the drop in resistance that occurs within an extremely small temperature range. When a material's resistance drastically vanishes at low temperature, this unique phenomenon is indicated as superconductivity.

4.2.2 Meissner Effect

In 1933, a German physicist Walther Meissner and his student Robert Ochsenfeld were able to discover one of the most fundamental properties of superconducting phenomenon: Meissner–Ochsenfeld effect. This property in quantum physics signifies an expulsion of magnetic induction (B) from the bulk of a superconductor when it is cooled below its critical temperature, i.e., $B = 0$ [8]. The Meissner–Ochsenfeld effect is able to attain this phenomenon by producing small currents within the surface of the superconductor that can effectively cancel out magnetic fields that come in contact with its material surrounding [9]. In terms of historical overview, Walther Meissner and Robert Ochsenfeld attempted to measure magnetic field distribution within a superconducting state in high purity Pb and Sn single crystals. The cooling vessel was filled with liquid helium and a lead plate was placed on it; furthermore, a light ferrite magnet was placed on top of the lead plate and it is intriguing to note that levitation was observed. As the lead was placed in liquid helium, its temperature dramatically decreased as it possessed superconducting properties which eventually repelled the magnetic force of the ferrite magnet. The fascinating property of the Meissner effect was very difficult to observe and conduct in laboratories due to the complications associated with handling of helium liquefier. However, the discovery of La-Ba-Cu-O high-temperature superconductor by Georg Bednorz and K. Alex Muller resulted in development of numerous oxide superconductors that obtain a T_c value higher than liquid nitrogen and this further enabled all scientists and academics to easily observe Meissner effect in their respective laboratories due to efficient handling procedures [10].

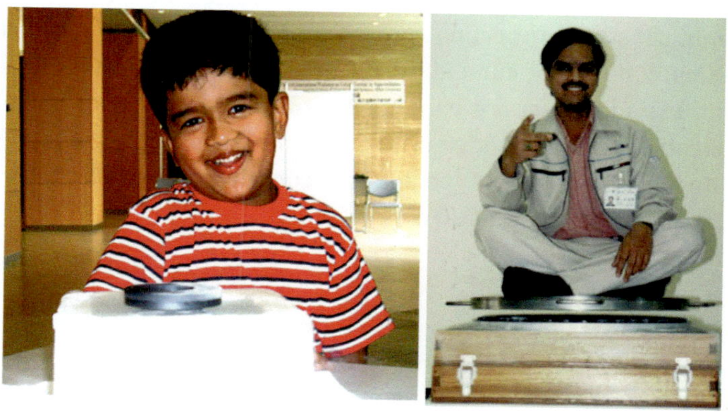

Figure 4.2 Displays levitation principle of high T_c superconducting material as the levitated object is magnet at liquid oxygen temperature (left). This is evidently the principle of the Meissner effect. Levitation of human beings is also possible due to strong pinning centers in YBCO (right).

4.2.3 Type I vs. Type II Superconductors

In theory, superconducting materials are further classified into two categories depending on their behavior towards external magnetic field (H) and they include Type I and Type II superconductors. Type I superconductors are considered to be conventional superconductors and are comprised of pure substances such as metals that possess conductive properties in its regular state along with a very low critical temperature [11]. In general, Type I superconductors lose its superconducting state as the external magnetic field exceeds temperature (T) or critical value (H_c) only when the following superconductor is somewhat higher than T_c [12]. Common examples of Type I superconductors include Hg, Zn, and Pb. The general attributes of Type I superconductors include diamagnetism and an extremely low critical magnetic field that eventually leads to limited technological applications [13]. In contrast, Type II superconductors are considered to be high-temperature superconductors and can range in the forms of alloys to ceramics. This classification of superconductors obtains high critical magnetic field (greater than 1 T), high critical temperature (greater than 10 K), and exhibits two critical magnetic field [14]. Type II superconductors are also known as hard superconductors and they partly follow the Meissner effect but not in full effect as magnetic flux can penetrate inside its bulk. The mixed state properties are evident in Type II superconductors as superconducting current surrounds it in a form of vortex state and these materials are commonly utilized in variety of technical applications due to its high critical magnetic field value [15]. Lastly, common examples of Type II superconductors are NbTi, Nb_3Sn, YBCO, BSCCO, and $LaBaCuO_4$.

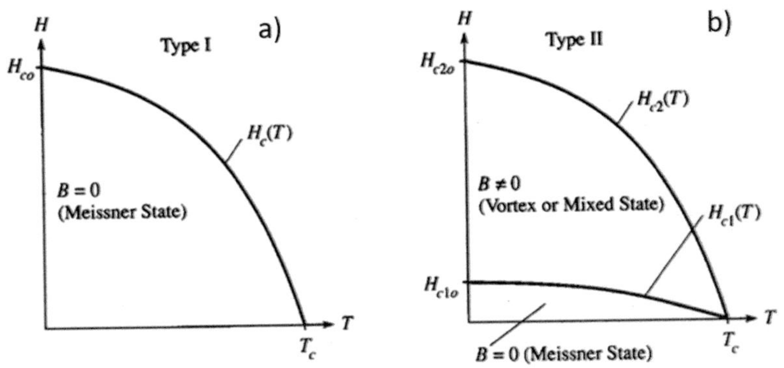

Figure 4.3 The differentiation of phase diagrams for Type I (left) and Type II (right) superconductors. It is observed that all high-temperature superconductors are Type II superconductors with a property of intermediate mixed state that enables magnetic flux within its mixed state; hence, it conveys that high-temperature superconductors are an attractive option for biomedical applications.

4.2.4 Josephson Effect

In 1962, another significant manifestation of the quantum nature of superconductivity—Josephson effects or weak superconductivity—was predicted and tested with numerous experiments. In fact, this phenomenon referred to two superconductors that were coupled together with a weak link and this was attainable due to a short constriction or a tunnel junction identified in a cross section area of a thin-film [16]. The following contact could be presented in numerous examples such as normal conducting layers and tunneling barriers that connect two superconducting electrodes. As a whole, the weak contact that is identified between two superconductors within a small arrangement of an area further suggests that superconducting contact is somewhat weakened [17]. In recent days, the Josephson effect is widely utilized in superconducting applications of sensors, electronics, and high frequency devices.

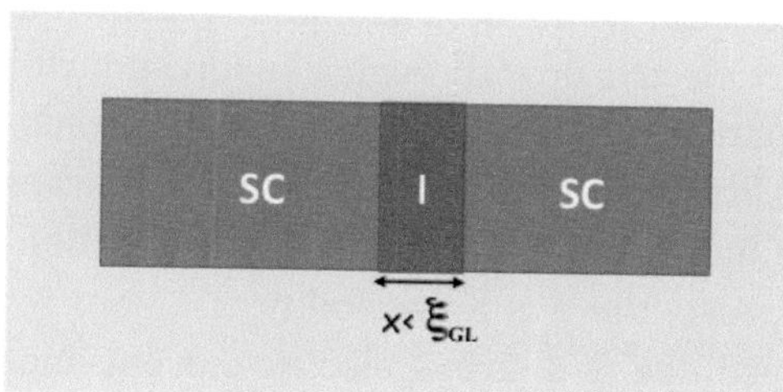

Figure 4.4 Displays Josephson effect. In fact, the coherence length is the largest insulating distance that can be tunneled by cooper pairs. Brian D. Josephson received the Nobel Prize in Physics in 1973 for his prediction of Josephson effect.

4.2.5 BCS Theory

BCS theory is also known as Bardeen–Cooper–Schrieffer theory as they are arranged in the order of physicist names and they include John Bardeen, Leon Cooper, and John Robert Schrieffer. This is the initial theory after 1911 discovery by Heike Kamerlingh Onnes that explains superconductivity in terms of microscopic effect due to the condensation of Cooper pairs. For several years, the law of conventional physics was not able to satisfactorily explain the theory of superconductivity and this theory was acknowledged with the Nobel Peace Prize in Physics during 1972 [18]. Furthermore, the BCS theory explained that electrons were able to move through a solid with zero resistance only due to two electrons emitting an attractive interaction within the same distance [19]. As a whole, when one electron shifts across the lattice, the neighboring nuclei are also attracted to it. The nuclei's motion can produce a transient hole that will draw the second electron towards the same direction as the first electron [20].

4.2.6 Critical Current

The critical current density can be considered as one of the significant characteristics of a superconductor viewing through the lens of practical applications as the maintenance of maximum electrical transport current density without producing any resistance would be highly valued for power transmission. The critical current of a superconductor can be denoted as I_c and is defined as a maximum value of direct electric flow through a superconductor without any energy dissipation [21]. Thus, if the current exceeds critical value then this will lead to superconducting material to transition into the normal state and further causing energy dissipation and external heating [22]. Also, there is a possibility of a zero resistance state to be destroyed due to a strong electric current imposed and this was referenced by Silsbee rule [23]. On the other hand, the critical current—I_c—for Type I superconductors can be defined as the current that creates the critical magnetic field (H_{cm}) around the surface of the sample. However, one can notice different parameters for cylindrical superconductor and can be defined as $I_c = cRH_{cm}/2$ where R is radius and c is constant; thus, when conditions pertain towards current $I > I_c$, the Type I superconducting materials would be indicated within intermediate state where one can observe constant alterations between normal and superconducting areas [24]. In contrast, Type II superconducting materials show that value of critical current is formulated by creating an initiation within the sample of vortices. The vortices tend to form at a sample surface that initiates current and can be noted as the first critical magnetic field: H_{c1}. Therefore, each vortex line has a core that indicates a quant of magnetic flux and this is only due to the Lorentz force that would act on the vortices towards the direction of current flow [25]. Actual samples show vortices are attached to sample boundaries and crystallite lattice defects and this further suggests that equilibrium was reached within magnetic flux pinning with production of small current [26]. When one increases the current of Lorentz force and at a situation where it exceeds critical value, the frustration of vortices could be observed due to its movement and generation of electric field in the sample [27]. During this stage, the matter in the sample maintains superconducting properties.

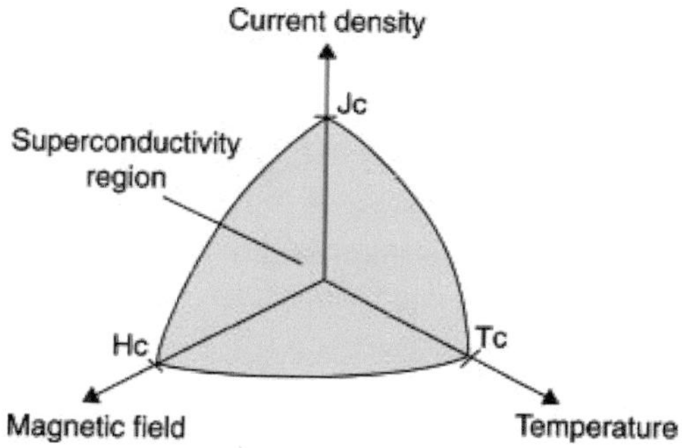

Figure 4.5 A schematic diagram of superconducting region of Type II superconductor where critical density, magnetic field, temperature should be maintained below critical surface for retaining superconducting properties.

4.2.7 The Cooper Effect

In 1956, an American physicist Leon Cooper made a groundbreaking discovery in the field of condensed matter physics as he first described Cooper Pair that is considered to be a pair of electrons which are bound together within low temperatures. Leon Cooper indicated that it is a small attraction between electrons in a metal that obtains the tendency to produce a paired state of electrons with lower energy compared to Fermi energy; thus, this suggests that the pair is bound [28]. Conventional superconductors possess this attraction only because of electron–phonon interaction. Cooper pairing can be classified as a quantum effect and it presents a foundational microscopic theory of superconductivity in regards to BCS model [29]. In addition to this, Cooper pair obtains an orbital moment that is equal in number to its harmonic as critical current of superconducting transition and binding energy would drastically decrease as orbital moment increases [30]. In accordance to a particular rule, the pairing would be filled up with low permissible value and summary spin of the couple would equal to a value of zero. Overall, the majority of superconducting materials possess Cooper pairs with zero orbital moment.

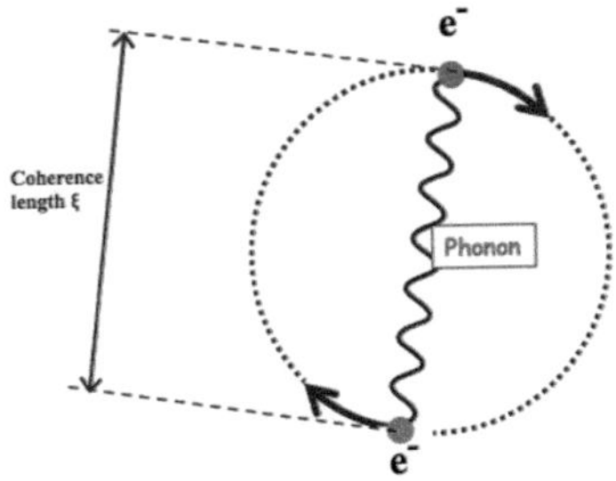

Figure 4.6 A model of Cooper pair attraction proposed by BCS theory.

4.2.8 London Penetration Depth

As a bulk superconductor is positioned within a weak external magnetic field H, the surrounding field would penetrate the superconducting material towards a short distance λ_L: London penetration depth. This is a defining property of superconducting materials as it characterizes a certain distance at which a weak magnetic field would penetrate into the bulk of a superconductor. For most superconductors, the penetration depth is considered to be at an order of 100 nm and typical values for λ_L are usually between 50 and 500 nm [31]. For instance, the penetration depth could be found by utilizing superfluid density as this is an essential parameter that is used to find T_c of high-temperature bulk superconductors. Furthermore, certain superconductors obtain a node at their energy gap and its penetration depth within 0 K would entirely be influenced by its magnetic field as superfluid density is dependent on it [32]. Due to the following reason, it is significant to obtain explicit measurements of absolute value of penetration depth within 0 K in order to have a strong comprehension in regards to the behavior of high-temperature superconducting materials [33].

4.2.9 Isotope Effect

In 1950, two notable, independent scientific groups had discovered the isotope effect as they were working on numerous mercury isotopes. In fact, the isotope effect had contributed to the crucial development of BCS theory because it had shown that T_c was proportional $M^{-1/2}$ where M is noted as the isotope mass for a given element [34]. Moreover, $M^{-1/2}$ would be proportional to vibration frequency of mass M; thus, this indicates superconductivity is an electronic phenomenon even when movements of electrons were observed in the vibration of crystal [35]. As a whole, it was clearly observed that the critical temperature of a superconductor would vary accordingly with isotopic mass. For example, the T_c would vary from 4.185 K to 4.146 K in mercury as average atomic mass M would vary within 4.186 K to 4.146 K atomic mass units [36]. Lastly, the transition temperature would also be altered when different isotopes of the same element would be mixed.

4.3 The Prominent Role of Superconductors in Biomedical Applications

4.3.1 Magnetic Resonance Imaging

The implementation of magnetic resonance imaging (MRI) into the healthcare system has revolutionized neuroscience by allowing medical doctors and researchers to visualize the diagnostic ability of a brain along with its soft tissue features enabling it to produce pictures without the aid of ultrasound or X-ray imaging. MRI obtains the ability of eradicating constant exposure to detrimental X-ray radiation and further providing strong accuracy and precision of diagnostic location; thus, the following advantages are encouraged towards a reduced need of exploratory surgery along with decreasing patient's length of stay.

In terms of science behind resonance imaging, the atom's nucleus functions as a small spinning magnet; however, when it is applied to a magnetic field, the nuclear spin—rotation around field direction with an attribute of frequency proportional to the value of field strength—would attempt to align [37]. Moreover, an applied pulse of optimal radio frequency causes the spinning nucleus to absorb energy from the pulse and further release it after many milliseconds later [38]. In analysis, it was evident that timing of relaxation was solely dependent on the chemical environment of an atom; however, its composition concluded to be different between diseased and healthy tissue in the framework of human physiology. In addition, the constant switching of magnetic field gradients on and off superimposes on the central area and further enables an accurate determination of precise location due to signals processed by a computer as it brings out non-familiar images of the human body [39].

Since its development from the 1970s and 1980s, the industry revolving around MRI grew to be an over 2 billion dollar market as it was recognized to be a versatile imaging technique. The MRI is prominently segmented in biomedical research and diagnostic medicine due to its capability of producing spatial images with physical and chemical data along with images of non-living organisms [40]. In the recent days, there are more than 25,000 MRI scanners stationed to be in use worldwide. On the other hand, it is important to note that a superconducting magnet functions as a central role within an MRI system; as a matter of fact, conventional magnets do not obtain the potential to produce typical field values that are required for MRI. Additionally, the effective utilization of superconducting materials is the only possibility of producing high stability of magnetic field and uniformity for attaining optimal precision, speed, and resolution towards economical clinical imaging [41].

Pursuing this topic further, functional magnetic resonance imaging (fMRI) is a notable extension technique of MRI that has been extensively utilized for mapping out alterations of regional cerebral blood flow rates and blood oxygenation volumes within a region of the brain. The fMRI has been proved to be a powerful tool to analyze activated brain regions that are responsible for essential functions of comprehension, speaking, vision, and movement of toes and fingers [42]. In the same way, MRI guidance imaging is another new technique that physicians utilize to accurately locate and remove tumors during surgical procedure. Consequently, magnetic resonance spectroscopy is utilized in a restricted manner for assessing neurological diseases, epilepsy, and brain tumors; in fact, spectroscopy provides valuable information concerning metabolic activity and chemical composition of brain tissue as it may assist towards evaluation of seizure activity [43].

After 30 years of remarkable progress and continued efforts invested towards the development of MRI, a great number of MRI systems have been installed worldwide. The enhancement observed with the current research in superconducting magnets enables further advancement of resolution and speed of MRI technologies. Without a doubt, superconducting materials played a powerful role with the ongoing evolution of diagnostic imaging which had shown enormous benefits towards patient welfare globally.

4.3.2 Ultra-Low Field Magnetic Resonance Imaging (ULF-MRI)

The conventional magnetic resonance imaging has indeed transformed the division of non-invasive imaging technique as this procedure has been commonly used in numerous diagnoses worldwide. As a matter of fact, a bulk-superconducting magnet produces high magnetic fields that play a central role towards the operation of MRI; in the same way, scientists in the current generation are also developing an alternative method called ultra-low-field MRI that focuses on utilizing low magnetic field that was produced from simple, low-cost magnets in which were constructed through copper wire [44]. In order to make up for the privation of high magnetic fields, it is crucial to utilize a sensitive superconducting detector noted as SQUID: superconducting quantum interference device. In fact, it portrays multiple benefits, as it is a low-cost system that could be widely implemented worldwide for initial screening purposes. As a notable case scenario, ULF-MRI provides a clear contrast for different tissue types such as prostate and breast tumors and this will aid towards an effective diagnosis [45]. Without a doubt, ULF-MRI is a valuable asset in healthcare since it is an enhanced instrument that reduces medical expenses along with improving its diagnostic screening [46]. In the near future, biomedical researchers and physicians hope to combine magneto-encephalography and ULF-MRI for a more promising and much environmentally friendlier instrumentation compared to high-field MRI systems [47].

4.3.3 Nuclear Magnetic Resonance

Nuclear magnetic resonance (NMR) has proved to be one of the most versatile spectroscopic devices in the 21st century due to its promising role within applications of modern science. The analytical nature of NMR led to Nobel Peace Prizes in Chemistry during 1991 and 2002 along with Nobel Peace Prize in Physics during 1952 [48]. Without a doubt, the basis of NMR spectroscopy in the biomedical sciences enabled physicians to recognize patients who are in danger of having heart disease through an analysis of concentration and size of lipoproteins [49]. Moreover, NMR technology is also utilized for assessing certain medications that enables medical doctors to choose optimal cholesterol medications for treating human patients in accordingly to their lipoprotein concentration and size [50].

In terms of the structure of NMR spectrometer, one can note that a superconducting magnet is placed at its heart as high homogeneity and high magnetic field could be produced to attain high-quality precision and resolution images for protein structure determination [51]. With constant progression in superconducting material research and development, an intriguing usage of materials could be observed from niobium titanium to niobium-tin conductors with a present emphasis towards a wide range of materials that are appropriate for ultra-high field applications in order to investigate highly complex molecules [52]. In recent days, biomedical researchers are developing power storage coils that are made out of MgB_2 superconducting wires due its advantages of lower production cost along with less cooling requirement [53].

Subsequently, one of the most fundamental methods of analyzing 3D structure of protein is through NMR. In fact, a protein is composed of thousands of amino acids that are connected to each other in a long chain; moreover, 20 amino acids tend to form a building block of the following chains and its order would be dictated by the information provided by genes [54]. These polymers serve essential activities for the sustainability of life such as production of electrochemical signals to stimulate neurotransmission and absorption of oxygen during breathing [55]. From this, it is evident that functional and structural properties of a protein would be portrayed as the frontier of medical science. However, the main challenge that medical researchers face is determining how the protein chain folds into its respective structure as it provides significant information about a protein's basic biochemical properties and function. For instance, obtaining information about these structures has enabled numerous pharmaceutical industries to develop medications that obtain the ability of precisely targeting specific proteins [56]. In other words, it is noted that basic protein fold is somewhat related to bioactivity since one basic protein fold may consist of various helices bundled together in a parallel manner that is usually involved in binding and recognition of RNA [57]. Thus, higher order function of a protein could be resulted from an integration of functions of 3D structure, molecular function, and basic protein folds.

Pursuing this topic further, in order to determine the 3D structure of a protein via NMR, a vast number of NMR measurements would be conducted. If only one NMR device was utilized, then a device must be adjusted multiple times as measurement conditions would be altered [58]. Additionally, a multiple magnetic system requires minimal necessary adjustments due to its effective determination of protein 3D structure could be analyzed within very few weeks rather than months [59]. In general, it is fascinating to note that obtaining information regarding the structure of a protein is a remarkable accolade that led to achieving numerous Nobel Prizes and NMR proved to be a resourceful device towards identification of protein structure and drug development.

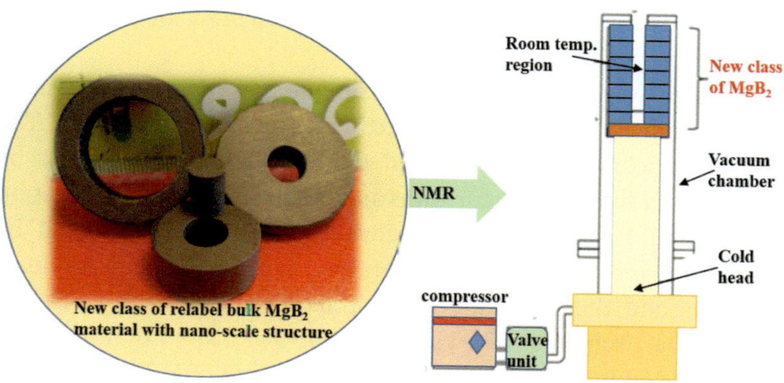

Figure 4.7 The schematic diagram of NMR set up where bulk MgB_2 superconducting material was utilized.

4.3.4 Diagnostic Techniques Using SQUIDs

Magnetocardiography (MCG) and magnetoencephalography (MEG) are distinguished examples of medical diagnostic testing that use SQUID. In the following systems, a wide variety of SQUID sensors aid in detecting magnetic signals from the human brain through non-invasive means.

First of all, a MEG unit would utilize several SQUID sensors to be placed on the scalp of a human subject for sensing the distribution of magnetic field, as this information is crucial to analyze the neural activity [60]. MEG showed numerous advantages compared to electroencephalography (EEG), as the spatial resolution was far superior due to magnetic permeability of human head is somewhat similar to vacuum with electrical conductivity to differ according to tissue [61]. In addition, another advantage is that MEG does not require contact with a human patient's skin since factors of skull thickness and moisture on the scalp can mislead the location of epilepsy source [62]. On the other hand, MEG has made rapid progress towards the analysis of unborn fetuses because this methodology attained the capability of delivering assessment of fetal neurological conditions, as it will aid medical surgeons during cases of high-risk pregnancies and detection of hemorrhage and hypoxia [63]. In fact, there are no other methodologies for conducting non-invasive evaluation of the fetal brain. However, the most pressing matter in regards to MEG is the challenge of implementing MEG units at a wider scale because problems arise with higher costs along with the need of databases to indicate relationship between surgery and magnetic source imaging (MSI) [64].

Second, MCG is a prominent technique that measures magnetic fields produced through the heart's electrical currents [65]. The multichannel instrument is used to estimate magnetic fields by utilizing mathematical algorithms and taking in account the location of a specific activity [66]. Moreover, MCG is widely utilized by physicians and researchers to accurately map out diagnosis for heart diseases and those include arrhythmia, coronary artery disease, and cardiac ischemia [67]. Various reports indicated that ischemia was detected with 100% accuracy with the use of MCG and played a powerful role in emergency departments and cardiology clinics worldwide [68].

Overall, the major challenge towards the development of MCG and MEG for the future relies on the need for further research towards superconductor development along with tackling performance costs and availability of helium. For instance, biomedical engineers would find a way of redesigning the structure of MCG and MEG to eliminate the need of helium cooling in the near future.

4.3.5 Magnetic Drug Delivery System

In the past 10 years, nanotechnology has been successfully implemented into biomedicine with an objective of delivering effective therapy and diagnosis to the public health and this has resulted into an innovative discipline known as nanomedicine. From nano-sized appliances capable of pinpointing specific targets in the human body to conjugating drugs, nanomedicine has shown to transform the fields of tissue engineering and drug delivery. Without a doubt, the usage of nano-sized vehicles to administer medicine with high local drug concentration to a diseased organ would be prone to minimal side effects along with minimal exposure to other areas of the human body [69]. As a whole, magnetic drug delivery system (MDDS) with the usage of superconducting magnet has been considered as a futuristic and attractive treatment method. This procedure aims to use particle motion control with an aid of magnetic force via DDS.

In terms of the conception of MDDS, the drug is transported through an ongoing stream of the blood and the magnet is placed on the surface of the human body that is near the branching point of the blood vessel [70]. Moreover, magnetic force is taken into account when the drug is transported in a certain direction and multiple repetitions are required to tackle the diseased tissue [71]. To illustrate, magnetic seeded drugs are incorporated into the human body through vein injection; further, drug travels from the vein into the heart's right atrium to pulmonary artery to right ventricle and will circulate within lungs, pulmonary vein, and heart's left atrium [72]. As a result, a superconducting magnet that is located outside of the human body would be utilized to navigate magnetic seeded drugs from the left ventricle while placing the magnet on diseased organs so that drug would be accumulated via magnetic force [73].

In practice, the unique principles of MDDS have been performed on numerous animals and they include pig, frog, rat, etc. In particular, this innovative application has been tested out on a Y-shaped blood vessel where abdominal aorta was divided into two sections of right and left ileum [74]. In this scenario, a strong magnet of 0.3 T was placed near the region of right ileum and magnetic capsules that consisted of the drug was administered into abdominal aorta; thus, this resulted in concentration of capsule to be increased twice due to the effect of a strong magnet [75]. In another case study, Kamei et al. [76], has pioneered a unique superconducting magnetic MDDS that had utilized a specialized material–ferubarbotran–to repair cartilage. In this investigation, a pig had cylindrical cartilage effect within the central patella. After 12 weeks following surgery, the results indicated that treatments utilization of superconducting magnet led to regeneration of new cartilage cells within a diseased area [77]. In essence, MDDS that utilizes superconducting materials indicated to be a promising therapy where patients are greatly benefited and would revolutionize medical and pharmaceutical industries.

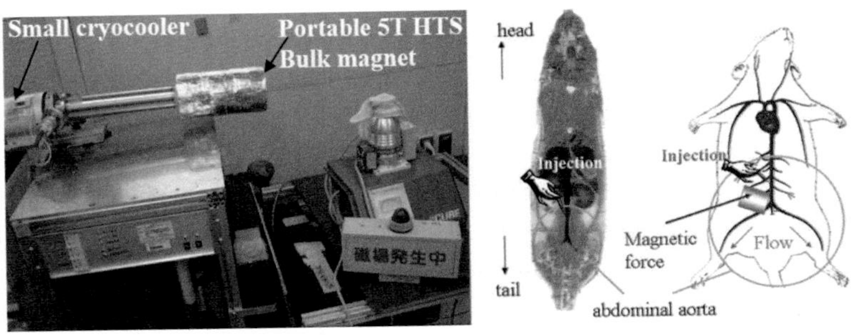

Figure 4.8 A new class of magnetic drug delivery system (MDDS) developed by utilizing a bulk Gd-123 superconducting material which can generate 5 T magnetic field (left). A portrayal of a navigation experiment tackling the branching point of a blood vessel in a rat (right).

4.3.6 Particle Beam Applications for Biomedical Diagnosis

In the 21st century, cancer is considered as the second leading cause of death in the global scale with an approximate estimation of 9.6

million deaths in 2018; hence, one can note that there are 1 in 6 deaths caused by cancer [78]. Further, one can notice that 70% of deaths observed in low- and middle-income nations are due to cancer and its economic impact is tremendously increasing [79]. Proceeding with the following complications, the utilization of radiology towards medical treatment and diagnosis has been commonly implemented for a century from the basis of X-ray discovery by Roentgen [80]. Numerous refinements and treatments had been developed as radiotherapy proved to be a promising treatment towards cancer as it obtained the potential of eradicating cancerous cells while maintaining healthy cells and tissues within the human body. More recently, it was noted that 90% of radiation therapy uses strong energy from Bremsstrahlung radiation that is provided from electron linear accelerators [81]. On the other hand, some notable examples of radiology include contrast media diagnostic method, fluorescent diagnosis, computerized topography, etc. It is essential to note that enhancement in applications of radiology would highly depend on development of compact electron accelerators and X-ray source technology.

As a matter of fact, superconducting magnetic technology enabled significant enhancements in cyclotron accelerators within hadron therapy. In this particular form of radiation therapy, medical doctors can strategically use ionized particle beams instead of X-ray for handling pediatric cancer tumors. Moreover, the execution of high-field superconducting cyclotrons suggested that it is a viable option due to low operating costs and compact infrastructure.

4.4 Relevance to Sustainable Development Goals

The United Nations SDGs can be regarded as an integrated, universal policy agenda with an aim of global prosperity and sustainable environment for our complex society [82]. It is essential to recognize SDG agenda attempts to encourage health policy workers to recognize the need of multisectoral collaboration and new accountability measures within primary health care systems. For instance, a strong collaboration between engineering and health care systems is required to through policy-making and research for addressing health-related outcomes in UN's SDGs. In essence, the application of superconducting engineering in the biomedical sciences would require a genuine engagement for the advancement of a wide array of medical applications to tackle health-related outcomes with the help of governmental platforms.

Figure 4.9 The upcoming future technological advancements require superconducting magnet technology to attain sustainable development goals (SDGs) through zero emission of carbon dioxide (left). The Government of Japan is investing efforts to signify and advertise the significance of SDGs within the local community (right).

4.5 Conclusion

In summary, superconducting technology has played a pivotal role in the field of biomedical applications. SQUID, MRI, and NMR are notable instruments that are globally used to aid physicians for patient diagnosis while accelerators are implemented in complex medical treatments. The recent developments in superconducting magnetic technology allowed a wide range of medical miracles to be feasible. However, an effective blueprint and continued interdisciplinary superconducting research in biomedicine are required to be performed multiacademically. Therefore, skilled biomedical engineers with an expansive training of therapeutic technologies and diagnostic treatment are urgently required with the advocacy of universities and government agencies to indirectly tackle UN's SDGs.

References

1. Ottersen, O.P., Engebretsen, E. COVID-19 puts the sustainable development goals center stage. *Nat Med* **26**, (2020) 1672–1673.
2. Luyckx, V.A., Al-Aly, Z., Bello, A.K. et al. Sustainable development goals relevant to kidney health: an update on progress. *Nat Rev Nephrol* **17**, (2020) 15–32.
3. Wang, Z., Tang, K. Combating COVID-19: health equity matters. *Nat Med* **26**, (2020) 458.
4. Vinuesa, R., Azizpour, H., Leite, I. et al. The role of artificial intelligence in achieving the sustainable development goals. *Nat Commun* **11**, (2020) 233.
5. Paglione, J., Greene, R. High-temperature superconductivity in iron-based materials. *Nat Phys* **6**, (2010) 645–658.
6. Larbalestier, D., Gurevich, A., Feldmann, D. et al. High-T_c superconducting materials for electric power applications. *Nature* **414**, (2001) 368–377.
7. Heike Kamerlingh Onnes, – Facts. NobelPrize.org. Nobel Media AB 2020. Fri. 27 Nov 2020.
8. Huse, D., Fisher, M., Fisher, D. Are superconductors really superconducting?. *Nature* **358**, (1992) 553–559.
9. da Silva, R., Milošević, M., Shanenko, A. et al. Giant paramagnetic Meissner effect in multiband superconductors. *Sci Rep* **5**, (2015) 12695.
10. Krusin-Elbaum, L., Tsuei, C., Gupta, A. High current densities above 100 K in the high-temperature superconductor $HgBa_2CaCu_2O_{6+\delta}$. *Nature* **373**, (1995) 679–681.
11. Gordeev, S., de Groot, P., Oussena, M. et al. Current-induced organization of vortex motion in type-II superconductors. *Nature* **385**, (1997) 324–326.
12. Nelson, D. A new age for type-II superconductors? *Nature* **385**, (1997) 675–676.
13. Vagov, A., Wolf, S., Croitoru, M.D. et al. Universal flux patterns and their interchange in superconductors between types I and II. *Commun Phys* **3**, (2020) 58.
14. Cardwell, D., Welch, D., Huebener, R., Uher, C., Porch, A., Gough, C., Tarte, E. Introduction to section A2. *Handbook of Superconducting Materials*, (2002) 171–205.

15. White, G. Thermal expansion. *Handbook of Applied Superconductivity*, Vol. 2, (1998) 1107–1119.

16. Kohlmann, J., Zorin, A.B. Applications in quantum metrology. *Applied Superconductivity: Handbook on Devices and Applications*; Wiley Online Library (2015) 807–842.

17. Vischi, F., Carrega, M., Virtanen, P. et al. Thermodynamic cycles in Josephson junctions. *Sci Rep* **9**, (2019) 3238.

18. Frye, R.C. High T_c superconducting interconnections in semiconductor - based electronic systems. *Heterostructures on Silicon: One Step Further with Silicon*, Springer (1989) 169–185.

19. Huang, W.M., Lin, H.H. Pairing mechanism in multiband superconductors. *Sci Rep* **10**, (2020) 7439.

20. Potočnik, A., Krajnc, A., Jeglič, P. et al. Size and symmetry of the superconducting gap in the f.c.c. Cs_3C_{60} polymorph close to the metal-Mott insulator boundary. *Sci Rep* **4**, (2014) 4265.

21. Hong, M., Maher, D., Ellington, M., Hellman, F., Geballe, T., Ekin, J., Holthuis, J. Further investigations of the solid-liquid reaction and high-field critical current density in liquid-infiltrated Nb-Sn superconductors. *IEEE Trans Magn* **21**:2, (1985) 771–774.

22. Pourrahimi, S., Thieme, C.L.H., Foner, S., Suenaga, M. Nb_3Sn(Ti) powder metallurgy processed high field superconductors. *Appl Phys Lett* **48**:26, (1986) 1808–1810.

23. Talantsev, E.F., Crump, W.P., Tallon, J.L. Universal scaling of the self-field critical current in superconductors: from sub-nanometre to millimetre size. *Sci Rep* **7**, (2017) 10010.

24. Cho, D., Bastiaans, K.M., Chatzopoulos, D. et al. A strongly inhomogeneous superfluid in an iron-based superconductor. *Nature* **571**, (2019) 541–545.

25. Naqib, S.H., Islam, R.S. Possible quantum critical behavior revealed by the critical current density of hole doped high-T_c cuprates in comparison to heavy fermion superconductors. *Sci Rep* **9**, (2019) 14856.

26. Tripathi, D., Dey, T.K. Scaling of pinning force density in (Bi,Pb)-2223 added MgB_2 superconductors. *J Superconduct Novel Magn* **28**:7, (2015) 2025–2032.

27. Luo, W., Huang, Z., Cai, X., Niu, R., Nie, R., Feng, Q., Wang, F., Gan, Z. Microstructure and superconducting properties of MgB_2 bulks prepared from Mg + B + $Mg(BH_4)_2$ composites. *Superconduct Sci Technol* **32**:8, (2019) 085006.

28. Hofstetter, L., Csonka, S., Nygård, J. et al. Cooper pair splitter realized in a two-quantum-dot Y-junction. *Nature* **461**, (2009) 960–963.

29. Diamantini, M.C., Postolova, S.V., Mironov, A.Y. et al. Direct probe of the interior of an electric pion in a Cooper pair superinsulator. *Commun Phys* **3**, (2020) 142.

30. Mi, S., Burset, P., Flindt, C. Electron waiting times in hybrid junctions with topological superconductors. *Sci Rep* **8**, (2018) 16828.

31. He, X., Gozar, A., Sundling, R., Božović, I. High-precision measurement of magnetic penetration depth in superconducting films. *Revi Sci Instrum* **87**:11, (2016) 113903.

32. Aurino, M., Di Gennaro, E., Di Iorio, F., Gauzzi, A., Lamura, G., Andreone, A. Discrete model analysis of the critical current-density measurements in superconducting thin films by a single-coil inductive method. *J Appl Phys* **98**:12, (2005) 123901.

33. Kogan, V.G. Meissner response of anisotropic superconductors. *Phys Rev B* **68**:10, (2003) 104511.

34. Chu, W., Cheng, J., Chu, S. et al. Iron isotope effect and local lattice dynamics in the (Ba, K)Fe_2As_2 superconductor studied by temperature-dependent EXAFS. *Sci Rep* **3**, (2013) 1750.

35. Huang, W.M., Lin, H.H. Anomalous isotope effect in iron-based superconductors. *Sci Rep* **9**, (2019) 5547.

36. Zech, D., Keller, H., Conder, K. et al. Site-selective oxygen isotope effect in optimally doped $YBa_2Cu_3O_{6+x}$. *Nature* **371**, (1994) 681–683.

37. Hernandez, D., Kim, K.-N. A review on the RF coil designs and trends for ultra high field magnetic resonance imaging. *Investig Magn Reson Imaging* **24**:3, (2020) 95.

38. Athalye, P.S., Šekeljić, N.J., Ilić, M.M., Tonyushkin, A.A., Notaroš, B.M. Subject-loaded quadrifilar helical-antenna RF coil with high $B_1{}^+$ field uniformity and large FOV for 3-T MRI. *Concepts Magn Reson B: Magn Reson Eng* **46B**:3, (2016) 106–117.

39. Lu, H., Shang, S., Wang, S. Parallel-plate waveguide for volume radio frequency transmission in MRI. *Magn Reson Med* **74**:5, (2015) 1482–1491.

40. López, M., Vázquez, F., Solís-Nájera, S., Rodriguez, A.O. Numerical study of remote detection outside the magnet with travelling wave magnetic resonance imaging at 3T. *J Phys: Conf Ser* **582**, (2015) 012037.

41. Bertora, F., Borceto, A., Viale, A., Fabbricatore, P. A target field approach to open MRI magnet design. *IEEE Trans Appl Superconduct* **22**:3, (2012) 4905204–4905204.

42. Kim, N., Woo, M.-K., Kang, C.-K. Compensation for z-directional non-uniformity of a monopole antenna at 7T MRI. *J Korean Phys Soc* **68**:11, (2016) 1352–1358.

43. Ganzetti, M., Wenderoth, N., Mantini, D. Intensity inhomogeneity correction of structural MR images: a data-driven approach to define input algorithm parameters. *Front Neuroinform* **10**, (2016) 10.

44. Ueno, S. Studies on magnetism and bioelectromagnetics for 45 years: from magnetic analog memory to human brain stimulation and imaging. *Bioelectromagnetics* **33**:1, (2012) 3–22.

45. Mäkinen, A.J., Zetter, R., Iivanainen, J., Zevenhoven, K.C.J., Parkkonen, L., Ilmoniemi, R.J. Magnetic-field modeling with surface currents. Part I. Physical and computational principles of bfieldtools. *J Appl Phys* **128**:6, (2020) 063906.

46. Zetter, R., Mäkinen, A.J., Iivanainen, J., Zevenhoven, K.C.J., Ilmoniemi, R.J., Parkkonen, L. Magnetic field modeling with surface currents. Part II. Implementation and usage of bfieldtools. *J Appl Phys* **128**:6, (2020) 063905.

47. Zevenhoven, K.C.J., Mäkinen, A.J., Ilmoniemi, R.J. Superconducting receiver arrays for magnetic resonance imaging. *Biomed Phys Eng Express* **6**:1, (2020) 015016.

48. Niki, H., Okuda, H., Oshiro, M., Yogi, M., Seki, I., Fukuhara, M. Proton nuclear magnetic resonance studies of hydrogen diffusion and electron tunneling in Ni-Nb-Zr-H glassy alloys. *J Appl Phys* **111**:12, (2012) 124308.

49. Bykovetz, N., Hoser, A., Klein, J., Lin, C.L., Seehra, M.S. Neutron scattering measurements in $RbMnF_3$: a test of spin-wave-region theories at low temperatures and critical behavior near T_N. *J Appl Phys* **111**:7, (2012) 07E145.

50. Bykovetz, N., Birang, B., Klein, J., Lin, C.L. NMR measurements of power-law behavior in the spin-wave and critical regions of ferromagnetic EuO. *J Appl Phys* **107**:9, (2010) 09E142.

51. Schwall, R. MRI-superconductivity in the marketplace. *IEEE Trans Magn* **23**:2, (1987) 1287–1293.

52. Mazur, A.S., Vovk, M.A., Tolstoy, P.M. Solid-state ^{13}C NMR of carbon nanostructures (milled graphite, graphene, carbon nanotubes, nanodiamonds, fullerenes) in 2000–2019: a mini-review. *Fullerenes, Nanotubes Carbon Nanostr* **28**:3, (2020) 202–213.

53. Noudem, J.G., Xing, Y., Bernstein, P., Retoux, R., Higuchi, M., Arvapalli, S.S., Muralidhar, M., Murakami, M. Improvement of critical current

density of MgB_2 bulk superconductor processed by Spark Plasma Sintering. *J Am Ceram Soc* **103**:11, (2020) 6169–6175.

54. Chen, M., Chen, X., Schafer, N.P. et al. Surveying biomolecular frustration at atomic resolution. *Nat Commun* **11**, (2020) 5944.
55. Chen, X., Rajasekaran, N., Liu, K. et al. Synthesis runs counter to directional folding of a nascent protein domain. *Nat Commun* **11**, (2020) 5096.
56. Liu, Q., He, Qt., Lyu, X. et al. DeSiphering receptor core-induced and ligand-dependent conformational changes in arrestin via genetic encoded trimethylsilyl ^{1}H-NMR probe. *Nat Commun* **11**, (2020) 4857.
57. Tang, Y., Huang, Y., Hopf, T. et al. Protein structure determination by combining sparse NMR data with evolutionary couplings. *Nat Methods* **12**, (2015) 751–754.
58. Shi, H., Rangadurai, A., Abou Assi, H. et al. Rapid and accurate determination of atomistic RNA dynamic ensemble models using NMR and structure prediction. *Nat Commun* **11**, (2020) 5531.
59. Shimada, I., Ueda, T., Kofuku, Y. et al. GPCR drug discovery: integrating solution NMR data with crystal and cryo-EM structures. *Nat Rev Drug Discov* **18**, (2019) 59–82.
60. Rossi, F., Luria, G., Sommariva, S., Sorrentino, A. Bayesian multi–dipole localization and uncertainty quantification from simultaneous EEG and MEG recordings. **EMBEC & NBC 2017**, (2018) 843–846.
61. Chowdhury, R.A., Zerouali, Y., Hedrich, T., Heers, M., Kobayashi, E., Lina, J.-M., Grova, C. MEG–EEG information fusion and electromagnetic source imaging: from theory to clinical application in epilepsy. *Brain Topogr* **28**:6, (2015) 785–812.
62. Hong, J.H., Ahn, M., Kim, K., Jun, S.C. Localization of coherent sources by simultaneous MEG and EEG beamformer. *Med Biol Eng Comput* **51**:10, (2013) 1121–1135.
63. Hong, J.H., Jun, S.C. Scanning reduction strategy in MEG/EEG beamformer source imaging. *J Appl Math* **2012**, (2012) 1–19.
64. Pinotsis, D.A., Miller, E.K. Differences in visually induced MEG oscillations reflect differences in deep cortical layer activity. *Commun Biol* **3**, (2020) 707.
65. Uchiyama, T., Ma, J. Development of pico tesla resolution amorphous wire magneto-impedance sensor for bio-magnetic field measurements. *J Magn Magn Mater* **514**, (2020) 167148.
66. Zuo, S., Schmalz, J., Ozden, M.-O., Gerken, M., Su, J., Niekiel, F., Lofink, F., Nazarpour, K., Heidari, H. Ultrasensitive magnetoelectric sensing

system for pico-tesla magnetomyography. *IEEE Trans Biomed Circ Syst* **14**:5, (2020) 971–984.

67. Zuo, S., Heidari, H., Farina, D., Nazarpour, K. Miniaturized magnetic sensors for implantable magnetomyography. *Adv Mater Technol* **5**:6, (2020) 2000185.

68. Inomata, N., Suwa, W., Toan, N.V., Toda, M., Ono, T. Resonant magnetic sensor using concentration of magnetic field gradient by asymmetric permalloy plates. *Microsyst Technol* **25**:10, (2019) 3983–3989.

69. Hernandez, F.J., Kavruk, M., Hernandez, L.I., Ozalp, V.C. Nanocapsules in biomedicine: promises and challenges. *Advanced Theranostics Materials*, John Wiley & Sons (2015) 209–233.

70. Schmidt, C.K., Medina-Sánchez, M., Edmondson, R.J. et al. Engineering microrobots for targeted cancer therapies from a medical perspective. *Nat Commun* **11**, (2020) 5618.

71. Eguchi, H., Umemura, M., Kurotani, R. et al. A magnetic anti-cancer compound for magnet-guided delivery and magnetic resonance imaging. *Sci Rep* **5**, (2015) 9194.

72. Qiu, Y., Tong, S., Zhang, L. et al. Magnetic forces enable controlled drug delivery by disrupting endothelial cell-cell junctions. *Nat Commun* **8**, (2017) 15594.

73. Sharifi, M., Hasan, A., Nanakali, N.M.Q. et al. Combined chemo-magnetic field-photothermal breast cancer therapy based on porous magnetite nanospheres. *Sci Rep* **10**, (2020) 5925.

74. Sato, I., Umemura, M., Mitsudo, K. et al. Simultaneous hyperthermia-chemotherapy with controlled drug delivery using single-drug nanoparticles. *Sci Rep* **6**, (2016) 24629.

75. Peer, D., Karp, J., Hong, S. et al. Nanocarriers as an emerging platform for cancer therapy. *Nat Nanotechnol* **2**, (2007) 751–760.

76. Guan, Y.Q., Zheng, Z., Huang, Z. et al. Powerful inner/outer controlled multi-target magnetic nanoparticle drug carrier prepared by liquid photo-immobilization. *Sci Rep* **4**, (2014) 4990.

77. Kim, J.H., Eguchi, H., Umemura, M. et al. Magnetic metal-complex-conducting copolymer core–shell nanoassemblies for a single-drug anticancer platform. *NPG Asia Mater* **9**, (2017) e367.

78. Anderson, A.S., Martin, R.M., Renehan, A.G. et al. Cancer survivorship, excess body fatness and weight-loss intervention—where are we in 2020?. *Br J Cancer* **124**, (2021) 1057–1065.

79. Higashi, A., Hashimoto, Y., Ohsawa, D., Shirai, T., Noda, K. Secondary electron emission measurement from Cr and Cu bombarded by an Ne^{10+} beam at 6 MeV/n. *Prog Theor Exp Phys* **2020**:3, (2020) 033I01.

80. Kumar, D., Englesbe, A., Parman, M., Stutman, D., Finkenthal, M. Charge state distribution and emission characteristics in a table top reflex discharge—effect of ion confinement and electrons accelerated across the sheath. *Phys Plasmas* **22**:11, (2015) 113504.

81. Kitagawa, A., Fujita, T., Muramatsu, M., Biri, S., Drentje, A.G. Review on heavy ion radiotherapy facilities and related ion sources (invited). *Rev Sci Instrum* **81**:2, (2010) 02B909.

82. Xu, Z., Li, Y., Chau, S.N. et al. Impacts of international trade on global sustainable development. *Nat Sustain* **3**, (2020) 964–971.

Chapter 5

Overview of Shaping $YBa_2Cu_3O_7$ Superconductor

Jacques G. Noudem
Normandy University, ENSICAEN, CRISMAT Lab, CNRS, Caen, 14000, France
jacques.noudem@ensicaen.fr

In this chapter, different routes to prepare the textured $YBa_2Cu_3O_7$ (YBCO) superconductor material will be described as: (i) 2D thick-film YBCO fabrics, (ii) 3D YBCO textured foams, and (iii) 3D multiple holes textured YBCO.

The properties of these three shapes will be compared with the conventional top seeded melt textured growth (TSMTG) YBCO.

Using a combination of standard ceramic processing and an infiltration technique, single grain-superconductor fabrics, foams, and multiple holes of $YBa_2Cu_3O_y$ (Y123) have been produced. The process, involving negligible shrinkage, results in an improved microstructure with a uniform distribution of submicron $Y_2Ba_1Cu_1O_5$ (Y211) inclusions in the Y123 matrix. The foam or artificial patterned holes structure could be highly favorable for good oxygenation, better cooling and consequently for an efficient heat extraction from the

High-T_c Superconducting Technology: Towards Sustainable Development Goals
Edited by Muralidhar Miryala

ISBN 978-981-4877-65-7 (Hardcover), 978-1-003-16468-5 (eBook)
www.jennystanford.com

superconducting element in view to overcome the typical flux jump and hot-spot phenomena commonly observed in the trapped field and, or superconducting fault current limiters respectively. In this chapter, the processing conditions and the properties are reported. The investigations reveal superconducting properties being similar to those of bulk melt processed materials. The advantages of the infiltration growth process and of the novel YBCO are discussed.

5.1 Introduction

Since the discovery of the high temperature superconductivity (HTS) in BaLaCuO [1], the oxide YBCO ceramics seems to be one of the promising candidates for superconducting applications. Huge effort has been engaged to the development of the processing techniques to prepare the material with interesting properties. The applications of superconductors for different systems require various geometries and forms as pellets, rods, thin and thick films, and complex bulk shapes.

The development of different melt processing methods has led to $YBa_2Cu_3O_7$ (Y123) ceramics with properties acceptable for many applications [2–5]. However, despite these properties some problems exist, e.g., (i) oxygenating the whole bulk material without inducing any micro-cracks during the tetragonal-to-orthorhombic phase transition [6], (ii) suppression of the hot spots, commonly observed during the transport current application or the flux jumps commonly observe during the magnetic trapped field measurements. To overcome these problems, the superconducting material processed as a fabric, foam structure or with the artificial drilled holes have been developed [7–11]. Using a combination of standard superconducting ceramic processing and an infiltration technique [12–14], mono-domain superconductor fabrics, foams, and perforated Y123 have been processed. An important feature of this process is that it offers the flexibility to produce large samples of near-net shape without distortions and cracks. In this process, the liquid source (barium rich liquid phase and copper oxide) infiltrates into porous Y_2O_3 cloth, Y_2BaCuO_5 (Y211) foams, preformed Y211 pellet, above the peritectic temperature (T_p), and reacts to form Y123. This can be expressed by the chemical reactions:

$$Y_2O_3 + (4BaCuO_2 + 2CuO) \rightarrow Y_2BaCuO_5 + (3BaCuO_2 + 2CuO) \rightarrow 2YBa_2CuO_7 \quad (5.1)$$

$$Y_2BaCuO_5 + (3BaCuO_2 + 5CuO) \rightarrow 2YBa_2Cu_3O_7 \quad (5.2)$$

The novel morphology of superconducting fabric, foam, or bulk superconductor with multiple holes material processed seems to be a good candidate for increasing interfacial flux pinning if the pores can be made sufficiently small. Many other prospects are related to this novel structure like, e.g.,

- more efficient heat transfer,
- faster oxygenation and less related micro-cracking, possibility of reinforcement, and of interlocking connections, etc.

Objectives of the present work are to give an overview of

(i) 2D thick-film YBCO fabrics;
(ii) mono-domain Y123 developed on 3D foam and artificial patterned multiple holes of Y211;
(iii) superconducting properties of the porous materials.

The review describes briefly in the first part, the developement of the shaping of 2D and 3D mono-domains of $YBa_2Cu_3O_7$ (Y123). The second part will focus on various characteristics of the prepared samples, basically, the detail of the superconducting properties.

5.2 Experimental Procedures and Results

5.2.1 2D Thick-Film YBCO Fabrics

The details of the processing conditions of $YBa_2Cu_3O_y$ (Y123) mono-domain fabrics are reported elsewhere [10]. The thick Y123 mono-domain fabrics were manufactured from commercially available Y_2O_3 fabrics by infiltration with a eutectic mixture of barium-cuprate and copper oxides (see Eq. (5.1), Section 5.1) during an adequate solidification process in presence of a higher melting seed crystal. Basically, the process described elsewhere [10] involves the conversion of commercially [15] available Y_2O_3 cloths, woven in different forms, square, satin, etc. (Fig. 5.1), into the Y123 phase by the infiltration of liquid phases during a series of distinct heat treatment stages. The various steps involved in the process are schematically shown in Fig. 5.2. The experimental configuration consists of a source of liquid phase(s) in contact with the Y_2O_3 cloth for infiltration. Basically, the thick Y123 mono-domain sample can be obtained by free standing or growth on top of the ceramic substrate. The resulting Y123 samples have the dimension of thick films ~100 μm thickness, while demonstrating microstructure features of mono-domain bulk Y123 samples. The growth mechanism was well described by E. S. Reddy et al. [10]. The macro- and microstructures obtained shown a typical mono-domain nature of the samples, analogous to top seeded melt-textured bulk Y123 material [9].

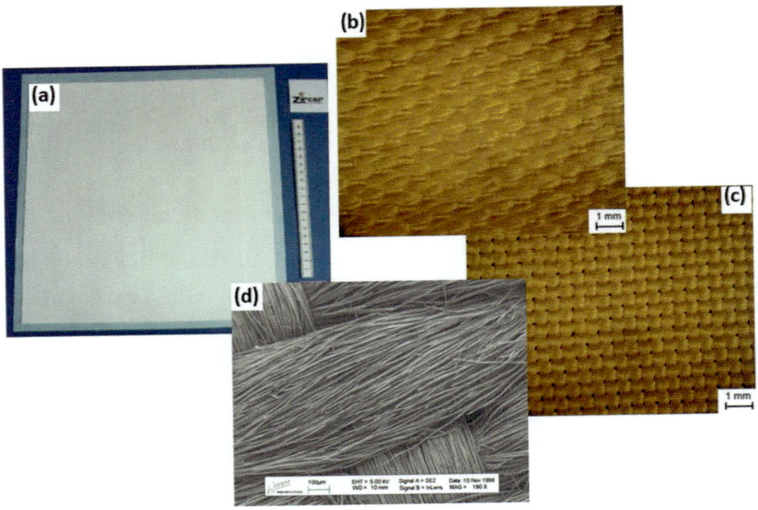

Figure 5.1 Commercially available Y_2O_3 fabric (a) main precursor [15]; (b) satin woven form; (c) square woven form; (d) high magnification of square form.

All the as-processed samples in this work were oxygenated at 450°C for 72–100 h in flowing oxygen. For microstructural examination, the samples were mounted using cold resin and polished on a rotating machine (Tegrapol-31, Struers Inc.) using varying grades of diamond particle solution until ¼ μm as a final. The polished surfaces were examined using an optical microscope (Olympus, BH2-UMA) and scanning electron microscope, SEM (Philips XL 30 FEG) with EDX.

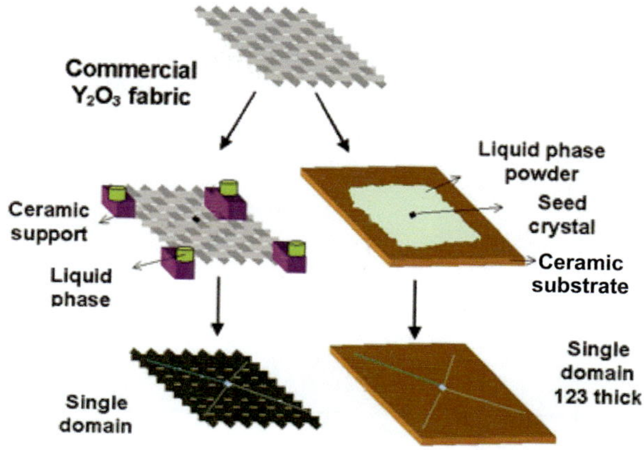

Figure 5.2 Various steps involved in the infiltration and growth process for the conversion of Y_2O_3 cloths into mono-domain Y123 fabrics [10].

Resistivity measurements were performed using the standard four-point contact method. Contacts between the sample and the current leads have been fabricated by painting silver paste (4929 Dupont Inc.) and subsequent annealing at 900°C for half an hour to ensure a good silver diffusion followed by oxygenation at low temperatures. A commercial PPMS (magnet power supply) apparatus and a SQUID (quantum design) magnetometer were used for resistivity and magnetization measurements.

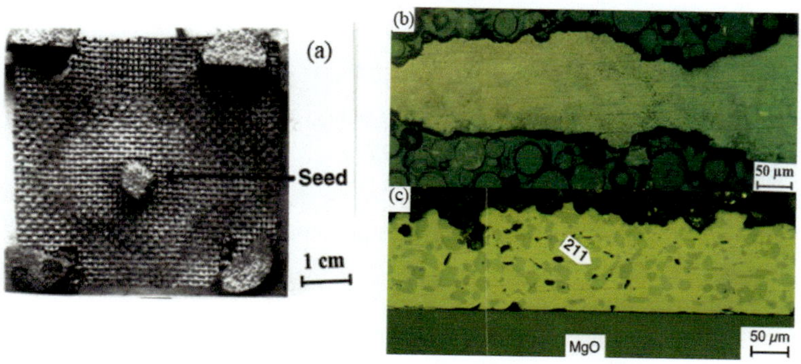

Figure 5.3 (a) Self-supporting mono-domain Y123 fabric prepared from square woven Y_2O_3 cloth. Polished surface of cross section showing the uniform Y211 particles distribution into the Y123 matrix; (b) self-supporting sample; (c) sample prepared on top of MgO substrate [10].

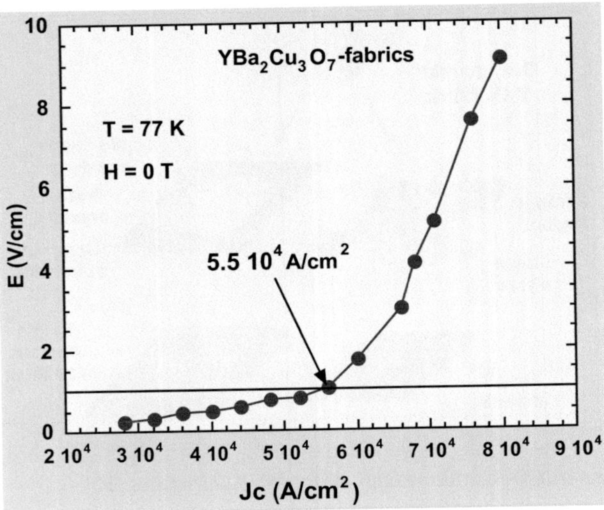

(a)

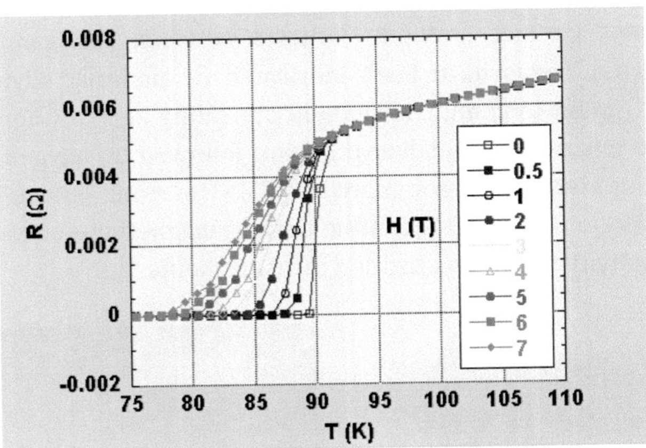

(b)

Figure 5.4 Transport properties of Y123 fabrics [18]. (a) Temperature dependence of the resistance at different applied magnetic fields. (b) DC transport at self-field and 77 K.

The controlled nucleation and growth of the Y123 mono-domain from a square woven Y_2O_3 fabric with a Nd123 seed is shown in Fig. 5.3a. The Y211 inclusions can be clearly observed in the self-supporting configuration and sample growth on the MgO substrate (Fig. 5.3b,c). In addition, the transport properties were investigated.

The temperature dependence of the resistance at different applied magnetic fields is shown on Fig. 5.4a. At self-field, the superconducting transition temperature is 90 K Y123 respectively with a typical transition width of 2 K. For fabric samples, transport J_c measurements in self-field showed the transition between the superconducting state and the flux-flow regime to occur at a critical current density (J_c) of 5.5×10^4 A/cm^2 at 77 K. This value is comparable to some reported [6, 16, 17] data related to the conventional melt textured of Y123.

5.2.2 3D YBCO Superconducting Foams

About the 3D $YBa_2Cu_3O_7$ (Y123), the Y_2BaCuO_5 Y211 precursor foams of different dimensions were fabricated as replica of raw material of polyurethane foam (Fig. 5.5) with a porosity of 10–100 pores per inch (ppi). Details of $YBa_2Cu_3O_y$ (Y123) foam preparation have been reported elsewhere [7, 8]. Figure 5.6a shows the schematic diagram of different steps involved in the preparation of Y123 foam by an infiltration-growth process [7].

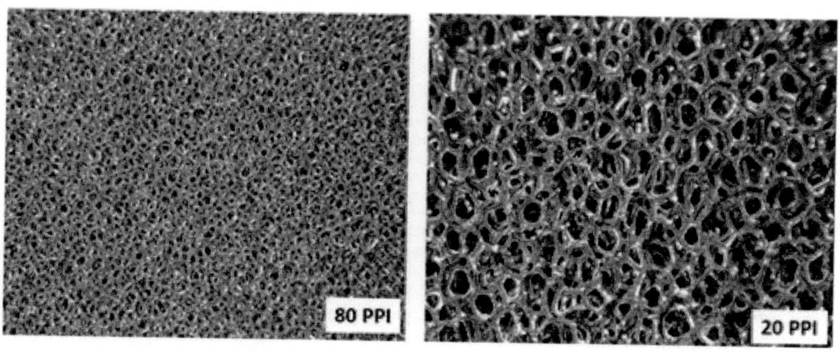

Figure 5.5 (a) Raw material of polyurethane foams of various pore sizes.

The liquid phase ($3BaCuO_2$ + 5CuO) or $Ba_3Cu_5O_8$ (Y035) was prepared by solid state milling of stoichiometric amounts of high purity $BaCO_3$ and CuO in the agate bowl of a planetary mill for 2 h. The mixed powder was then heat-treated at 870°C, 24 h with intermediate grindings. The preparation of Y211 foams involves the coating/impregnation of polyurethane foam of desired porosity with optimized water based Y211 slurry. This slurry was prepared by mixing commercially available Y211 fine (~2 μm) powder in water with 5 wt% polyvinyl-alcohol (PVA) as binder. The process of coating/impregnation of polyurethane foams involves dipping the foams into the Y211 slurry and drying at room temperature. Repeating the step several times results in a thick Y211 coating. The slow heating at a rate of 50°C/h to 600°C for 6 h in air burnt the organic components PVA and polyurethane. The final Y211 foam was sintered in air at 1100°C, 20 h for densification. The final green Y211 pre-form was transformed into single-grain Y123 superconducting foam by infiltration growth process reported elsewhere [7, 12–14] and using the configuration shown in Fig. 5.6a. The sample was rapidly heated to 1050°C for 30 min. At this stage the liquid phase infiltrates into the interstitial spaces between Y211 particles due to the capillary action. The peritectic reaction occurring between the green Y211 reservoir and the liquid phase source Y035 during slow cooling stage (1010°C to 980°C at 0.3°C/h) from the peritectic temperature results in the growth of Y123 grain. The samples, after grain growth, from 980°C were cooled to room temperature in 6 h. The nucleation and orientation of Y123 phase is controlled by placing a *c*-axis oriented seed- either $NdBa_2Cu_3O_y$ (Nd123) on top of the foam with a Y_2O_3 cloth [15] as support for the seed and to facilitate a homogeneous grain growth into the foam (Fig. 5.6b).

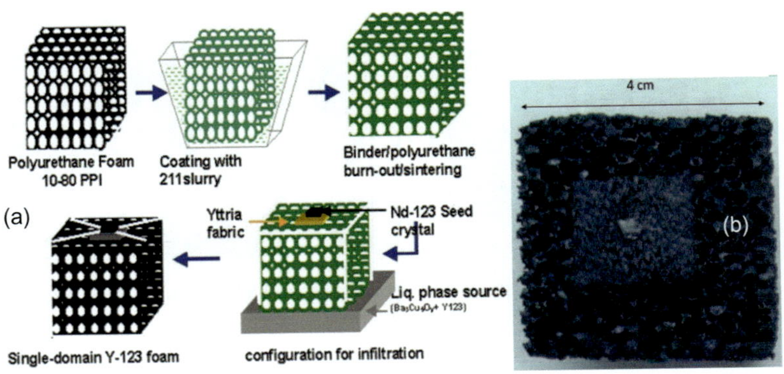

Figure 5.6 (a) Schematic diagram showing the processing of a single domain $YBa_2Cu_3O_{7-x}$ foam. (b) A mono-domain Y123 superconducting foam [7, 8].

The sample has an open porosity of 20 ppi. The alveoli morphology is evidence with a strut thickness of about 300 μm (Fig. 5.7a). The final Y123 structure should be favorable to coolant connectivity (liquid nitrogen bath) during the quenching stage due to the large surface area. Polished surface (Fig. 5.7b) of the struts reveal the typical microstructural features known from melt processed bulk materials, with parallel platelet gaps and trapped Y211 particles in a Y123 matrix. From the microstructures, the volume content of Y211 particles is estimated to be around 40 volume percent with an average particle size around 2–5 μm.

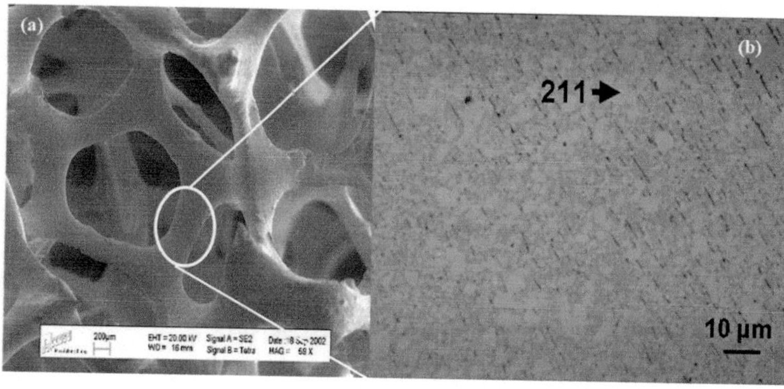

Figure 5.7 (a) Honeycomb structure of the superconducting Y123 foam. (b) Polished surface of the struts reveals residual Y211 particle (white) inclusions in the Y123 matrix [8].

The zero field cooled magnetization curve, Fig. 5.8, shows a very narrow transition ($\Delta T \approx 1.5$ K) with an offset ≈92 K, comparable to value taken from inset resistivity measurements. This abrupt transition characterizes the presence of quasi-pure $YBa_2Cu_3O_7$ composition confirming that the initial polyurethane foam and organic compounds were fully burnt into the initial Y211 precursor leaving no residual contamination. The magnetic critical current density, $J_c = 40$ A/cm^2 at 77 K and self-field was measured [8] and seems at the range of the values usually obtained on the mono-domain of Y123 superconductors.

5.2.3 3D Multiple Holes Textured YBCO

On the other hand, the details of seeded infiltration growth (SIG) and multiple holes process of $YBa_2Cu_3O_y$ (Y123) are reported elsewhere [9, 19, 20]. We can cite some benefits of this new design among which:

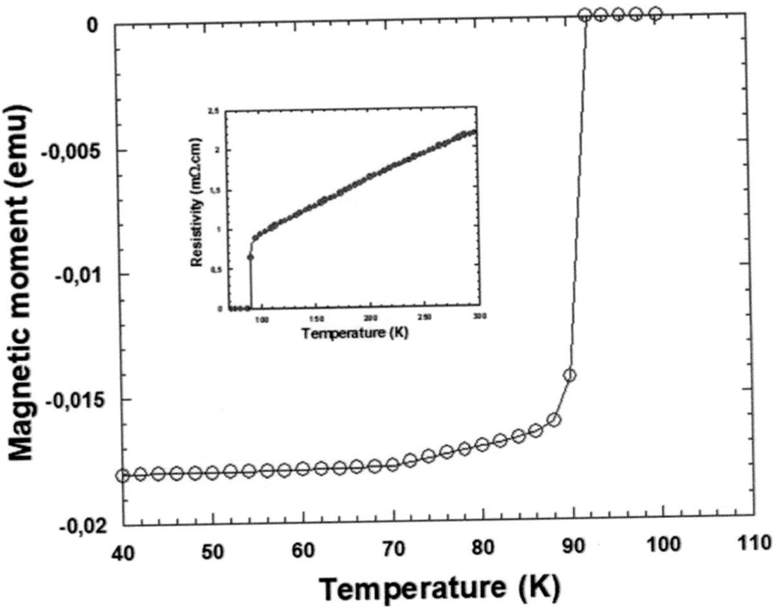

Figure 5.8 Temperature dependence of magnetic moment and electrical resistivity (inset).

- Decreased porosity in large samples as reported in some studies [6, 21]
- Better oxygenation (faster, more homogeneous)
- Small increase of J_c (increase of the surface pinning)
- Large increase of the magnetic trapped field up to 32% in comparison to the plain sample
- Large exchange surface important: (i) heat transfer (ii) oxygenation, and (iii) cooling

On Fig. 5.9, the $Y_2Ba_1Cu_1O_y$ (Y211) block or pellet is placed on top of the liquid phase ($3BaCuO_2$ + $2CuO$) or $Ba_3Cu_5O_8$ (Y035) to allow the liquid infiltration driving by capillarity force. A $NdBa_2Cu_3O_7$ (Nd123) seed as a nucleation center is placed on the top of the Y211 preform.

The holes into the preform are performed by drilling cylindrical cavities with diameter of 1–2 mm through the disk or square shaped sample. The holes are arranged in a regular network in the plane of the samples (Fig. 5.9b). On the other hand, the thin wall and plain single domain of $YBa_2Cu_3O_y$ are grown by using the top seeding melt growth process. Details are given elsewhere [9].

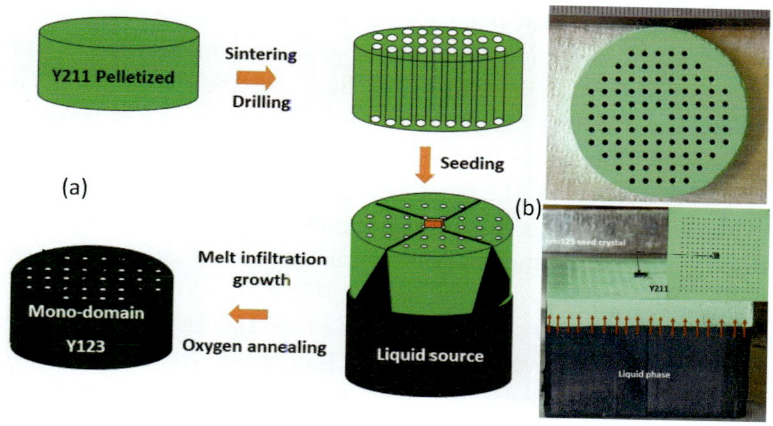

Figure 5.9 (a) Schematic diagram showing the processing of mono-domain 123. (b) Artificially patterned Y211 with multiple holes [22].

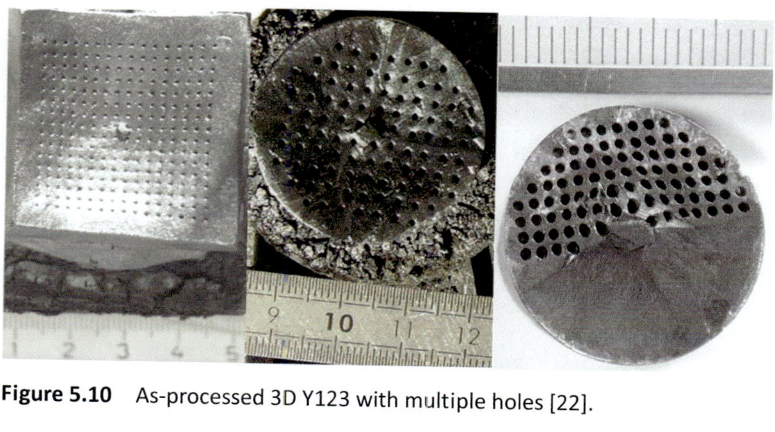

Figure 5.10 As-processed 3D Y123 with multiple holes [22].

The macroscopic top view of multiple holes 3D Y123 samples are presented in Fig. 5.10. We can observe that the resulting Y123 samples are mono-domains with the crystalline growth from the seed position to the edge of the sample. Another way to demonstrate the growth of the mono-domain from the multiple holes structure is shown by Fig. 5.10 (right). The growth lines of faceted growth on the surface of the half-perforated mono-domain are not clearly observed but exist compared to the half-plain part. This shows that the pre-formed with holes do not seem to disturb the growth of the mono-domain, which is confirmed in Ref. [9], where the growth was monitored for both the plain and sample with multiple holes.

Optical macrographs of some as grown samples with 0.8 mm holes are shown in Fig. 5.11 [9]. It illustrates the cross sections of plain and perforated samples. The porosity is drastically reduced for the drilled sample. For the plain sample, a large porosity and crack zones are noticeable. The mono-domain character of the sample was evidenced by the flux mapping measurements. This was also used to verify the homogeneity of the samples and to investigate the field ability [9].

The three-dimensional distribution of the trapped field (as shown in Fig. 5.12) was scanned with Hall sensor at ~0.5 mm above the top surface of the sample magnetized at 77 K. The trapped field value is higher in the multiple holes Y123 pellet (454 mT) than in the plain one (330 mT). This represents an increase of 37% for the multiple holes sample than for the plain one, in agreement with other reports [23].

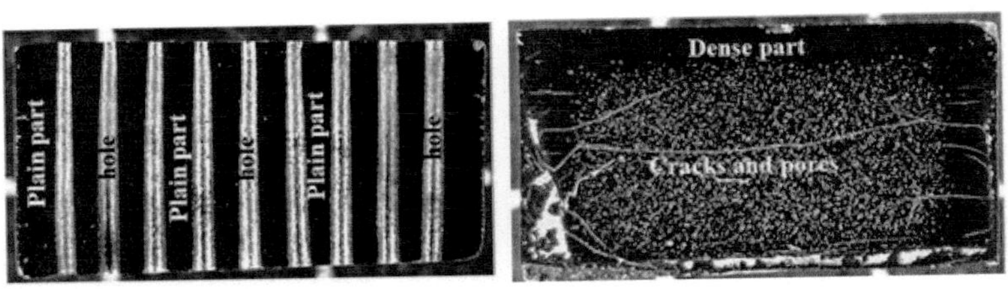

Figure 5.11 Cross section of Y123 with multiple holes and plain pellet [9].

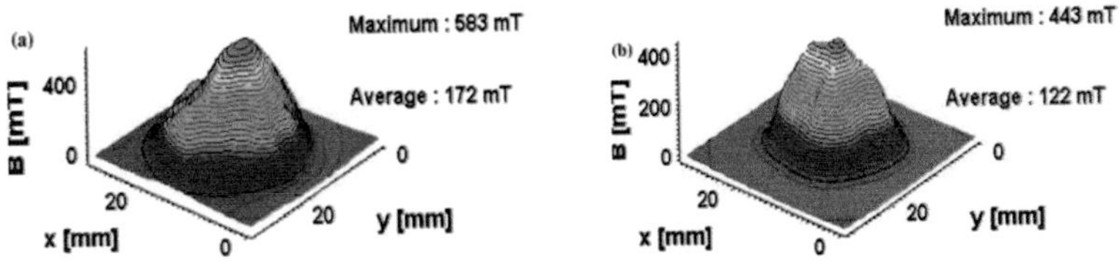

Figure 5.12 Magnetic trapped measurements on a (left) multiple holes and (right) plain mono-domain samples [9].

5.3 Conclusion

The mono-domain YBCO (Y123) with various shapes have been successfully prepared from (i) commercially available Y_2O_3 cloth, (ii) polyurethane as a raw material, and (iii) Y211 with artificial patterned holes. In all cases, the SIG process has been used. The mono-domain nature and microstructure resulted in properties similar to those of bulk materials. The superconducting properties are comparable to the conventional melt textured Y123 superconductors. Regarding the functional properties, the ability of the Y123 material with multiple holes to trap a high field has been demonstrated. At 77 K, the magnetic trapped field value for the multiple holes sample is 32% larger than in the plain ones.

The use of Y123 foam or multiple holes sample with a large surface area could allow one to effectively extract the heat or to overcome the magnetic flux jump generated into the bulk superconductor.

References

1. Bernorz, J. G. and Müller, K. A. (1986). Possible high T_c superconductivity in the Ba-La-Cu-O system, *Z. Phys. B: Condens. Matter*, **64**, pp. 189–193.
2. Paul, W., Lakner, M., Rhyner, J., Unternahner, P., Baumann, Th., Chen, M., Widenhorn, L. and A. Guérig, A. (1997). Test of 1.2 MVA high-T_c superconducting fault current limiter, *Supercond. Sci. Technol.*, **10**, p. 914.
3. Obradors, X., Puig, T., Mendoza, E., Plain, J., Figueras, J., Granados, X., Carillo, A. E., Vares, E., Sandiumengue, F. and Tixador, P. (2000). Tuning the critical currents in bulk MTG YBCO for current limiting devices, *Supercond. Sci. Technol.*, **13**, p. 879.
4. Noudem, J., Barbut, J. M., Belmont, O., Sanchez, J., Tixador, P. and Tournier, R. (1998). Current limitation at 1080 A under 1100 V with bulk Bi-2223, *IEEE Trans. Appl. Supercond.*, **9**, p. 664.
5. Tomita, M. and Murakami, M. (2003). High-temperature superconductor bulk magnets that can trap magnetic fields of over 17 tesla at 29 K, *Nature*, **421**, p. 517.
6. Isfort, D., Chaud, X., Tournier, R. and Kapelski, G. (2003). Cracking and oxygenation of YBaCuO bulk superconductors: application to *c*-axis elements for current limitation, *Physica C*, **390**, p. 341.

7. Reddy, E. S. and Schmitz, G. J. (2002). Superconducting foams, *Supercond. Sci. Technol.*, **15**, pp. 21–24.
8. Noudem, J. G., Reddy, E. S. and Schmitz, G. J. (2003). Magnetic and transport properties of $YBa_2Cu_3O_y$ superconductor foams, *Physica C*, **390**, p. 286.
9. Chaud, X., Meslin, S., Noudem, J., Harnois, C., Porcar, L., Chateigner, D. and Tournier, R. (2005). Isothermal growth of large YBaCuO single domains through an artificial array of holes, *J. Cryst. Growth*, **275**, p. 855.
10. Reddy, E. S., Noudem, J. G., Tarka, M. and Schmitz, G. J. (2000). Mono-domain $YBa_2Cu_3O_y$ superconductor fabrics prepared by an infiltration process, *Supercond. Sci. Technol.*, **13**, pp. 716–721.
11. Reddy, E. S., Hari Babu, N., Shi, Y., Cardwell, D. A. and Schmitz, G. J. (2005). Processing of large grain Y-123 superconductors with pre-defined porous structures, *Supercond. Sci. Technol.*, **18**, p. 15.
12. Cao, H., Chaud, X., Noudem, J. G., Zhang, C., Hu, R., Li, J. and Zhou, L. (2010). Mechanism of seeded infiltration growth process analysed by magnetic susceptibility measurements and in situ observation, *Ceram. Int.*, **36**, p. 1383.
13. Cao, H., Moutalbi, N., Harnois, C., Hu, R., Li, J., Zhou, L. and Noudem, J. G. (2010). Novel configuration of processing bulk textured $YBa_2Cu_3O_{7-x}$ superconductor by seeded infiltration growth method, *Physica C*, **470**, p. 68.
14. Meslin, S. and Noudem, J.G. (2004). Infiltration and top seeded growth mono-domain $YBa_2Cu_3O_{7-x}$ bulk superconductor, *Supercond. Sci. Technol.*, **17**, p. 1324.
15. Zircar Products Inc., www. Zircarzirconia.com, Florida NY, USA.
16. Braithwaite, D., Bourgault, D., Sulpice, A., Barbut, J. M., Tournier, R., Monot, I., Lepropre, M., Provost, J. and Desgardin, G. (1993). Scaling properties of the anisotropic critical current density in bulk textured YBaCuO. Evidence toward a 3D flux line lattice, *J. Low Temp. Phys.*, **91**, p. 1–11.
17. Porcar, L., Bourgault, D., Barbut, J. M., Barrault, M., Germi, P., and Tournier, R. (1997), High critical transport currents of melt textured YBCO up to 6000 A, *Physica C*, **275**, pp. 293–298.
18. Noudem, J. G., Guilmeau, E., Reddy, E. S., Noe, M. and Schmitz, G. J. (2003). Transport properties of hot-forming textured Bi:2223 and single domain $YBa_2Cu_3O_y$ fabric materials, *Physica C*, **386**, pp. 202–205.

19. Reddy, E. S., Herweg, M. and Schmitz, G. J. (2003). Processing of Y_2BaCuO_5 foams, *Supercond. Sci. Technol.*, **16**, pp. 608–612.

20. Noudem, J. G., Meslin, S., Harnois, C., Chateigner, D. and Chaud, X. (2004). Melt textured $YBa_2Cu_3O_{7-x}$ bulk with artificially patterned holes: a new way of processing c-axis fault current limiter meander, *Supercond. Sci. Technol.*, **17**, pp. 931–936.

21. Diko, P., Fuchs, G. and Krabbes, G. (2001). Influence of silver addition on cracking in melt-grown YBCO, *Physica C*, **363**, pp. 60–66.

22. Meslin, S., Harnois, C., Chubilleau, C., Horvath, D., Grossin, D., Reddy. E. S. and Noudem, J. G. (2006). Shaping and reinforcement of melt textured $YBa_2Cu_3O_{7-\delta}$ superconductors, *Supercond. Sci. Technol.*, **19**, pp. 585–589.

23. Noudem, J. G., Meslin, S., Horvath, D., Harnois, C., Chateigner, D., Ouladdiaf, B., Eve, S., Gomina, M., Chaud, X. and Murakami, M. (2007). Infiltration and top seed growth-textured YBCO bulks with multiple holes, *J. Am. Ceram. Soc.*, **90**, pp. 2784–2790.

Chapter 6

Development of MgB_2 Superconducting Super-Magnets: Its Utilization towards Sustainable Development Goals

Muralidhar Miryala
Materials for Energy and Environmental Laboratory,
Superconducting Materials Laboratory,
Graduate School of Science and Engineering,
Shibaura Institute of Technology, Koto City, Tokyo 135-8548, Japan
miryala1@shibaura-it.ac.jp

The objective of this review chapter is to improve the flux pinning and critical current density (J_c) of bulk MgB_2 materials for day-to-day life applications and to highlight the aim of United Nation's sustainable goals. We prepared several sets of bulk MgB_2 materials from commercial high-purity powders of Mg metal and carbon-coated amorphous B using a single-step solid-state reaction process. Special focus was given toward optimizing high performance of bulk MgB_2 by utilizing (i) the sintering temperature, (ii) the sintering time, (iii) the silver addition, (iv) MgB_4 addition, (v) carbon-encapsulated boron, and (vi) excess Mg with silver addition. All samples

High-T_c Superconducting Technology: Towards Sustainable Development Goals
Edited by Muralidhar Miryala

ISBN 978-981-4877-65-7 (Hardcover), 978-1-003-16468-5 (eBook)
www.jennystanford.com

were characterized utilizing X-ray diffraction, scanning electron microscopy, atomic force microscopy, and trapped field experiments from 20 K to 40 K. The superconducting performance, T_c and J_c at 20 K and 25 K was recorded by superconducting quantum interference device (SQUID) magnetometer. The J_c values in Mg-rich MgB_2 material with 4 wt% of Ag were higher than those of silver-free MgB_2 bulks. The sample with 4 wt% Ag combined with 1.5 wt% of carbon-coated B exhibited the highest J_c of 430 kA/cm^2 at 20 K and self-field. The bulk MgB_2 sample sintered for 3 h and 1 wt% of MgB_4 addition showed superior J_c values and a sharp T_c at 38.1 K, which is twice as high compared to pure sample. The highest trapped value of 1.51 at 20 K was achieved in 20 mm in size bulk MgB_2 sample sintered at 775 °C. The results demonstrate a strong correlation between the microstructure achieved, the resulting pinning performance, and trapped field values. The present results demonstrate that bulk MgB_2 magnets produced with a simple sintering method have potential for several industrial applications including in Magnetic Resonance Imaging (MRI), Nuclear Magnetic Resonance (NMR), Magnetic Drug Delivery System, Maglev Trains, Power Cable Technology, Water Cleaning Systems, etc. that are crucial aspects in addressing United Nation's (UN) sustainable development goals.

6.1 Introduction

Discovery of high T_c = 39 K in a binary metal MgB_2 has instigated tremendous research and development activities [1]. MgB_2 consists of hexagonally arranged boron atoms with a base diameter of 3.5 Å, which is nearly equal to the *c*-spacing creating a sizable room to accommodate an Mg atom. The planar boron layer is sandwiched between hcp type arrangements of the Mg atoms in the AlB_2 type structure with space group P6/mmm. The boron layer plays the crucial role as the holes in the Fermi level are constituted by the contribution of holes from the boron atoms in this plane [2–4]. MgB_2 is a two-gap superconductor as has been found experimentally using the Raman spectroscopy [5], point contact spectroscopy [6], tunneling spectroscopy [7], and angle-resolved photoelectron spectroscopy (ARPES) [8]. Its Fermi surface consists of nearly 2D, distinctly cylindrical topology wrapped along the direction perpendicular to the boron layer. The 2D behavior of the Fermi surface is due to the weak interlayer coupling between the boron layers. The two gaps consists of, the smaller gap which is due to 3D π-band holes with a gap Δ_π = 2.7 meV, while the larger gap is due to the 2D σ-band holes with a gap of Δ_σ = 6.2 meV [8]. The uniquely high T_c (= 39 K) is due to the 2D character of its electronic density of states and the strong electron-phonon coupling in MgB_2. This is assisted by the low mass of boron $\left(M \propto \frac{1}{\omega^{1/2}}\right)$, in analogy to the exotic thought that metallic hydrogen could possess remarkably high and sizable T_c even up to the room temperature [9]. It was found that the Eliashberg function in MgB_2 is maximum at around ~600 cm^{-1} [10,11], clearly demonstrating that the Raman zone center peak [12] with the full width half maxima of ~300 cm^{-1} found at 600 cm^{-1} is responsible for the coupling with holes for the Cooper pair formation. The BCS mechanism of superconductivity in MgB_2 is strongly supported by the boron isotope effect found in MgB_2 [13].

There have been numerous attempts to increase the T_c in MgB_2 by using many techniques such as pressure, substitution of various elements including C, Fe, Ag, Mn, Al and some co-doping techniques. [12,14–17]. The concept co-doping in MgB_2 was proposed with the view that expansion of the unit cell volume could result into higher T_c [18,19]. However, experimentally it is often found that the any attempt to modify the crystal structure in turn increases disorder in the system. Disorder in the crystal lattice can propagate the phenomena such as band filling, inter band scattering, and anharmonicity in the system [4,20]. Which have a detrimental effect on the T_c and thus so far, the entire attempt to increase T_c have met with a disappointment. But not all is gloomy, as the magnetic field that this superconductor could withstand after doping at liquid hydrogen temperatures (20 K) has been exceptional [21]. There has been tremendous improvement in the superconducting properties, mainly due to various doping studies and the processing techniques that has come up in the recent years. The ability of MgB_2 to store large magnetic field in the superconducting state has been an attractive activities for the researchers recently. The large J_c and H_{c2} are mainly due to the large coherent length that this superconductor possesses. MgB_2 provides multiple benefits as compared to other classes of superconductors. MgB_2 is free from the problem of weak links in the grain connectivity, the almost negligible magnetic anisotropy, simple stoichiometry and furthermore, there is no need for any oxygenation in the processing of MgB_2, thus this tremendously scales down the time needed for the synthesis. Also, it finds application in the liquid hydrogen temperature which other metallic superconductors, such as $Nb_3(Ge, Sn)$ do not offer. The possibility of the large-scale synthesis at a relatively lower cost, basically due to the cheap starting materials and the ease of fabrication has been an attractive characteristic of this class of superconductors. Above all, the relatively high mechanical strength of MgB_2 even though with relatively low weight further multiplies it area of applications [22,23]. MgB_2 has been successfully synthesized in various forms, thin film, bulk, wire, and tape with exceptional superconducting properties [21,24,25]. The robust superconductivity, high mechanical strength, and the large T_c have led to a number of applications of MgB_2 in our daily life, including in the field of medical science. It can be used as super-magnets, for energy production and energy storage, fault current limiters, noncontact bearings for liquid pumping, magnetic shielding screens, nuclear magnetic resonance (NMR), and the magnetic resonance imaging (MRI) [26,27].

Variety of growth and processing techniques have been adopted to enhance the superconducting properties along with its mechanical properties, keeping in mind the need for the ease of large-scale synthesis for commercial and industrial applications. The conventional method of powder metallurgy for the synthesis of MgB_2 is modified over the years and many workers have adopted to techniques that promises higher densities such as, high pressure sintering technique, hot isostatic pressing technique, hot compaction technique, spark plasma sintering technique, etc., to achieve high J_c [28–31]. Another simple method of in-situ synthesis with single step synthesis has been adopted by some groups for the large-scale production of high quality samples with high J_c [25, 32,33]. Giunchi et al. have adopted an appealing approach of the reactive Mg liquid infiltration process; using this technique, samples of various size and shapes were synthesized [34]. Hollow cylindrical samples with dimensions of the inner diameter 7 mm, outer diameter 9 mm, and the cylindrical length of 26 mm was successfully fabricated. In addition, it was shown that various shapes such as solid cylinders, rectangular rods, hollow, and solid discs of various sizes can be prepared with the ability of each large sized MgB_2 to hold the trap field of magnitude greater than 1 T [27]. A good-quality MgB_2 sample with superior superconducting properties requires that the sample possess good grain connectivity, high T_c, higher density, and higher mechanical strength so as to support larger current capability and also be able to withstand mechanical pressures without losing its superconductivity. These requirement have resorted MgB_2 community to various means such as chemical doping [35] and irradiation [36] and refining the initial particle size during the synthesis by ball milling [37], optimizing and choosing the sintering duration, and the sintering temperatures, etc. For instance, with optimized sintering conditions Muralidhar et al. have found that $J_c > 10^5$ A/cm^2 for the field of 1 T and at 20 K can be easily achieved without any need for the external chemical doping [38]. It was suggested that beyond the sintering temperature >900 °C, formation of unwanted secondary phases such as MgB_4 assisted flux flow and hence resulted into lower J_c in the system. Likewise, with the sintering temperature <700 °C, the unreacted Mg considerably reduced J_c of the system. It was further shown that a single-step sintering temperature of ~ 775 °C and the sintering duration of 3 h resulted in the best J_c [32,39]. We believe that with further refining of the particle size, the J_c can be significantly improved. The correlation between the J_c and the microstructure is inevitable; the normal precipitate in the superconducting matrix and the superconducting MgB_2 particle with sizes equivalent to the coherence length enhances flux pinning thereby increasing the flux trapping ability of the superconductors. While a reverse effect is seen when the size of the precipitate do not match with the coherence length, a robust superconductor requires a proper choice of the grain size and the ruling out of any possible secondary phase formation. This has been recently shown by Muralidhar et al. that with the optimized grain size of ~100 nm or less the value of J_c can be increased considerably (up to 250 KA/cm^2) in MgB_2 superconductors even for the field of 1 T and at 20 K [25]. Recently, a trapped field of 3 T was reported by Durrel et al. at 17 K for the sample of 3 cm size [40]. Apart from the tuning of the sintering conditions and grain sizes, researchers have adopted number of doping techniques and the choice of dopants to improve the superconducting properties in MgB_2. This has shown phenomenal success in achieving MgB_2 samples with superior quality. It was demonstrated that the carbon-doped MgB_2 showed a high J_c > 33 KA/cm^2 at self-field and at 5 K with considerable improvement in the upper critical field [41,42]. Furthermore, various sources of carbon were used, for example, oxygen-free pyrene gas as a source of carbon improved J_c to as high as 1 MA/cm^2 [43]. Studies on the carbon doping via carbon nanotubes addition during the synthesis revealed an exceptional J_c = 10 KA/cm^2 at 20 K in the presence of the high field of 4 T [44]. Carbohydrate has been used as a source of doping carbon with the motivation of achieving dopant homogeneity at relatively less reactive time and to rule out the possibility of the formation of secondary phases. A remarkably high J_c close to 1 MA/cm^2 at a field of 1 T and at temperature of 20 K is reported [33]. Recently, carbon encapsulated boron has been used by Liu et al., which showed remarkable increase in the properties of carbon doped MgB_2 [45]. Use of nanoparticle-sized SiC as a source of carbon doping has also helped to improve J_c [46,47]. The success of these appealing approaches of using carbon via different sources for extensive doping in MgB_2 is because of the ability of carbon to replace boron with ease. The atomic radius of carbon (0.77 pm) is less than the atomic radius of boron (0.88 pm), thus, substitution of carbon into the boron sites is rather easy and thereby shrinking the unit cell size. Other dopants such as Mn, Fe, Si, and Ag have been successfully doped to improve J_c [48]. Muralidhar et al. demonstrated high J_c with 4 wt% Ag additions, the J_c increased to 200 KA/cm^2 and further increasing of the doping percentage to 10 wt% Ag resulted into the formation of unwanted secondary phases such as $AgMg_3$ [25]. Surprisingly, the increase in the Ag doping percentage resulted in the increase of the bending strength of the bulk MgB_2 samples, which is attributed to the fact that Ag nanoparticles played the crucial role of fillers, filling the porous MgB_2 matrix, and thereby increasing its mechanical strength. In addition to the single element doping, many researchers have attempted successfully the codoping methods [35, 49]. The pressure-dependent studies in MgB_2 have shown that with decrease in the volume of the unit cell the T_c decreases and therefore, increase of the volume of the unit cell was expected to increase the T_c, however, to the contrary, the various codoping methods, which increases the unit cell volume has always shown a detrimental effect on the T_c, although the current carrying capacity of MgB_2 was improved.

In this chapter, we present variety of bulk MgB_2 samples produced by sintering technique at 775 °C for 3 h in Ar atmosphere. All samples were characterized using a scanning electron microscope (SEM), transmission electron microscope (TEM), and a superconducting quantum interface device (SQUID) magnetometer, along with trapped field measurements in a temperature range of 20 K to 40 K. We have found that trapped field values are improved with increasing the size of the bulk and improved with decreasing temperature. Further, this value was dramatically improved by nano boron utilized bulk MgB_2 samples, which is crucial for the production high performance of bulk MgB_2 samples for superconducting super-magnet applications in order to advocate several sustainable development goals (SDGs) proposed by the United Nations for a blueprint to *"achieve a better and more sustainable future for all"*.

6.2 Experimental

Bulk polycrystalline MgB_2 samples were made with in-situ solid-state reaction. High-purity commercial powders of Mg (99.9% purity, 200 mesh), amorphous B powder (99% purity, 300 mesh), nano amorphous B powder (98% purity, 200 nm), carbon-encapsulated B powder (98% purity, 200 nm), and high purity metallic silver (99.99%) were used for the production of bulk MgB_2 super-magnets. All combinations were mixed in a nominal composition of Mg:B = 1:2. Also, produced several set of samples utilizing an optimized silver (4 wt%) and Mg rich, bulk MgB_2 samples to improve the flux pinning and mechanical performance. The starting powders were thoroughly ground in a glove box under Ar atmosphere. The powder mixture was pressed into pellets of 20 mm and 30 mm diameter using an isostatic pressing machine. Eventually, Ti sheets were wrapped around the consolidated pellets that were subjected to sintering in Ar atmosphere in a tubular furnace. The samples were prepared with a single-step heat treatment at 775 °C, which was found to be the optimum in the previous experiments [39]. The temperature of the furnace was raised to the target temperature in 5 h. During the sintering in flowing Ar gas the temperature was kept constant for 3 h, and then reduced to room temperature with a cooling rate of 100 °C/h. After sintering, the Ti sheets were removed from the MgB_2

bulks. Finally, the samples were encapsulated into stainless steel rings as shown in Fig. 6.1, which was effective in preventing cracks and maintaining the temperature at both top and bottom surfaces during trapped field measurements.

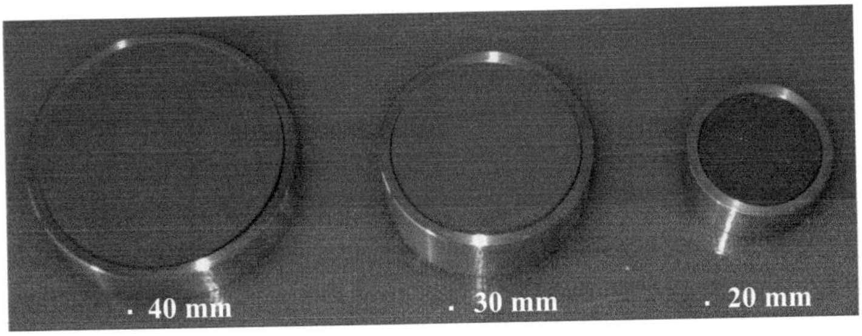

Figure 6.1 Optical photographs of MgB_2 samples 20 mm and 30 mm in diameter produced with sintering process at 775 °C for 3 h in Ar atmosphere. All the samples were reinforced with stainless steel rings. Note the samples are of good quality with no appreciable cracks.

6.2.1 Characterization

The microstructure of these samples was studied by means of scanning electron microscopy (SEM) and transmission electron microscopy (TEM). T_c was defined as the onset temperature at which diamagnetic signals were first observed. Magnetization hysteresis loops (MHL) were measured at 7 K to 38 K using a SQUID magnetometer with the maximum applied field of 5 T. The magnetic J_c values were estimated based on the extended Bean's critical state model using the relation $J_c = 2\Delta m/[a^2d(b - a/3)]$ where d is the sample thickness, a, b are cross-sectional dimensions ($b \geq a$), and Δm is the difference of the magnetic moments during increasing and decreasing field processes in the M-H loop [50].

The trapped magnetic fields were measured by magnetizing the bulk MgB_2 sample in a 10 T superconducting magnet in a temperature range of 20 K–40 K. The cryostat consists of a cryocooler capable of cooling a MgB_2 disk to temperatures below 20 K. MgB_2 bulk samples 20 mm and 30 mm in diameter and 7 mm in thickness were investigated. A standard Pt sensor was mounted inside the top and bottom copper plates to monitor temperature. The transverse Hall sensor, which was directly glued at the center top surfaces of the bulk, was 0.5 mm distant from the sample surface.

The bulk MgB_2 superconductor was activated using field-cooling magnetization in a hybrid superconducting magnet composed of NbTi and Nb_3Sn coils cooled by a cryocooler. The hybrid magnet could generate a magnetic field of 10 T in a room temperature bore of 100 mm diameter. The cryostat with the sample was placed into the superconducting hybrid magnet. The magnet was powered to 3 T, and then the bulk MgB_2 sample was inserted and cooled in the field-cooling process to below 20 K. The trapped magnetic field was recorded after reducing magnetic field to zero and a waiting period of 1 h. To stabilize the temperature, we always employed a waiting time of 30 min at each temperature before recording the data.

6.3 Innovative Activities to Improve The Performance of Bulk MgB_2: Sintering Temperature

Clearly, the low density of MgB_2, homogeneity of the dopant species, and the low J_c as compared to the thin film counterpart has been a long standing issue yet to be addressed. One of the solutions to this is a proper choice of sintering conditions, sintering time, dopant, and the choice for the source of dopant that can be effectively used to serve the purpose of dopant homogeneity, enhancing density and thereby help increasing the J_c and H_{c2} of the bulk MgB_2 samples.

In this study, we will elucidate on the fabrication and characterization of disk-shaped bulk MgB_2 superconductors with high performance as superconducting super-magnets. Several samples were fabricated using a solid-state reaction in pure Ar atmosphere from 750 to 950 °C to obtain the highest critical current density as well as large trapped field values. More details on processing conditions are reported in the experimental section. The sintering process decreases the porosity and increases the connectivity between the particles in bulk MgB_2. However, the optimum sintering temperature is very crucial in MgB_2 material to improve the performance and to avoid the volatility and oxidation of Mg at high temperature. As a result, one can expect the improvement in trapped field values or critical current density (J_c). Recent data clearly indicated that the high J_c values were recorded for the MgB_2 samples sintered at 775 and 800 °C as compared to those sintered at the medium sintering temperature or high sintering temperature (see Fig. 6.2). For the samples sintered at 775 and 800 °C, the highest remnant value of J_c was around 181 kA/cm^2. In order to improve the performance, further optimization of sintering temperature around 775 to 800 °C is crucial and made several samples between these temperatures.

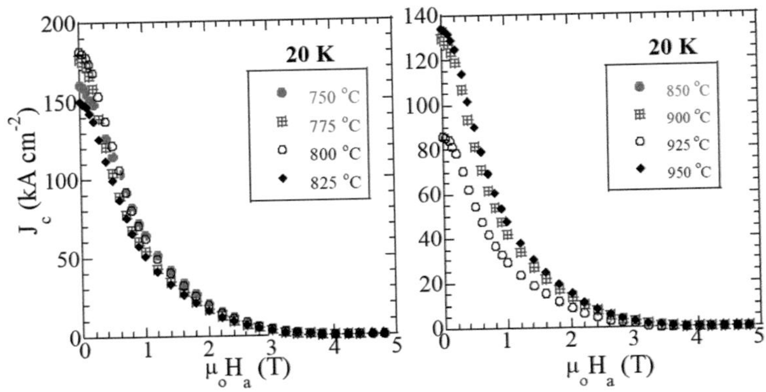

Figure 6.2 Field dependence of the critical current densities (T = 20 K) for MgB_2 superconductor with varying sintering temperatures, prepared in Ar atmosphere.

The critical current behavior of the bulk MgB_2 material at 20 K, sintered for 3 h at 775 °C and 805 °C in steps of 5 °C in argon atmosphere is presented in Fig. 6.3 (left). It was notable that the critical current density was increased with increasing the sintering temperature from 775 °C to 805 °C. Further, the self-field critical current density was improved from 170 kA/cm^2 to 245 kA/cm^2 for the sample sintered at 775 °C to 805 °C. These values are quite high as compared to the recently reported bulk MgB_2 material processed via the solid-state sintering method [51]. Note that the highest critical current densities of 245 kA/cm^2 and 120 kA/cm^2 at 20 K, in self-field and 1 T was recorded for the sample sintered for 3 h at 805 °C without using high-energy milling, high-pressure apparatus, or doping material [38]. To understand more improved performance in MgB_2 sample sintered for 3 h at 805 °C, made microstructure analysis by AFM and results are presented in Fig. 6.3.

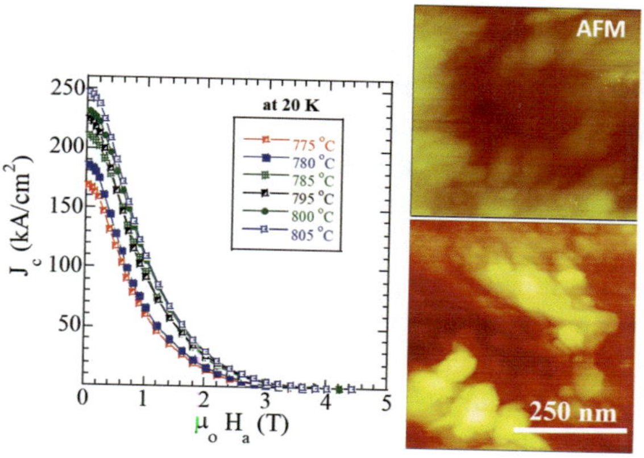

Figure 6.3 Atomic force microscope (AFM) images of sintered bulk MgB_2 samples sintered 3 h at 775 °C (right top) and 3 h at 805 °C prepared in Ar atmosphere (right bottom).

Figure 6.3 (right) presents an atomic force microscopy (AFM) image for the sample sintered for 3 h at 775 °C (top right) and 805 °C (bottom right). As can be seen in the figure that MgB_2 grains are within the nanometer range around 100 nm, which may be responsible for the improved critical current density in these materials. In our earlier study clarified that sintering temperature changes the grain size as well as decrease the impurity phases in the MgB_2 matrix [51]. The optimum grain size is very important to control the grain size and its connectivity. The optimized gains will be offering the grain boundary pinning together with refining the connectivity. The low-temperature sintering mean optimum sintering is crucial which might open a new path to batch production of bulk MgB_2 materials several industrial applications.

6.4 Innovative Activities to Improve the Performance of Bulk MgB_2: Sintering Time

In this study, the effect of sintering time on the flux pinning and superconducting properties of bulk MgB_2 were investigated. The results presented in Section 6.3 indicated that samples sintered moderate temperatures around 775–800 °C gives the best values of critical current and trapped filed at 20 K. The X-ray diffraction conformed that all samples produced moderate temperature is single phase. To further improve the flux pinning and to optimize the sintering time for the bulk MgB_2 material, several samples were produced at 775 °C for 1 h to 10 h and the resulting X-ray diffraction results are presented in Fig. 6.4, left. It is noticeable that the sample sintered even at 775 °C for 1 h showed single phase MgB_2 along with very little trace amounts of MgO impurity. It is possible that the oxygen trapped during the sample pressing into pellets and could contribute to the formation of MgO. Further, it is also likely to happen that the Mg is good oxygen scrubber, and it reacts easily with oxygen even in air during the samples wrapping process in titanium foils before loading the samples into the furnace.

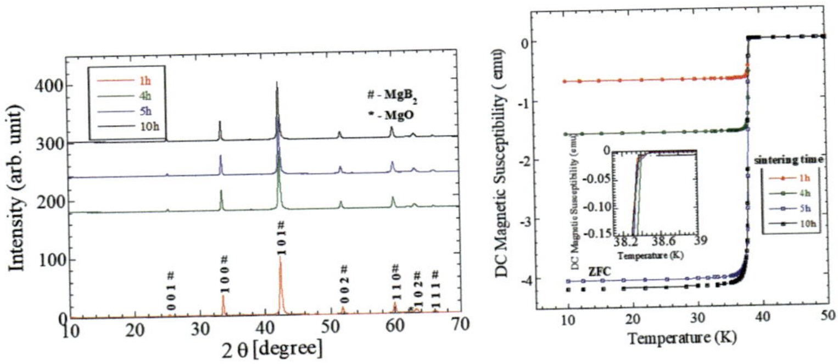

Figure 6.4 X-ray diffraction spectra of bulk MgB_2 samples produced sintering process at 775 °C for 1 h to 10 h (bottom to top) in argon atmosphere (left figure). Temperature dependence of magnetization for MgB_2 samples produced sintering process at 775 °C for 1 h to 10 h in argon atmosphere (right figure).

It was observed that all samples showed high critical current densities as compared to the recently reported bulk MgB_2 material processed via the solid-state sintering method [51]. Figure 6.4 (right) shows how the superconducting transition temperature is affected by the sintering time in bulk MgB_2 material produced at 775 °C in Ar atmosphere. All samples have an onset superconducting transition in the range 38.2 K to 38.8 K with a sharp superconducting transition. The high T_c and the sharp transition indicate that all sintered MgB_2 samples exhibit good crystallinity and homogeneity. These results suggest that superconducting transition temperature in the superconducting matrix in bulk MgB_2 materials is not so sensitive to sintering time.

Figure 6.5 shows the critical current behavior of the bulk MgB_2 material at 20 K, sintered for 1 h, 4 h, 5 h, and 10 h at 775 °C. The critical current density was decreased with increasing the sintering temperature from 1 h to 10 h.

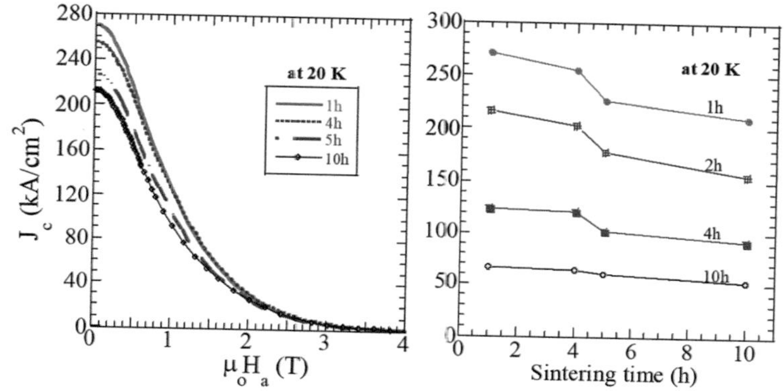

Figure 6.5 Field dependence of the critical current densities (T = 20 K) for MgB_2 superconductor with varying sintering times at 775 °C, prepared in argon atmosphere (left); Sintering time versus the critical current densities (T = 20 K) for MgB_2 superconductors prepared in argon atmosphere (right).

These results clearly demonstrate that the sintering time is crucial to obtain the high critical current density in bulk MgB_2 material. Samples synthesized at 775 °C for 1 h turn out to be the most promising ones and the short sintering time is effective to obtain the high J_c in bulk MgB_2 material. Note that the highest critical current density of 270 kA/cm^2 at 20 K and self-field was recorded for the samples sintered for 1 h (see Fig. 6.5, left). To understand the sintering time dependence of the critical current density, we have measured the critical current density performance at 20 K in self-field, 0.5 T, 1 T, 1.5 T, and 2 T and the corresponding J_c values are plotted in Fig. 6.5 (right). It is clear from the figure that the sample sintered at 775 °C for 1 h exhibits the best performance. Further, the critical current density was decreased with increasing the time, which again indicates that an optimum time is very important to increase the bulk MgB_2 performance. It was reported that single phase MgB_2 material one can obtained even sintering time for 1 h. By employing the optimum sintering time will be control the grain size, which was established that grain bounders act as pinning centers and that optimum control of the grin size will improve the pinning force [51]. On the other hand, the critical current density enhancement in MgB_2 superconducting wire is recently reported on the bases of the structural imperfections caused by the sintering temperatures. As the sintering temperature becomes lower, the structural imperfections of the MgB_2 material are increased, as reflected by detailed X-ray refinement and the normal state resistively. Further, the low sintering temperature keeps the grain size is small, which strongly improve the pinning. Therefore, the small grain size and its dispersion is very important to improve the critical current density of the bulk MgB_2 materials similar the tapes and wires. To control of the microstructural imperfections, especially the grain size down to nano scale looks crucial in bulk MgB_2 material. The recent macrostructure results obtained by AFM and EBSD observations indicated that the number of nanometer-sized grains is higher in samples processed at lower sintering temperatures around 800 °C as compared to the samples sintered at high temperatures [38]. The preset results indicate that there is still room to improve the performance of sintered bulk MgB_2 material by controlling the grain size to the nanoscale level, which is crucial and targeted to the large-scale applications at the boiling temperature of liquid hydrogen.

6.5 Innovative Activities to Improve the Performance of Bulk MgB_2: Silver Addition

Superconducting super-magnet applications, it is required to produce good-quality bulk MgB_2 material with high critical current density (J_c) and an acceptable mechanical performance. To improve the critical current density of MgB_2 material, a variety of processing techniques have been developed. As a result, the transport J_c values at liquid helium temperature and 10 T for silicon-doped MgB_2 samples reached the level of 10^5 A/cm^2. However, the critical current density values of MgB_2 materials still need to be further improvement for high magnetic field applications. Already achieved current values are still much lower than those of Nb_3Sn material [52]. The silver addition to MgB_2 will be one of the solutions to overcome this problem and improve the mechanical performance. The addition of Ag to MgB_2 powders during the sintering process has been found the reduction of loss of Mg, which is commonly observed in the sintering process. Further, the superconducting critical temperature remains almost unchanged with silver addition similar to the Y-123 system. Interestingly, silver does not react chemically with MgB_2 and is believed to reduce the defect scattering density [53]. This chapter ponders upon systematic study on silver addition combined with optimum processing conditions on MgB_2 bulk material is crucially important to further improvement of the critical current density (J_c) and optimization of the mechanical performance to utilize this material for several industrial applications. X-ray diffraction (XRD) patterns recorded from the silver-added bulk MgB_2 materials sintered at 775 °C are clarified that all samples contain the main phase was MgB_2. All the peaks corresponding to pure MgB_2 material in the case of silver-free sample, while the silver-added samples have additional peaks at (2Θ) of 26.78°, 38.18°, 42.74°, and 57.17° [25]. These additional peaks are close to the Ag_2O as compared to the $AgMg_3$ respectably. Note that the intensity of the Ag-based peaks is increased systematically with increasing the silver content from 0 to 10 wt% in the MgB_2 material. This trend is quite good, and indicate that silver was not reacted the MgB_2 phase similar to the $YBa_2Cu_3O_y$ material. XRD analyses suggest that silver cannot react with MgB_2 main phase, while keeping the lower sintering temperature with optimum sintering time.

On the other hand, the temperature dependence of the DC magnetic susceptibility in magnetic field of 1 mT for the MgB_2 sample with varying content of silver addition showed onset superconducting transition in the range 38.5 to 38.7 K with sharp transition less than 0.7 K (see Fig. 6.6, left). The high T_c and sharp transition show that all sintered MgB_2 samples exhibit good crystallinity and homogeneity. The observed superconducting transition temperatures are similar to that of recently reported bulk MgB_2 material fabricated with a standard solid-state process. XRD and DC magnetic susceptibility results suggest that silver cannot react with MgB_2 main phase, while keeping the lower sintering temperature with optimum sintering time and superconducting transition temperature in the superconducting matrix in bulk MgB_2 materials is not destroying to the silver content similar to the RE-123 materials respectively.

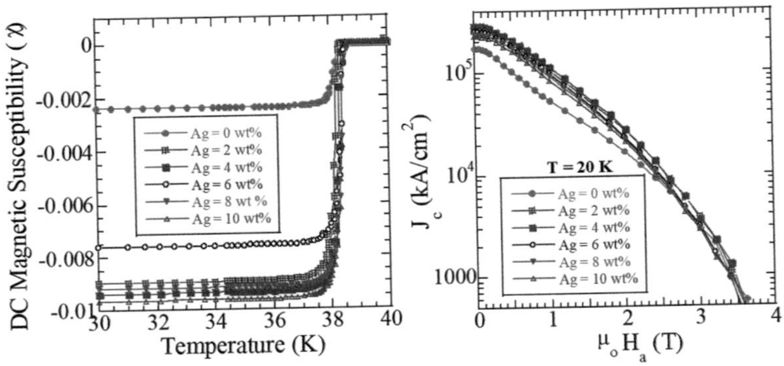

Figure 6.6 Temperature dependence of magnetic susceptibility (left) and field dependence of the critical current densities at $T = 20$ K (right) for MgB_2 sample with 0–10 wt% Ag, sintered at 775 °C for 3 h in Ar atmosphere.

Figure 6.6 (right) shows the critical current density (J_c) at 20 K as a function of applied field for pure as well as Ag-added bulk MgB_2 materials, obtained from M (H) curves using Bean's critical state model. Note that the J_c values are systematically increased with silver content up to 4 wt% and decreased thereafter. The critical current density (J_c) values in whole field range are 30% higher than the silver free martial in the case of 4 wt% of silver-added sample. For the sample added 4 wt% silver, the highest remnant value of J_c was around 292 kA/cm^2, which is somewhat higher than reported values for silver-added bulk MgB_2 materials [54]. The J_c values at 20 K in fields were 210 kA/cm^2 at 0.5 T and 115 kA/cm^2 at 1 T. These results suggested that the small amount of Ag addition along with optimum sintering temperature is crucial to improve the performance of the bulk MgB_2 material, without using additional source of Mg or adapting any special technique to produce the bulk MgB_2 material. The improved critical currents must be attributed to the mainly two reasons, i) formation of high density of normal conducting $AgMg_3$ regions in MgB_2 matrix [53,54], ii) controlling the grain size by adopting the optimum sintering temperature [51], and iii) improving the connectivity between the MgB_2 grain by silver particles. The recorded value of critical current densities obtained in the present work is higher than the recently reported mostly in the Ag-added MgB_2 bulk material [53].

To understand the improved critical current density (J_c) on silver-added samples, the microstructural studies are vital. Further, recent results on bulk MgB_2 material clearly indicated that to precisely control the grain size, reduction of B-rich MgB_2, and MgO phases will play an important role to improve the current-carrying capacity of the bulk MgB_2 material. The most important aspect of any engineering material is challenging the control the microstructure but at the same time, the most important task in the optimization of the processing parameters of engineering materials.

High magnification compositional SEM images for the polished cross section of MgB_2 samples showed a several futures, i) MgB_2 grain structure, ii) Ag-based Mg phases, and iii) large silver oxide particles. Notably, the large size silver oxide particles are trapped in the pore region, which were observed in all samples. Furthermore, at high temperatures one can find small white particles, which are identified through EPMA analysis to be Ag-based Mg phase with the composition close to $AgMg_3$ [54]. These results are consistent with X-ray diffraction studies (XRD). To see the clear morphology, size, and dispersion of the Ag-based Mg phase in the MgB_2 matrix, collected SEM micrographs in very high magnifications, and selective samples specially for the silver content of 4 wt% and 8 wt% are presented in Fig. 6.7. The samples with lower content of silver addition showed the uniform dispersion of nano meter size Ag-Mg particles in the MgB_2 matrix, which are quit significant, as they can act as flux pinning centers. Further, the particle density was dramatically improved with increasing the silver content. The size of the Ag-based Mg grains in sintered MgB_2 materials varied from 10 to approximately 20 nm [54]. It is also reported that Ag-enriched phase distributed in the MgB_2 matrix had the composition close to $AgMg_3$, and EDX analysis with TEM clarified that Ag-rich phases 10–20 nm in size may occupy in MgB_2 matrix [54]. XRD analysis also supported the result of the SEM analysis that the $AgMg_3$ formation in MgB_2 matrix, which content and dispersion is depended on the initially added silver content. Regarding the particle size and dispersion, 4 wt% of silver addition seem to be optimum for the production of high-performance bulk MgB_2 samples with good quality. The presence of silver-based Mg particles active as a flux pinning media to improve the critical current density (J_c) of MgB_2 material. Further, the mechanical properties of silver-free and silver-added MgB_2

bulks have been estimated through the three-point bending test at room temperature clarified that the bending strength of the Ag-added MgB_2 materials are superior as compared to the silver-free material. Note that the average bending strength for the 10 wt% of silver-added material is 10 times high as compared to the silver-free material [25]. This result makes it clear that a small quantity of the silver is significantly influenced by their packing ratio or grain connectivity of bulk MgB_2 materials, which will be useful for the several industrial applications to utilize bulk material as superconducting super-magnets.

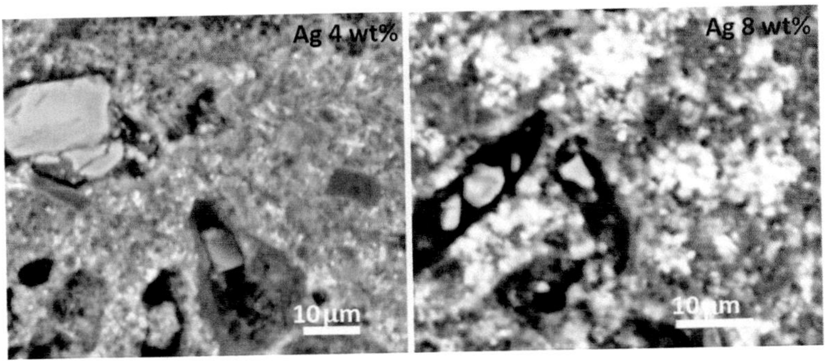

Figure 6.7 SEM images of polished cross sections for bulk MgB_2 samples sintered at 775 °C for 3 h in Ar atmosphere: (left) 4 wt% of silver; and (right) 8 wt% of silver.

6.6 Superconducting Properties of Bulk MgB_2 with MgB_4 Addition

In the trails to enhance the current performance of bulk MgB_2, carbon doping using CNTs, boron carbide, various carbohydrates, SiC, etc., were briefly studied [45–47]. Although, this led to an improvement in superconducting J_c, but posed a serious challenge of nonuniform carbon distribution in the matrix. In our previous studies, we employed usage of carbon-coated boron to solve this issue, where carbon was uniformly distributed [55–59]. Although it is all good till here, this technique is not commercially viable because of its high cost. Therefore, we tried to search for other alternatives to improve superconducting performance by a commercially viable method.

In this study, we thought of using MgB_4, because of its similarity in chemical composition with MgB_2 and low-cost synthesis.

First, MgB_4 was synthesized separately via sintering of Mg and B powder mixture taken in ratio of Mg:B = 1:4. This mixture was thoroughly ground in a glove box in argon atmosphere. The powder mixture was then pressed into pellets using a uniaxial press. The consolidated pellets were then wrapped in titanium foils and sintered in a tube furnace at 1050 °C for 2 h in argon atmosphere. To remove any possible MgO, the MgB_4 pellets were ground into powder and immersed in a bath of 1 Molar HNO_3 solution for an hour. The powder was then cleaned by ethanol to wash out the acid followed by a distilled water wash. After that, the powder was cleaned in acetone and dried. Now, we synthesize MgB_2, while using freshly prepared MgB_4 as additives in various amounts. MgB_2 was produced from high-purity Mg metal (99.9% purity, 200 meshes) and amorphous B powder (99% purity, 300 meshes) along with MgB_4 (0, 1, 2, 3, 4, 5, and 10 wt%) are mixed in the nominal ratio Mg:B = 1:2.

X-ray diffraction and SEM showed MgB_2 to be the main matrix phase with a little amount of MgB_4, as expected [60]. The intensity of the MgB_4 peak matched with that of MgB_4 added. Peaks with concentrations/low intensity lower than 5 wt% were not considered. A minute amount of MgO was detected, which is formed during transferring bulk to furnace. In addition, lattice parameter did not change by much, which tells us that there is no influence from MgB_4 addition on lattice [61]. All the samples exhibited a sharp superconducting transition or small T_c transition width along with T_c onsets and offsets (see Fig. 6.8, left). The transition width, ΔT_c was small for all the samples, except the sample with a high amount of MgB_4. This tells us that addition of MgB_4 in huge amount is detrimental to the MgB_2 matrix properties. The transition temperature of sample with 10 wt% of MgB_4 slightly decreases, which is 37.85 K, slightly lower than pure sample. The results presented above demonstrate that a small quantity of MgB_4 does not disturb the superconducting behavior of the MgB_2 material by much.

Figure 6.8 (right) displays the critical current density, J_c, of the MgB_2 bulks with 0, 1, 2, 3, 4, 5 and 10 wt% of MgB_4 via M-H loops measures at 25 K, H//c-axis. J_c at both low and high magnetic fields increased tremendously in the bulk with 1 wt% of MgB_4 when compared to the reference/pure bulk. The remnant J_c value was around 225 kA/cm^2 and J_c at 0.4 Tesla reached 185 kA/cm^2. Results show that with further increase in MgB_4 content the critical current density starts dropping. AT 25 K, MgB_2 bulk with 10 wt% of MgB_4 had the remnant J_c value close to 183 kA/cm^2. High magnification SEM images (Fig. 6.9) revealed grain refinement upon addition of MgB_4. Compared to the reference/pure sample, high density of very small MgB_4 particles in size of 20–300 nm are observed in the bulk with 1 wt% MgB_4 (see Fig. 6.9, left). However, the bulk with 10 wt% MgB_4 contains less tiny refined particles, in accordance with the J_c performance. Hence, we expect that addition of external MgB_4 can enhance critical current density by creating a well-connected grain morphology and nanometer size grains, which will effectively increase the grain boundary pinning strength [61].

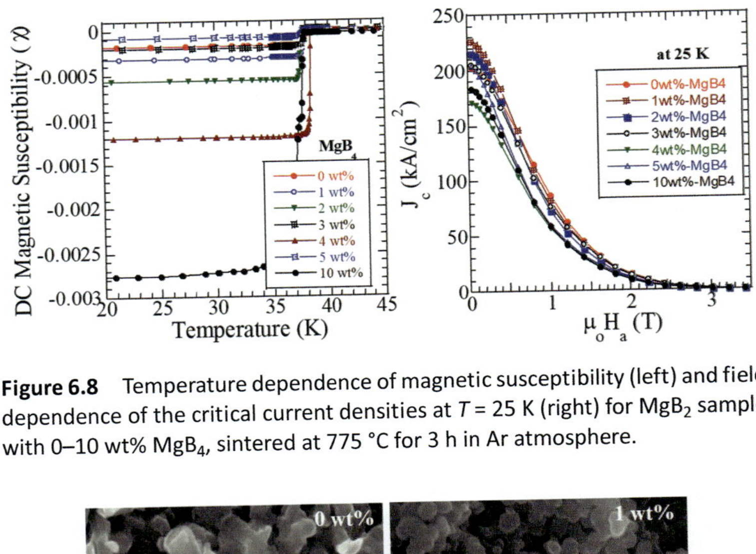

Figure 6.8 Temperature dependence of magnetic susceptibility (left) and field dependence of the critical current densities at T = 25 K (right) for MgB_2 sample with 0–10 wt% MgB_4, sintered at 775 °C for 3 h in Ar atmosphere.

Figure 6.9 High-magnification SEM images of pure MgB_2, and MgB_2 with 1 wt% MgB_4. In the sample with 1 wt% MgB_4 several fine grains of size 20–300 nm can be seen, which are responsible for high self-field J_c.

6.7 Innovative Activities to Improve the Performance of Bulk MgB_2: Utilizing Carbon-Encapsulated Boron

For superconducting super-magnet applications, it is necessary to produce a good-quality and high-performance bulk MgB_2 material with high J_c and low production cost. To improve the critical current density of the MgB_2 material, a variety of processing techniques have been developed including chemical doping [62], refining of the initial particle size by ball milling [63], and irradiation [64]. As a result, the transport J_c values at liquid helium temperature and 10 T reached the level of 10^5 A/cm^2 for Si-doped MgB_2 samples. Similar improvements were also observed with additions of carbon, boron carbide, carbon nanotubes, carbohydrates or hydrocarbons, graphene oxide, and Ni-Co-B nanoparticles [65,66]. However, the critical current density values of MgB_2 materials still need a further improvement for high magnetic field applications. Recent microstructural observations by AFM and SEM indicated that a proper sintering time is very important to control the grain size of the MgB_2 material [67]. X-ray diffraction, DC magnetization, and SEM analyses proved that single-phase and homogenous MgB_2 bulks were produced by sintering at 775 °C and 800 °C, which exhibited the highest critical current density recorded so far. The samples sintered at 775 °C for 1 h showed J_c of 270 kA/cm^2 at 20 K [56]. A trapped field value of 1.51 T at 20 K was achieved in such MgB_2 samples of 20 mm diameter and 7 mm thickness. The simple sintering route is attractive for mass production of bulk MgB_2 materials. However, a further improvement in J_c is needed for such a simple technology. Utilizing the carbon-encapsulated boron might help, since carbon doping of MgB_2 superconducting wires had very positive results, record critical current densities being reported in such tapes and wires [38]. Recent reports on bulk MgB_2 samples doped by carbon-encapsulated boron showed superior critical current density as compared to pure samples [68,69]. Processing optimization is still needed for bulk MgB_2 material to reach uniform carbon dispersion in the bulk MgB_2 material. One of the points is to find an optimum content of the carbon-encapsulated boron in the compound.

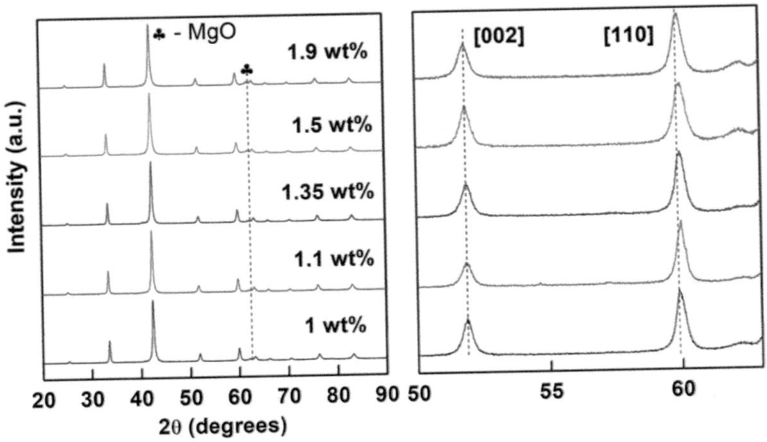

Figure 6.10 (Left) XRD of bulk MgB_2 synthesized with various carbon wt% ccb. (Right) Shift in [110] XRD peaks of bulk MgB_2 synthesized with various carbon wt% ccb.

In the present section, we report on superconducting properties, X-ray diffraction, scanning electron micrographs, and chemical analysis of the carbon-coated boron bulk MgB_2 materials fabricated by a simple sintering method, exhibiting a high J_c and a homogenous carbon distribution in the whole sample. In addition, we optimized the carbon encapsulated boron (CEB) amount for obtaining the possible highest J_c. In order to do so, we tested a series of bulk MgB_2 was synthesized by in situ sintering, with various concentrations of CEB such as 1, 1.1, 1.35, 1.5, and 1.9 wt% carbon coating. The precursors were commercial powders (Furu-uchi Chemical Corporation) of amorphous Mg powder (99.9% purity, 200 meshes) and carbon-encapsulated boron powder (PAVZYUM). From X-ray diffractograms (Fig. 6.10, left) we can see that major phase id MgB_2, with minor MgO peaks. But upon closer view (Fig. 6.10, right), we can see a slight shift in (110) peak toward right with increasing carbon wt%. This is due to the carbon substitution into the lattice, which causes reduction in bond length due to the lower atomic radius. Hence, the shift is proportional to the amount of carbon substituted. Lattice parameter calculations using Bragg's law revealed that lattice parameter a (equivalent to b) drops with increasing carbon content, up to 1.5 wt%, but above 1.5 wt% it starts to increase and lattice parameter c remains unchanged (standard lattice parameters of pure MgB_2

are $a \sim b \sim 3.086$ Å and $c \sim 3.524$ Å). The lattice parameter a was calculated by using (110) peak, parallel to ab-plane, while c was calculated using (002) peak, perpendicular to ab-plane. Because of the layered hexagonal arrangement of MgB_2, the carbon substitution only affects ab-plane and hence there is no shift in lattice parameter c. Precise values of lattice parameters are listed in the Table 6.1.

Table 6.1 Lattice parameters a, c of MgB_2 bulks

Carbon coated boron (ccb) wt%	a (nm)	c (nm)
1	30.854	35.24
1.1	30.849	35.24
1.35	30.826	35.24
1.5	30.798	35.24
1.9	30.830	35.24

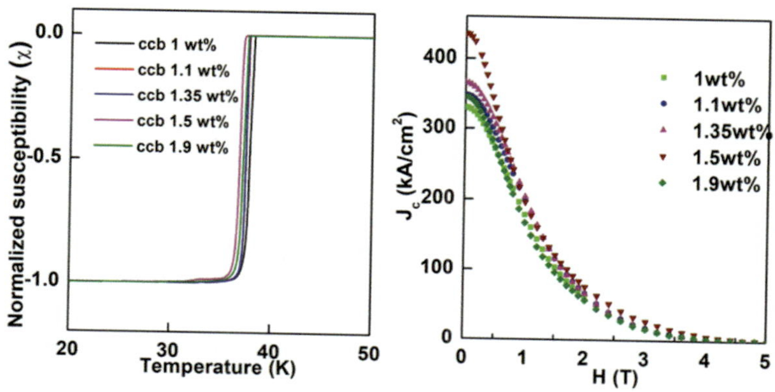

Figure 6.11 T_c (left) and J_c (right) for the bulk MgB_2 synthesized with various carbon wt% ccb.

M-T curves (Fig. 6.11, left) show sharp transition, which indicates high-quality MgB_2 bulk formation. T_c starts to decrease with increasing CEB wt% till 1.5 wt% and starts to raise till 1.9 wt%. Detailed results can be found elsewhere [70]. T_c reduction with increasing C doping is well known from previous studies. The main reasons are carbon substitution into the lattice and the associated lattice strain. This phenomenon has been observed in virtually every case of dopants. The onset superconducting temperature ($T_{c,\mathrm{onset}}$) values follow the similar trend to that of lattice parameter shift. This synchrony of these two results suffices as evidence for saying that maximum carbon doping has happened when 1.5 wt% CEB was used.

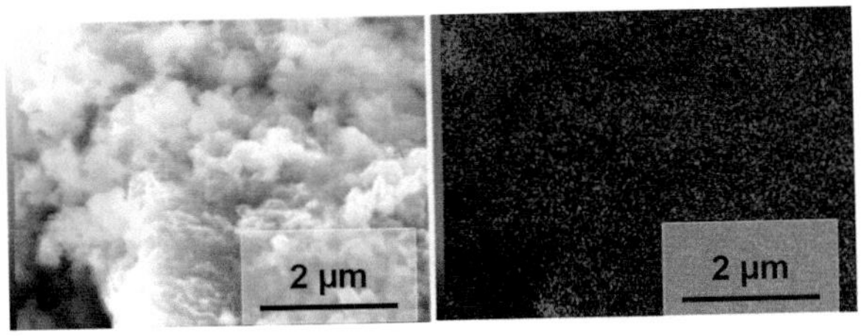

Figure 6.12 Low-magnification SEM image (left) and EDX (right) for the bulk MgB_2 synthesized with 1.5 carbon wt% ccb.

Figure 6.13 High-magnification SEM images for the bulk MgB_2 synthesized with 1.5 carbon wt% ccb.

The critical current densities of various bulks were measured from M-H loops. All the bulks exhibited superior properties when compared to the regular bulk. Especially, the bulk with 1.5 wt% CEB exhibited the highest value among all other bulks, 435 kA/cm^2 at self-field, 20 K (Fig. 6.11, right). At low fields close to self-field as well as high fields, J_c of the 1.5 wt% CEB-based bulk was higher than the rest. This increment is because of the defects and nano-sized lattice strain zones that are formed because of carbon lattice substitution, which acted, as flux pinning centers. SEM and EDX (Figs. 6.12 and 6.13) characterization was done to ensure uniform carbon distribution into the matrix as well as to observe the grain size. EDX spectrum reveals uniform carbon distribution, while the high-magnification images depict fine particles that are around 50–200 nm. These results show that the carbon addition as CEB, especially 1.5 wt% is crucial in particle size refinement as well as impurity scattering through the matrix which improves J_c. Hence, this technique aids in synthesizing stable high-performance bulk MgB_2 and therefore more reliable for use in commercial applications.

6.8 Role of Excess Mg in Enhancing Superconducting Properties of Ag-Added Carbon-Coated Boron-Based Bulk MgB_2

Recently, silver/Ag addition has received attention as it has no impact on T_c while improving the J_c, particularly in YBCO bulk system [71,72]. Similar studies were made on MgB_2 to observe if similar effect can be imitated [25]. In addition, MgB_2 being highly porous and less dense, Ag addition seems interesting as it can increase toughness, density, and other mechanical properties which MgB_2 lack. Some research shows that Gibbs energy of the Ag-Mg phase formation is negative which tells us the definite/spontaneous formation. In addition, unless the sintering temperature is above 1200 °C, the Gibbs energy of Ag-Mg formation is superior to that of MgB_2 formation [73]. This suggests that a minute amount of Mg is consumed by the Ag during the sintering before formation of MgB_2. So, in order to maintain the 1:2 stoichiometric ratio of precursors Mg:B, we improvised adding excess Mg being sufficient for both Ag-Mg as well as MgB_2 formation. In our prior investigations where we, it is found that Ag-Mg based intermetallic secondary phases are formed helping in improving superconducting properties. Here, we will study the excess Mg required to solve the Mg-deficiency problem and its impact on superconducting properties of MgB_2 bulk.

To study the optimum amount of excess Mg precursor required for balancing the loss from Ag-Mg phases formation, a series of bulk MgB_2 is synthesized by an in situ solid-state reaction. Several amounts of Mg had been varied such as 5, 7.5, 10, 12.5 and 15 wt% excess or x = 1.05, 1.075, 1.1, 1.125, 1.15 in x:2 as in Mg:B, also represented in further discussions as Mg_xB_2. The precursors used were commercial powders (Furu-uchi Chemical Corporation) of amorphous Mg powder (99.9% purity, 200 meshes), carbon-encapsulated boron powder (1.5% carbon) and 4 wt% Ag. These powders were then sintered like any other MgB_2 bulk samples in previous chapters.

From XRD (Fig. 6.14), we can observe that the main phase is MgB_2 along with formation of Ag-Mg phases and scarce MgO. It also illustrates that at $x = 1.075$ bulk seems to have the highest Ag-Mg phase fraction or large peak intensity of Ag-Mg phase. In bulks, where $x < 1.075$, the excess Mg is forming Ag-Mg phase but only minute amount. Whereas in bulks with $x > 1.075$, the excess Mg seems to be contributing to other phases such as MgO which there by reducing the amount of Ag-Mg phase which can be deduced with the peak intensity of MgO. From Rietveld phase fraction analysis calculated using MAUD software, we found that highest Ag-Mg phase fraction percentage of around 2% is observed in $x = 1.075$ bulk, while in the highest excess Mg-added bulk showed only 0.3 wt% Ag-Mg phase. Detailed data can be found elsewhere [60, 74].

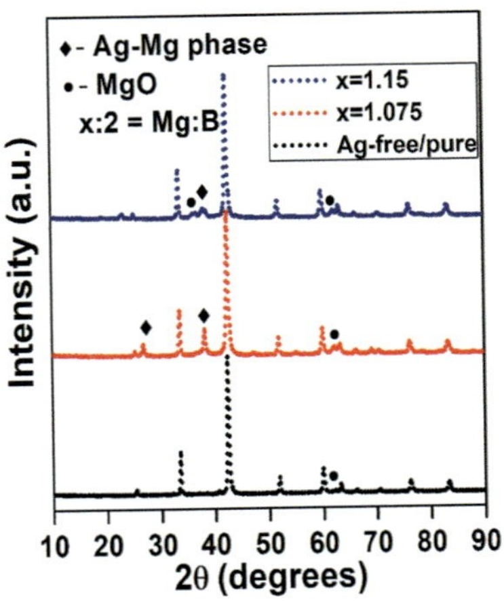

Figure 6.14 XRD of best sample 7.5 wt% excess Mg along with pure and the highest 15 wt% excess Mg-added MgB_2 bulks.

To study the effect of excess Mg addition on the superconducting properties of bulk Mg_xB_2. Superconducting transition temperature (T_c) (Fig. 6.15, left) shows that all the bulks have a sharp superconducting transition or ΔT_c is approximately < 1 K. However, the ΔT_c is increasing with the increase in excess Mg addition, which is probably because of the high MgO content. Onset T_c of all bulks is near to 37.5 K (slightly lower than 39 K), which is expected from a carbon doping in form of carbon-encapsulated boron [56]. This reduction is mainly due to carbon but not Ag, because Ag does not substitute into the lattice at current experimental conditions [73,75]. $T_{c(onset)}$ increases with increase in excess Mg, which tells us that sufficient Mg was form high percentage/quality of MgB_2 bulk superconducting phase. High J_c values are observed in $x = 1.075$ bulk at 20 K (Fig. 6.15, right) which is 347 kA/cm^2 at self-field (approx. 50% improvement) and 244 kA/cm^2 at 0.5 T while the values of pure bulk is around 220 kA/cm^2 at self-field. High magnification SEM images show nano-particle layers of Ag-Mg phases near grain boundaries (Fig. 6.16, left). These nano structure are around 20–40 nm close to the range of MgB_2's coherence length, which make them good pinning centers. These results elucidate us that this 7.5 wt% excess Mg addition, along with defects (from carbon substitution) and other high field pinning centers (Ag-Mg nano phases) is a contributing factor for improving the high field J_c. While EDX (Fig. 6.16, right) elemental mapping shows that these layers are formed throughout the matrix uniformly. Therefore, we think that slightly excess Mg addition is necessary and beneficial in obtaining optimum pinning effect from Ag addition in sintered bulk MgB_2 system.

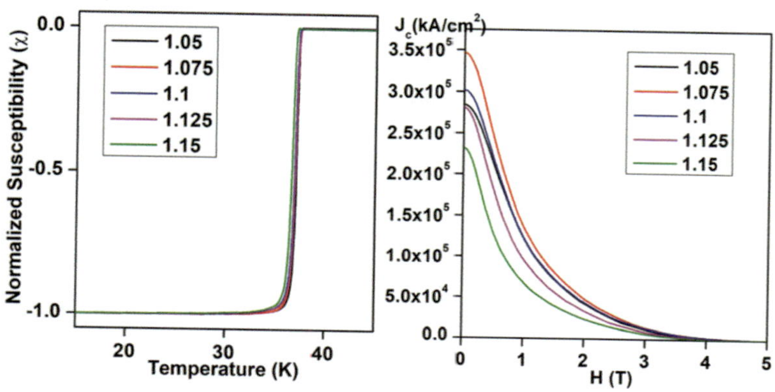

Figure 6.15 T_c (left) and J_c (right) for the Ag added ccb 1.5 wt% based MgB_2 bulk synthesized with various amounts of excess Mg.

Figure 6.16 SEM (left) and EDX (right) for the Ag-added ccb 1.5 wt% based MgB_2 bulk synthesized with various amounts of excess Mg.

6.9 Realizing High-Trapped Field MgB_2 Bulk Magnets for Addressing SDGs

So far, we have discussed various systems that majorly focused on improving superconducting critical current density (J_c), even at the cost of critical transition temperature in some cases. This shows the importance of J_c in superconductors especially when considering bulks. This is because the J_c is a direct factor/parameter in determining the trapped field. Practical applications of bulk MgB_2 superconducting magnets are akin to those of melt-textured YBCO [76]. These bulks are used as trapped field magnets in applications such as NMR and MRI, fault current limiters, noncontact bearings for liquid pumping, and for magnetic shielding screens [76,77]. However, all the mentioned applications need high critical current density bulks with good quality, large size. Some reports have showed the MgB_2 bulk magnet's capability for applications operated at 20 K [40, 78,79]. Various research groups have already reported decent trapped field performances of bulk MgB_2 materials. High-trapped field values close to 1 T in MgB_2 bulk superconductors at 20 K were demonstrated by Yoo and Murakami in 2003 [80]. Viznichenko et al. [81] later reported better trapped field values in a uniform polycrystalline MgB_2 bulk with a 28 mm diameter and 11 mm thickness, created via high pressure of 2 GPa about 1.5 T and 2.3 T at 20 K and 6 K, respectively. For most of the engineering applications, a simple processing route is desirable for mass production of bulk

MgB_2 materials. Hence, in our work we show good superconducting performances and trapped field values for MgB_2 bulk materials synthesized via conventional sintering technique, with considering mass production on mind.

First, MgB_2 pellets of 40 mm, 30 mm, and 20 mm in diameter and 7 mm in thickness were prepared using conventional sintering technique. The trapped field was then measured after magnetizing the bulk sample with a superconducting magnet (hybrid superconducting magnet composed of NbTi and Nb_3Sn coils cooled by a cryocooler that can generate 10 T) in the temperature range between 20 K and 40 K. A Hall sensor was directly placed at the top and bottom surfaces of the MgB_2 bulk, and the distance between probe and sample surface is 0.5 mm. The trapped magnetic field was recorded after reducing the external magnetic field to zero with a waiting interval of 1 h. To ensure the temperature stabilization, we waited 30 min at each temperature step prior to the trapped field measurements.

The trapped fields of bulk MgB_2 samples sintered at various temperatures such as 700, 750, 775, 800, 850, 875, 900, and 950 °C were measured at 20–40 K. All the samples had the same dimensions of 20 mm diameter and 7 mm thickness and the TF values are measured at a distance of 0.5 mm above the sample surface. Prior to magnetic measurements, we waited for 1 h after reducing the external field from 3 T to zero. The maximum trapped field values increased with decreasing temperature from 40 to 20 K (see Fig. 6.17).

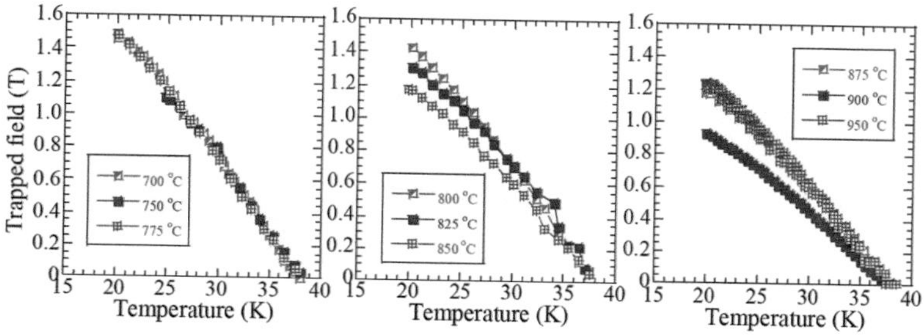

Figure 6.17 Temperature dependence of the trapped fields for bulk MgB_2 samples 20 mm in diameter and 7 mm in thickness, prepared in Ar atmosphere with a sintering method.

Bulks sintered at lower temperatures (<1073 K) exhibit relatively high trapped fields around 1.4–1.51 T at 20 K (see 6.17, left), while the others (> 1073 K) show around 0.92–1.2 T (see Fig. 6.17, middle and right). Of all the bulks, the highest TF value was 1.51 T at 20 K recorded in the sample sintered at 775 °C or 1048 K, which also exhibited highest J_c value among all bulks. This high performance on the bulk sintered at 775 °C is mostly due to the optimum grain size obtained at the perfect sintering temperature. Microstructural analysis revealed a grain size in the order of 300–400 nm (see Fig. 6.18). Further, superior TF values were also observed in bulk sintered at 950 °C showed trapped fields higher than that of the sample sintered at 900 °C. High-magnification SEM images (see Fig. 6.18, left) revealed presence of submicron-sized particles which were analyzed by EDX to be MgO particles and B-rich submicron-sized particles in the MgB_2 matrix. These particles may be responsible for improved field-trapping ability of the bulk via improved flux pinning. The present results demonstrate that MgB_2 bulk magnets produced with a simple sintering method have potential for industrial applications. It is also true that trapped fields could be further improved through the optimization of processing conditions. Fine dispersion of suitable secondary phase pinning centers in the MgB_2 matrix will also be effective in increasing trapped field values. In the case of bulk superconductors, the enlargement of bulk diameter leads to the enhancement of field trapping ability so long as the trapped field is less than the irreversibility field. These bulk can be equipped in MRI and other applications that can help in maintaining health care, sustainable energy with zero-energy losses, etc., that are important aspect in addressing SDGs.

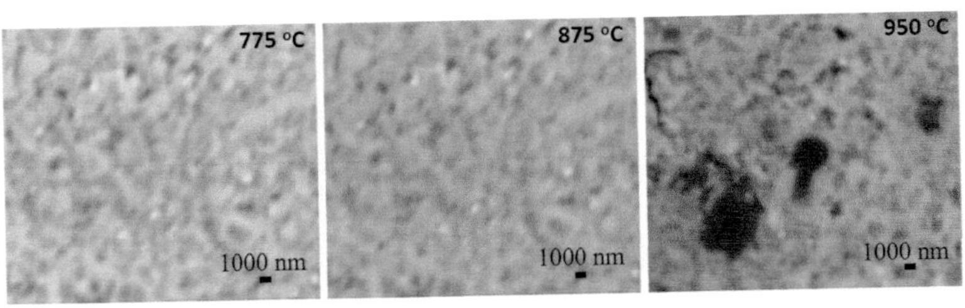

Figure 6.18 High-magnification SEM images of polished cross-sections for bulk MgB_2 samples sintered for 3 h in Ar atmosphere at 775 °C (left), 875 °C (middle), and 950 °C (right).

6.10 Concluding Remarks

The bulk MgB_2 superconducting material is a solid candidate for trending and upcoming magnet applications. Especially, the above-mentioned dopants and enhancements to fabrication techniques can greatly aid in transforming material suitable for usage in real-life applications. With popular $REBa_2Cu_3O_y$ superconductors having huge fabrication time and brittle nature, MgB_2 hails as a leading choice primarily due to its easy synthesis and machinability. Superconductors are important part to create a sustainable energy manipulation that addresses some of the SDGs. Moreover, the applications of this class of materials range from high-trapped field magnets in medical field to high-speed maglevs in transportation, which makes it a perfect solution to meet SDG goals. Hence, we think that MgB_2 especially with improved performance can be utilized in our paths to solve SDG goals.

Acknowledgments

This work was partly supported by Shibaura Institute of Technology (SIT) Research Center for Green Innovation and Grant-in-Aid FD research budget code: 112282.

References

1. Nagamatsu J, Nakagawa N, Muranaka T, Zenitani Y, and Akimitsu J, Superconductivity at 39 K in magnesium diboride, *Nature*, **410** (2001) pp. 63–64.
2. Kortus J, Mazin I, Belashchenko KD, Antropov V, and Boyer L, Superconductivity of metallic boron in MgB_2, *Phy. Rev. Lett.*, **86**, (2001) pp. 4656–4659.
3. An J, and Pickett W, superconductivity of MgB_2: covalent bonds driven metallic, *Phy. Rev. Lett.*, **86**, (2001) pp. 4366–4369.
4. AY Liu, I Mazin, and J Kortus, Beyond Eliashberg superconductivity in MgB_2: Anharmonicity, two-phonon scattering, and multiple gaps, *Phy. Rev. Lett.*, **87**, (2001) 087005 pp. 4.
5. Chen X, Konstantinović M, Irwin J, Lawrie D, and Franck J, Evidence for two superconducting gaps in MgB_2, *Phy. Rev. Lett.*, **87**, (2001) 157002 pp. 4

6. Gonnelli R, Daghero D, Ummarino G, Stepanov V, Jun J, Kazakov S, and Karpinski J, Direct evidence for two-band superconductivity in MgB_2 single crystals from directional point-contact spectroscopy in magnetic fields, *Phy. Rev. Lett.*, **89**, (2002) 247004 pp. 4.
7. Choi HJ, Roundy D, Sun H, Cohen ML, and Louie SG, The origin of the anomalous superconducting properties of MgB_2, *Nature*, **418**, (2002) pp. 758–760.
8. Souma S, Machida Y, Sato T, Takahashi T, Matsi H, Wang S-C, Ding H, Kaminski A, Campuzano *JC*, Sasaki S, and Kadowaki, K, The origin of multiple superconducting gaps in MgB_2, *Nature*, **423**, (2003) pp. 65–67.
9. Ashcroft NW, Metallic hydrogen: a high-temperature superconductor?, *Phy. Rev. Lett.*, **21**, (1968) pp. 1748.
10. Osborn R, Goremychkin E, Kolesnikov A, and Hinks D, Phonon density of states in MgB_2, *Phy. Rev. Lett.*, **87**, (2001) 017005 pp. 4.
11. Shukla A, Matteo C, Matteo d'A, Michele L, Francesco M, Christophe B, Michael K, Karpinski J, Kazakov SM, Jun J, Daghero D, and Parlinski K, Phonon dispersion and lifetimes in MgB_2, *Phy. Rev. Lett.*, **90**, (2003) 95506 pp. 4.
12. Li W, Li Y, Chen R, Zeng R, Dou S, Zhu M, and Jin H, Raman study of element doping effects on the superconductivity of MgB_2, *Phy. Rev. B*, **77**, (2008) 094517 pp. 7.
13. Bud'Ko S, Lapertot G, Petrovic C, Cunningham C, Anderson N, and Canfield P, Boron isotope effect in superconducting MgB_2, *Phy. Rev. Lett.*, **86**, (2001) pp. 1877–1880.
14. Xu S, Moritomo Y, Kato K, and Nakamura A, Mn-substitution effects On Mgb_2 superconductor, *J. Phys. Soc. Japan*, **70**, (2001) pp. 1889–1891.
15. Dou SX, Soltanian S, Zhao Y, Getin E, Chen Z, Shcherbakova O, and Horvat J, The effect of nanoscale Fe doping on the superconducting properties of MgB_2, *Superconduct. Sci. Technol.*, **18**, (2005) pp. 710–715.
16. Li J, Li L, Liu F, Dong C, Xiang J, and Zhao Z, Superconductivity, superstructure, and structure anomalies in $Mg_{1-x}Al_xB_2$, *Phy. Rev. B*, **65**, (2002) 132505 p. 3.
17. Kunc K, Loa I, Syassen K, Kremer R, and Ahn K, MgB_2 under pressure: phonon calculations, Raman spectroscopy, and optical reflectance, *J. Phys.: Condens. Matter*, **13**, (2001) pp. 9945–9962.
18. Shi L, Zhang H, Zhou S, Zhao J, and Zuo J, Superconductivity and the disorder effect in Ag and Al double doped MgB_2, *J. Appl. Phys.*, **100**, (2006) 023905 p. 6.

19. Mustapić M, Horvat J, Hossain MS, Skoko Z, and Dou SX, Enhancing superconducting properties of MgB_2 pellets by addition of amorphous magnetic Ni–Co–B nanoparticles, *Superconduct. Sci. Technol.*, **26**, (2013) 075013 p. 9.

20. Yildirim T, Gülseren O, Lynn JW, Brown CM, Udovic TJ, Huang Q, Rogado N, Regan KA, Hayward MA, Slusky JS, He T, Haas MK, Khalifah P, Inumaru K, and Cava RJ, Giant anharmonicity and nonlinear electron-phonon coupling in MgB_2: a combined first-principles calculation and neutron scattering study, *Phy. Rev. Lett.*, **87**, (2001) 037001 p. 4.

21. Eom C, et al., High critical current density and enhanced irreversibility field in superconducting MgB_2 thin films, *Nature*, **411**, (2001) pp. 558–560.

22. Buzea C and Yamashita T, Review of the superconducting properties of MgB_2, *Superconduct. Sci. Technol.*, **14**, (2001) R115–R146.

23. Noudem J, Aburras M, Bernstein P, Chaud X, Muralidhar M, and Murakami M, Development in processing of MgB_2 cryo-magnet superconductors, *J. Appl. Phys.*, **116**, (2014) 163916 p. 4.

24. Jin S, Mavoori H, Bower C, and Van Dover R, High critical currents in iron-clad superconducting MgB_2 wires, *Nature*, **411**, (2001) pp. 563–565.

25. Muralidhar M, Inoue K, Koblischka MR, Murakami A, and Murakami M, Effects of silver addition on critical current densities and mechanical properties in bulk MgB_2, *J. Adv. Eng. Mater.*, **17**, (2015) pp. 831–838.

26. Vinod K, Kumar RA, and Syamaprasad U, Prospects for MgB_2 superconductors for magnet application, *Superconduct. Sci. Technol.*, **20**, (2006) R1–R13.

27. Perini E and Giunchi G, Field cooling of a MgB_2 cylinder around a permanent magnet stack: prototype for superconductive magnetic bearing, *Superconduct. Sci. Technol.*, **22**, (2009) 045021 p. 6.

28. Tampieri A, Celotti G, Sprio S, Caciuffo R, and Rinaldi D, Study of the sintering behaviour of MgB_2 superconductor during hot-pressing, *Phys. C: Superconduct.*, **400**, (2004) pp. 97–104.

29. Shields T, Kawano K, Holdom D, and Abell J, Microstructure and superconducting properties of hot isostatically pressed MgB_2, *Superconduct. Sci. Technol.*, **15**, (2002) pp. 202–205.

30. Gümbel A, Eckert J, Fuchs G, Nenkov K, Müller K-H, and Schultz L, Improved superconducting properties in nanocrystalline bulk MgB_2, *Appl. Phys. Lett.*, **80**, (2002) pp. 2725–2727.

31. Prikhna T, Eisterer M, Weber H W, Gawalek W, Kovylaev V, Karpets M, Moshchil V, Kozyrev A, Basyuk T, Chaud X, Goldacker W, Sokolovsky V, Noudem J, Borimskiy A, Sverdun V, and Prisyazhnaya E, Temperature–pressure induced nano-structural inhomogenities for vortex pinning in bulk MgB_2 of different connectivity, *Phys. C: Superconduct.*, **503**, (2014) pp. 109–112.

32. Muralidhar M, Kenta N, Koblischka MR, and Murakami M, High critical current densities in bulk MgB_2 fabricated using amorphous boron, *Phys. Status Solidi (a)*, **212**, (2015) pp. 2141–2145.

33. Zhou S, Pan A V, Wexler D, and Dou SX, Sugar coating of boron powder for efficient carbon doping of MgB_2 with enhanced current-carrying performance, *Adv. Mater.*, **19**, (2007) pp. 1373–1376.

34. Giunchi G, Ripamonti G, Cavallin T, and Bassani E, The reactive liquid Mg infiltration process to produce large superconducting bulk MgB_2 manufacts, *Cryogenics*, **46**, (2006) pp. 237–242.

35. W Li, R Chen, X Xu, Y Hu, M Zhu, Y Li, and S Dou, Enhancement of critical current of SiC and malic acid codoped MgB_2/Fe wires, *Int. J. Modern Phys, B*, **29**, (2015)1542032 p. 5.

36. Bugoslavsky Y, Cohen L, Perkins G, Polichetti M, Tate T, Gwilliam R, and Caplin A, Enhancement of the high-magnetic-field critical current density of superconducting MgB2 by proton irradiation, *Nature*, **411**, (2001) pp. 561–563.

37. Xu X, Kim JH, Shahriar-Al Hossain M, Park J, Zhao Y, Dou SX, Yeoh WK, Rindfleisch M, and Tomsic M, Phase transformation and superconducting properties of MgB2 using ball-milled low purity boron, *J. Appl. Phys.*, **103**, (2008) 032912 p. 7.

38. Muralidhar M, Nozaki K, Kobayashi H, Zeng XL, Koblischka-Veneva A, Koblischka MR, Inoue K, and Murakami M, Optimization of Sintering Kinetics of Bulk MgB_2 for Highest Flux Pinning, *J Alloys Compounds*, **649**, (2015) pp. 833–842.

39. Muralidhar M, Inoue K, Koblischka MR, and Murakami M, Fabrication of MgB_2 bulk magnets with high critical currents, *Adv. Sci. Technol.* (Trans Tech Publ), **95**, (2014) pp. 196–201.

40. Durrell JH, Dancer CEJ, Dennis A, Shi Y, Xu Z, Campbell AM, Hari Babu N, Todd RI, Grovenor CRM, and Cardwell DA, A trapped field of >3 T in bulk MgB_2 fabricated by uniaxial hot pressing, *Supercond. Sci. Technol.*, **25**, (2012) 112002 p. 5.

41. Masui T, Lee S, and Tajima S, Carbon-substitution effect on the electronic properties of single crystals, *Phys. Rev. B*, **70**, (2004) 024504 p. 7.

42. Pogrebnyakov AV, and Xi XX, Properties of thin films with carbon doping, *Appl. Phys. Lett.*, **85**, (2004) pp. 2017–2019.

43. Maeda M, et al., In field J_c improvement by oxygen free pyrene gas diffusion into highly dense MgB_2 superconductor, *J Appl. Phys.*, **109,** (2011) 023924 p. 4.

44. Dou SX, Yeoh W, Horvat J, and Ionescu M, Effect of carbon nanotube doping on critical current density of superconductor, *Appl. Phys. Lett.*, **83**, (2003) pp. 4996–4998.

45. Y Liu, F Lan, Z Ma, N Chen, H Li, S Barua, D Patel, M Shahriar, Al Hossain, S Acar, JH Kim, and SX Dou, Significantly enhanced critical current density in nano-MgB_2 grains rapidly formed at low temperature with homogeneous carbon doping, *Superconduct. Sci. Technol.*, **28**, (2015) 055005 p. 7.

46. Dou SX, Braccini V, Soltanian S, Klie R, Zhu Y, Li S, Wang X, and Larbalestier D, Nanoscale-SiC doping for enhancing J_c and H_{c2} in superconducting MgB_2, *J. Appl. Phys.*, **96**, (2004) pp. 7549–7555.

47. Da Silva L, Serrano G, Serquis A, Metzner V, and Rodrigues D Jr, Artificial pinning centers in MgB_2 superconducting bulks, *Superconduct. Sci. Technol.*, **28**, (2015) 025008 p. 7.

48. Zhang S, Kopas C, Wagner B, Queen D, and Newman N, In-situ electron paramagnetic resonance studies of paramagnetic point defects in superconducting microwave resonators, *Appl. Phys. Lett.*, **109**, (2016) 122602 p. 5.

49. S Barua, D Patel, N Alzayed, M Shahabuddin, JM Parakkandy, Shahabuddin Shah M, Z Ma, M Mustapić, MS Al Hossain, and JH Kim, Correlation between in-field J_c enhancement and grain connectivity in co-doped MgB_2 superconductor, *Mater. Lett.*, **139,** (2015) pp. 333–335.

50. Chen D-X and Goldfarb RB, Kim model for magnetization of type-II superconductors, *J. Appl. Phys.*, **66**, (1989) pp. 2489–2500.

51. Muralidhar M, Inoue K, Koblischka MR, Tomita M, and Murakami M, Optimization of processing conditions towards high trapped fields in MgB_2 bulks, *J. Alloys Compounds*, **608**, (2014) pp. 102–109.

52. Lee PJ and Larbalestier DC, Microstructural factors important for the development of high critical current density Nb_3Sn strand, *Cryogenics*, 48, (2008) pp. 283–292.

53. Grivel J-C, Abrahamsen A, and Bednarcik J, Effects of Cu or Ag additions on the kinetics of MgB_2 phase formation in Fe-sheathed wires, *Supercond. Sci. Technol.*, **21**, (2008) 035006 pp5.

54. Kumar D, Pennycook SJ, Narayan J, Wang H, and Tiwari A, Role of silver addition in the synthesis of high critical current density MgB_2 bulk superconductors, *Supercond. Sci. Technol.*, **16**, (2003) pp. 455–458.

55. Muralidhar M, Hguchi M, Jirsa M, Diko P, Kokal I, and Murakami M, Improved critical current densities of bulk MgB_2 using carbon-coated amorphous boron, *IEEE Trans. Appl. Supercond.*, **27**, (2017) 7801872 p. 4.

56. Muralidhar M, Higuchi M, Kitamoto K, Koblischka MR, Jirsa M, and Murakami M, Enhanced critical current density in bulk MgB_2, *IEEE Trans. Appl. Supercond.*, **28**, (2018) 8000405 p. 5.

57. Muralidhar M and Murakami M, Progress in critical current density (J_c) in sintering MgB_2 bulk, *J. Supercond. Nov. Magn.*, **31**, (2018) pp. 2677–2681.

58. Kumar D, Muralidhar M, Rao MS, and Masato M, Raman spectroscopy of carbon doped MgB_2 prepared using carbon encapsulated boron as precursor, *J. Alloys Compounds*, **723**, (2017) pp. 751–756.

59. Arvapalli SS, Muralidhar M, and Murakami M, Optimization of Mg precursor concentration to obtain high J_c in MgB_2 synthesized with Ag addition and carbon encapsulated boron, *IEEE Trans. Appl. Supercond.*, **29**, (2019) 8000904 p. 4.

60. Ishiwata J, Muralidhar M, Inoue K, and Murakami M, Effect of MgB_4 addition on the superconducting properties of polycrystalline MgB_2, *J. Phys. Proc.*, **65**, (2015) pp. 69–72.

61. Miryala M, Arvapalli SS, Diko P, Jirsa M, and Murakami M, Flux pinning and superconducting properties of bulk MgB_2 with MgB_4 addition, *J. Adv. Eng. Mater.*, **22**, (2019) 1900750 pp. 1–7.

62. Kirat G, Kizilaslan O, Aksan MA, and Yakinci ME, Effect of coaddition on the critical current density of MgB_2 superconductor, *Phys. B Condens. Matter.*, **445**, (2014) pp. 24–27.

63. Bohnenstiehl SD, Susner MA, Yang Y, Collings EW, Sumption MD, Rindfleisch MA, and Boone R, Carbon doping of MgB_2 by toluene and malic-acid-in-toluene, *Phys. C Supercond. its Appl.*, **471**, (2011) pp. 108–111.

64. De Silva KSB, Gambhir S, Wang XL, Xu X, Li WX, Officer DL, Wexler D, Wallace GG, and Dou SX, The effect of reduced graphene oxide addition on the superconductivity of MgB_2, *J. Mater. Chem.*, **22**, (2012) pp. 13941–13946.

65. Dou SX, Shcherbakova O, Yoeh WK, Kim JH, Soltanian S, Wang XL, Senatore, C, Flukiger, R, Dhalle, M, Husnjak, O, and Babic, E, Mechanism

of enhancement in electromagnetic properties of MgB_2 by nano SiC doping, *Phys. Rev. Lett.,* **98**, (2007) 097002 p. 4.

66. Hossain MSA, Senatore C, Flükiger R, Rindfleisch MA, Tomsic MJ, Kim JH, and Dou SX, The enhanced J_c and Birr of insitu MgB_2 wires and tapes alloyed with $C_4H_6O_5$ (malic acid) after cold high pressure densification, *Supercond. Sci. Technol.,* **22**, (2009) 095004 p. 8.

67. Higuchi M, Muralidhar M, Diko P, Jirsa M, and Murakami M, Record critical current density in bulk MgB_2 using carbon-coated amorphous boron with optimum sintering conditions, *J. Phys. Conf. Ser.,* **871,** (2017) 012056 p. 6.

68. Bhagurkar AG, Yamamoto A, Anguilano L, Dennis AR, Durrell JH, Hari Babu N, and Cardwell DA, A trapped magnetic field of 3 T in homogeneous, bulk MgB_2 superconductors fabricated by a modified precursor infiltration and growth process, *Supercond. Sci. Technol.,* **29**, (2016) 035008 p. 8.

69. Kumar D, Muralidhar M, Rao MSR, and Murakami M, Effect of Ag addition on the surface topography and the vibrational dynamics of MgB_2, *J. Supercond. Nov. Magn.,* **31**, (2018) pp. 2033–2038.

70. Arvapalli SS, Muralidhar M, Pinmangkorn S, and Murakami M, *J. Adv. Eng. Mater.* **22**, (2020) 2000478 p. 7

71. Muralidhar M, and Murakami M, Improved superconducting properties due to Ag_2O addition in NEG-123, *Phys. C Supercond.,* **360,** (2001) pp. 657–660.

72. Farbod M, and Batvandi MR, Doping effect of Ag nanoparticles on critical current of $YBa_2Cu_3O_7$-δ bulk superconductor, *Phys. C Supercond.,* **471**, (2011) pp. 112–117.

73. Sun T, Zhang XP, Zhao YG, Shen R, Wang K, Zhang LW, Cao BS, Xiong YH, Li PJ, and Wen HH, Study of the reactions between MgB_2 and Ag at high temperature, *Phys. C Supercond. its Appl.,* **382**, (2002) pp. 367–372.

74. Arvapalli SS, Miryala M, and Murakami M, Beneficial impact of excess Mg on flux pinning in bulk MgB_2 synthesized with Ag addition and carbon encapsulated boron, *J. Adv. Eng. Mater.,* **21**, (2019) 1900497, pp. 1–7.

75. Kimishima Y, Takami S, Uehara M, Numa S, Sugiyama Y, and Kuramoto T Pinning properties of Ag/MgB_2 bulk system, *Phys. C Supercond. its Appl.,* **468,** (2008) pp. 1181–1184.

76. Murakami M, Processing and applications of bulk RE–Ba–Cu–O superconductors, *Int. J. Appl. Ceram. Technol.,* **4**, (2007) pp. 225–241.

77. Malozemoff AP, World academy of ceramics, in: LJ Gauckler (Ed.), *Techna Group – Monog. Mater. Soc.*, **8**, (2009) 279.

78. Perini E, Giunchi G, Saglietti L, Albisetti A, Matrone A, and Cavaliere A, Magnetic field trapping in MgB2 bulks and inserts, *IEEE Trans. Appl. Supercond.*, **21**, (2011) pp. 2690–2693.

79. Muralidhar M, Ishihara A, Suzuki K, Fukumoto Y, Yamamoto Y, and Tomita M, Optimization of the fabrication process for high trapped field MgB_2 bulks, *Phys. C*, **494,** (2013) pp. 85–88.

80. Yoo SI and Murakami M, Presented at 11th International Workshop on Critical Currents in Superconductors (IWCC11), Nihon University, Tokyo, Japan 28–31, (2003).

81. Viznichenko RV, Kordyuk AA, Fuchs G, Nenkov K, Müller KH, Prikhna TA, and Gawalek W, Temperature dependence of trapped magnetic field in MgB_2 bulk superconductor, *Appl. Phys. Lett.*, **83**, (2003) pp. 4360–4362.

Chapter 7

Powder Technology of Magnesium Diboride and Its Applications

Soo Kien Chen,[a] Oon Jew Lee,[b] and Muralidhar Miryala[c]
[a]*Department of Physics, Faculty of Science, Universiti Putra Malaysia, 43400 Serdang, Selangor, Malaysia*
[b]*Faculty of Science and Marine Environment, Universiti Malaysia Terengganu, 21030 Kuala Nerus, Terengganu, Malaysia*
[c]*Superconducting Research Laboratory, Graduate School of Science and Engineering, Shibaura Institute of Technology, 3-7-5 Toyosu, Koto-ku, Tokyo 135-8548, Japan*
chensk@upm.edu.my

This chapter reviews strategies for enhancing critical current density, J_c of MgB_2 superconductor from the perspective of powder technology. Focus is on the powder processing and characterization of the material. Discussion is mainly based on phase formation, crystal lattice, microstructure and transport electrical resistivity properties. The effects of the structural and electrical properties on critical temperature, T_c, J_c, upper critical field, H_{c2} and irreversibility field, H_{irr} are highlighted. The strategies are categorized into intrinsic and extrinsic ones. Intrinsically, properties of MgB_2 can be tuned by

High-T_c Superconducting Technology: Towards Sustainable Development Goals
Edited by Muralidhar Miryala

ISBN 978-981-4877-65-7 (Hardcover), 978-1-003-16468-5 (eBook)
www.jennystanford.com

manipulating its constituent elements, i.e. magnesium and boron. Extrinsic approach can be done by externally introduce chemical dopants during the sample preparation procedure. For this, addition of Dy_2O_3 is chosen for discussion as the dopant enhances both J_c and H_{c2} significantly without changing much the crystal structure and T_c. This chapter is divided into three main sections: (i) tuning magnesium and boron in undoped samples (ii) addition with rare earth oxides, and (iii) some large-scale applications of MgB_2. Finally, the importance of powder technology for MgB_2 and its relationship with some of the United National Sustainable Development Goals are concluded.

7.1 Introduction

The three important critical parameters, namely critical current density (J_c), upper critical field (H_{c2}), and critical temperature (T_c) define application range of superconductors. In this regard, powder technology has played a paramount role in not only the synthesis but also the enhancement of electromagnetic properties of magnesium diboride (MgB_2). Meanwhile, it is equally crucial to ensure such technology advancement is aligned with the United National Sustainable Development Goals (UN SDGs).

With its relatively higher T_c [1] (compared to the Nb-based low-T_c superconductors) and weak link free grain boundaries [2] (as opposed to the oxide high-T_c superconductors (HTSCs)), MgB_2 appears to be a suitable choice of material for various usage. Hitherto, the most common techniques of using magnesium and boron powders for synthesis of MgB_2 are the in situ and ex situ approaches [3]. The former yields porous samples but higher J_c while the latter leads to denser sample density. Since MgB_2 is a binary compound, this allows somewhat better control of sample preparation as compared with the quaternary oxide HTSCs. Magnesium is a volatile element. Magnesium loss due to evaporation escalates at higher temperatures above its melting point. Conversely, boron has a high melting point above 2000 °C at standard atmospheric pressure. Thus, the distinct chemical properties of magnesium and boron can be utilized to prepare MgB_2 with enhanced J_c and H_{c2}. This involves manipulation of magnesium and boron in many ways such as usage of boron powder form (crystalline or amorphous), starting particle size of the powders and nominal magnesium content. Preparing MgB_2 samples in this way is very practical, simple and cost-effective without relying on additional chemicals. Nonetheless, addition with chemical dopants represent another simple method to increase current carrying performance of MgB_2 without employing further materials processing or complicated procedures like ball milling, hot isotactic pressing and spark plasma synthesis.

7.2 Tuning Magnesium and Boron in Undoped Samples

7.2.1 Boron Powders

Boron powders influence J_c of MgB_2 considerably. Hence, it is essential to check purity, distribution of particle size, and form of the powders before using them to prepare samples. Commercial boron powders are available in crystalline, semi-crystalline, and amorphous form [4]. Quality of the powders varies depending on their manufacturing process. The methods used to prepare boron are mainly fused electrolysis, metallic reduction, hydrogen reduction, and decomposition of boron compounds [5, 6]. To prepare amorphous boron, the common method is by reduction of B_2O_3 using magnesium (metallic reduction) followed by purification using acid leaching process [7]. It was reported that the use of innovative method produced fine amorphous boron which yielded high J_c in the synthesized MgB_2 samples [8]. Nevertheless, the presence of boron oxide, B_2O_3 (probably as residue of metallic reduction) will give an adverse effect on the final MgB_2 products by contributing towards higher fraction of oxide phases. Moreover, oxides are the source of oxygen to react with Mg forming MgO. The presence of MgO is common in commercial Mg since Mg is a good oxygen scrubber that reacts easily with oxygen even in air.

Different impurities can be formed in MgB_2 samples depending on the source of boron powders. A study on the effects of different boron powders on the behaviour of J_c was conducted whereby four types of boron powders were used to prepare samples [4]. Table 7.1 shows the specifications of the boron powders according to the information on the labels of the respective chemical containers. X-ray diffraction (XRD) showed semi-crystalline feature of the A9597 boron powder despite being labelled as amorphous. C99 and A9597 showed dual particle size distribution as measured by particle size analyser (Table 7.1). XRD showed that B_2O_3 is the main impurity in the amorphous boron powders probably because they had been produced by Mg reduction of B_2O_3. However, it is not easy to quantify accurately the amount of B_2O_3 in the amorphous and crystalline boron powders.

Table 7.1 Specifications of the boron powders as specified in the labels of their chemical containers

Boron powder identity	Manufacturer	Form	Purity (%)	Particle size
C98	Alfa Aesar	Crystalline	98	325 mesh (21.10 μm)
C99	FluroChem	Crystalline	99	<40 μm (0.56, 11.42 μm)
A9597	Fluka	Amorphous	95–97	(0.56, 2.42 μm)
A9999	Alfa Aesar	Amorphous	99.99	325 mesh (0.54 μm)

Note: The parentheses refer to the peak values of particle size distribution determined using a Mastersizer E particle size analyser from Malvern Instruments [4]

While MgO was detected as the main impurity in the MgB_2 samples made of the amorphous and semi-crystalline borons, more impurities like B_2O, $B_{13}O_2$, $Mg_3(BO_3)_2$, and some unidentified phases were found in the samples prepared using the crystalline borons [4].

The finding can be explained based on reactivity of the boron powders. The amorphous boron powders are much more reactive than that of the crystalline ones leading to enhanced reaction rate in forming MgB_2 [5]. On the other hand, the reaction between the crystalline boron powders and magnesium is relatively slow. This is supported by the fact that unreacted Mg was found in the samples prepared using one of the crystalline borons [4]. Consequently, the longer reaction time led to the formation of $(Mg)B_xO_y$ phases. The amorphous boron powder reacted rapidly with magnesium to form MgB_2. For B_2O_3, its reaction with magnesium forms MgO and B.

Table 7.2 shows that the *c*-axis lattice parameter does not vary much across the samples. The sample made of A9597 boron powder has the smallest *a*-axis. It is interesting that the scanning electron microscopy (SEM) images revealed larger grains of the MgB_2 samples prepared using the crystalline borons compared with those made of amorphous borons consistent with the smaller particle sizes of the amorphous boron powders. Similarly, bimodal grain sizes were observed in the MgB_2 samples prepared using borons with bimodal particle sizes.

T_c varies across the samples from 37.9 K to 38.8 K (Table 7.2) without a clear indication of improved T_c with boron purity in contrast to previous finding [9]. However, the T_c transitions (Table 7.2) for the samples made of crystalline boron is broader than that of the amorphous boron samples suggesting greater sample inhomogeneity and/or poorer grain coupling of the former.

Table 7.2 Lattice parameters, T_c and ΔT_c of the samples prepared using the different boron powders mentioned in Table 7.1 [4]

Boron powder	Lattice parameters (Å) *a*	*c*	T_c (K)	ΔT_c (K)
C98	3.0826 ± 0.0007	3.520 ± 0.001	38.8	2.00
C99	3.0820 ± 0.0010	3.524 ± 0.002	37.9	3.00
A9597	3.0789 ± 0.0006	3.521 ± 0.001	38.2	0.75
A9999	3.0822 ± 0.0005	3.525 ± 0.001	38.2	1.00

As shown in Fig. 7.1, the samples prepared using the amorphous boron also show higher J_c. For example, the field-dependent critical current density, $J_c(H)$ at 6 K and 20 K of the best samples prepared using the amorphous boron is at least an order of magnitude larger than that prepared using the crystalline boron. J_c decreased more rapidly with magnetic field for the samples prepared using the lower purity crystalline boron in agreement with the XRD data showing that the samples contained a significant amount of oxide impurities which reduce cross-sectional area for supercurrent.

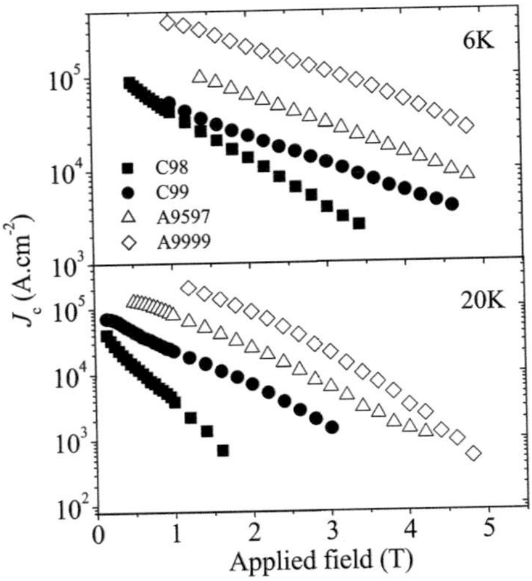

Figure 7.1 Field dependence of J_c at 6 K and 20 K for the samples prepared using different boron powders [4].

The highest J_c in the sample prepared using the highest purity amorphous boron is consistent with previous finding though the reported value of J_c there is lower [10]. The high $J_c(H)$ of the sample is believed to be mainly due to grain boundary pinning as revealed by the SEM images showing its smallest grain size among the samples. The presence of more grain boundaries (as a result of smaller grain size) may be reflected in the high room temperature resistivity values (178 and 225 μΩ·cm measured for 2 pellets) approaching that of the samples prepared using the crystalline borons (275 and 344 μΩ·cm measured for 2 pellets). Since magnesium is a very volatile element at elevated temperature, it is also possible that the higher $J_c(H)$ is partly contributed by the presence of Mg non-stoichiometry in the MgB_2 grains at a level insufficient to cause a degradation of T_c. Accordingly, H_{c2} would be increased by enhanced scattering.

7.2.2 Mixed Boron Precursors

To reduce the use of high-purity boron powders which are unsurprisingly expensive, mixture of impure boron powders (which are cheaper) as a precursor for reaction with magnesium was shown to yield MgB_2 with enhanced J_c compared to the samples prepared using the respective impure boron powders alone [11]. Comparison was made between the two types of mixed boron powders, namely (i) C99 and A9597 and (ii) C99 and A9999 (Table 7.1).

For (i), T_c remained largely unchanged by mixing C99 and A9597 boron powders for reaction with magnesium. $J_c(H)$, at both 6 K and 20 K was enhanced (Fig. 7.2). For example, J_c (6 K, 1.4 T) is 2.3×10^5 A/cm^2 for the mixed 10 wt.% C99 and 90 wt.% A9597 sample compared to 9.8×10^4 A/cm^2 for the sample made using purely A9597 boron and 3.6×10^4 A/cm^2 for the sample made using purely C99 boron. It is clear that $J_c(H)$ is approximately doubled compared to the best single precursor sample (purely A9597). The enhancement of $J_c(H)$ is effective up to 4.6 T at 6 K and 3 T at 20 K.

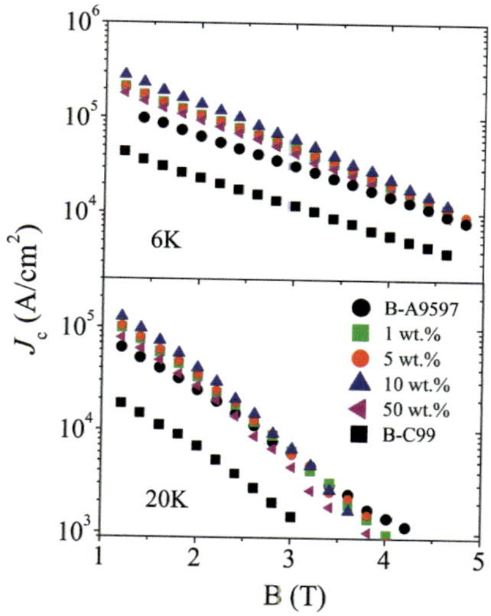

Figure 7.2 J_c versus field at 6 K and 20 K for the samples prepared from mixed boron powders of C99 and A9597. B-A9597 and B-C99 are samples made from the boron precursor of A9597 and C99 alone, respectively [11].

For (ii), no enhancement nor degradation of $J_c(H)$ was observed for up to 10 wt.% C99 mixed with 90 wt.% A9999 as shown in Fig. 7.3. Like the case in (i), T_c did not vary much by mixing C99 with A9999. However, the transition width was broadened as the fraction mixture of C99 was increased. By analysing (i) and (ii), it was argued that the enhancement of $J_c(H)$ of the former is due to its retention of high strain level compared to the latter. Such a strain is mainly a result of bimodal grain structure which retains due to the fact that both C99 and A9597 have bimodal particle size. In addition, the strain is believed to be contributed by the presence of impurities and disorder due to non-uniformity of the reaction rate (the samples contained crystalline and amorphous borons with different chemical reactivity). The scenario for the samples made of purely A9999 is different. Its strain is originated from its fine grains giving rise to high density of grain boundaries. Upon mixing with C99, grain size of the samples increased resulting in reduced grain boundary pinning.

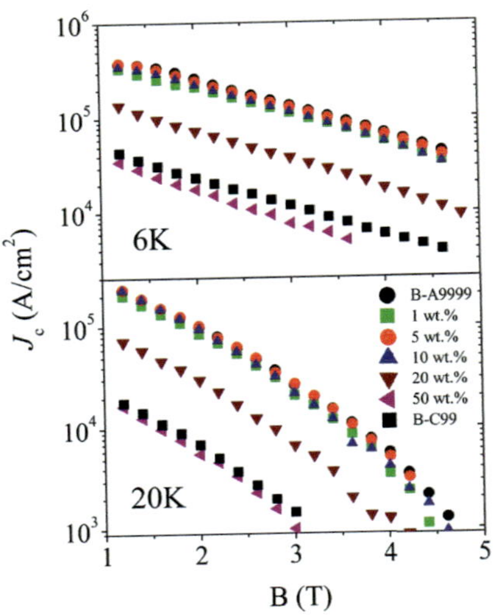

Figure 7.3 J_c versus field at 6 K and 20 K for the samples prepared from mixed boron powders of C99 and A9999. B-A9999 and B-C99 are samples made from the boron precursor of A9999 and C99 alone, respectively [11].

Nonetheless, for the two cases, mixing of excessive C99 leads to degradation in J_c due to the formation of more impurities as shown in the samples prepared using purely C99 [4].

7.2.3 Nominal Magnesium Non-Stoichiometry

By varying the nominal magnesium content, x in Mg_xB_2 (x = 0.95–1.5), the resulting structural changes and their effects on the superconducting properties were demonstrated [12]. The structural changes were evidenced by powder XRD, Raman spectroscopy and temperature dependence of electrical resistivity. The nominal Mg deficient samples ($x < 1.0$) showed enhanced J_c versus field, H_{c2} and H_{irr} compared to the near-stoichiometric samples. However, the true x values in Mg_xB_2 are lower than the nominal mixing ratios due to magnesium loss (high volatility of magnesium at elevated temperature and formation of MgO).

As shown in Table 7.3, both nominal Mg-non-stoichiometric samples of $x = 0.95$ and 1.5 have lower connectivity as indicated by the active area fraction, A_F. While the lower A_F for $x = 0.95$ is likely to be due to the presence of MgB_4, it is unclear what causes such a lower value of A_F in the $x = 1.5$ sample since XRD did not show any peaks of Mg (excess magnesium) and MgB_4. The smaller *RRR* and larger ρ_A(40 K) values indicate more intragrain scattering for the samples with $x \leq 1$. The *RRR* values are within those of the "dirty" samples and also lower than that of the single crystal (~5) [13, 14]. Despite the increase of MgB_4 fraction, the relatively constant values of *RRR* and ρ_A(40 K) for $x < 1$ suggest that the higher boride phase did not contribute to the intragrain scattering.

Table 7.3 T_c and electrical resistivity properties of the samples [12]

x (Mg_xB_2)	T_c [K]	ρ(40 K) [μΩ·cm]	RRR	Active area fraction, A_F	ρ_A(40 K) [μΩ·cm]
0.95	37.5	157.25	2.06	0.04	6.29
0.98	37.5	33.21	2.07	0.19	6.31
1.00	37.5	32.60	1.96	0.23	7.50
1.20	37.5	47.89	2.56	0.10	4.79
1.50	38.0	32.38	3.57	0.09	2.91

In terms of lattice parameters, Mg deficiency led to higher value of c-axis. The values of the c-axis are in the range of 3.5260–3.5277 Å across the samples. These values are actually higher than most others reported in the literature. A typical low c-axis is ~3.5187 Å [15], whereas typical high values are ~3.524 Å [1, 16] and 3.525 Å [17]. In general, high c values have been observed in the samples prepared without excess magnesium in the reaction environment [4, 18].

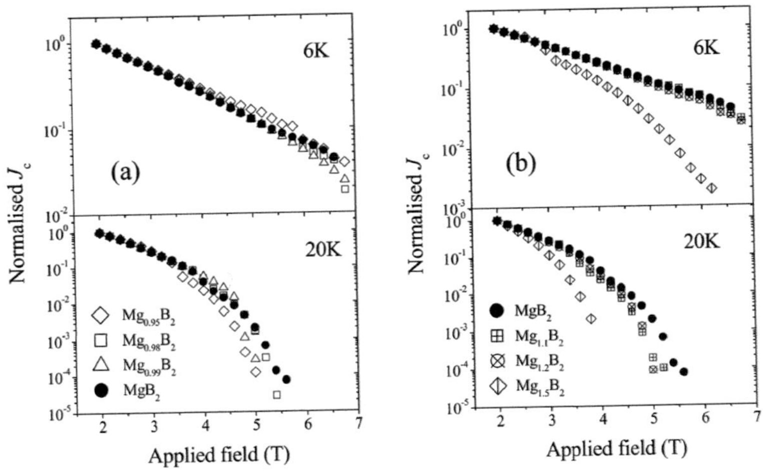

Figure 7.4 Normalized J_c at 6 K and 20 K versus field for (a) $0.95 \leq x \leq 1.0$ and (b) $1.0 \leq x \leq 1.5$ of Mg_xB_2 samples. J_c was normalized to J_c(2 T) [12].

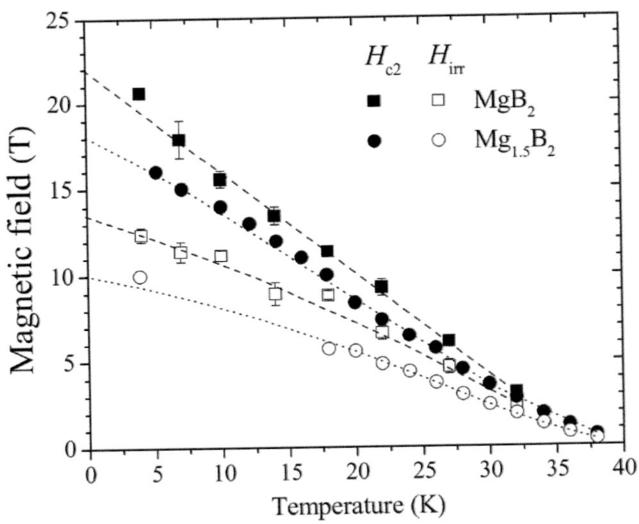

Figure 7.5 Temperature dependence of H_{irr} and H_{c2} for x = 1.0 and 1.5 of Mg_xB_2 samples [12].

Structural change as a result of variation in the nominal Mg content in Mg_xB_2 was also observed through the measured Raman spectra. MgB_2 with its P6mmm space group shows a dominant peak at 600 cm^{-1} which has been ascribed to the E_{2g} mode [19]. In the Raman spectra measured across the samples, two additional shoulders at around 400 cm^{-1} and 730 cm^{-1} were noticeable. Similar finding had also been reported in Al-doped samples and the phenomenon was associated with structural disorder [20, 21]. As induced by the disorder, the appearance of more than one phonon peak indicates the violation of Raman selection rules resulting in a modified Raman spectrum [20, 21]. As x decreased, the intensity of the 730 cm^{-1} peak increased relative to the 600 cm^{-1} peak.

Taken all, the data show the presence of structural disorder as induced by Mg deficiency which in turn enhances dependency of J_c on field (Fig. 7.4) and H_{c2} and H_{irr} (Fig. 7.5).

7.2.4 MgB_4 as Precursor for Reaction with Mg

7.2.4.1 Influence of heat treatment conditions

Instead of the in situ reaction between elemental magnesium and boron powders, MgB_2 can be synthesized via reaction between the higher boride phase of MgB_4 and Mg [22]. In fact, reversibility of the reaction $MgB_4 + Mg \rightarrow MgB_2$ has been proven [23, 24]. By such method, it was reported that the MgB_2 samples showed denser microstructure with high J_c and H_{c2} [23-25]. It is possible to synthesize MgB_2 by reaction of MgB_4 and Mg using a much reduced sintering time, for instance, 4 to 8 h [22] compared to 5 days at 750 °C [26]. To prepare MgB_4 in argon gas flow atmosphere, the challenge is to reduce MgO phase. Through acid leaching on the synthesized MgB_4 powders, weight fraction of MgO was reduced from 16.5 ± 0.5% to 7.0 ± 0.9% compared to ~28% as reported previously by Yan et al. [24].

Figures 7.6 and 7.7 show weight fraction (%) of the phases formed via reaction between MgB_4 and Mg in the temperature range of 650 °C–950 °C for 4 and 8 h, respectively. The weight fraction was estimated based on the XRD data using the software X'pert HighScore Plus. MgB_2, MgB_4, MgO, and Mg are the phases which were detected across the samples. For the samples sintered for 4 and 8 h, MgB_2 phase with the highest weight fraction of about 54% was obtained at 750 °C above which it decreased drastically. In addition, broadening in the peaks was observed suggesting degraded crystallinity of the MgB_2 with the increasing sintering temperature. For MgB_4 phase, its weight fraction increased with sintering temperature for both sets of samples sintered for 4 and 8 h. This can be understood by rapid vaporization of magnesium at higher sintering temperature leading to reduced rate of conversion from MgB_4 to MgB_2. As a result, some MgB_4 powder remained unreacted. Moreover, decomposition of MgB_2 into MgB_4 and Mg at high sintering temperature increased the amount of the MgB_4 phase [14]. Transformation of MgB_2 into MgB_4 is possible between 610 °C and 862 °C [27]. For MgO, the heat treatment condition did not have a very pronounced effect on its weight fraction. The weight fraction of MgO is around 14–21% and 15–19% for the samples sintered for 4 and 8 h, respectively.

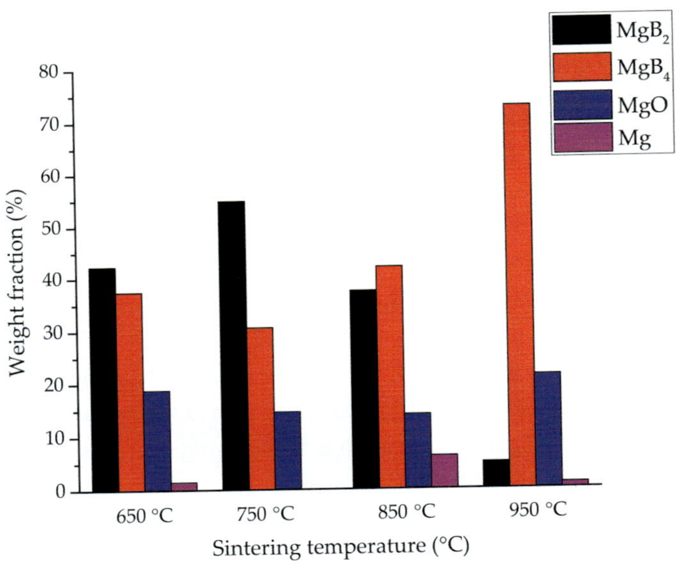

Figure 7.6 Weight fractions of phases formed for the samples with 4 h of sintering [22].

Variation of lattice parameters for the samples sintered for 4 and 8 h are shown in Figs. 7.8 and 7.9, respectively. The *a*-axis decreased with sintering temperature. However, no specific trend of evolution in the *c*-axis with sintering temperature was observed.

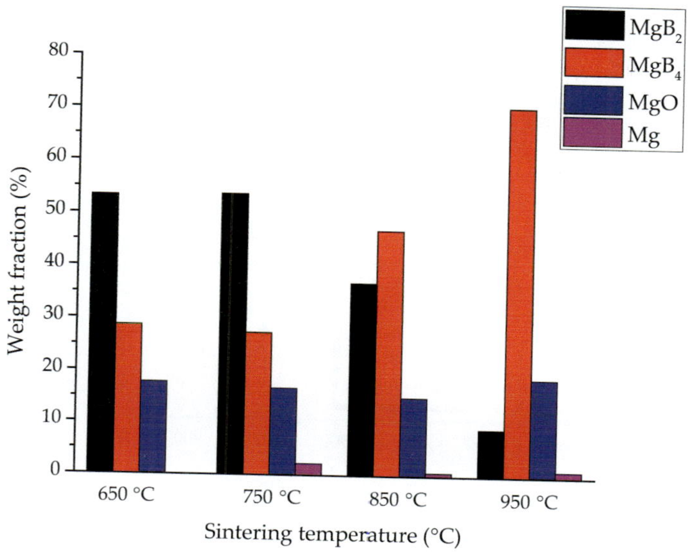

Figure 7.7 Weight fractions of phases formed for the samples with 8 h of sintering [22].

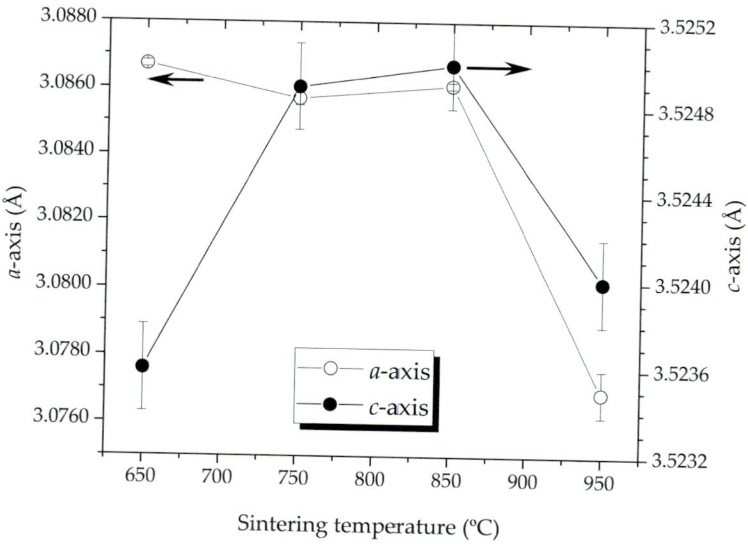

Figure 7.8 Lattice parameters for the samples with 4 h of sintering [22]. Lines are guide for eyes.

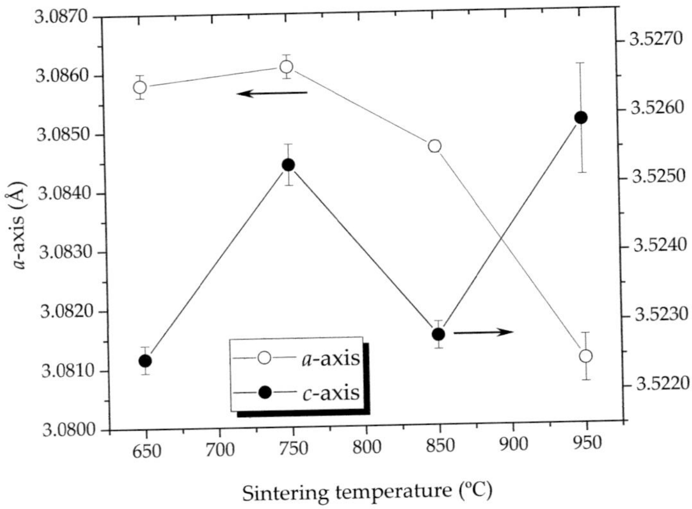

Figure 7.9 Lattice parameters for the samples with 8 h of sintering [22]. Lines are guide for eyes.

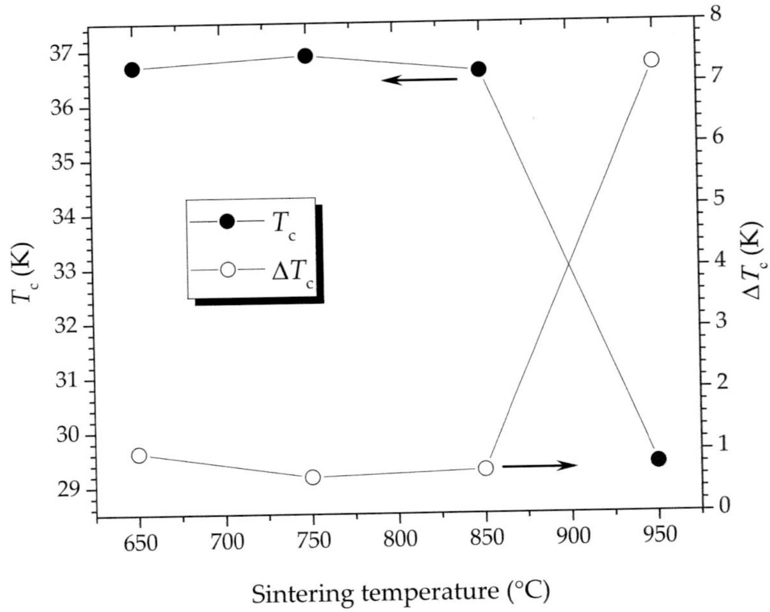

Figure 7.10 T_c and ΔT_c for the samples with 4 h of sintering [22]. Lines are guide for eyes.

Density of the samples was calculated according to the weight and volume of the pellet samples. The density values are in the range of 1.22 g/cm^3 to 1.41 g/cm^3. This value is slightly lower compared with the values of 1.36 g/cm^3 to 1.67 g/cm^3 and 1.2 g/cm^3 to 1.7 g/cm^3 as obtained by Tan et al.[23] and Yan et al. [24], respectively. The reported density for the pure MgB_2 prepared using in situ reaction of magnesium and boron is around 1.2 g/cm^3 [28].

Values of T_c for the samples sintered for 4 and 8 h are shown in Figs. 7.10 and 7.11, respectively. T_c was found to decrease upon increasing the sintering temperature above 850 °C. The decrease in T_c was associated with lattice distortion and the presence of small weight fraction of MgB_2 phase (Figs. 7.6 and 7.7). All the samples sintered at or below 850 °C showed a sharp T_c transition (ΔT_c). A larger ΔT_c in the samples sintered at 950 °C indicated that high sintering temperature caused more severe sample inhomogeneity.

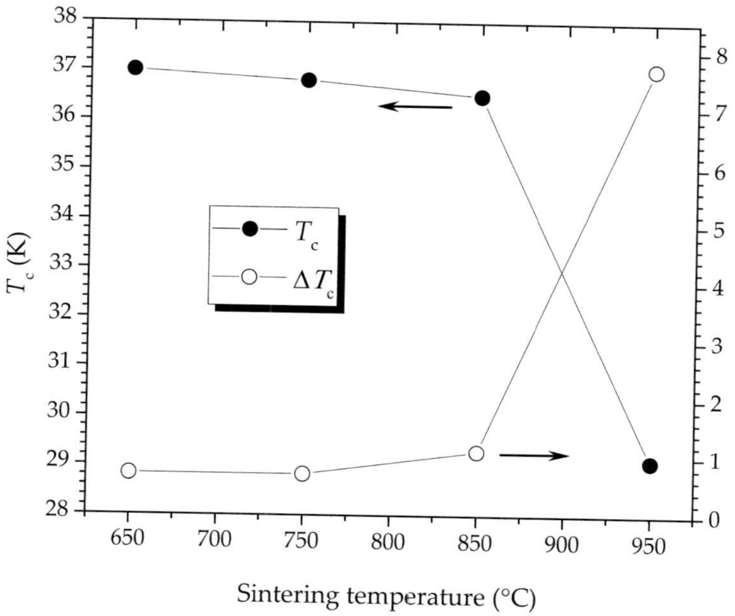

Figure 7.11 T_c and ΔT_c for the samples with 8 h of sintering [22]. Lines are guide for eyes.

Figure 7.12 shows J_c at 5 K and 1 T for the samples sintered from 650 °C to 950 °C for 4 and 8 h. In general, the behaviour of the J_c

can be explained based on the weight fraction of MgB_2 (Figs. 7.6 and 7.7). Cross-sectional areas for supercurrent is likely to be higher for the samples with larger fraction of superconducting MgB_2 phase. As shown in Figs. 7.6 and 7.7, samples sintered at 650 °C and 750 °C (4 and 8 h) have higher weight fraction of MgB_2. Hence, J_c is higher for these 4 samples as expected (Fig. 7.12). However, the sample sintered at 650 °C for 4 h had J_c which is comparable to others yet its weight fraction of MgB_2 (42%) is the lowest among the four samples. This suggests that the pinning strength in this sample is enhanced, possibly because of poorer crystallinity as it was sintered at lower temperature and for shorter time. As the weight fraction of MgB_2 decreased with the sintering temperature from 850 °C to 950 °C, J_c decreased accordingly. J_c at 5 K and 1 T for the sample sintered at 750 °C for 4 h was approximately 1.0×10^5 A/cm^2.

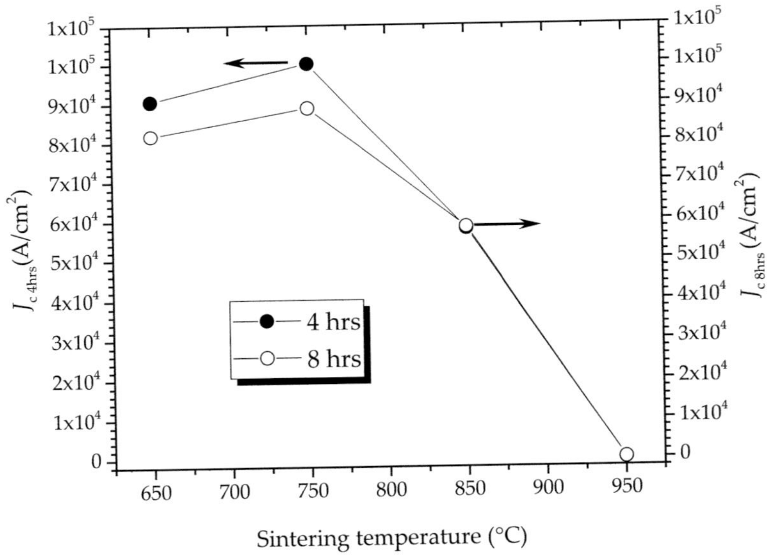

Figure 7.12 J_c at 5 K and 1 T for the samples sintered from 650 °C to 950 °C for 4 (left) and 8 h (right) [22]. Lines are guide for eyes.

7.2.4.2 Influence of nominal Mg content and heat treatment conditions

Influence of nominal Mg content, x on the Mg_xB_2 samples prepared by reaction between MgB_4 and Mg was studied [29]. The nominal

Mg content, x of Mg_xB_2 was varied according to $x = 1.2$, 1.5, and 1.7. The samples were sintered at 650 °C and 750 °C, respectively, for 8 h. As shown in Figs. 7.13 and 7.14, excess Mg ($x > 1.0$) increased the weight fraction of MgB_2 compared with that of the stoichiometric ($x = 1.0$) samples of about 54% [22]. The weight fraction of MgB_2 for the samples sintered at 750 °C is higher compared to that at 650 °C. Thermodynamically, the reaction between Mg and MgB_4 was intensified at 750 °C giving rise to the formation of more MgB_2 phase. This coincides with a decrease in the weight fraction of unreacted Mg (Figs. 7.13 and 7.14). For the sample $Mg_{1.2}B_2$ which has the largest fraction of MgB_2 phase, its MgB_4 weight fraction is also the lowest among others. It was noticed the weight fraction of MgB_2 for both sets of samples decreased for $x > 1.2$. As the weight fraction of MgB_2 increased with the increase of sintering temperature from 650 °C to 750 °C, this indicated reduction in the availability of Mg for oxidation. Hence, weight fraction of MgO for the samples sintered at 750 °C is lower.

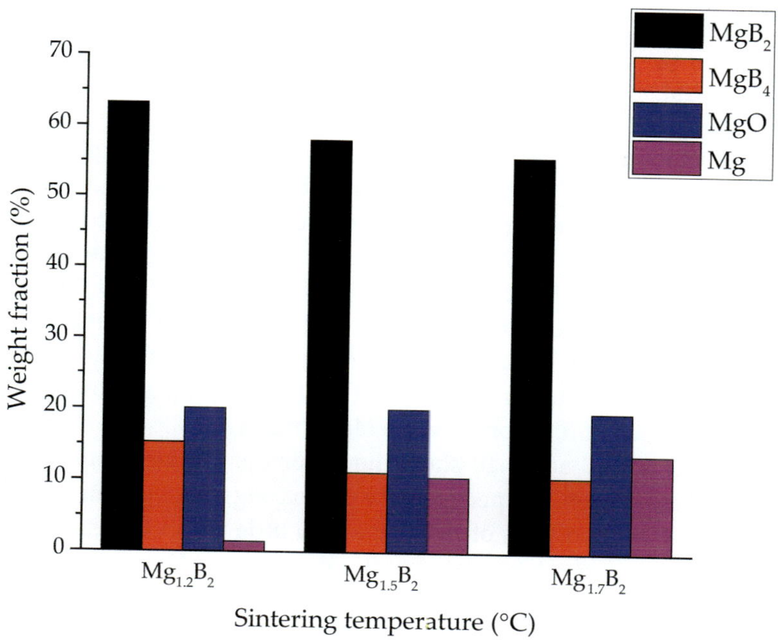

Figure 7.13 Weight fractions of phases formed for the samples sintered at 650 °C [29].

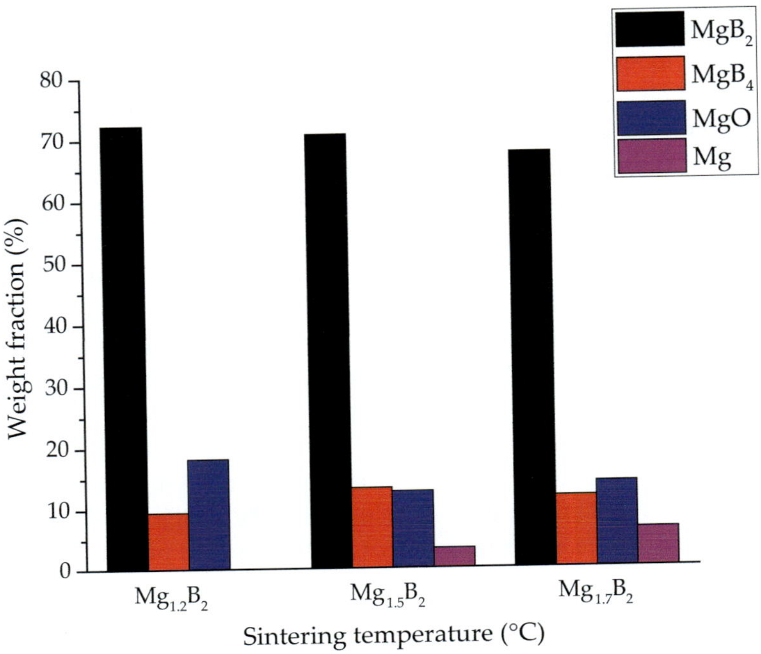

Figure 7.14 Weight fractions of phases formed for the samples sintered at 750 °C [29].

Evolution of lattice parameters with sintering temperature of 650 °C and 750 °C is shown in Figs. 7.15 and 7.16, respectively. The *c*-axis for both sets of samples decreased with *x* in agreement with previous report for Mg excess samples [12]. For the *a*-axis, it decreased and increased slightly with *x* at 650 °C and 750 °C, respectively. The density of the samples (calculated as the mass per unit volume) is in the range of 1.43–1.46 g cm^{-3}. This value is also higher than the stoichiometric samples ($x = 1$) prepared using the same method of reaction between MgB_4 and Mg [22].

Figures 7.17 and 7.18 show the T_c for the samples sintered at 650 °C and 750 °C, respectively. By comparing the two figures, it is noticed that the values of T_c change very little with *x* and sintering temperature except for the sample $Mg_{1.7}B_2$. For $Mg_{1.7}B_2$ sintered at 650 °C, it has the lowest T_c (27.5 K) most likely due to lattice distortion caused by the unreacted Mg [30]. Moreover, a large ΔT_c (11.5 K) suggests a more severe phase inhomogeneity [31] in that sample possibly caused by the large amount of unreacted Mg (weight fraction 14% (Fig. 7.13)).

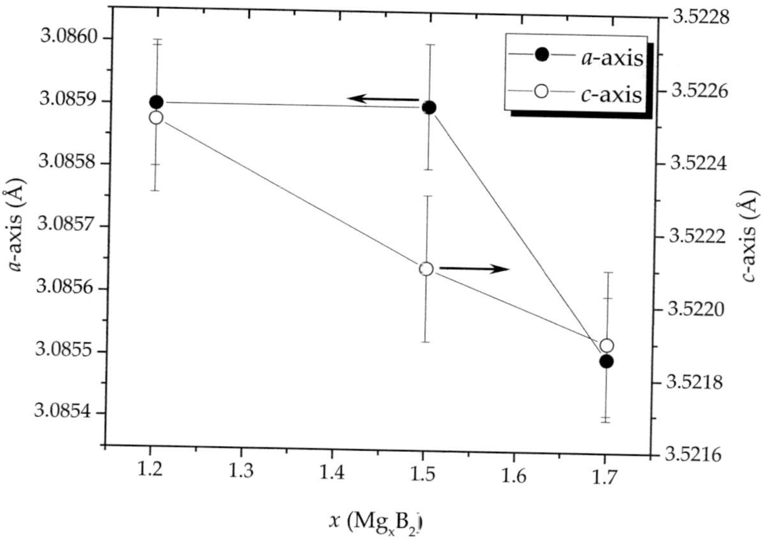

Figure 7.15 Lattice parameters for the samples sintered at 650 °C [29]. Lines are guide for eyes.

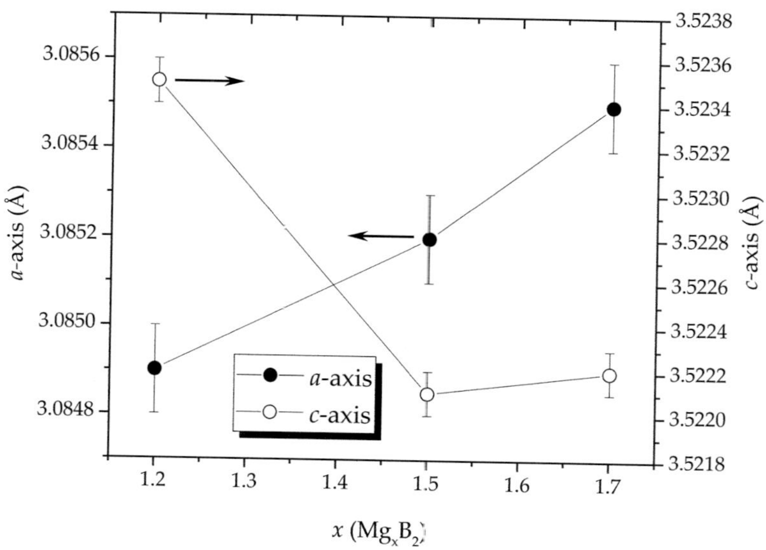

Figure 7.16 Lattice parameters for the samples sintered at 750 °C [29]. Lines are guide for eyes.

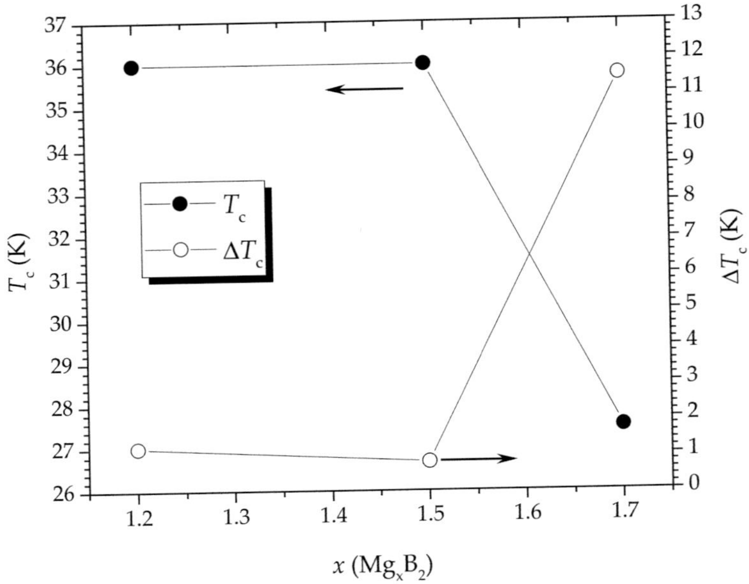

Figure 7.17 T_c and ΔT_c for the samples sintered at 650 °C [29]. Lines are guide for eyes.

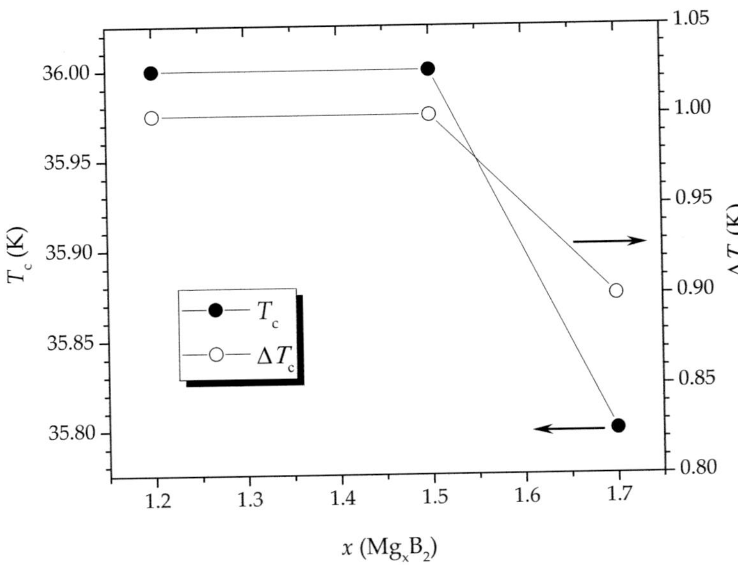

Figure 7.18 T_c and ΔT_c for the samples sintered at 750 °C [29]. Lines are guide for eyes.

Figure 7.19 shows the field dependence of J_c at 5 K for the samples sintered at 650 °C and 750 °C. No significant enhancement of J_c was observed for the samples sintered at 650 °C implying the influence of excess nominal Mg content is negligibly small probably due to the relatively low sintering temperature around the melting point of magnesium (~650 °C). However, J_c varied considerably with excess nominal Mg for the samples sintered at 750 °C even though their MgB_2 weight fractions (Fig. 7.14) are close to each other, indicating the presence of different current-carrying limiting factors.

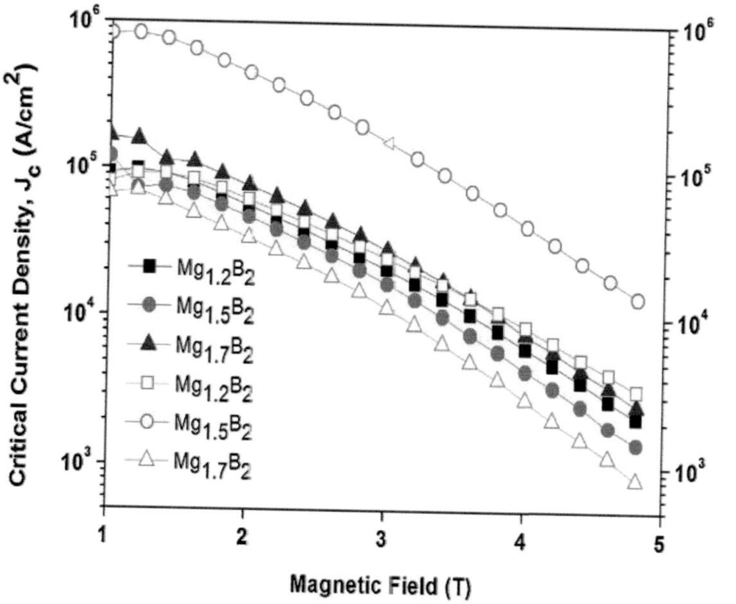

Figure 7.19 J_c versus field at 5 K for the Mg_xB_2 samples sintered at 650 °C (closed symbols) and 750 °C (open symbols) [29].

The sample with $x = 1.5$ sintered at 750 °C shows the highest J_c among others. However, $Mg_{1.5}B_2$ sintered at 650 °C falls within the Mg-rich region [12] as indicated by a more rapid drop of J_c with field and the presence of a large amount of unreacted Mg (10%). The amount of unreacted Mg decreased as the J_c increased tremendously at 750 °C and the sample also has the least amount of MgO (Fig. 7.14). As shown in Fig. 7.19, the form of dependency of J_c on the field remained unchanged, suggesting that the increased magnitude of J_c is a result of combined effect of improved grain connectivity due to excess Mg [32] and enhanced flux pinning by unreacted Mg [30]. At 5 K, 2 T, J_c is increased by almost one order of magnitude to 4.5×10^5 A/cm^2.

7.3 Addition with Rare Earth Oxides

To increase the critical parameters of MgB_2, various oxide phases have been attempted as additives for in situ reaction with magnesium and boron [33]. Among the oxides, rare-earth (RE) oxides have been reported to enhance the electromagnetic properties of MgB_2 greatly [34, 35]. Most of the RE oxides were found to show insignificant substitution in either magnesium or boron site. This may due to the large difference between the ionic size of RE and magnesium giving rise to limited solubility of RE in MgB_2 [34, 35]. As a result, T_c of MgB_2 reacted with RE oxides does not degrade much as compared with carbon substitution at boron site [36]. For instance, T_c was decreased by 1 K only with 15 wt.% of Y_2O_3 addition [37] and insignificant change of T_c for 10 wt.% Tb_4O_7 addition [38]. The formation of nano precipitates is essential for charge carrier scattering and to serve as flux pinning centres for increasing H_{c2} and high field J_c. These precipitates can be MgO, Mg-B-O, RE-Mg/O, RE-B/O, and unreacted RE oxides. Among the reaction products, formation of REB_4 and REB_6 is common [34]. To pin the magnetic flux lines effectively from moving, it is desirable to have the size of the precipitates to be close to the coherent length of MgB_2. Nevertheless, addition of excessive RE oxides may degrade J_c due to the distribution of the precipitates at the grain boundaries thus reduces the effective area for supercurrent path.

Addition of Dy_2O_3 into mixture of magnesium and boron during heat treatment allows dysprosium to react with boron forming DyB_4 [39]. Other impurities are MgO and unreacted Dy_2O_3. The latter was evidenced from the XRD patterns when the additive was excessive. Preventing formation of MgO is rather difficult due to the presence of oxygen in Dy_2O_3 and during the sample preparation process. Fortunately, MgO itself is also one of the flux pinners [40, 41]. Based on the SEM imaging, grain size of the samples was refined with addition of Dy_2O_3 [39]. The decreasing trend of room temperature resistivity suggests improved grain coupling as a result of Dy_2O_3

addition. T_c was not affected much by addition of Dy_2O_3. The zero-field-cool magnetization versus temperature measurement using the SQUID magnetometer showed that the T_c was decreased by 0.5 K only with 5 wt.% of Dy_2O_3 addition. Such small decrease in T_c is consistent with the insignificant change of lattice parameters (Fig. 7.20). For the sample with 0.5 wt.% addition of Dy_2O_3, J_c (1 T) at 6 K and 20 K is 6.5×10^5 and 3.5×10^5, respectively (Fig. 7.21). These values are higher by a factor of more than four compared to that of the pure sample. Such enormous enhancement of J_c is attributed to the distribution of nano precipitates of DyB_4 and MgO within the grains acting as effective flux pinning centres as shown by high-resolution TEM imaging. Majority of these precipitates are ~10 nm or smaller while some with the size of ~70 nm were also observed. It was reported that the irreversibility field, H_{irr} at 20 K (as defined at the field where J_c (20 K) drops to 100 Acm^{-2}) was not improved with Dy_2O_3 addition compared to that of the pure sample [34].

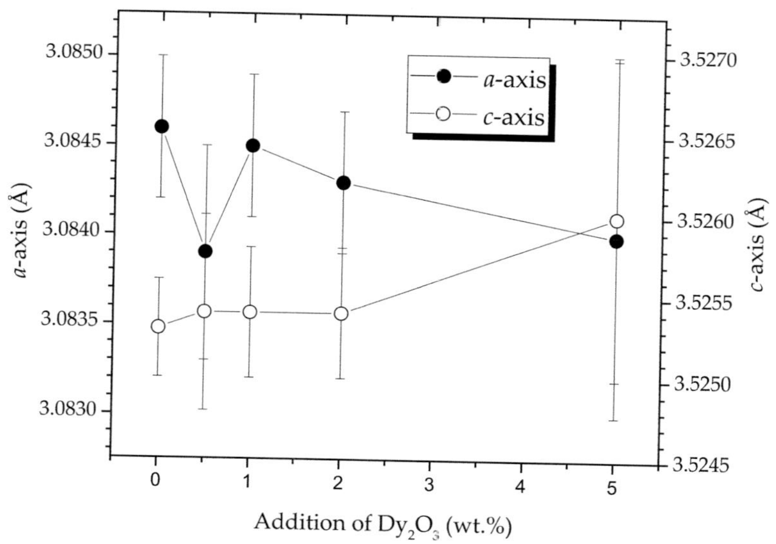

Figure 7.20 Lattice parameters for the samples added with different weight percentages of Dy_2O_3 [39]. Lines are guide for eyes.

Recently, the positive role of Dy_2O_3 as additive on the superconducting properties of MgB_2 was ascertained again [42, 43]. For the MgB_2 samples prepared using the high temperatures and under pressure (HTP) processing technique [42], the addition of Dy_2O_3 did not change the lattice parameters and T_c. The observed nano precipitates which were formed within the grains enhanced the flux pinning properties. Dy_2O_3 was also found to favour the formation of MgB_2 phase for the samples prepared using the modified internal Mg diffusion (MIMD) process with the use of carbon coated boron powder [43]. Addition of 2 wt.% of Dy_2O_3 enhanced J_c at 4.2 K (≤7 T) and 20 K (≤4 T).

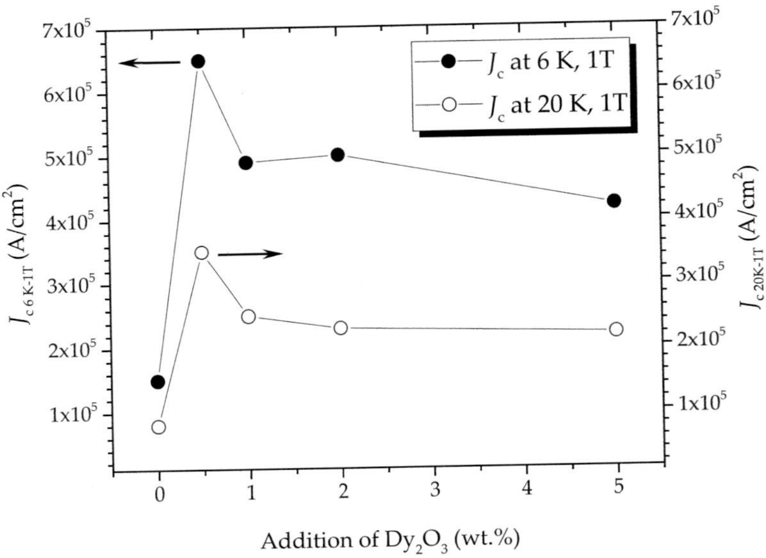

Figure 7.21 J_c (1 T) at 6 K and 20 K for the samples added with different weight percentages of Dy_2O_3 [39]. Lines are guide for eyes.

7.4 Large-Scale Applications

Progress in fabrication of wires and tapes based on the powder-in-tube (PIT) method has benefited tremendously from the development of the powder technology of MgB_2. In situ and ex situ PIT are the most common techniques in preparing mono and multifilamentary conductors [44, 45]. MgB_2 can be considered as an alternative material to the conventional Nb-based low-T_c superconductors. It is anticipated that the price competitiveness of the former will overtake the latter in the near future. In general, application of MgB_2 conductors can be divided into two categories namely (i) low magnetic field in cryogen-free environment around 20 K and (ii) high magnetic field at liquid helium temperature [46]. Commercial applications of Nb-based materials are NbTi at low field like MRI instruments and Nb_3Sn at high field [45]. Some commercial applications in the temperature range 4–30 K are possible with the price performance of MgB_2 wires [47]. At the moment, the primary competitors to MgB_2 in 4–30 K are BSCCO and YBCO [45]. Instead of relying on the expensive liquid helium, cooling using cryogen free system is highly desirable for MgB_2. Unlike the HTSCs, cooling can be done using the much cheaper liquid nitrogen.

To date, kilometres length of MgB_2 wires have been fabricated successfully [47, 48]. Applications of MgB_2 wires are mainly within the large-scale range including magnetic resonance imaging (MRI), fault current limiter (FCL), solenoid, transformers, motors and generators, magnetic separation, adiabatic demagnetization refrigerators (ADR), superconducting magnetic energy storage (SMES), MagLev trains, and usage in high energy physics (HEP) [46, 47].

For magnets made of MgB_2 conductors, their potential applications are klystron [49], particle accelerator in HEP [50], tokamak magnet for the International Thermonuclear Experimental Reactor (ITER) Project [51, 52], MRI, nuclear magnetic resonance (NMR) system, and superconducting quantum interference devices (SQUIDs). The ultra-lightweight MgB_2 wires are especially useful for space application [53]. In this regard, magnet made of Ti-MgB_2 tape was designed and fabricated as magnetic shield to protect astronauts from cosmic radiation during long space missions [54]. The project was supported under the framework of the European Space Radiation Superconducting Shielding (SR2S).

To date, some applications of MgB_2 have been realized successfully. In 2006, ASG Superconductors, Columbus Superconductors, and Paramed designed and commissioned the first open MRI system with cryogen-free superconducting MgB_2 magnets generating a central field of 0.5 T in a free patient gap of about 60 cm [48]. The magnets are made of multifilamentary ex situ MgB_2 tapes. The MRI system operating at temperatures of 15–20 K has the advantage of a lower operational cost and an enhanced stability compared to Nb-Ti. The cost of the system is also lower in terms of materials compared to the use of HTSCs. For power transmission, 30 m hybrid energy transfer line with MgB_2 superconducting cable operating at liquid hydrogen was demonstrated [55]. At 21 K, critical current of the cable reached about 3500 A. The developed 30 m hybrid energy system is capable of delivering up to 135 MW of chemical and electrical power in total. Another achievement along this line is the successful commissioning of a helium gas-cooled transfer line incorporating 2×20 m long cables operating at a record DC current of up to 20 kA at 24 K. The cables are made of superconducting MgB_2 round wire [48]. Such electrical system was developed under the framework of The High Luminosity LHC (HL-LHC) Project at CERN.

For MgB_2 bulks, a trapped magnetic field of 3 T at 5 K was demonstrated [56]. The MgB_2 bulks, fabricated through the modified precursor infiltration and growth process, were reported to show homogenous microstructure and its density reaches 2.4 g/cm^3 [57]. By using the similar approach, carbon doped MgB_2 bulks were shown to exhibit a maximum trapped field of 4.5 T at 7.5 K [58]. Another development which is worth mentioning is the fabrication of MgB_2 cup via magnesium infiltration method for magnetic shielding [59]. This is crucial for detecting very small or weak signal for highly sensitive system. One such example is detection of weak magnetic field using SQUID magnetometer.

7.5 Concluding Remarks

The progress in the powder technology is clearly an innovation to ensure long-term sustainable development in the industry of MgB_2 (UN SDG 9). With such advancement, the electromagnetic performance of MgB_2 will be further enhanced. Concurrently, improvement in the material synthesis method will lead to cheaper cost of production for MgB_2 powder. Cheaper cost of material can be translated into cheaper and more affordable energy especially for power transmission lines (UN SDG 7). Like any other superconductors, MgB_2 is a material for green technology hence clean energy due to its zero DC electrical resistance below the T_c (UN SDG 7). Accordingly, reduction in CO_2 emission is expected with the use of MgB_2 superconductor for the related applications (UN SDG 13).

Acknowledgements

S. K. Chen acknowledges the financial support of the Ministry of Education Malaysia under the Fundamental Research Scheme Grant (FRGS/1/2018/STG07/UPM/02/1).

References

1. Nagamatsu, J., Nakagawa, N., Muranaka, T., Zenitani, Y., and Akimitsu, J. (2001). Superconductivity at 39 K in magnesium diboride, *Nature*, **410**, pp. 63–64.
2. Larbalestier, D. C., Cooley, L. D., Rikel, M. O., Polyanskii, A. A., Jiang, J., Patnaik, S., Cai, X. Y., Feldmann, D. M., Gurevich, A., Squitieri, A. A., Naus, M. T., Eom, C. B., Hellstrom, E. E., Cava, R. J., Regan, K. A., Rogado, N., Hayward, M. A., He, T., Slusky, J. S., Khalifah, P., Inumaru, K., and Haas, M. (2001). Strongly linked current flow in polycrystalline forms of the superconductor MgB_2, *Nature*, **410**, pp. 186–189.
3. Yamamoto, A., Tanaka, H., Shimoyama, J.-I., Ogino, H., Kishio, K., and Matsushita, T. (2011). Towards the realization of higher connectivity in MgB_2 conductors: In-situ or sintered ex-situ?, *Jpn. J. Appl. Phys.*, **51**, pp. 010105.
4. Chen, S. K., Yates, K. A., Blamire, M. G., and MacManus-Driscoll, J. L. (2005). Strong influence of boron precursor powder on the critical current density of MgB_2, *Supercond. Sci. Technol.*, **18**, pp. 1473–1477.
5. Zhou, J., and Bai, P. (2015). A review on the methods of preparation of elemental boron, *Asia-Pac. J. Chem. Eng.*, **10**, pp. 325–338.
6. Zhigach, A. F., and Stasinevich, D. C. (1977) *Boron and Refractory Borides*, eds. Matkovich, V. I., "Methods of preparation of amorphous boron," (Springer, Berlin, Heidelberg) pp. 214–226.
7. Seifolazadeh, A., and Mohammadi, S. (2016). Synthesis and characterization of nanoboron powders prepared with mechanochemical reaction between B_2O_3 and Mg powders, *Bull. Mater. Sci.*, **39**, pp. 479–486.
8. Vignolo, M., Romano, G., Martinelli, A., Bernini, C., and Siri, A. S. (2012). A novel process to produce amorphous nanosized boron useful for MgB_2 synthesis, *IEEE Trans. Appl. Supercond.*, **22**, pp. 6200606–6200606.
9. Ribeiro, R. A., Bud'ko, S. L., Petrovic, C., and Canfield, P. C. (2003). Effects of boron purity, Mg stoichiometry and carbon substitution on properties of polycrystalline MgB_2, *Phys. C*, **385**, pp. 16–23.

10. Zhou, S., Pan, A. V., Horvat, J., Qin, M. J., and Liu, H. K. (2004). Effects of precursor powders and sintering processes on the superconducting properties of MgB_2, *Supercond. Sci. Technol.*, **17**, pp. S528–S532.
11. Chen, S. K., and MacManus-Driscoll, J. L. (2009). Enhanced critical current densities in MgB_2 by mixing relatively impure boron powders, *Supercond. Sci. Technol.*, **22**, pp. 065009.
12. Chen, S. K., Serquis, A., Serrano, G., Yates, K. A., Blamire, M. G., Guthrie, D., Cooper, J., Wang, H., Margadonna, S., and MacManus-Driscoll, J. L. (2008). Structural and superconducting property variations with nominal Mg non-stoichiometry in Mg_xB_2 and its enhancement of upper critical field, *Adv. Funct. Mater.*, **18**, pp. 113–120.
13. Eltsev, Y., Lee, S., Nakao, K., Chikumoto, N., Tajima, S., Koshizuka, N., and Murakami, M. (2002). Anisotropic superconducting properties of MgB_2 single crystals probed by in-plane electrical transport measurements, *Phys. Rev. B*, **65**, pp. 140501.
14. Masui, T., Yoshida, K., Lee, S., Yamamoto, A., and Tajima, S. (2002). Phonon contribution to the transport properties of single-crystal MgB_2, *Phys. Rev. B*, **65**, pp. 214513.
15. Kazakov, S. M., Puzniak, R., Rogacki, K., Mironov, A. V., Zhigadlo, N. D., Jun, J., Soltmann, C., Batlogg, B., and Karpinski, J. (2005). Carbon substitution in MgB_2 single crystals: Structural and superconducting properties, *Phys. Rev. B*, **71**, pp. 024533.
16. Mickelson, W., Cumings, J., Han, W. Q., and Zettl, A. (2002). Effects of carbon doping on superconductivity in magnesium diboride, *Phys. Rev. B*, **65**, pp. 052505.
17. Tarantini, C., Braccini, V., Ferdeghini, C., Gatti, F., Manfrinetti, P., Marre, D., Palenzona, A., Pallecchi, I., Bernini, C., Tumino, A., Sheikin, I., Aebersold, H. U., Lehmann, E., and Putti, M. (2005). Critical field of magnesium diboride in substituted and irradiated samples, *IEEE Trans. Appl. Supercond.*, **15**, pp. 3223–3226.
18. Xiao, H., Peng, W., Song, W. H., Ma, R. C., Zhang, L., Du, J. J., and Sun, Y. P. (2003). Influence of Mg deficiency on the properties of MgB_2, *Phys. C*, **386**, pp. 648–652.
19. Kunc, K., Loa, I., Syassen, K., Kremer, R. K., and Ahn, K. (2001). MgB_2 under pressure: phonon calculations, Raman spectroscopy, and optical reflectance, *J. Phys.: Condens. Matter*, **13**, pp. 9945–9962.
20. Di Castro, D., Cappelluti, E., Lavagnini, M., Sacchetti, A., Palenzona, A., Putti, M., and Postorino, P. (2006). Raman spectra of neutron-irradiated and Al-doped MgB_2, *Phys. Rev. B*, **74**, pp. 100505.

21. Monni, M., Ferdeghini, C., Putti, M., Manfrinetti, P., Palenzona, A., Affronte, M., Postorino, P., Lavagnini, M., Sacchetti, A., Di Castro, D., Sacchetti, F., Petrillo, C., and Orecchini, A. (2006). Role of charge doping and lattice distortions in codoped $Mg_{1-x}(AlLi)_xB_2$ compounds, *Phys. Rev. B*, **73**, pp. 214508.

22. Tan, K. L., Tan, K. Y., Lim, K. P., Shaari, A. H., and Chen, S. K. (2012). Optimization of phase formation and superconducting properties in MgB_2 prepared by phase transformation from MgB_4, *J. Electron. Mater.*, **41**, pp. 673–678.

23. Tan, K. S., Jun, B.-H., and Kim, C.-J. (2009). Superconducting properties of a MgB_2 bulk formed by using a MgB_4 + Mg mixture, *J. Korean Phys. Soc.*, **54**, pp. 1626–1629.

24. Yan, S. C., Yan, G., Liu, C. F., Lu, Y. F., and Zhou, L. (2007). Experimental study on phase transformation between MgB_2 and MgB_4, *J. Am. Ceram. Soc.*, **90**, pp. 2184–2188.

25. Nardelli, D., Matera, D., Vignolo, M., Bovone, G., Palenzona, A., Siri, A. S., and Grasso, G. (2013). Large critical current density in MgB_2 wire using MgB_4 as precursor, *Supercond. Sci. Technol.*, **26**, pp. 075010.

26. Schmitt, R., Glaser, J., Wenzel, T., Nickel, K. G., and Meyer, H. J. (2006). A reactivity study in the Mg–B system reaching for an improved synthesis of pure MgB_2, *Phys. C*, **436**, pp. 38–42.

27. Brutti, S., Ciccioli, A., Balducci, G., Gigli, G., Manfrinetti, P., and Palenzona, A. (2002). Vaporization thermodynamics of MgB_2 and MgB_4, *Appl. Phys. Lett.*, **80**, pp. 2892–2894.

28. Dou, S. X., Soltanian, S., Horvat, J., Wang, X. L., Zhou, S. H., Ionescu, M., Liu, H. K., Munroe, P., and Tomsic, M. (2002). Enhancement of the critical current density and flux pinning of MgB_2 superconductor by nanoparticle SiC doping, *Appl. Phys. Lett.*, **81**, pp. 3419–3421.

29. Tan, K. L., Lim, K. P., Halim, A. S., and Chen, S. K. (2013). Enhanced critical current density in MgB_2 prepared by reaction of MgB_4 and Mg, *Phys. Status Solidi A*, **210**, pp. 616–622.

30. Sinha, B. B., Kadam, M. B., Mudgel, M., Awana, V. P. S., Kishan, H., and Pawar, S. H. (2010). Synthesis and characterization of excess magnesium MgB_2 superconductor under inert carbon environment, *Phys. C*, **470**, pp. 25–30.

31. Wenzel, T., Nickel, K. G., Glaser, J., Meyer, H.-J., Eyidi, D., and Eibl, O. (2003). Electron probe microanalysis of Mg–B compounds: stoichiometry and heterogeneity of superconductors, *Phys. Status Solidi A*, **198**, pp. 374–386.

32. Zeng, R., Lu, L., Wang, J. L., Horvat, J., Li, W. X., Shi, D. Q., Dou, S. X., Tomsic, M., and Rindfleisch, M. (2007). Significant improvement in the critical current density of in situ MgB_2 by excess Mg addition, *Supercond. Sci. Technol.*, **20**, pp. L43–L47.

33. Chen, S. K., Maeda, M., Yamamoto, A., and Dou, S. X. (2017) *Vortices and Nanostructured Superconductors*, eds. Crisan, A, Chapter 3 "Chemically and mechanically engineered flux pinning for enhanced electromagnetic properties of MgB_2," (Springer, Cham) pp. 65–108.

34. Katsura, Y., Shimoyama, J., Yamamoto, A., Horii, S., and Kishio, K. (2007). Effects of rare earth doping on the superconducting properties of MgB_2, *Phys. C*, **463–465**, pp. 225–228.

35. Pan, X. F., Cheng, C. H., and Zhao, Y. (2011). Effect of rare-earth oxides doping on the superconductivity and flux pinning of MgB_2 superconductor, *J. Supercond. Novel Magn.*, **24**, pp. 1611–1616.

36. Yeoh, W. K., and Dou, S. (2007). Enhancement of H_{c2} and J_c by carbon-based chemical doping, *Phys. C*, **456**, pp. 170–179.

37. Wang, J., Bugoslavsky, Y., Berenov, A., Cowey, L., Caplin, A. D., Cohen, L. F., MacManus Driscoll, J. L., Cooley, L. D., Song, X., and Larbalestier, D. C. (2002). High critical current density and improved irreversibility field in bulk MgB_2 made by a scaleable, nanoparticle addition route, *Appl. Phys. Lett.*, **81**, pp. 2026–2028.

38. Vinod, K., Neson, V., Syamaprasad, U., Shipra and Sundaresan, A. (2008). Structural and superconducting properties of bulk MgB_2 with added nano Tb_4O_7, *Supercond. Sci. Technol.*, **21**, pp. 025003.

39. Chen, S. K., Wei, M., and MacManus-Driscoll, J. L. (2006). Strong pinning enhancement in MgB_2 using very small Dy_2O_3 additions, *Appl. Phys. Lett.*, **88**, pp. 192512.

40. Kováč, P., Hušek, I., Melišek, T., Grivel, J. C., Pachla, W., Štrbík, V., Diduszko, R., Homeyer, J., and Andersen, N. H. (2004). The role of MgO content in ex situ MgB_2 wires, *Supercond. Sci. Technol.*, **17**, pp. L41.

41. Jiang, C. H., Hatakeyama, H., and Kumakura, H. (2005). Effect of nanometer MgO addition on the in situ PIT processed MgB_2/Fe tapes, *Phys. C*, **423**, pp. 45–50.

42. Yang, Y., Sumption, M. D., and Collings, E. W. (2016). Influence of metal diboride and Dy_2O_3 additions on microstructure and properties of MgB_2 fabricated at high temperatures and under pressure, *Sci. Rep.*, **6**, pp. 29306.

43. Brunner, B., Rosová, A., Kováč, P., Reissner, M., and Dobročka, E. (2016). Effect of Dy_2O_3 doping on phase formation and properties of MgB_2

wires made by the modified internal magnesium diffusion process, *Supercond. Sci. Technol.*, **30**, pp. 025004.

44. Flükiger, R., Suo, H. L., Musolino, N., Beneduce, C., Toulemonde, P., and Lezza, P. (2003). Superconducting properties of MgB_2 tapes and wires, *Phys. C*, **385**, pp. 286–305.

45. Vinod, K., Kumar, R. G. A., and Syamaprasad, U. (2007). Prospects for MgB_2 superconductors for magnet application, *Supercond. Sci., and Technol.*, **20**, pp. R1–R13.

46. Penco, R., and Grasso, G. (2007). Recent development of MgB_2-based large scale applications, *IEEE Trans. Appl. Supercond.*, **17**, pp. 2291–2294.

47. Tomsic, M., Rindfleisch, M., Yue, J., McFadden, K., Phillips, J., Sumption, M. D., Bhatia, M., Bohnenstiehl, S., and Collings, E. W. (2007). Overview of MgB_2 superconductor applications, *Int. J. Appl. Ceram. Technol.*, **4**, pp. 250–259.

48. Ballarino, A., and Flükiger, R. (2017). Status of MgB_2 wire and cable applications in Europe, *J. Phys.: Conf. Ser.*, **871**, pp. 012098.

49. Tanaka, H., Suzuki, T., Kodama, M., Koga, T., Watanabe, H., Yamamoto, A., and Michizono, S. (2020). Performance of MgB_2 superconductor developed for high-efficiency klystron applications, *IEEE Trans. Appl. Supercond.*, **30**, pp. 1–5.

50. Cooley, L. D., Eom, C. B., Hellstrom, E. E., and Larbalestier, D. C. (2001). Potential application of magnesium diboride for accelerator magnet applications, *Proc. Part. Accel. Conf.*, PACS2001, pp. 203–207.

51. Qiu, W., Jie, H., Patel, D., Lu, Y., Luzin, V., Devred, A., Somer, M., Shahabuddin, M., Kim, J. H., Ma, Z., Dou, S. X., and Hossain, M. S. A. (2016). Improvement in the transport critical current density and microstructure of isotopic $Mg^{11}B_2$ monofilament wires by optimizing the sintering temperature, *Sci. Rep.*, **6**, pp. 36660.

52. Coppi, B., Airoldi, A., Albanese, R., Ambrosino, G., Bombarda, F., Bianchi, A., Cardinali, A., Cenacchi, G., Costa, E., Detragiache, P., De Tommasi, G., DeVellis, A., Faelli, G., Ferrari, A., Frattolillo, A., Frosi, P., Giammanco, F., Grasso, G., Lazzaretti, M., Mantovani, S., Migliori, S., Pierattini, S., Pironti, A., Ramogida, G., Rubinacci, G., Sassi, M., Tavani, M., Tumino, A., and Villone, F. (2013). New developments, plasma physics regimes and issues for the Ignitor experiment, *Nucl. Fusion*, **53**, pp. 104013.

53. Kováč, P., Hušek, I., Rosová, A., Kulich, M., Kováč, J., Melišek, T., Kopera, L., Balog, M., and Krížik, P. (2018). Ultra-lightweight superconducting wire based on Mg, B, Ti and Al, *Sci. Rep.*, **8**, pp. 11229.

54. Musenich, R., Nardelli, D., Brisigotti, S., Pietranera, D., Tropeano, M., Tumino, A., Cubeda, V., Calvelli, V., and Grasso, G. (2016). Ti–MgB_2 conductor for superconducting space magnets, *IEEE Trans. Appl. Supercond.*, **26**, pp. 1–4.

55. Vysotsky, V. S., Antyukhov, I. V., Firsov, V. P., Blagov, E. V., Kostyuk, V. V., Nosov, A. A., Fetisov, S. S., Zanegin, S. Y., Rachuk, V. S., and Katorgin, B. I. (2015). Cryogenic tests of 30 m flexible hybrid energy transfer line with liquid hydrogen and superconducting MgB_2 cable, *Phys. Procedia*, **67**, pp. 189–194.

56. Bhagurkar, A. G., Yamamoto, A., Anguilano, L., Dennis, A. R., Durrell, J. H., Hari Babu, N., and Cardwell, D. A. (2016). A trapped magnetic field of 3 T in homogeneous, bulk MgB_2 superconductors fabricated by a modified precursor infiltration and growth process, *Supercond. Sci. Technol.*, **29**, pp. 035008.

57. Bhagurkar, A. G., Yamamoto, A., Babu, N. H., Durrell, J. H., Dennis, A. R., and Cardwell, D. A. (2014). Synthesis of dense bulk MgB_2 by an infiltration and growth process, *Supercond. Sci. Technol.*, **28**, pp. 015012.

58. Bhagurkar, A. G., Yamamoto, A., Wang, L., Xia, M., Dennis, A. R., Durrell, J. H., Aljohani, T. A., Babu, N. H., and Cardwell, D. A. (2018). High trapped fields in C-doped MgB_2 bulk superconductors fabricated by infiltration and growth process, *Sci. Rep.*, **8**, pp. 13320.

59. Huang, Z., Luo, W., Cai, X., Guo, Z., Pan, X., Xi, D., Wu, Q., Niu, R., Yang, C., Li, Y., Linghu, K., Gan, Z., Nie, R., Feng, Q., and Wang, F. (2019). Magnetic shielding capability of MgB_2 cup made by liquid infiltration process, *IEEE Trans. Appl. Supercond.*, **29**, pp. 1–5.

Chapter 8

Ultrasonication: A Cost-Effective Way to Synthesize High-J_c Bulk MgB_2

Arvapalli Sai Srikanth
Materials for Energy and Environmental Laboratory,
Superconducting Materials Laboratory,
Graduate School of Engineering and Science,
Shibaura Institute of Technology, Koto City, Tokyo 135-8548, Japan
arvapallisaisrikanth5@gmail.com, miryala1@shibaura-it.ac.jp

8.1 Introduction

MgB_2 is a trendy material suitable for several superconductor applications such as bulk magnets for compact magnetic resonance imaging (MRI) and nuclear magnetic resonance (NMR), electric motors, etc. Although T_c is slightly lower (39 K) [1] compared to superconducting cuprates (~90 K), the processing time, easy fabrication, and cost parameters make it attractive. In general, liquid helium cooling is used in MgB_2 based practical applications which is expensive as well as difficult to obtain. However, cryogen-free cryocoolers [2–4], liquid hydrogen [5], and neon are capable

High-T_c Superconducting Technology: Towards Sustainable Development Goals
Edited by Muralidhar Miryala

ISBN 978-981-4877-65-7 (Hardcover), 978-1-003-16468-5 (eBook)
www.jennystanford.com

of reaching the required cooling temperature without expensive and technically complicated use of liquid helium. This promotes a rapid research on MgB_2 superconducting material. The light weight is another factor improving efficiency of the devices and increasing the range of applications [6–11], especially in space. Till today, conventional low-temperature superconductors such as Nb_3Sn and NbTi make up the majority of superconducting market because of good qualities, such as decent critical current density, ductility, strength, easiness to fabricate into long wires, etc. Efforts are being made to replace them with high temperature superconductors (HTS) superconductors and MgB_2 superconductors. A few main reasons for the industrial sector to show interest in MgB_2 are production, light weight and density (2.55 g/cm^3), cost effectiveness (both Mg and B are abundant and cheap), robustness of material, etc. From previous discussion we know that MgB_2 is anisotropic, however is very less when compared to trending HTS. Hence, MgB_2 unlike HTS does not require an extra texturing step in processing. Certain sectors, such as space application, have huge interests because of how light MgB_2 can be. All above-mentioned features and advantages make MgB_2 a strong candidate for making superconductors that can be readily used in real-world applications. It is always advised that for optimum performance the superconductors must be used quite below T_c. In case of MgB_2 the most suitable operating temperature is around 20–30 K which can be achieved easily using present cryocooler technology or liquid hydrogen which can avoid the use of very expensive liquid helium. MgB_2 is the most promising material for high-field magnet application in place of Nb_3Sn and NbTi as the technology for them is materialized and MgB_2 can just fit in. Apart from the magnet applications, such as NMR, MRI, water cleaning, etc., MgB_2 superconductor has potential uses in superconducting transformers, rotors, and transmission cables, etc. One main drawback that hinders this scenario is moderate J_c and trapped field values of MgB_2 bulk.

Several research groups have been trying to improve trapped field and critical current density in MgB_2 bulks. Most of the trials were related to optimization of synthesis parameters [12,13], novel synthesis techniques [14–16], refining the raw precursors, doping, additions [17,18], and fabrication of films [19]. The primary synthesis method for bulk MgB_2 is the solid state sintering. This process is highly scalable and by maintaining uniformity in synthesis parameters, one can obtain uniformity in product's properties, such as elemental distribution, density, J_c, etc. In addition, this sintering method also ensures a uniform trapped field, which is crucial for levitation and super-magnet device fabrication. Some researchers have tried manipulating the sintering temperature, sintering duration, multi-step heating [13,20–24], etc. In our previous research, we have optimized the sintering process such 775–800 °C for 3 h for best performance bulk MgB_2 [12]. One small disadvantage is the need of an inert atmosphere during the sintering, as Mg is highly reactive with oxygen. Anyway, it does not hinder massive production of bulk MgB_2 material. Different techniques and synthesis methods were also practiced such as spark plasma sintering [25–27], diffusion method [16], infiltration growth [28], and chemical routes (combustion, pyrolysis, precipitation, etc.). All these methods have their advantages and disadvantages. One way that can improve superconducting performance is to introduce pinning centers to pin vortex lattice motion and thereby reach a high performance up to high magnetic fields. The coherence length of MgB_2 materials is quite high when compared to other HT superconductors. Hence, defects with bigger sizes can thus act as pinning centers. Some of the effective pinning centers used are nonsuperconducting inclusions, defects, grain boundaries, voids, etc. Such pinning centers can be MgB_4 [29,30], MgO [31,32], metals [33,34], metal oxides [35,36], grain refiners, rare-earth elements [37], dislocations, defects created by irradiation [38], carbon doping [39–46], etc.

It has been proven that increase of grain boundary area results in increase of J_c, in particular at low-magnetic fields. In one of our previous works dealing with MgB_2 we used a commercial nano-amorphous boron [47] to enhance vortex pinning. The results were outstanding, but the powder was expensive, making the whole fabrication process costly. Ball milling is a technique proved to effectively decrease powder size, however is not scalable and has problems like poor T_c, B_2O_3 formation and contamination by other impurities [48,49]. Here, we utilized "high-energy ultrasonication," a novel technique that is scalable and results in uniformly refined particles [50,51]. An important advantage of this method is that the process can be controlled and optimized via tuning frequency and power. However, this process is new, the interaction between dispersant and dispersion is crucial. Optimum conditions for the best results need to be established. The present work tries to elucidate on this cheap ultrasonication technique and how cheap commercial boron refined in ethanol medium results in increase of grain boundaries in the final microstructure and therefore produce high performance bulk MgB_2 while preserving the T_c. Overall, this work emphasized the structure–property relation in superconducting bulk MgB_2 when fabricated in conjunction with novel and cheap ultrasonication technique.

8.2 Experimental

8.2.1 High-Energy Ultrasonication

The working principle of this technique is that it generates powerful waves using vibration from a metal probe. These high-energy waves cause turbulence in the medium and pass the high energy to the particles, which then bombard each other and the container walls, resulting in a sharp breakage. Tiny air bubbles are formed, releasing destructive forces inside the solution, which aid in splintering the particles. An important advantage of this method is that the waves' generation can be controlled and optimized by tuning frequency and power. In this, we employed Mitsui UX-300 Ultrasonic Homogenizer.

8.2.2 Boron Ultrasonication

In order to develop a cheap, scalable fabrication process, we ultrasonicated cheap commercial boron (Furu-uchi chemicals, 300 mesh, 99% purity) using ultrasonication (Mitsui ultrasonic Homogenizer). The working principle comprises generation of high power waves in the beaker containing boron powder dispersed in ethanol. In this experiment, we systematically varied the ultrasonication time such as 30, 60, 90, and 120 min, maintaining power constant, at 50% (150 W) and frequency at 20 kHz. The power was switched ON and OFF alternatively every 30 s. This ensured that the metal tip generating pulses got enough time to cool off the entire heat produced by the particles bombardment. Thus the real ultrasonication times were 15, 30, 45, and 60 min. Immediately after the ultrasonication, the powder was heated at 100 °C for 1 h in a muffle furnace to remove ethanol. Later, the powders were characterized by scanning electron microscope (SEM) to study the particle characteristics.

8.2.3 Synthesis of MgB_2

The precursors were commercial powders (Furu-uchi Chemical Corporation) of amorphous Mg powder (99.9% purity, 200 meshes) and an ultrasonicated boron powder. One gram of MgB_2 was synthesized using 0.529 g of Mg and 0.471 g of B (ultrasonicated) in the molar ratio of 1:2. The powders were rigorously mixed and ground in a glove box and heat treated after pelletizing as discussed in Chapter 2. The pellets were then removed out from the furnace and the outer surfaces are polished to avoid any surficial MgO formed. From here on we address the various MgB_2 bulks as Bulk1, Bulk2, Bulk3, Bulk4, and Bulk5 corresponding to the boron precursor used which is ultrasonicated for 0, 15, 30, 45, and 60 min, respectively. The bulks were later characterized for the analysis.

8.2.4 Characterization of MgB_2

The constituent phases of the samples were identified with a high-resolution automated Rigaku smart-lab X-ray powder diffractometer (RINT2200-) with a step size of 0.01° from 10° to 90°, using Cu-K_α radiation generated at 40 kV and 30 mA. The microstructures of both the ultrasonicated boron powders and bulk MgB_2 samples were studied using a field emission scanning electron microscope (FE-SEM).

Superconducting critical temperature (T_c) and magnetization hysteresis loops (M-H) were measured using SQUID Magnetometer (Quantum Design, model MPMS5). Specimens for SQUID measurements, with approximate dimensions of 1 × 1 × 0.75 mm^3, were cut from bulk MgB_2 samples. J_c was calculated from the M-H loops using the extended Bean critical state model formula for finite rectangular samples,

$$J_c = 20\ \Delta m/[a^2c\ (b - a/3)], \tag{8.1}$$

where a, b are cross-sectional dimensions, $b > a$, and c is thickness of the specimen (a, b, c in mm). Δm (in emu units, 1 emu = 10^{-3} Am^2) is the difference of magnetic moments during increasing and decreasing field in the *M-H* loop. The T_c, J_c trends are discussed in correlation with microstructural results.

8.3 Results and Discussion

8.3.1 Boron XRD

Ball milling as a standard technique for boron powder refinement is not suitable for industrial use because of the impurities, mainly B_2O_3 created during this process. XRD analysis revealed that the amount of B_2O_3 raised with increasing milling time, which resulted in a decrease of T_c of the MgB_2 bulk [49]. The ultrasonication process does not show such a drawback. XRD results proved absence of B_2O_3 diffraction peaks even after 1 h of ultrasonication (100% intensity peak- usually at approx. $2\theta = 27.8°$), as shown in Fig. 8.1. Note that any oxides in the precursor promote formation of MgO, which is detrimental to the superconducting properties.

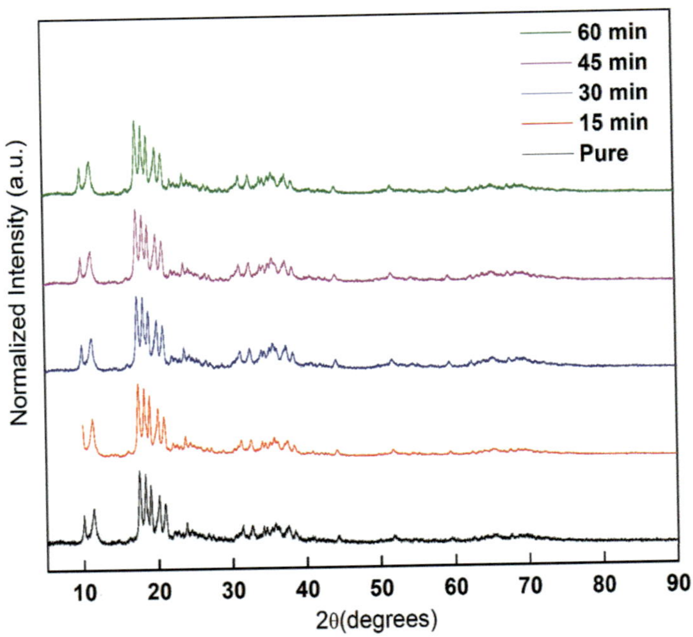

Figure 8.1 XRD of the ultrasonicated boron (0, 15, 30, 45, 60 min). All diffraction patterns showed no signs of B_2O_3 formation.

8.3.2 Microstructural Analysis of Ultrasonicated Boron

The particles size was studied by means of SEM. Fig. 8.2 shows the SEM micrographs of 15, 30, 45, and 60 min ultrasonicated boron. The particles size was significantly reduced with ultrasonication as when compared to pure bulk shown in Fig. 8.3. In addition, clustering or agglomeration can be observed in boron powder ultrasonicated for longer than 15 min, especially in 30, 45, and 60 min samples as pointed in the micrographs in red circles. Refinement up to a few tens of nanometers can be observed. This implies that although there is size refinement with increase in ultrasonication, longer intervals over 15 min can result in clustering. This can be possible because ultrasonication involves collision between particles. We propose that these particles after 15 min of bombardment end up with irregular surfaces and smaller sizes which when subjected to further collisions can result in interlocking of particles.

Figure 8.2 SEM micrographs of 15, 30, 45, and 60 min ultrasonicated boron.

Another possible explanation can be agglomeration, which is mainly due to the tendency of the system to minimize surface energy [52]. Similar phenomenon was reported recently, when boron was subjected to ball milling for size reduction [49]. The authors concluded that after 2 h of ball milling, the performance goes down because of volume reduction in MgB_2 phase. They also observed formation of unreacted Mg and B_2O_3. In the case of ultrasonication, very minute amount of MgO was observed and T_c was close to pure MgB_2. Ultrasonication occurs to be better than ball milling in terms of balanced quality, performance, and processing time.

Figure 8.3 SEM micrograph of nonultrasonicated or reference boron.

8.3.3 MgB_2 XRD

From the XRD of MgB_2 bulks shown in Fig. 8.4, it is evident that there are no contaminants present in the matrix apart from a scare quantity of MgO, which is formed during the transfer from glove box to furnace and the pressing step. The intensity of MgO peak ($2\theta \sim 62.3°$) can be seen, which is very small when compared to other peaks. All the remaining peaks belong to MgB_2.

8.3.4 Superconducting Performance Measurements

For clear depiction of $T_{c,onset}$ we have plotted the first derivative of DC magnetic susceptibility against temperature (Fig. 8.5). All the samples show high $T_{c,onset}$ such as 38.5 K. The width of the peak represents ΔT_c, which is around 0.7 K for all bulks that depicts the high quality of MgB_2 fabricated. These results also point to a high purity, as most of the secondary phases present in a superconducting material result in degradation of critical temperature. Additives and dopants such as carbon in various forms, *Ti*, Cu, Fe, SiC, and others resulted in great decrease in T_c [46,53–57].

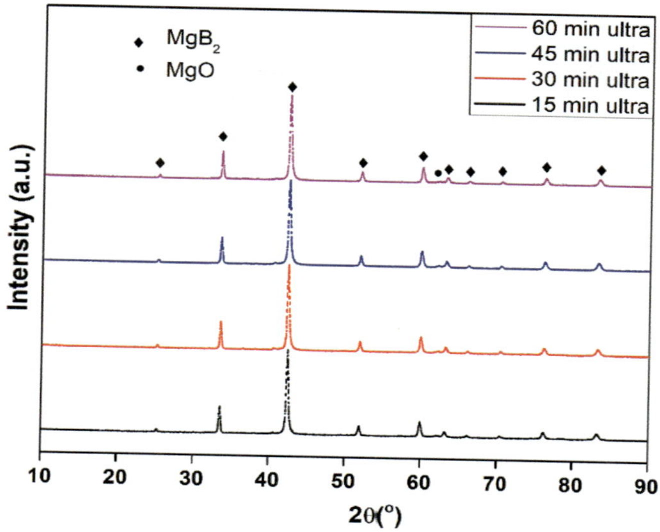

Figure 8.4 XRD of the MgB_2 bulk prepared from ultrasonicated boron (15, 30, 45, 60 min).

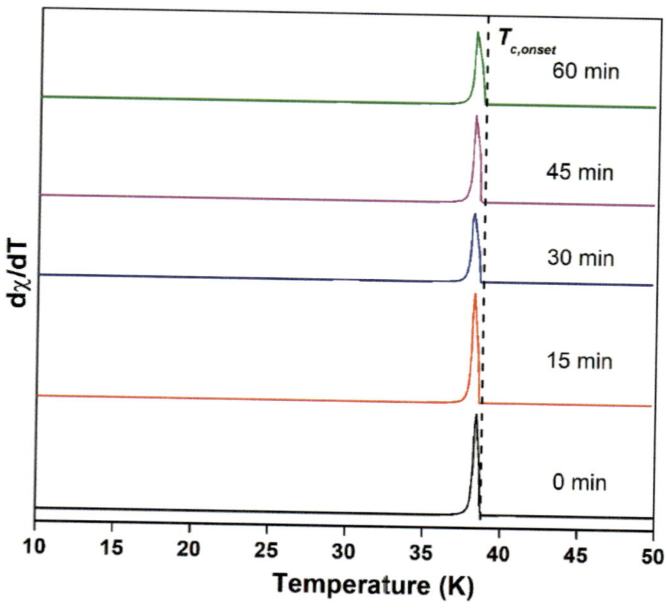

Figure 8.5 First derivative of DC susceptibility vs. temperature of the MgB_2 bulk prepared from ultrasonicated boron (0, 15, 30, 45, 60 min).

Critical current density measurements via M-H loops presented in Fig. 8.6 show that Bulk2 shows high performance such as high self-field J_c of 300 kA/cm^2 at 20 K, while other bulks Bulk3, Bulk4, and Bulk5 exhibited self-field J_c close to 270, 275, and 260 kA/cm^2 respectively. Thus, there is about 36% improvement in Bulk2, 22% improvement for Bulk3, Bulk4, and 18% improvement Bulk5 when compared to Bulk1 (220 kA/cm^2).

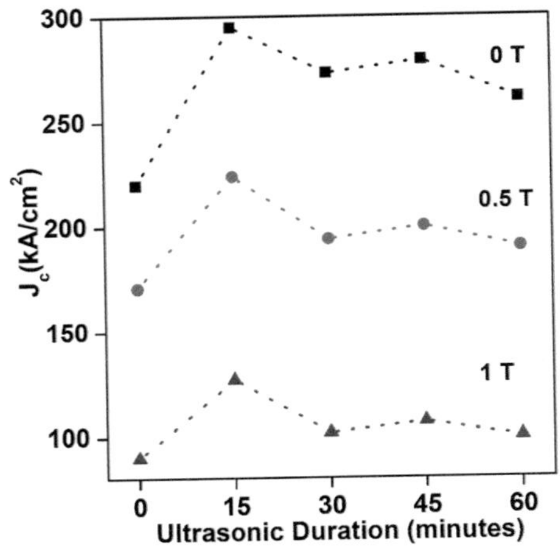

Figure 8.6 Critical current densities of various MgB_2 bulks prepared using ultrasonicated boron powder. All the values are measured at 20 K.

In order to understand this improvement, we carried out several microstructural studies using FE-SEM as well as estimated average grain size and distribution using ImageJ software. In general, during the liquid-solid reaction of Mg and B, fine MgB_2 particles dissolve in the Mg melt and contribute to growth of bigger crystals [58]. From Figs. 8.7 to 8.11, we can observe that the particles in microstructure corresponding to Bulk1, Bulk2, Bulk3, Bulk4, and Bulk5, respectively. The particles in Bulk2 are much finer than other bulks which is due to the refined boron precursor powder. However, the particles in Bulk3, Bulk4, and Bulk5 are bigger, resulting less J_c than Bulk2. This is because the clustered boron particles merged to form larger MgB_2 grains which in turn reduced the grain boundary area. As we know, grain boundaries are primary pinning centers in bulk MgB_2 superconducting system. Hence the J_c reduction with the longer durations of ultrasonication is a result of agglomeration or clustering of nano-B particles. This conclusion can be further supported by average particle sizes of Bulk2 which is 260 nm, while Bulk1, Bulk3, Bulk4, and Bulk5 have 480, 320, 350, and 370 nm, respectively. The J_c observed was in proportion with the measured average particle sizes for these bulks. From these results we can comprehend that this system reaches optimum at 15 min, from the current analysis. It can also be said that further ultrasonication might lead to an increase in boron cluster size and number, which will imply that the bulk microstructure consists larger grains, reducing the number of grain boundaries and therefore J_c. Similar technique can be added to wires and tapes system, to harvest high performance at cheap means.

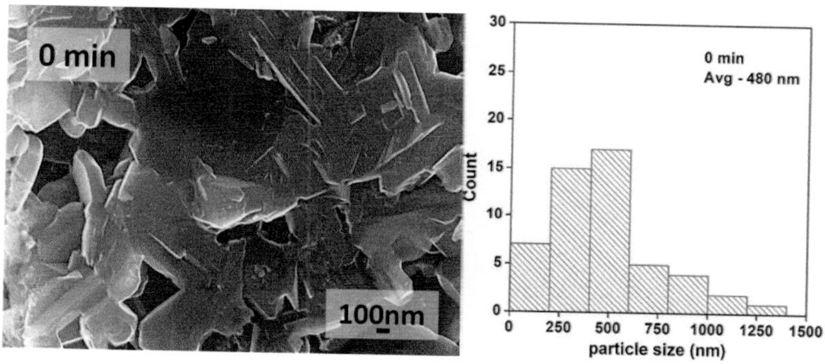

Figure 8.7 Microstructures and particle size distribution of the MgB_2 bulk prepared from commercial boron (nonultrasonicated).

Figure 8.8 Microstructures and particle size distribution of the MgB_2 bulk prepared from 15 min ultrasonicated boron (in ethanol medium).

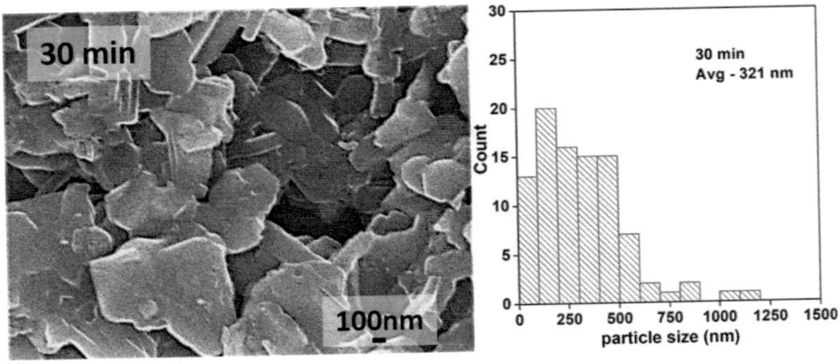

Figure 8.9 Microstructures and particle size distribution of the MgB_2 bulk prepared from 30 min ultrasonicated boron (in ethanol medium).

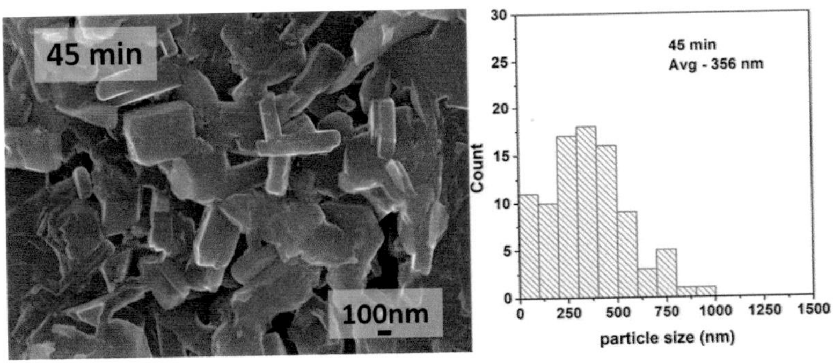

Figure 8.10 Microstructures and particle size distribution of the MgB_2 bulk prepared from 45 min ultrasonicated boron (in ethanol medium).

Figure 8.11 Microstructures and particle size distribution of the MgB_2 bulk prepared from 60 min ultrasonicated boron (in ethanol medium).

8.4 Conclusion

Cheap and novel high-energy ultrasonication technique was used on cheap commercial boron precursor to increase superconducting performance of sintered bulk MgB_2. Boron refinement up to nanometer regime was achieved without formation of B_2O_3. Longer ultrasonication resulted in agglomeration of boron particles that decreased the J_c values. XRD and magnetic studies of bulk MgB_2 fabricated with this powder revealed that the bulk is primarily high quality ($T_c \sim$ 38.5 K; $\Delta T_c \sim$ 0.7 K) single phase, with scare MgO. According to our study, 15 min of boron ultrasonication in ethanol is optimal for the best MgB_2 performance and reaches saturation thereafter because of formation of larger grains in the matrix. J_c increment of around 36% was observed in this optimized regime of MgB_2 preparation. The ultrasonication is a cost-effective, impurity-free, and scalable technique and hence a suitable way to synthesize bulk MgB_2 for practical applications.

References

1. Nagamatsu J, Nakagawa N, Muranaka T, Zenitani Y and Akimitsu J. (2001). Superconductivity at 39K in magnesium diboride, *Nature.* **410**, 63.
2. Duband L and Collaudin B. (1999). Sorption coolers development at CEA-SBT, *Cryogenics.* **39**, pp. 659–63.
3. Kim Y, Park I and Jeong S. (2013). Experimental investigation of two-stage active magnetic regenerative refrigerator operating between 77 K and 20 K, *Cryogenics.* **57**, pp. 113–21.
4. Pan C, Zhang T, Zhou Y and Wang J. (2016). A novel coupled VM-PT cryocooler operating at liquid helium temperature, *Cryogenics.* **77**, pp. 20–4.
5. Kajikawa K and Nakamura T. (2009). Proposal of a fully superconducting motor for liquid hydrogen pump with MgB_2 wire, *IEEE Trans. Appl. Supercond.* **19**, pp. 2–6.
6. Rafieazad M, Balcı Ö, Acar S and Somer M. (2017). Review on magnesium diboride (MgB_2) as excellent superconductor: Effects of the production techniques on the superconducting properties, *Bor Derg.* **2**, pp. 87–96.

7. Baig T, Al Amin A, Deissler R J, Sabri L, Poole C, Brown R W, Tomsic M, Doll D, Rindfleisch M, Peng X, Mendris R, Akkus O, Sumption M and Martens M. (2017). Conceptual designs of conduction cooled MgB_2 magnets for 1.5 and 3.0 T full body MRI systems, *Supercond. Sci. Technol.* 30 043002 (32 pp).

8. Yao W, Bascuñán J, Hahn S and Iwasa Y. (2010). MgB_2 coils for MRI applications, *IEEE Trans. Appl. Supercond.* **20**, pp. 756–9.

9. Eisterer M. (2007). Magnetic properties and critical currents of MgB_2, *Supercond. Sci. Technol.* **20**, pp. R47–73.

10. Tomsic M, Rindfleisch M, Yue J, Mcfadden K, Phillips J, Sumption M D, Bhatia M, Bohnenstiehl S and Collings E W. (2007). Overview of MgB_2 superconductor applications, *Int. J. Appl. Ceram. Technol.* **4**, pp. 250–9.

11. Larbalestier D, Gurevich A, Feldmann D M and Polyanskii A. (2001). High-T_c superconducting materials for electric power applications, *Nature.*

12. Muralidhar M, Inoue K, Koblischka M R, Tomita M and Murakami M. (2014). Optimization of processing conditions towards high trapped fields in MgB_2 bulks, *J. Alloys Compd.* **608**, pp. 102–9.

13. Muralidhar M, Nozaki K, Kobayashi H, Zeng X L, Koblischka-Veneva A, Koblischka M R, Inoue K and Murakami M. (2015). Optimization of sintering conditions in bulk MgB_2 material for improvement of critical current density, *J. Alloys Compd.* **649**, pp. 833–42.

14. Mackinnon I D R, Shahbazi M, Alarco J A and Talbot P C. (2017). Low temperature decomposition of metal borohydride drives autogenous synthesis of MgB_2, *Supercond. Sci. Technol.* 30 055004.

15. Bhagurkar A G, Yamamoto A, Dennis A R, Durrell J H, Aljohani T A, Nadendla H B and Cardwell D A. (2017). Microstructural evolution in infiltration-growth processed MgB_2 bulk superconductors, *J. Am. Ceram. Soc.* **100**, pp. 2451–60.

16. Rosová A, Kulich M, Kováč P, Brunner B, Scheiter J and Häßler W. (2017). The effect of boron powder on the microstructure of MgB_2 filaments prepared by the modified internal magnesium diffusion technique, *Supercond. Sci. Technol.* 30 055001 (12pp)

17. Arvapalli S, Miryala M and Murakami M. (2019). Optimization of Mg precursor concentration to obtain high J_c in MgB_2 synthesized with Ag addition and carbon encapsulated boron, *IEEE Trans. Appl. Supercond.* **29**, pp. 1–4.

18. Arvapalli S, Miryala M and Murakami M. (2019). Beneficial impact of excess Mg on flux pinning in bulk MgB_2 synthesized with Ag addition and carbon encapsulated boron, *Adv. Eng. Mater.* 21 1900497

19. Kauffmann-Weiss S, Hassler W, Guenther E, Scheiter J, Denneler S, Glosse P, Berthold T, Oomen M, Arndt T, Stocker T, Hanft D, Moos R, Weiss M, Weis F and Holzapfel B. (2017). Superconducting properties of thick films on Hastelloy prepared by the aerosol deposition method with ex situ MgB_2 powder, *IEEE Trans. Appl. Supercond.* **27**, pp. 2–5.

20. Cai Q, Liu Y, Guo Q, Ma Z and Li H. (2016). Variation of MgB_2 growth mode and enhancement of critical current density by instantaneously 800 °C-triggered low-temperature sintering, *Scr. Mater.* **124**, pp. 184–8.

21. Zou X, Zhang W, Wang Q, Zheng L, Yu X, Yu Z, Zhang H, Zhao Y and Zhang Y. (2019). Preparation of MgB_2 superconducting wires by the rapid heating and quenching method, *Mater. Lett.* **244**, pp. 111–4.

22. Ma Z, Liu Y, Cai Q, Jiang H and Yu L. (2013). Excellent J_c in the low-temperature sintered MgB_2 superconductors consisted of uncompleted MgB_2 phase and residual Mg, *Mater. Chem. Phys.* **141**, pp. 378–82.

23. Ma Q, Peng J, Ma Z, Cheng F, Lan F, Li C, Yang Z, Liu C and Liu Y. (2018). Improved grain connectivity and critical current density in ex-situ MgB_2 superconductors prepared by two-step sintering, *Mater. Chem. Phys.* **204**, pp. 62–6.

24. Aldica G, Burdusel M and Badica P. (2014). Trapped magnetic field in a (NdFeB)-(MgB_2) pair-type bulk magnet, *Phys. C Supercond. its Appl.* **505**, pp. 19–23.

25. Noudem J G, Aburras M, Bernstein P, Chaud X, Muralidhar M and Murakami M. (2014). Development in processing of MgB_2 cryo-magnet superconductors, *J. Appl. Phys.* 116 163916.

26. Murakami A, Iwamoto A and Noudem J G. (2018). Mechanical properties of bulk MgB_2 superconductors processed by spark plasma sintering at various temperatures, *IEEE Trans. Appl. Supercond.* 28 8400204.

27. Giunchi G, Ripamonti G, Cavallin T and Bassani E. (2006). The reactive liquid Mg infiltration process to produce large superconducting bulk MgB_2 manufacts, *Cryogenics.* **46**, pp. 237–42.

28. Ishiwata J, Muralidhar M, Inoue K and Murakami M. (2015). Effect of MgB_4 addition on the superconducting properties of polycrystalline MgB_2, *Phys. Procedia.* **65**, pp. 69–72.

29. Peng J, Cai Q, Cheng F, Ma Z, Li C, Xin Y and Liu Y. (2017). Enhancement of critical current density by a "MgB_2-MgB_4" reversible reaction in self-sintered ex-situ MgB_2 bulks, *J. Alloys Compd.* **694**, pp. 24–9.

30. Prikhna T, Eisterer M, Goldacker W, Gawalek W, Sokolovsky V, Weber H W, Kozyrev A V, Moshchil V E, Sverdun V B, Kovylaev V V, Karpets

M V, Basyuk T V and Shaternik A V. (2015). Effect of nanostructural inhomogeneities on the superconducting characteristics of MgB_2 with enhanced grain connectivity, *IEEE Trans. Appl. Supercond.* 25 6200404.

31. Prikhna T A, Eisterer M, Weber H W, Gawalek W, Kovylaev V V, Karpets M V, Basyuk T V and Moshchil V E. (2014). Nanostructural inhomogeneities acting as pinning centers in bulk MgB_2 with low and enhanced grain connectivity, *Supercond. Sci. Technol.* 27 044013 (9pp).
32. Zeng R, Lu L, Wang J L, Horvat J, Li W X, Shi D Q, Dou S X, Tomsic M and Rindfleisch M. (2007). Significant improvement in the critical current density of in situ MgB_2 by excess Mg addition, *Supercond. Sci. Technol.* **20**, pp. 1–6.
33. Ansari I A. (2019). Study of dynamic behaviors for nano Fe-doped MgB_2 superconductor via ac-susceptibility measurements, *Ceram. Int.* **45**, pp. 1523–7.
34. Sudesh, Kumar N, Das S, Bernhard C and Varma G D. (2013). Effect of graphene oxide doping on superconducting properties of bulk MgB_2, *Supercond. Sci. Technol.* 26 095008 (8pp).
35. Ansari I A, Shahabuddin M, Ziq K A, Salem A F, Awana V P S, Husain M and Kishan H. (2007). The effect of nano-alumina on structural and magnetic properties of MgB_2 superconductors, *Supercond. Sci. Technol.* **20**, pp. 827–31.
36. Pan X F, Cheng C H and Zhao Y. (2011). Effect of rare-earth oxides doping on the superconductivity and flux pinning of MgB_2 superconductor, *J. Supercond. Nov. Magn.* **24**, pp. 1611–6.
37. Bugoslavsky Y, Cohen L F, Perkins G K and Polichetti M. (2001). Enhancement of the high-magnetic-field critical current density of superconducting MgB_2 by proton irradiation, *Nature.* **411**, pp. 561–3.
38. Konduru P, Singh A V P, Kandasami A, Dinakar K and Sreehari Sastry S. (2019). Enhancement of superconducting parameters of MgB_2 by low energy carbon ion implantation, *Nucl. Instruments Methods Phys. Res. Sect. B Beam Interact. with Mater. Atoms.* **438**, pp. 42–7.
39. Zhou S, Pan A V, Wexler D and Dou S X. (2007). Sugar coating of boron powder for efficient carbon doping of MgB_2 with enhanced current-carrying performance, *Adv. Mater.* **19**, pp. 1373–6.
40. Bohnenstiehl S D, Susner M A, Yang Y, Collings E W, Sumption M D, Rindfleisch M A and Boone R. (2011). Carbon doping of MgB_2 by toluene and malic-acid-in-toluene, *Phys. C Supercond. its Appl.* **471**, pp. 108–11.

41. Wilke R H T, Bud'ko S L, Canfield P C, Finnemore D K, Suplinskas R J and Hannahs S T. (2004). Systematic effects of carbon doping on the superconducting properties of $Mg(B_{1-x}C_x)_2$, *Phys. Rev. Lett.* **92**, pp. 2–5.

42. Ye S J, Matsumoto A, Zhang Y C and Kumakura H. (2014). Strong enhancement of high-field critical current properties and irreversibility field of MgB_2 superconducting wires by coronene active carbon source addition via the new B powder carbon-coating method, *Supercond. Sci. Technol.* 27, 085012 (10pp).

43. Jun B H, Park S D and Kim C J. (2012). Refinement and carbon incorporation effects on the superconducting properties of MgB_2 through wet milling process of low purity boron powder, *J. Alloys Compd.* **535**, pp. 27–32.

44. Herbirowo S, Sofyan N, Saragih R, Imaduddin A, Sebleku P, Herbirowo S, Sofyan N, Saragih R and Imaduddin A. (2017). Properties of carbon nanotubes-doped Fe-sheath MgB_2 for superconducting wires, *Am. Inst. Phys.* 1826 020007.

45. Muralidhar M, Higuchi M, Jirsa M, Diko P, Kokal I and Murakami M. (2017). Improved critical current densities of bulk MgB_2 using carbon-coated amorphous boron, *IEEE Trans. Appl. Supercond.* **27**, pp. 18–21.

46. Arvapalli S S, Muralidhar M and Murakami M. (2018). High-performance bulk MgB_2 superconductor using amorphous nano-boron, *J. Supercond. Nov. Magn.* **32**, pp. 1891–5.

47. Xu X, Kim J H, Yeoh W K, Zhang Y and Dou S X. (2006). Improved J_c of MgB_2 superconductor by ball milling using different media, *Supercond. Sci. Technol.* **19**, pp. L47–50.

48. Kang M O, Joo J, Jun B H, Park S, Kim C S and Kim C. (2019). Effect of boron milling on phase formation and critical current density of MgB_2 bulk superconductors, *Prog. Supercond. Cryog.* **21**, pp. 18–24.

49. Palomino R L, Miró A M B, Tenorio F N, Jesús F S De, Escobedo C A C and Ammar S. (2016). Ultrasonics sonochemistry sonochemical assisted synthesis of $SrFe_{12}O_{19}$ nanoparticles, *Ultrason. – Sonochemistry.* **29**, pp. 470–5.

50. Sreekumar V M, Babu N H, Eskin D G and Fan Z. (2015). Materials Science & Engineering A Structure – property analysis of in-situ Al – $MgAl_2O_4$ metal matrix composites synthesized using ultrasonic cavitation, *Mater. Sci. Eng. A.* **628**, pp. 30–40.

51. Nakaso K, Shimada M, Okuyama K and Deppert K. (2002). Evaluation of the change in the morphology of gold nanoparticles during sintering, *J. Aerosol Sci.* **33**, pp. 1061–74.

52. Vives Díaz N E, Hosmani S S, Schacherl R E and Mittemeijer E J. (2008). Nitride precipitation and coarsening in Fe-2.23 at.% V alloys: XRD and (HR)TEM study of coherent and incoherent diffraction effects caused by misfitting nitride precipitates in a ferrite matrix, *Acta Mater.* **56**, pp. 4137–49.

53. Yang F, Feng J Q, Xiong X M, Liu G Q, Zhang P X, Li C S, Wang Q Y, Feng Y, Pang Y C, Li S Q and Yan G. (2015). The effect of high-energy ball milling on the microstructure and properties of Ti-doped MgB_2 bulks and wires, *Phys. Procedia.* **65**, pp. 157–60.

54. Rui X F, Sun X F, Xu X L, Zhang L and Zhang H. (2005). Doping Effect of Nano- YBCO Additive on MgB_2, *Int. J. Mod. Phys. B.* **19**, pp. 375–7.

55. Aldica G, Plapcianu C, Badica P and Groza J R. (2008). Superconducting layered composites of Fe-MgB_2 obtained by field assisted sintering. A global road map for ceramic materials and technologies: forecasting the future of ceramics, *International Ceramic Federation - 2nd International Congress on Ceramics, ICC 2008, Final Programme.*

56. Pan X F, Shen T M, Li G, Cheng C H and Zhao Y. (2007). Doping effect of Pr_6O_{11} on superconductivity and flux pinning of MgB_2 bulk, *Phys. Status Solidi(a).* **204**, pp. 1555–60.

57. Ma Z and Liu Y. (2012). Sintering process and its mechanism of MgB_2 Superconductors Sintering of Ceramics - New Emerging Techniques, **ISBN:978-953-51-0017-1**, pp. 469–98.

58. Glowacki B A, Majoros M, Vickers M, Evetts J E and Shi Y. (2001). Superconductivity of powder-in-tube MgB_2 wires, *Supercond. Sci. Technol.* **14**, pp. 193–9.

Chapter 9

New Potential Family of Iron-Based Superconductors towards Practical Applications: $CaKFe_4As_4$ (1144)

Shiv J. Singh[a,b] **and Andrzej Morawski**[a]
[a]*Institute of High Pressure Physics, Polish Academy of Sciences, ul. Sokolowska 29/37, Warsaw, 01-142, Poland*
[b]*Clarendon Laboratory, Department of Physics, University of Oxford, Parks Road, Oxford OX1 3PU, UK*
sjs@mail.unipress.waw.pl

Most of iron-based superconductors depict the superconductivity by a suitable doping and control of the actual composition in these superconductors is always a challenge. The presence of few stoichiometric families of iron-based superconductors provides a unique opportunity to understand the superconductivity without any substitution disorder by doping. Among them, recently discovered $CaKFe_4As_4$ (1144) is a unique stoichiometric iron-based superconductor with a remarkably high critical temperature of 35 K, owing to its very high upper critical field ($\mu_0 H_{c2}$~100 T), low anisotropy and high critical current density (up to J_c~10^7–10^8 A/cm^2).

High-T_c Superconducting Technology: Towards Sustainable Development Goals
Edited by Muralidhar Miryala

ISBN 978-981-4877-65-7 (Hardcover), 978-1-003-16468-5 (eBook)
www.jennystanford.com

This high-T_c superconducting family, having the highest J_c of any other iron-based superconductor, has been placed as a strong contender for magnet applications, although more research works are required in this direction. In present chapter, we will review about the superconducting properties and current progress of this new 1144 family with other families of iron-based superconductors.

9.1 Introduction

Superconductivity is a very interesting phenomenon which shows zero resistance and expel magnetic fields below a certain temperature (called "critical transition temperature T_c"). This property was first observed in Mercury (Hg) metal in 1911 by H. K. Onnes [1]. Following this pioneer work, various superconductors have been discovered especially at low temperature [2]. Interestingly, new superconducting materials are always discovered during some interval of times and works as a catalyst for this research field. However, still searching a room temperature superconductor is not achieved and a lot of efforts are being made in this direction. In the past 100 years, seven Nobel prizes have been awarded in this superconducting field, which indicates that the research communities are always working seriously in this fantastic field. A large number of elements, intermetallic compounds, and other types of compounds have been discovered as superconductors and some of them are suitable for use in devices [3–5]. Prominent members of the conventional (phonon mediated) superconductors include Nb_3Sn (T_c = 18 K) and NbTi (T_c = 9.8 K) which are currently using in the superconducting solenoids in laboratory electromagnets and MRI machines operated with liquid Helium temperature [6]. In 2001, a binary compound MgB_2 was discovered as a superconductor and has the highest T_c of around 39 K for conventional superconductors at atmospheric pressure [7]. This transition temperature is close to the limit suggested by Bardeen–Cooper–Schrieffer (BCS) theory for phonon-mediated superconductivity [2]. The boron isotope effect confirms that MgB_2 is a phonon-mediated superconductor [7].

In 1986, the superconductivity in layered copper oxide compounds was discovered as first high T_c family of superconductor by G. Bednorz and Alex Muller [8], and pushed transition temperatures over 100 K [9,10]. It has attracted great attention since its discovery because T_c can exceed the boiling temperature of liquid nitrogen

(77 K), which is much higher than the supposed limit of $T_c \sim 40$ K derived according to the BCS theory of conventional superconductivity [2]. In these cuprate superconductors, phonons do not appear to play a role to form superconducting pairing (cooper pairing), but the specific mechanism remains elusive [11]. This high-T_c superconductors have a layered crystal structure consisting of superconducting CuO_2 planes separated by charge reservoir layers. The parent compound does not show the superconducting properties; however, it may be doped by electrons or holes type into the CuO_2 planes [12,13]. On holes doping, the antiferromagnetic Mott insulating phase of the parent compounds disappears and superconductivity emerges. The reported superconducting phase diagram shows a dome-like shape behaviour of T_c as a function of doping and the maximum transition temperature T_c is obtained around 16% doped per CuO_2 plaquette [11,13]. A similar phase diagram is also seen on electrons doping, albeit with a more robust antiferromagnetic phase and a lower T_c. While in the case of hole-doped side, there exists an enigmatic state above T_c called the pseudogap, where the electron density of states within certain momentum region is suppressed [11–13]. Superconducting properties of the representative member of cuprate superconductors are shown in Table 9.1. Despite their promising characteristics such as high T_c and high H_{c2} for technology, adoption of this superconductor family has been hindered by materials challenges such as brittleness and anisotropies nature.

In 2008, second high T_c family, i.e. iron-based superconductors, was discovered [14] and intensely studied during the last 12 years [15–18]. Again, phonons are not believed to play a dominant role in the superconducting pairing similar to cuprate superconductors. On the basis of crystal structure of the parent compounds, the iron-based superconductors can be divided into six categories, as shown in Fig. 9.1: 1111 type (REFeAsO), 122 ($AEFe_2As_2$), 11 (FeCh), 111 (LiFeAs), and 1144 ($AeAFe_4As_4$). They have unique physics of pairing mechanism [11] and have high-transition temperature (T_c) up to 58 K [19] with small anisotropy [15,17,20]. Table 9.1 shows superconducting parameters for various superconductors. The upper critical field H_{c2} of MgB_2 [4,7] and conventional metallic superconductors Nb-Ti and Nb_3Sn is below 25 T at liquid helium temperature [3], whereas iron-based superconductors have very high upper critical fields (H_{c2}), and large critical current density (J_c) [15,20].

Table 9.1 Summary of superconducting parameters of $CaKFe_4As_4$ (1144) with other iron-based superconductors, MgB_2, and Nb-based superconductors

Parameters	1111	BaK122	BaCo122	111	11	1144	Bi-2223	REBCO	MgB_2	Nb_3Sn	NbTi
T_c (K)	58	37–38	22	18	16	33–36	108	92	39	18	9
$\mu_0 H_{c2}(0)$(T)	>100	80	>100	40	40	>100	>100	>100	18	25	11.5
$J_c^{4.2K}$ (A/cm^2)	~10^6	~5×10^6	~2.6×10^5	~10^6	~10^5	~10^8	~10^6	~10^7	~10^6	~10^6	~4×10^5 (5 T)
ξ_{ab} (nm)	1.8–2.3	1.5–2.4	2.45	4.8	1.2	1.1	1.5	1.5	6.5	3	4
ξ_c (nm)			1.48	1.7		1.7					
γ_H (around T_c)	4–5	1.5–2	2.7	2.5	1.1–1.9	2.5	50–90	5–7	2–2.7	Negligible	Negligible

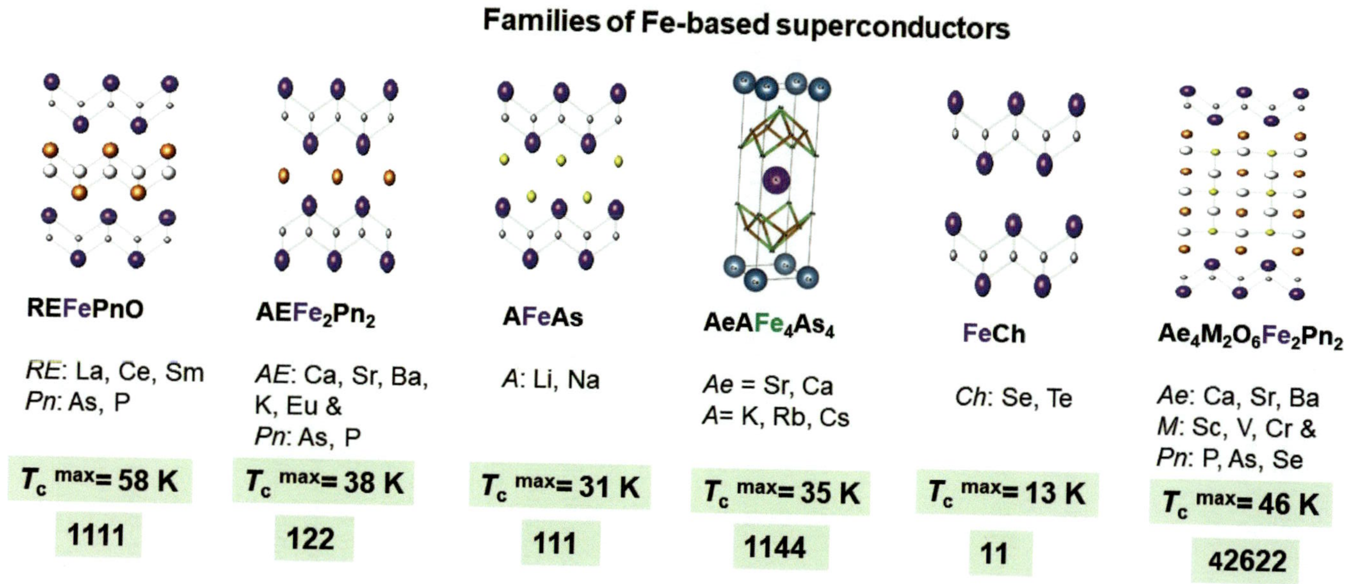

Figure 9.1 Various categories of iron-based superconductors based on their crystal structure.

The second family of high-T_c superconductor has also much lower anisotropy than that of first one (cuprates: $Bi_2Sr_2Ca_{n-1}Cu_nO_{2n+4}$ (BSCCO), $REBa_2Cu_3O_7$ (REBCO)). Some families such as 122 are nearly isotropic at low temperature which suggests high irreversibility fields (H_{irr}) of iron-based superconductors close to its upper critical field (H_{c2}). These properties advise that iron-based superconductors have strong potential for high-field applications comparable to that of cuprates. Furthermore, in cuprate (REBCO), the intergrain critical current decreases exponentially with increasing grain boundary angle when it is larger than 3–5°, whereas for 122, the reported critical angle is around 9° [21]. This is named as weak link effects, i.e. granular nature. This effect is well studied by the remanent magnetisation study, which demonstrates the coupled effect of inter- and intragranular behaviour due to the origin of weak Josephson coupling between grains [22–24]. Due to the large grain boundary angle, the iron-based superconducting wires and tapes can be expected to have a large transport current [16]. It can be fabricated by simple and low-cost powder-in-tube (PIT) method rather than the complex-coated conductor technique. These mentioned properties propose that the iron-based superconducting wires can be a good candidate to design high magnetic field well beyond the capacities of conventional Nb-Ti based magnets [15,17] which are desired for the next generation of nuclear magnetic resonance (NMR) spectrometers, accelerators, and fusion magnets, and also for cryocooled medical magnetic resonance imaging (MRI) systems operating at intermediate fields.

Among various iron-based superconductors, 122 ($AEFe_2As_2$, AE = Ba, Sr) family has been widely investigated because of its excellent properties such as very high upper critical field, relatively small anisotropy and an advantageous grain-boundary nature [15–17]. The parent compound of this family does not show the superconductivity; however, by suitable doping or by applying pressure, the superconductivity can appear. The highest superconducting properties have been reported for this family by K doping. Among the 122 materials, $Ba_{1-x}K_xFe_2As_2$ possesses the highest T_c up to 38 K [25] and large upper critical fields (H_{c2}) over 100 T with low anisotropy (γ) ~1–2. These properties are advantageous for high-field applications. In $Ba_{1-x}K_xFe_2As_2$ (BaK122), superconductivity is induced by substituting Ba with K (hole doping), where Ba and K randomly occupy the same crystallographic site in an arbitrary ratio of doping content (x). Therefore, their superconducting properties significantly depend on the doping contents (x) [26]. Note that the significant doping dependence of superconducting properties is almost similar to other 122 materials with different dopant elements. As a result, a fine adjustment of doping level (x) is required to achieve better properties of bulks and thin films. Furthermore, it is always challenging to control the actual doping level.

A new family 1144 of iron-based superconductors has been discovered in November 2016 and reported that some pairs of alkali and alkaline earth elements of different size leads to stoichiometric $AeAFe_4As_4$ (Ae = Ca, Sr and A = K, Rb, Cs) [27]. The stoichiometric compound gives the superconducting transition (T_c) between 31 and 36 K, and allows working with a system where unwanted effects caused by chemically substitutions which play an important role for other families. The 1144 family is particularly suitable for the studies of the fundamental superconducting properties, exhibiting clean-limit behaviour, and having a fairly high-critical temperature [28]. The recent reports based on the single crystals of $CaKFe_4As_4$ depict that all superconducting properties are almost similar or better than 122 family [20,29] and also no worries about the doping level. It has been also established that this family has very high critical current density of the order of 10^8 A/cm^2 [20]. Its field and temperature dependence are better than 122 family and other iron-based superconductors. Recent studies show that 1144 has huge potential for the practical application [20,29].This chapter is especially focused on $CaKFe_4As_4$ which is a unique representative member of 1144 family. We will review its superconducting properties and the possibilities of its application. Furthermore, the understanding of superconductivity mechanism in a doped composition is quit challenging due to different electronic phases. Thus, the presence of few stoichiometric iron-based superconductors such as 1144 provides a unique opportunity to study the phenomenon of superconductivity with much high accuracy.

9.2 Structural Properties of 1144

The 122 parent compound $AEFe_2As_2$ (AE = Ca, Ba, Sr) has $ThCr_2Si_2$-type structure (Space group: *I*4/*mmm*) [25,26]. Substitution of alkali metal such as K, Na, Rb, Cs at AE sites could be possible in this compound and *I*4/*mmm* space group remains unchanged, where the doped alkali metal randomly occupies crystallographically equivalent sites in AE layers having the same structural type. Recently discovered $CaKFe_4As_4$ (CaK1144) is one of most important members of AEA1144 structural family of iron-based superconductors, here AE = Alkaline-earth and A = alkali metals [27]. Due to the large difference between ionic radii of AE and A, AEA1144 crystallises through alternate stacking of the AE and A layers across the Fe_2As_2 layer along c-axis as shown in Fig. 9.2(a) and (b) for $CaKFe_4As_4$. The introduction of a new K-layer in 122 type structure changes the space group from *I*4/*mmm* to *P*4/*mmm*, i.e. a tetragonal structure of $CaKFe_4As_4$ has Ca and K layers stack alternatively across the Fe_2As_2 layer along the c-axis [30]. In 122, the distance of the As-atoms above and below the Fe-atoms is equivalent, as shown in Fig. 9.2(c) for $CaFe_2As_2$, whereas for $CaKFe_4As_4$, FeAs-layers are shifted away from the high-symmetry points, also the distance of the As-atoms is different for the upper and the lower case, as shown in Fig. 9.2(d). Also, the bond angle As-Fe-As in upper and lower superconducting FeAs layers is around 107.2° and 109.1°. The comparative structural parameters for 122 and 1144 are listed in Table 9.2. These studies suggest that 1144 has slightly unsymmetrical tetragonal structure due to the introduction of an extra alkali layer in comparison to 122 families.

Some important points have to be noted between 122 and 1144 structures. First, the tetragonal lattice centring of 122 structure drops from body centred to primitive lattice in the 1144 structure, i.e. *I*4/*mmm* changes to *P*4/*mmm*. Because of this, some more diffraction peaks generate in the x-ray patterns which become the identification of this new 1144 phase. Second, due to the alternating layers of Ca and K in 1144, the n-glide symmetry across the Fe-planes is broken and furthermore, equivalence of the As sites on either side of the Fe plane in $CaFe_2As_2$ (122) is disturbed in the case of $CaKFe_4As_4$. It can be easily observed by examining their immediate As environment in the 1144 structure. These inequivalent As sites

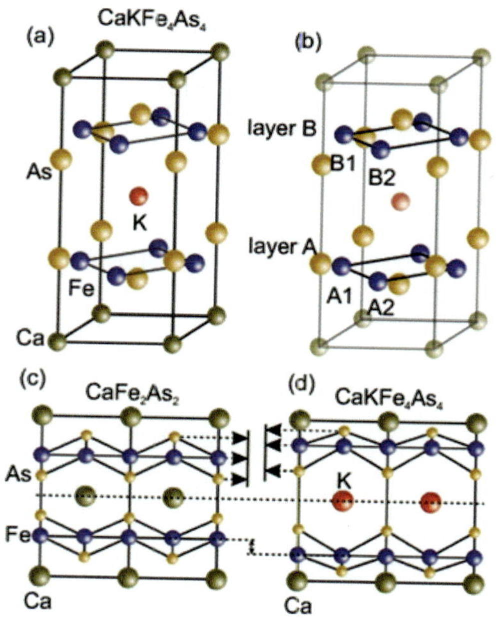

Figure 9.2 **(a)** Structure of $CaKFe_4As_4$ **(b)** structure of 1144 with two layers of FeAs, defined as layer A and layer B. Space symmetries of atomic positions of **(c)** $CaFe_2As_2$ (122) and **(d)** $CaKFe_4As_4$ (1144) are shown. In the case of 122, Fe atoms are in high symmetry points. The distance of the As-atoms above and below from Fe-atoms is equivalent. While in case of 1144, due to extra layer of K, FeAs layers are shifted away from the high symmetry points and also the distance of As-atoms is different for the upper (layer B) and lower case (layer A). Reproduced with permission from [30].

Table 9.2 List of structural parameters of 122 and 1144 families

122 family	**1144 family**
Tetragonal	Tetragonal
I4/mmm	P4/mmm
Lattice parameter: a = 3.9612 (Å)	a = 3.866 (Å)
Lattice parameter: c = 13.006 (Å)	c = 12.817 (Å)
Bond angle:	
$\alpha_{As(1)\text{-}Fe\text{-}As(1)} = 109.6°$	$\alpha_{As(1)\text{-}Fe\text{-}As(1)} = 107.2°$
$\alpha_{As(2)\text{-}Fe\text{-}As(2)} = 109.6°$	$\alpha_{As(2)\text{-}Fe\text{-}As(2)} = 109.1°$

(*Continued*)

Table 9.2 (*Continued*)

122 family	1144 family
The distance of the As-atoms above and below from Fe-atoms $h_{As(1)} = h_{As(2)} = 1.346$ Å	$h_{As(1)} = 1.429$ Å $h_{As(2)} = 1.381$ Å

are no longer constrained to sit at the same height from the Fe plane, as shown in Fig. 9.2d. Because of this arrangement, the Fe-site symmetry is reduced from 42 m which is common in most iron-based superconductors, to 2 mm by broken symmetry across the plane [15].

9.3 Transition Temperature (T_C) and Upper Critical Field (H_{C2})

Various studies based on the transport and thermodynamics properties such as electrical resistivity, thermoelectric power, Hall effect, and specific heat have been reported for $CaKFe_4As_4$ [28]. These detailed studies clearly show the superconducting transition T_c is around 34–35 K, as reported for the single crystal and polycrystalline sample [20,28,31]. This transition temperature is very closed to that of MgB_2 [7] and 122 superconductors [25]. The transition temperature (T_c) is shown in Table 9.1 with respect to other stoichiometric superconductors and 122. No phase transition either structural or magnetic transition is observed up to the lowest temperature for 1144 similar to stoichiometric MgB_2 and LiFeAs, whereas 122 shows a first-order structural phase transition from tetragonal to orthorhombic with the simultaneous onset of long-range antiferromagnetic order around 140 K [25]. Furthermore, the temperature dependence of Hall resistivity and thermoelectric power measurements on 1144 crystals confirm the positive charge carrier, playing a dominant role in normal state of 1144 [28]. The calculated carrier concentration is around 7.4×10^{21} cm^{-3} at 40 K and 1.3×10^{22} cm^{-3} at 200 K. The overall behaviours of the Hall resistivity and thermoelectric power are similar to those reported for optimally or slightly over-doped $Ba_{1-x}K_xFe_2As_2$ [32]. In $Ba_{1-x}K_xFe_2As_2$ (BaK122), superconductivity is induced by substituting Ba with K (hole doping), where Ba and K randomly occupy the same crystallographic site in an arbitrary ratio x. Therefore, the superconducting properties of (Ba,K)Fe_2As_2 significantly depend on doping contents (x) [26].

To understand the magnetic field response of a superconductor, the study of upper critical field (H_{c2}) with temperature dependence is one of the most important parameters which enables us to understand the mechanism of superconductivity. Upper critical field is an intrinsic property of a type II superconducting material in which the resistive transition (ρ-T) width increases with increasing magnetic field. Normally, the upper critical field (H_{c2}) can be estimated by the magneto-transport studies and the H_{c2} at 0 K can be calculated using the one band Werthamer–Helfand–Hohenberg (WHH) formula [33]: $H_{c2}(0) = -0.693T_c[dH_{c2}/dT]_{Tc}$, here, $[dH_{c2}/dT]_{Tc}$ is obtained from the slope of the curve H_{c2} versus T at T_c. It is noted that H_{c2} values extrapolated using the WHH model are not valid for the low-temperature range. The slope of upper critical field from transport data measured in the magnetic field up to 16 T gives much larger magnetic fields, $\mu_0H_{c2}(0)$= ~82 T and ~168 T for $H//c$ and $H//ab$, respectively, for single crystal [28], whereas in the case of polycrystalline 1144, the reported $\mu_0H_{c2}(0)$ value is around 138 T [31]. The polycrystal sample has random grain orientations and generally, it can show the average superconducting properties of both orientations of single crystal one. The estimated upper critical field μ_0H_{c2} based on these low field experiments is anisotropic and exceeds 100 T at low temperatures similar to other superconductors such as 122 [34].

Furthermore, measurements in high magnetic fields are very important which can give quantitative values of the upper critical field ($H_{c2}(T \rightarrow 0)$). A recent report has studied the magneto-resistivity of $CaKFe_4As_4$ up to 90 T and down to 4.2 K to produce the complete phase diagram of this stoichiometric superconductors [35]. It is shown in Fig. 9.3 and confirm more accurate value (around 100 T) of the upper critical field. Fig. 9.3 also depicts the high field data for $Ba_{0.6}K_{0.4}Fe_2As_2$ single crystal up to 60 T and confirm the H_{c2} value around 70 T [36]. These values are much higher than the Pauli paramagnetic limit of $1.85T_c$ which is estimated around 56 T and 65 T for a single band superconductor for 122 and 1144, respectively. This phenomenon has been also found in optimally doped iron-based superconductors such as $FeSe_{0.5}Te_{0.5}$ [17,37]. Hence, it suggests the dominant role of Pauli paramagnetic effects in iron-based superconductors.

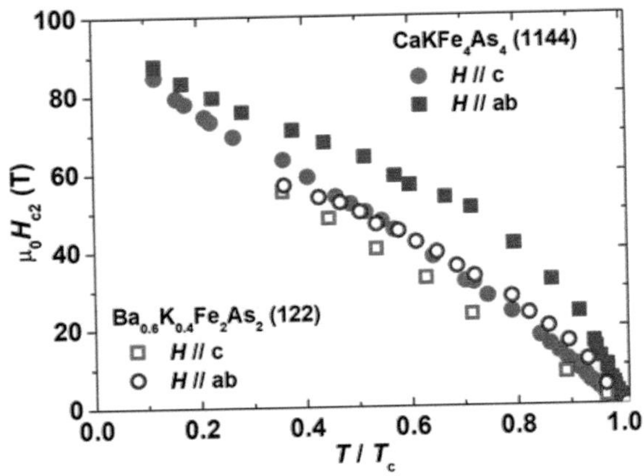

Figure 9.3 The temperature dependence behaviour of the upper critical field H_{c2} is shown for single crystal $CaKFe_4As_4$ (1144) (T_c~35 K) and $Ba_{0.6}K_{0.4}Fe_2As_2$ (122) (T_c~28.2 K) in H//ab and H//c. The data used have been taken from the reference [35] for 1144 and Reference [36] for 122.

The ratio of the upper critical field for different orientations $\gamma = H_{c2}^{ab} / H_{c2}^{c}$ generally defines the anisotropy of a superconductor which decreases from 3.5 to 1 for 1144 with decreasing temperatures, as shown in Fig. 9.4. The temperature dependence behaviour of anisotropy for 122 is also shown in Fig. 9.4 which indicates that both family follow the same behaviour in low temperature [35,36]. These studies confirm the isotropic superconducting behaviour of $CaKFe_4As_4$ in the low-temperature limits. Furthermore, penetration-depth studies based on the plasma frequencies give the calculated anisotropy around 4.5 near the T_c similar to the measured anisotropy from the other measurements [38].

In a superconductor, an arbitrarily small attractive interaction among the electrons at the Fermi surface binds them into pairs that is called Cooper pairs. The length scale, known as coherence length (ξ^{GL}), is another characteristic parameter for a superconductor [2,17]. This characteristic parameter depicts the spatial extent to which the interactions of electrons in a Cooper pair occurs. Using the upper critical field values, we can extract the coherence length.

In the vicinity of the transition temperature T_c, the Ginzburg–Landau coherence lengths are ξ_{ab}^{GL} = 1.66(4) nm and ξ_{c}^{GL} = 0.53(2) nm and in the low-temperature region, the coherence lengths are ξ_{ab}^{GL} = 1.83(4) nm and ξ_{c}^{GL} = 1.87(6) nm for single crystal $CaKFe_4As_4$. These coherence length values also confirm the isotropic nature of this $CaKFe_4As_4$ in the low-temperature region [20,35]. The small coherence length of $CaKFe_4As_4$ is compatible with the presence of large upper critical field (H_{c2}). Since the coherence is given by $\xi = \hbar v_F / 2\pi k_B T_c$, so it consistent with small Fermi velocities (v_F) and high T_c value of $CaKFe_4As_4$. Even larger upper critical field can be imagined in clean surface superconductivity having a thin layer of thickness ξ and also, the calculated the mean free path of ℓ = 26.6 nm. Since this mean free path due to elastic scattering from impurities is much larger than the coherence length, $CaKFe_4As_4$ can be describe as a clean superconductor [28].

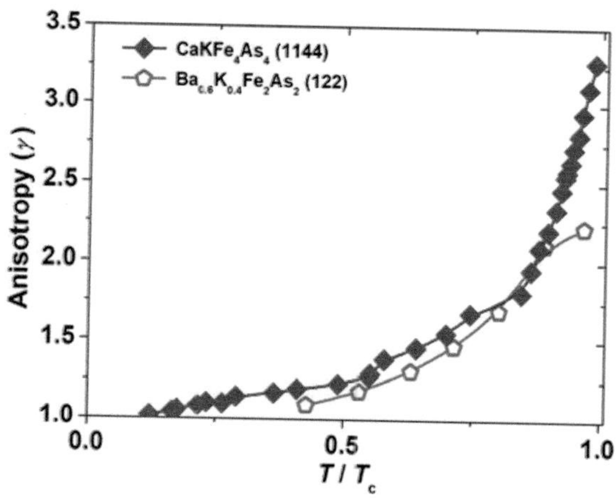

Figure 9.4 The anisotropy parameter γ plotted as a function of temperature for $CaKFe_4As_4$ (1144) and $Ba_{0.6}K_{0.4}Fe_2As_2$ (122) (T_c~28.2 K). The parameter γ decreases with decreasing temperature and approaches to 1 as T tends to 0, indicating isotropic superconductivity. The data used have been taken from Reference [35] for 1144 and Reference [36] for 122.

9.4 Critical Current Properties

A superconductor has a unique property of lossless electric transport. The current carrying capacity of a superconductor can be defined by the critical current density which is determined by the material's ability to trap vortices, namely, vortex pinning. Defect structure inside a superconductor, generally, plays an important role for the vortex pinning behaviours. The critical current density, J_c, is an essential property of a superconductor in order to identify the best candidate for practical applications. Magnetisation hysteresis loops (M-H) at different temperatures allow us to extract the magnetic J_c as a function of magnetic field using the extended Bean's critical state model [39]. For a rectangular-shaped samples, the magnetic field dependence of the critical current density, J_c (in units of A/cm^2) is given by the relation $J_c = 2\Delta M/[a(1\text{-}a/3b)]$ (for $a < b$); where $\Delta M = M_{up} - M_{dn}$, M_{up} and M_{dn} are the magnetisation while magnetic fields are decreasing and increasing, respectively. The magnetisation ΔM expressed in A/cm is obtained from the magnetic moment divided by the sample volume, $V = a \times b \times c$, where a and b are in-plane sample dimensions perpendicular to the applied magnetic field and c is sample dimension, parallel to the applied magnetic field. For the cylindrical sample in order to estimate the critical current, we use the expression $J_c = 3\Delta M/d$, where d is the diameter of the cylinder.

Bean model [39] estimates the critical current density (J_c) of a superconductor assuming that supercurrents flow with a density equal to the critical current density $J_c(H)$, which remains a constant independent of the magnetic field (H). Several other critical state models were proposed by Kim et al. or Kim-Anderson et al. [40], to extend the critical state model for various applications and to incorporate the field dependence of critical current density. However, Bean model is most common and widely used method to estimate the critical current density from M-H curves in which the value of $J_c(T, H)$ does not vary much with the magnetic field for low temperature isotherms. Bean model gives a good estimation of the critical current density and provides a simple, intuitive framework in which data could be analysed, despite the need for approximations when applying the model to samples of finite dimensions.

The initial studies represented that 1144 family has all superconducting properties almost similar to 122 family[27,28]. However, in 2017, extremely large critical current density and its unusual temperature dependence in 1144 single crystal had been reported [20]. After that many papers have been reported on this family [29]. Before that, the following four families of iron-based superconductors have been mainly focused on for the critical current properties and the relevant vortex physics: 1111 (REFeAsO), 122 ($AEFe_2As_2$), 111 (AFeAs), and 11 (FeSe). Most of the experiments were carried out on single-crystal samples and J_c was determined by magnetic hysteresis loop measurements by using Bean model [39], as discussed above. Conclusively, the results can be summarised as: (1) Generally, high J_c is obtained of the order of 10^6 A/cm^2 at low temperature and field, for example, $J_c \sim 2 \times 10^6$ A/cm^2 at 5 K and 0 T for SmFeAs(O,F) (1111) [41], ~2×10^6 A/cm^2 for (Ba,K)Fe_2As_2 (122) [42], ~0.1×10^6 A/cm^2 for LiFeAs (111) [43], and ~0.2×10^6 A/cm^2 for Fe(Te,Se) (11), (2) Strong pinning behaviour from the magnetisation hysteresis measurements and a power-law decay of J_c at the low-field region [44]. (3) The second magnetisation peak (SMP) observed in the *M-H* loops, suggesting high J_c values under high magnetic field, (4) Role of different pinning source, for examples (i) dense vortex pinning nanostructures due to inhomogeneous distributions of dopant atoms [45], (ii) charged dopant atoms [46], and (iii) structural or magnetic domain boundaries [34]. (5) Controversial reports on doping dependence of J_c, for example one report of BaCo122 (Ba(Fe,Co)$_2$As$_2$) [47] showed that J_c and T_c follow the same doping dependence, whereas in other reports for BaCo122 [48] and BaK122 [49], the doping dependence of J_c does not follow the T_c and reaches the maxima for the underdoped region. P doped Ba122 [50] system suggests that J_c is determined by some extrinsic factors rather than by the doping concentration. Hence, the present iron-based superconductors have always an issue about the exact content of doping and due to this, a final agreement on the superconducting J_c behaviour has not been reached yet.

To study the critical current properties of 1144, magnetic hysteresis *M*(*H*) loops have been reported at constant temperatures between 2 and 34 K in an applied magnetic field up to 16 T for the single crystal in both directions *H*//ab and *H*//c, and also for the polycrystalline 1144 sample [20]. High-quality single crystals are used to find the maximum possible superconducting properties of $CaKFe_4As_4$, whereas polycrystals are used to study the granular effect on the superconducting properties. Single crystal 1144 shows the SMP below 26 K for *H*//c but no such peak is observed in *H*//ab and polycrystalline sample. Similar effect is also observed for other superconducting crystals such as BaK122 [42], MgB_2 [4],

LiFeAs [43], and Nb_3Sn [3]. Although the exact reasons behind the SMP (fishtail effect) are not clear yet, it is commonly associated with order-disorder transition and is strongly dependent on the sample orientation in externally applied magnetic field, being diminished when H//(ab). However, in a polycrystalline sample as compared with single crystals, the distribution of grain sizes as well as the grain boundaries is likely to smear out any features, normally, seen in single crystal samples, as found in many iron-based superconductors. The obtained J_c value for H//ab has ultra-high values of the order of 10^8 A/cm^2 which is an order of magnitude larger than the value obtained when H//c, as shown in Fig. 9.5. However, the polycrystalline sample has a low J_c which is at least three orders of magnitude lower than that of single crystal. The low J_c value of the polycrystalline sample is possible due to its extrinsic factors, such as grain morphology, surface roughness, intergrain voids, and interfaces. Hence, the improved synthesis process will be needed for the polycrystalline samples that can reduce these extrinsic factors by adopting high-pressure synthesis, a ball-milling process, resintering, and the optimisation of both the temperature and length of the preparation process [17,22,31], etc.

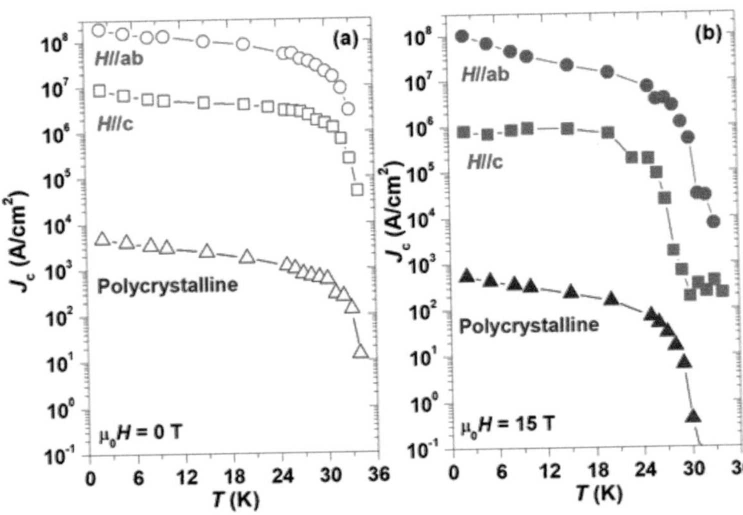

Figure 9.5 The temperature dependence of critical current density J_c for single crystal along c-axis and ab-plane and for polycrystalline $CaKFe_4As_4$ at **(a)** μ_0H = 0 T **(b)** μ_0H = 15 T. The data used have been taken from the reference [20]. Reproduced with permission from [20].

To emphasise this important aspect, Fig. 9.5 shows the temperature dependence of the reported J_c value for single crystal and the polycrystal sample at low and high magnetic fields, i.e. 0 and 15 T. It depicts that J_c shows almost temperature independent behaviour up to 25 K even with increasing magnetic fields up to 15 T. For example, a J_c value of ~1.9(2) × 10^8 A/cm^2 for H//(ab) for 1 T and only dropping to ~0.7(2) × 10^8 A/cm^2 by 15 T [20]. These are some of the highest values obtained for any iron-based superconductor single crystal (Table 9.1). Furthermore, using the coherence length, the estimated depairing current density (J_d) for $CaKFe_4As_4$ is obtained as 1.6 × 10^8 A/cm^2 at low temperature [35]. Such a high critical current value supports strong vortex pinning behaviours in this stoichiometric compounds and suggests this high-T_c superconductor as a potential candidate for practical applications.

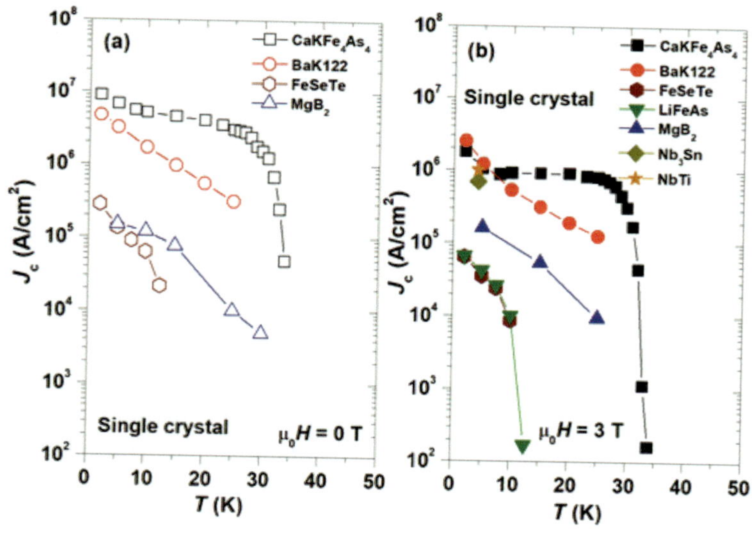

Figure 9.6 The temperature dependence of the critical current density of $CaKFe_4As_4$ single crystal compared with other superconductors such as (Ba,K)Fe_2As_2 (BaK122), Fe(Se,Te) (11), LiFeAs (111), FeSeTe (11), and MgB_2, Nb_3Sn, NbTi in magnetic field **(a)** H = 0 T and **(b)** H = 3 T for H//c. The data used have been taken from the reference [20]. Reproduced with permission from [20].

To make a more clear view about such a large J_c value, Fig. 9.6 shows a comparison of the temperature dependence of critical current behaviour of this 1144 superconductor with other single crystals of iron-based superconductors including MgB_2 and conventional superconductors for H//c direction at magnetic field of 0 and 3 T. It is evident that $CaKFe_4As_4$ have an ultra-high self-field J_c of the order of 10^7 A/cm^2 for H//c and these characteristics remain robust up to 31 K in 3 T (up to a value of 10^6 A/cm^2). Generally, defects inside a superconductor may act as effective pinning centres, so, high density of defects play an important role to achieve high values of the critical current density. One can consider many defects in a doped superconductor. However, $CaKFe_4As_4$ is a stoichiometric compound and a rather clean superconductor, but it exhibits very high critical current values, which suggest another interesting source of pinning rather than doping concentration. It should be noted that in experiment, it is rather difficult to determine unambiguously the type of main pinning centres.

A high J_c of the order of 10^4–10^5 A/cm^2 even under high magnetic field around 10 T is found in many optimally doped iron-based superconductors [16,17]. The critical current properties generally depend on the high T_c, vortex motion and thermal fluctuation, similar to the high-T_c cuprates. Due to these phenomena, the vortex dynamics show interesting effects, such as flux creep, thermally activated flux flow, and the presence of a SMP or the fishtail effect. The ability to carry the critical current is governed by the vortex pinning strength and motion [26]. Different kinds of pinning determine the elastic or plastic motion of vortex lattice, formation of vortex glass, vortex melting, and order-disorder phase transitions [20]. As vortex motion is strongly affected by the effects of disorders in superconductors. In the case of 122, the critical current J_c and SMP effect significantly depend on the type of dopant element and their concentration [26]. On the other hand, the largest J_c for all dopants, as stated earlier, is reported at the slightly underdoped composition, i.e. near the phase boundary between the superconducting phase and the antiferromagnetic-orthorhombic (AFO) phase.

The extremely high critical current density of $CaKFe_4As_4$ suggests that the vortex core energy of the flux lines can be enhanced in this stoichiometric superconductor and the reduced symmetry of tetragonal structure could play an important role of pinning in magnetic field. The fundamental measure of vortex pinning is the pinning force per unit volume F_P, and this dictates J_c. The scaling analysis of the normalised pinning force density f_p(h) depicts the peak position of $f_P(h)$ and takes a value of 0.40–0.45 for the highest-J_c samples [49] ($f_P = F_P/F_P^{max}$; $h = H/H_{irr}$), suggestive of a common mechanism, i.e. point pinning mechanism which plays a dominant role and yields high J_c value. In case of stoichiometric 1144, the flux pinning force density, f_P(h), collapses onto on a single curve with a peak position of $h_{max} = 0.28$ [20] and suggests the existence of both surface and the point core pinning of the vortices.

However, the presence of disordered vortex lattice has been confirmed up to 8 T by the recent STM studies [51] where the vortices are pinned at the locations where the superconducting order parameter is strongly suppressed due to pair breaking. With increasing the magnetic field, the vortex core size decreases which suggests to very large upper critical fields. At the low temperature, the coherence length, i.e. the vortex core size estimated from upper critical field study is ~1.9 nm [35], in reasonable agreement with the value determined by STM of ~1.3 nm [51], which demonstrates the importance of point pinning in this material. Furthermore, the surface topography measured by STM depicts steps on the surface due to structural defects, surface reconstruction and small stripes, which as possible sources of surface pinning. Hence, the STM studies fully support the pinning force analysis of $CaKFe_4As_4$, as discussed above. Other types of two-dimensional surface pinning may be influenced by the anisotropic layered structure and stacking faults that may occur between Fe layers along the *c*-axis [29].

STM studies have also been done for other families of iron-based superconductors where disorder in the vortex lattice has been observed and studied. Interestingly, a varied behaviour has been observed in doped superconductors during experiments similar to the case of J_c. For example, a disordered vortex lattice is obtained in vortex lattice images for 25% K-doped $SrFe_2As_2$ [52] and 10% Co-doped $BaFe_2As_2$ [53], whereas 40% K-doped $BaFe_2As_2$ [54] depicts a more ordered lattice at the same values of the applied magnetic fields. These experimental results clearly indicate that charge doping for these materials by itself cannot explain the origin of vortex pinning. It has been suggested that the different pinning strength found in these compounds might be possible due to the electronic inhomogeneity due to ion size mismatch between substituted elements. For example, the mismatch between K and Ba is five times smaller than between K and Sr [55]. Furthermore, in the case of stoichiometric FeSe, vortices are preferentially positioned near twin boundaries where superconductivity is strongly suppressed. In bulk Fe(Se,Te) irradiated by heavy ions, the amorphous regions at the columnar defects created by irradiation are worked as a pinning centres for vortices which suggest to a complicated pinning landscape [56]. The stoichiometric pnictide LiFeAs has no twin boundaries or any substitution disorder but the observed vortex lattice is a very disordered at large magnetic field which can be due to lattice softening due to high field [57]. The STM studies in stoichiometric $CaKFe_4As_4$ uniquely show the link between the locations with a finite density of states at the Fermi level and the vortex positions [51]. This stoichiometric 1144 studies clearly conclude no any role of the charge doping in the origin of vortex pinning in iron-based superconductors. Also, it suggests that the observed disordered vortex lattice characteristic of iron-based superconductors is due to order parameter suppression at defects whose size is of the order of the vortex core size [51].

Furthermore, the observed high J_c might be related to the intrinsic effects which are somehow related to asymmetry tetragonal structure of 1144. The ultra-high critical current densities and the vortex phase diagram of $CaKFe_4As_4$ [20] place this stoichiometric high-T_c superconducting family as a realistic contender for practical applications.

9.5 Development towards Application

9.5.1 Polycrystalline Sample

Extremely high critical current density of the order of 10^8 A/cm^2 at low temperature, its rather robust field dependence and isotropic upper critical field up to H_{c2}~100 T suggest highly improved superconducting properties [20] as compared with other iron-based superconductors. These experimental findings recommend the 1144 family as a possible strong candidate for superconducting magnet application where high quality wires or tapes are required. As we know, wires and tapes for bulk applications are always based on polycrystalline materials and we need a large amount and high-quality polycrystalline sample. Hence, an optimisation of the synthesis method of 1144 powder is required, so that a robust synthesis process can be established.

Iyo et al. [27] synthesised powder $CaKFe_4As_4$ by solid state reaction method but found that these samples can form also impurity phases of $CaFe_2As_2$ and KFe_2As_2. One should keep in mind that 122 phase is more stable phase than 1144 phase. $CaFe_2As_2$ and KFe_2As_2 (122) are two most common phases that can appear as impurity phases during the formation of $CaKFe_4As_4$. Furthermore, the single crystal growth phase diagram of $CaKFe_4As_4$ [58] shows that the 1144 phase can be only formed in the temperature range between 920 °C and 980 °C, and has a strong sensitivity to impurity phase formation outside this range. Taking clues from this phase diagram, the optimisation of synthesis process of polycrystalline 1144 phase has been recently reported [31] by preparing various polycrystalline samples in a wide temperature range from 900 °C to 1100 °C with very small temperature interval (~25 °C). These samples are characterised by various measurements such as XRD, magnetisation, and transport to reach a final conclusion. These studies find that the synthesis temperature of 955 °C for 6 h is the optimum condition for obtaining pure polycrystalline $CaKFe_4As_4$ with the highest onset T_c value of 34.2 K and a sharp transition width of 2 K. The block diagram of the optimisation process is shown in Fig. 9.7. The sample prepared at 900 °C is dominated by impurity phases formed mainly of KFe_2As_2 (~50%) and $CaFe_2As_2$ (~25%). With increasing synthesis temperature, both impurity phases decreased significantly and no impurity phase is detected in the $CaKFe_4As_4$ synthesised at 955 °C for 6 h. For higher synthesis temperature, the 1144 phase coexists with a dominant impurity phase $CaFe_2As_2$ evident in the structural analysis (~25% for the growth at 1000 °C). A small amount (~1%-2%) of KFe_2As_2 is also observed as reported in the XRD patterns [31]. Furthermore, the heating time dependence had also been checked for the best synthesis temperature (955 °C), which shows that samples prepared at 955 °C for a longer period of time of 17 h have the tendency to start forming $CaFe_2As_2$ impurity phases.

The variation of transition temperature for these polycrystalline 1144 samples with respect to the synthesis temperature is shown in Fig. 9.8(a) which clearly depicts a high T_c for the optimum synthesis conditions (955 °C for 6 h). By increasing the synthesis temperature above 955 °C, the superconducting transition decreased very rapidly and the transition to superconducting state broadens. Furthermore, the critical current J_c behaviour is also shown in Fig. 9.8(b) where the highest values of J_c is obtained for the 1144 sample prepared in optimum synthesis conditions (955 °C, 6 h), it suggests that this sample has better grain connectivity and quality compared to other samples. The field dependence of the calculated J_c for this sample at various temperatures shows that the J_c value at 2 K is larger than 10^4 A cm^{-2} even at 5 T, and improved compared with other 1144 polycrystalline samples [20]. Such a large value has also been observed in other polycrystalline sample of iron-based superconductors [16] and it is much larger than that observed in specifically processed C-doped MgB_2 [4,7].

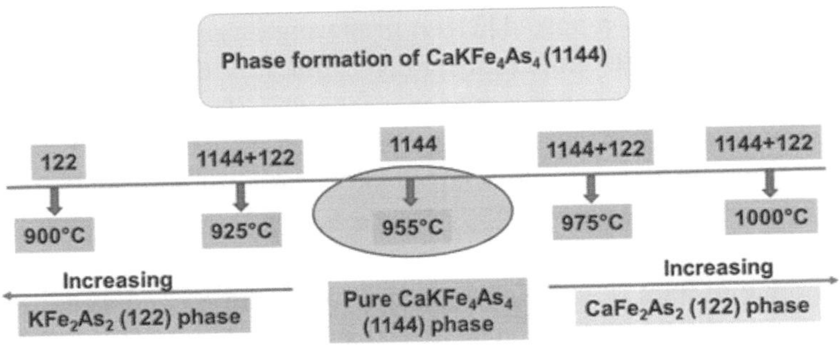

Figure 9.7 Block diagram of the optimisation of synthesis process of $CaKFe_4As_4$ (1144) phase. This diagram is drawn based on the reported studies in [31]. KFe_2As_2 and $CaFe_2As_2$ (122) were appeared as an impurity phase during 1144 synthesis process.

Furthermore, the sample prepared at 955 °C for longer time (17 h), the transition temperature is almost similar to the optimum sample but the superconducting volume fraction is reduced [31]. This may be caused by a small amount of the observed impurity phases. These studies suggest that the pure phase of polycrystalline samples of $CaKFe_4As_4$ could be formed in a very narrow temperature window and for a short heating duration, i.e. 955 °C for 6 h. The pure 1144 can be formed around 955 °C and slightly difference in temperature such as 20–25 °C can generate the impurity phases, as shown in Fig. 9.7. This optimised synthesis temperature lies in middle of the optimised temperature range (between 920 °C and 980 °C) for the phase pure single crystals of $CaKFe_4As_4$. The reproducibility of this optimum synthesis process has been checked by preparing 6 g sample in one batch [31]. Hence, this method can be used to prepared 40–50 g powder sample which generally need for the fabrication of superconducting wires and tapes for the initial stage.

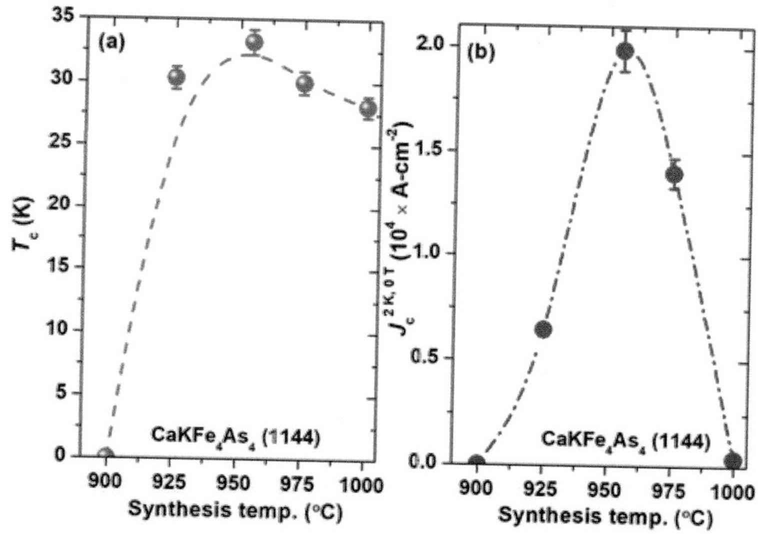

Figure 9.8 The synthesis temperature dependence of **(a)** transition temperature (T_c) and **(b)** the critical current density (J_c) at 2 K and 0 T magnetic field. The data have been taken from the reference [31].

9.5.2 Superconducting Wires and Tapes

Iron-based superconductor cannot be deformed plastically due to its relatively hard and tough phase. So, a filamentary configuration can be obtained by packing powders in a metallic tube. Therefore, the PIT method is a very convenient way for producing superconducting wires and tapes in a laboratory, and can also be used for iron-based superconductors [16]. Also, the PIT process uses very simple deformation techniques and low material costs, which is very attractive from the application aspect. Indeed, PIT techniques is well-established techniques and already used for brittle superconductors such as Nb_3Sn [59], MgB_2 [60], and ceramic Bi-cuprate superconductors [61].

PIT method already has been used for other iron-based superconductors such as 1111, 122, and 11 [16]. This process consists mainly of three steps: (1) filling of a metallic sheath with the precursor powder of the selected superconductor; (2) cold deformation of the so-obtained composite tube down to a wire or a flat tape; (3) final heat treatment in controlled atmosphere for the synthesis or sintering of the superconducting phase core. Normally, there are two different processes for PIT technique: in-situ and ex-situ process. In first case, precursors are mixed according to the final superconducting composition and packed into a metal tube. After deformation, the final reaction performed within the final wires or tapes during the sintering process. In the latter case, the final prepared superconducting materials are sealed in the metal tube and then do the deformation process.

In 2018, after details study of the critical current properties of single crystal 1144 [20], the first 1144 tapes had been reported [62] by using ex-situ PIT method. The prepared 1144 powder was mixed with the small amount of Sn addition and loaded into silver tube with outer diameter ~8 mm and inner diameter ~5 mm. Normally, Sn additions help to improve the grain connectivity of the superconducting materials during the annealing process as reported for other families of iron-based superconductors [16,63]. The Ag-tube with 1144 powder was placed in another metal tubes of different sheath materials Ag, SS/Ag and Cu/Ag which is used for the mechanical strength. Then, these tubes were drawn in form of different thickness of tapes. For sintering, HIP (hot isostatic pressing) furnace was used up to 150 MPa at 600 °C for 1 h. Small length (~3 cm) of the prepared tapes were used for the transport critical current measurements by four-probe method in a magnetic field up to 14 T at 4.2 K. The transport J_c behaviour of these tapes with magnetic field is shown in Fig. 9.9 at 4.2 K with different sheath materials. The transport J_c of the Ag-sheathed tape (prepared at 600 °C, 1 h, ambient pressure) at self-field achieves 7×10^4 A cm^{-2} while it is one order magnitude is lower (~7×10^3 A cm^{-2}) at 4.2 K after the application of 10 T. This value is almost three times larger than that of the Ag-sheathed tape annealed at 500 °C [62]. Hence, the great improvement of the transport J_c was obtained just by the enhancement of the sintering temperature by 100 °C; however, at the same time, impurity phases were also appeared. Additionally, the J_c of the SS/Ag-sheathed tape (prepared at 600 °C, 1 h, ambient pressure) is further improved, as shown in Fig. 9.9. The self-field J_c for this sample reaches 10^5 A cm^{-2}, and even achieves 10^4 A cm^{-2} at 10 T. Microstructural analysis suggests that the improvement mainly comes from the diminished intergranular voids provided by the higher strength outer sheath such as SS-tube. Such high transport J_c of 10^4 A cm^{-2} at the high magnetic field suggests the practicality of this 1144 material in high fields. By applying a pressure of 150 MPa to the superconducting core, the Cu/Ag-sheathed tapes reach the highest performance. The transport J_c reach to 2.1×10^5 A cm^{-2} at self-field and 2.2×10^4 A cm^{-2} at 10 T, indicating that this system is promising for practical application. However, these tapes contain the impurity phases such as KFe_2As_2 and CaAgAs, and had been suggested that silver sheath reacts with core 1144 powder during the sintering process of tapes.

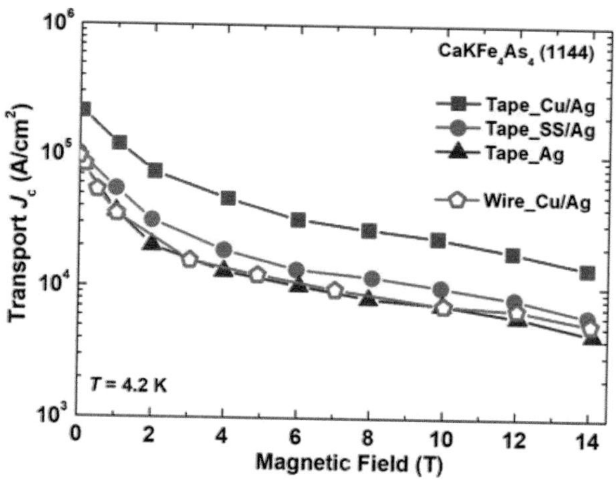

Figure 9.9 Field dependence of transport critical current density J_c at 4.2 K for 1144 wires and tapes where the field is parallel to the tape surface. The data have been taken from the reference [62] and [64].

In the direction of superconducting wire fabrication, one paper had been reported based on the ex-situ PIT method [64] similar to that of the 1144 tape. Ag-tube having outer diameter 4.5 mm and inner diameter 3 mm was filled with the prepared 1144 powder. These tubes were drawn into a round shape wires of diameter of 1.2 mm using dies with circular holes. Similar to 1144 tape process,

HIP techniques were used for the sintering of these wires at 700 °C for 0.5–4 h in an Ar atmosphere with pressures of 0.1–175 MPa. Transport current J_c has been measured at 4.2 K in presence of magnetic field and is shown in Fig. 9.9. The obtained transport J_c of the best wire (prepared at 700 °C, 1 h, 175 MPa pressure) was 9.6×10^4 A-cm^{-2} under self-field and 7.6×10^3 A-cm^{-2} at 10 T. These studies clearly depict that the sintering pressure plays an important role for the J_c value and the core density. At the same time, impurity phases such as KFe_2As_2, FeAs, and CaO were also detected in the core. The presence of these impurity phases reduced the connections of grains which might be a reason of low transport J_c. Fig. 9.9 depicts the transport J_c behaviours of best 1144 wire. Interestingly, the 1144 wires (prepared at 700 °C, 4 h, 175 MPa pressure) and tapes (prepared at 600 °C, 1 h, ambient pressure) with Ag sheath show almost similar behaviour of the transport J_c at 4.2 K. However, the higher transport J_c of the 1144 tapes with Cu-sheath (prepared at 700 °C, 1 h, 150 MPa pressure)) suggests the better grain connectivity than that of other 1144 tapes and wires, besides the impurity phases.

Overall, the reported 1144 wires and tapes had impurities; because of this, the obtained transport J_c was almost three orders of magnitude lower than that of the reported value of the single crystal 1144. However, the reported 1144 wires and tapes indicate that a high-core density and the resulting suppression of weak links by high-pressure sintering play a key role in achieving a practical J_c, as also obtained for K-doped $AEFe_2As_2$ pressed tapes [16]. Furthermore, the sintering of 1144 powder during wires and tapes fabrication at higher temperature above 500 °C degrade the 1144 phase which is similar to the reported post-sintering effects on 1144 polycrystalline sample above 500 °C [31]. Interestingly, the transport J_c of 1144 tapes is one order of magnitude higher than the magnetic J_c for the best 1144 polycrystalline samples that was obtained by the best synthesis method, i.e. 955 °C and 6 h (Fig. 9.8b).

Over the past decades, the 122 family has been the subject of extensive research in the fabrication process of wires and tapes. In recent years, rapid progress was made on the critical current density (J_c) of the 122 wires and tapes based on a PIT technique. Encouraging breakthroughs were made after almost 10 years, including a high transport J_c exceeding the practical level of 10^5 A-cm^{-2} (at 4.2 K, 10 T) and the first 100 meter-class wire [16]. One should be noted, 122 phase is very much stable and has easy synthesis route. After the phase formation, the 122 phase generally does not degrade during sintering or annealing process up to ~850–900 °C. However, 1144 phase has very narrow window for the phase formation and even after phase formation, the sintering or annealing process destroy the 1144 phase above 500 °C. Generally, the sintering or annealing process plays an important role in the fabrication process of high-quality wires and tapes. Further improvement of grains and grain boundaries in the 1144 wires and tapes requires, besides high-quality starting material, advanced processing techniques such as high-impact ball milling, high-pressure synthesis. Above discussion suggests that future studies based on the fabrication process of 1144 wires and tapes will be needed to find a robust method by using the high-quality 1144 powder.

9.6 Conclusions

1144 iron-based superconductors have the critical temperature, T_c, around 35 K without the need for additional doping. Recent studies on 1144 single crystals reveal highly improved superconducting properties as compared with those reported for the 122 family. It includes a high critical current density of the order of 10^8 A-cm^{-2} at low temperature, a rather robust field dependence as well as an isotropic upper critical field of up to $\mu_0 H_{c2}$ = 100 T. These experimental findings place the 1144 family as a strong contender for magnet applications at around 20 K, where the niobium-based superconductors cannot play a role owing to their lower T_c. In the direction of wire and tapes, the optimisation of synthesis process of polycrystalline sample finds the best synthesis conditions and shows that 1144 phase has a very narrow temperature window to the formation of pure phase and small temperature differences of 20–25 °C generates impurity phases in the form of 122 phases. The reported 1144 tapes and wires have been prepared by PIT method and show the transport J_c up to 10^5 A/cm^2 with lots of impurity phases. These studies suggest that further progress is needed for the development and the fabrication of the high-quality tapes and wires of this promising material. We believe, very high-pressure HIP especially prepared hot extrusion with the quenching, can conserve the 1144 phase during wires and tapes fabrication.

References

1. Onnes, H. K., The resistance of pure mercury at helium temperatures, *Comm. Phys. Lab Uni. Leiden,* vol. 120b, p. 124c, 1911.
2. Kittel, C., *Introduction to Solid State Physics*, Hoboken, NJ: John Wiley & Sons, 2005.
3. Godeke, A., "A review of the properties of Nb_3Sn and their variation with A15 composition, morphology and strain state," *Supercond. Sci. Technol.,* vol. 19, p. R68–R80, 2006.
4. Eisterer, M., "Magnetic properties and critical currents of MgB_2," *Supercond. Sci. Technol.,* vol. 20, p. R47–R73, 2007.
5. Zhou, D. et al., "An overview of rotating machine systems with high-temperature bulk superconductors," *Supercond. Sci. Technol.,* vol. 25, p. 103001 (12 pp.), 2012.
6. Bray, J. W., "Superconductors in applications; some practical aspects," *IEEE Trans. Appl. Superconductivity,* vol. 19, pp. 2533–2539, 2009.
7. Buzea, C., Yamashita, T., "Review of the superconducting properties of MgB_2," *Supercond. Sci. Technol.,* vol. 14, p. R115, 2001.
8. Bednorz, J. G., Müller, K. A., "Possible high-T_c superconductivity in the Ba–La–Cu–O system," *Zeitschrift für Physik B Condensed Matter,* vol. 64, pp. 189-193, 1986.
9. Schilling, A., Cantoni, M., Guo, J. D. Ott, H. R., "Superconductivity above 130 K in the Hg–Ba–Ca–Cu–O system," *Nature,* vol. 363, pp. 56–58, 1993.
10. Przybylski, K; Morawski, A., "Preparation of extra high T_c materials-microstructure and properties of Hg-Ba-Ca-Cu-O and Tl-Ba-Ca-Cu-O single crystals and films," in *Advanced Studies on Superconducting Engineering*, Budapest, Hungary, The Supertech Consortium, 2001, pp. 47-80.
11. Alexandrov, A. S., "High-temperature superconductivity: the explanation," *Phys. Scr.,* vol. 83, p. 038301, 2011.
12. Shen, K. M.; Seamus-Davis, J. C., "Cuprate high-T_c superconductors," *Mater. Today,* vol. 11, no. 9, pp. 14-21, 2008.
13. Lee, P. A.; Nagaosa, N.; Wen, X. G., "Doping a Mott insulator: physics of high-temperature superconductivity," *Rev. Mod. Phys.,* vol. 78, p. 17, 2006.
14. Kamihara, Y.; Watanabe, T.; Hirano, M.; Hosono, H., "Iron-based layered superconductor $La[O_{1-x}F_x]FeAs$ (x = 0.05–0.12) with T_c = 26 K," *J. Am. Chem. Soc.,* vol. 130, p. 3296, 2008.

15. Hosono, H.; Yamamoto, A.; Hiramatsu, H; Ma, Y., "Recent advances in iron-based superconductors toward applications," *Mater. Today,* vol. 21, no. 3, pp. 278-302, 2018.

16. Yao, C.; Ma, Y., "Recent breakthrough development in iron-based superconducting wires for practical applications," *Supercond. Sci. Technol.,* vol. 32, p. 023002, 2019.

17. Singh, Shiv J.; Mele, P., "Future potential of new high T_c iron-based superconductors," in *Superconductivity,* Springer, 2020, pp. 243–268.

18. Shimoyama, J., "Potentials of iron-based superconductors for practical future materials," *Supercond. Sci. Technol.,* vol. 044002, p. 27, 2014.

19. Singh, S. J.; Shimoyama, J.; Yamamoto, A.; Ogin, H.; Kishio, K., "Transition temperature and upper critical field in $SmFeAsO_{1-x}F_x$ synthesized at low heating temperatures," *IEEE Trans. Appl. Superconductivity,* vol. 23, no. 3, p. 7300605, 2013.

20. Singh, S. J.; Bristow, M.; Meier, W. R.; Taylor, P.; Blundell, S. J.; Coldea, A. I., "Ultrahigh critical current densities, the vortex phase diagram, and the effect of granularity of the stoichiometric high-T_c superconductor $CaKFe_4As_4$," *Phys. Rev. Mater.,* vol. 2, p. 074802, 2018.

21. Katase, T.; Ishimaru, Y.; Tsukamoto, A.; Hiramatsu, H.; Kamiya, T.; Tanabe, K.; Hosono, H., "Advantageous grain boundaries in iron pnictide superconductors," *Nat. Commun.,* vol. 2, p. 409, 2011.

22. Iida, K.; Hanisch, J.; Yamamoto, A., "Grain boundary characteristics of Fe-based superconductors," *Supercond. Sci. Technol.,* vol. 33, p. 043001, 2020.

23. Singh, Shiv J.; Shimoyama, J.; Yamamoto, A.; Ogino, H.; Kishio, K., "Significant enhancement of the intergrain coupling in lightly F-doped SmFeAsO superconductors," *Supercond. Sci. Technol.,* vol. 26, p. 065006, 2013.

24. Singh, Shiv J.; Beck, R.; Wurmehl, S.; Hess, C.; Buechner, B., "Granular behavior observed in the polycrystalline superconducting LiFeAs," *Supercond. Sci. Technol.,* vol. 28, p. 025006, 2015.

25. Rotter, M.; Tegel, M.; Johrendt, D., "Superconductivity at 38 K in the iron arsenide $Ba_{1-x}K_xFe_2As_2$," *Phys. Rev. Lett.,* vol. 101, p. 107006, 2008.

26. Ishida, S.; Song, D.; Ogino, H.; Iyo, A.; Eisaki, H.; Nakajima, M.; Shimoyama, J.; Eisterer, M., "Doping-dependent critical current properties in K, Co, and P-doped $BaFe_2As_2$ single crystals," *Phys, Rev. B,* vol. 95, p. 014517, 2017.

27. Iyo, A.; Kawashima, K.; Kinjo, T.; Nishio, T.; Ishida, S.; Fujihisa, H.; Gotoh, Y.; Kihou, K.; Eisaki, H.; Yoshida, Y., "New-structure-type Fe-based

superconductors: $CaAFe_4As_4$ (A = K, Rb, Cs) and $SrAFe_4As_4$ (A = Rb, Cs)," *J. Am. Chem. Soc.,* vol. 138, pp. 3410-3415, 2016.

28. Meier, W. R.; Kong, T.; Kaluarachchi, U. S.; Taufour, V.; Jo, N. H.; Drachuck, G.; Böhmer, A. E.; Saunders, S. M.; Sapkota, A.; Kreyssig, A.; Tanatar,, M. A.; Prozorov, R.; Goldman, A. I.; Balakirev, F. F.; Gurevich, A.; Bud'ko, S. L.; Canfield, P. C., "Anisotropic thermodynamic and transport properties of single-crystalline $CaKFe_4As_4$," *Phys. Rev. B.,* vol. 94, p. 064501, 2016.

29. Ishida, S.; Iyo, A.; Ogino, H.; Eisaki, H.; Takeshita, N.; Kawashima, K.; Yanagisawa, K.; Kobayashi, Y.; Kimoto, K.; Abe, H.; Imai, M.; Shimoyama, J.; Eisterer, M., "Unique defect structure and advantageous vortex pinning properties in superconducting $CaKFe_4As_4$," *Npj Quantum Mater.,* vol. 4, p. 27, 2019.

30. Lochner, F.; et al., "Electronic properties, low-energy Hamiltonian, and superconducting instabilities in $CaKFe_4As_4$," *Phys. Rev. B,* vol. 96, p. 094521, 2017.

31. Singh, Shiv J.; Cassidy, Simon J.; Bristow, M.; Blundell, Stephen J.; Clarke, Simon J.; Coldea, Amalia I., "Optimization of superconducting properties of the stoichiometric $CaKFe_4As_4$," *Supercond. Sci. Technol.,* vol. 33, p. 025003, 2020.

32. Ohgushi, K.; Kiuchi, Y., *Phys. Rev. B,* vol. 85, p. 064522, 2012.

33. Werthamer, N. R.; Helfand, E.; Hohenberg, P. C., "Temperature and purity dependence of the superconducting critical field, H_{c2}," *Phys. Rev.,* vol. 147, p. 295, 1966.

34. Sun, D. L.; et al., "Comparative study of upper critical field H_{c2} and second magnetization peak Hsp in hole- and electron doped $BaFe_2As_2$ superconductors," *Phys. Rev. B,* vol. 80, p. 144515, 2009.

35. Bristow, M.; Knafo, W.; Reiss, P.; Meier, W.; Canfield, P. C.; Blundell, Stephen J.; Coldea, Amalia I., "Competing pairing interactions responsible for the large upper critical field in a stoichiometric iron-based superconductor $CaKFe_4As_4$," *Phys. Rev. B,* vol. 101, p. 134502, 2020.

36. Yuan, H. Q.; Singleton, J.; Balakirev, F. F.; Baily, S. A.; Chen, G. F.; Luo, J. L.; Wang, N. L., "Nearly isotropic superconductivity in $(Ba,K)Fe_2As_2$," *Nature,* vol. 457, pp. 565–568, 2009.

37. Serafin, A.; Coldea, A. I.; Ganin, A. Y.; Rosseinsky, M. J.; Prassides, K.; Vignolles, D.; Carrington, A., "Anisotropic fluctuations and quasiparticle excitations in $FeSe_{0.5}Te_{0.5}$," *Phys. Rev. B,* vol. 82, p. 104514, 2010.

38. Hashimoto, K.; Serafin, A.; Tonegawa, S.; Katsumata, R.; Okazaki, R.; Saito, T.; Fukazawa, H.; Kohori, Y.; Kihou, K.; Lee, C. H.; Iyo, A.; Eisaki,

H.; Ikeda, H.; Matsuda, Y.; Carrington, A.; Shibauchi, T., "Evidence for superconducting gap nodes in the zone-centered hole bands of KFe_2As_2 from magnetic penetration depth measurements," *Phys. Rev. B,* vol. 82, p. 014526, 2010.

39. Bean, C. P., "Magnetization of high-field superconductors," *Rev. Mod. Phys.,* vol. 36, p. 31, 1964.

40. Anderson, P. W.; Kim, Y. B., "Hard superconductivity: theory of the motion of Abrikosov flux lines," *Rev. Mod. Phys. 36, 39,* vol. 36, p. 39, 1964.

41. Fang, L.; Jia, Y.; Mishra, V.; Chaparro, C.; Vlasko-Vlasov, V. K.; Koshelev, A. E.; Welp, U.; Crabtree, G. W.; Zhu, S.; Zhigadlo, N. D.; Katrych, S.; Karpinski, J.; Kwok, W K., "Huge critical current density and tailored superconducting anisotropy in $SmFeAsO_{0.8}F_{0.15}$ by low-density columnar-defect incorporation," *Nat. Commun.,* vol. 4, p. 2655, 2013.

42. Yang, H.; Luo, H.; Wang, Z.; Wen, H. H., "Fishtail effect and the vortex phase diagram of single crystal $Ba_{0.6}K_{0.4}Fe_2As_2$," *Appl. Phys. Lett.,* vol. 93, p. 142506, 2008.

43. Pramanik, A. K.; Harnagea, L.; Nacke, C.; Wolter, A. U. B.; Wurmehl, S.; Kataev, V.; Büchner, B., "Fishtail effect and vortex dynamics in LiFeAs single crystals," *Phys. Rev. B,* vol. 83, p. 094502, 2011.

44. van der Beek, C. J. et al., "Flux pinning in $PrFeAsO_{0.9}$ and $NdFeAsO_{0.9}F_{0.1}$ superconducting crystals," *Phys. Rev. B,* vol. 81, p. 174517, 2010.

45. Yamamoto, A.; et al., "Small anisotropy, weak thermal fluctuations, and high field superconductivity in co-doped iron pnictide $Ba(Fe_{1-x}Co_x)_2As_2$," *Appl. Phys. Lett.,* vol. 94, p. 062511, 2009.

46. van der Beek, Cornelis J.; et al., "Quasiparticle scattering induced by charge doping of iron-pnictide superconductors probed by collective vortex pinning," *Phys. Rev. Lett.,* vol. 105, p. 267002, 2010.

47. Shen, B.; et al., "Flux dynamics and vortex phase diagram in $Ba(Fe_{1-x}Co_x)_2As_2$ single crystals revealed by magnetization and its relaxation," *Phys. Rev. B,* vol. 81, p. 014503, 2010.

48. Prozorov, R.; et al., "Intrinsic pinning on structural domains in underdoped single crystals of $Ba(Fe_{1-x}Co_x)_2As_2$," *Phys. Rev. B,* vol. 80, p. 174517, 2009.

49. Song, D.; et al., "Distinct doping dependence of critical temperature and critical current density in $Ba_{1-x}K_xFe_2As_2$ superconductor," *Sci. Rep.,* vol. 6, p. 26671, 2016.

50. Demirdiş, S.; et al., "Disorder, critical currents, and vortex pinning energies in isovalently substituted $BaFe_2(As_{1-x}P_x)_2$," *Phys. Rev. B,* vol. 87, p. 094506, 2013.

51. Fente, A.; et al., "Influence of multiband sign-changing superconductivity on vortex cores and vortex pinning in stoichiometric high-T_c $CaKFe_4As_4$," *Phys. Rev. B,* vol. 97, p. 134501, 2018.

52. Song, C. L.; et al., "Dopant clustering, electronic inhomogeneity, and vortex pinning in iron-based superconductors," *Phys. Rev. B,* vol. 87, p. 214519, 2013.

53. Yin, Y.; et al., "Scanning tunneling spectroscopy and vortex imaging in the iron pnictide superconductor $BaFe_{1.8}Co_{0.2}As_2$," *Phys. Rev. Lett.,* vol. 102, p. 097002, 2009.

54. Shan, L.; et al., "Observation of ordered vortices with Andreev bound states in $Ba_{0.6}K_{0.4}Fe_2As_2$," *Nat. Phys.,* vol. 7, pp. 325–331, 2011.

55. Song, C. L. et al., "Dopant clustering, electronic inhomogeneity, and vortex pinning in iron-based superconductors," *Phys. Rev. B,* vol. 87, p. 214519, 2013.

56. Song, C. L.; et al., "Suppression of superconductivity by twin boundaries in FeSe," *Phys. Rev. Lett.,* vol. 109, p. 137004, 2012.

57. Hanaguri,, T.; et al., "Scanning tunneling microscopy/spectroscopy of vortices in LiFeAs," *Phys. Rev. B,* vol. 85, p. 214505, 2012.

58. Meier, W. R.; et al., "Optimization of the crystal growth of the superconductor $CaKFe_4As_4$ from solution in the FeAs-$CaFe_2As_2$-KFe_2As_2 system," *Phys. Rev. Mater.,* vol. 1, p. 013401, 2017.

59. Kunzler, J. E.; et al., "Superconductivity in Nb_3Sn at high current density in a magnetic," *Phys. Rev. Lett.,* vol. 6, p. 89, 1961.

60. Flükiger, R.; et al., "Superconducting properties of MgB_2 tapes and wires," *Physica C: Superconductivity,* vol. 385, pp. 286-305, 2003.

61. Hikata, T.; et al., "Ag-sheathed Bi-Pb-Sr-Ca-Cu-O superconducting wires with high critical current density," *Jpn. J. Appl. Phys.,* vol. 28, p. L82, 1989.

62. Cheng, Z.; et al., "Effects of core density and impurities on the," *Supercond. Sci. Technol.,* vol. 32, p. 105014, 2019.

63. Singh, S. J.; et al., "Enhancement of intergranular current density of Sm-based oxypnictide superconductors with Sn addition," *Supercond. Sci. Technol.,* vol. 27, p. 085010, 2014.

64. Pyon, S.; et al., "Fabrication and characterization of $CaKFe_4As_4$ round wires sintered at high pressure," *Appl. Phys. Express,* vol. 11, p. 123101, 2018.

Chapter 10

Quasi 1D Layered $Nb_2Pd_xS_y$ Superconductor for Industrial Applications

Reena Goyal, Masato Murakami, and Muralidhar Miryala
Materials for Energy and Environmental Laboratory,
Superconducting Materials Laboratory,
Graduate School of Engineering and Science,
Shibaura Institute of Technology, Koto City, Tokyo 135-8548, Japan
reenagoyal88@gmail.com, miryala1@shibaura-it.ac.jp

This chapter provides the reader with a general overview of $Nb_2Pd_xS_y$ superconducting material. An overview of the electrical and magnetic properties and its various compositions and fabrication processes are discussed. This provides factors affecting superconducting properties of $Nb_2Pd_xS_y$ sample. It shows high upper critical field and flexible nature of superconducting fibers. The possible reasons for large upper critical field in this system are also described here. Due to the large value of upper critical field and the flexible nature of

High-T_c Superconducting Technology: Towards Sustainable Development Goals
Edited by Muralidhar Miryala

ISBN 978-981-4877-65-7 (Hardcover), 978-1-003-16468-5 (eBook)
www.jennystanford.com

$Nb_2Pd_xS_y$ superconducting material could be employed for practical applications and for the development of sustainable goals. These studies paved a way toward the technological development in the field of high-power transmission cables and electronic devices.

10.1 Introduction

Superconductivity, a fascinating research field with outstanding superconducting properties, has inspired the researchers because of its wide range of potential applications. Finding new superconductors to add into the superconducting family has long been a topic of interest in condensed matter physics [2,34,35,51]. After the search for new superconductors, researchers try to find various superconducting industrial applications. The theoretical and experimental studies concluded that there are two main parameters which decide superconducting applications of any superconductors. These are superconducting transition temperature (T_c) and the upper critical field (H_{c2}). Researchers always look for high transition temperature and high upper critical field. In general, high transition temperature suggests low operating cost while, on the other hand, high upper critical field claims the robustness of any superconductor against the magnetic field. These two parameters are beneficial for finding practical applications of any superconductor, but for some applications like high power transmission cable, researchers always have challenged to make flexible materials. Among various superconductors discovered yet, the discovery of cuprates and iron-based superconductors has sparked enormous interest in the scientific community because of various interesting properties [3,8,19]. From the previous experimental studies, it was observed that the highest transition temperature achieved in the case of cuprates and iron superconductors is 133 K and 56 K, respectively [2,28,29,42,53]. Structurally, cuprate and iron-based superconductors can be characterized as a sandwiched "hamburger" model [46]. It is known, the demand for YBCO with high transition temperature rises into industrial sectors for better performance and it is because of the requirement of low cost to achieve its superconducting properties [16]. The low-temperature superconductors are also gathered interest of researchers to study widely because of various other exciting properties belong to these superconductors. For example, Sr_2RuO_4 is a low-T_c compound, but due to the possibility of Cooper pairs pairing in a spin triplet configuration always motivates researchers to explore it vastly [21,32]. $LaFePO_3$, another superconductor with low T_c, prompted researchers at it led to the discovery of high transition temperature iron-based superconductors [23]. Various other low-temperature superconductors such as Li_2Pd_3B, $BaTi_2Sb_2O$, $LaNiGa_2$, and $IrxCh_2$ (Ch = Se and Te) have been discovered [11,15,18,48] and their discovery attracts community because of their scientific importance and potential applications.

By keeping the importance of these two parameters, $Nb_2Pd_xS_y$, a low-dimensional Pd-based ternary chalcogenide superconductor discovered by Q. Zhang et al. in 2013, motivated researchers to find its various exciting properties [57]. Structurally, Nb_2PdS_5 is a low-dimensional compound with a 1D chain-like structure. Although the transition temperature is ~6.6 K, the upper critical field of this superconductor crosses the Pauli paramagnetic limit. It has drawn much attention because of the high upper critical field and easy-to-grow flexible fibers because of its low-dimensional behavior. By using the ultrasonically cleaving technique, a low-dimensional superconducting wire can be easily cut into rectangular shaped wires with uniform width. It is important to have a flexible nature of superconducting material for its practical uses [36]. It is a challenge for researchers to make any flexible superconductors with outstanding properties. The first long flexible fibers of $Nb_2Pd_xS_y$ superconductor were prepared by Yu et al. with excellent upper critical field value and low transition temperature [54]. Interestingly, the discovery of long flexible fibers with the suitable size provided us a new route to high-power transmission cables and electronic devices [54]. After exploring the exciting superconducting properties of this compound, lots of isostructural superconductors have also synthesized, including $Nb_2Pd_xSe_5$, $Ta_2Pd_xS_5$, Ta_2PdSe_5, etc. [14,24,31,47,56].

Therefore, this chapter summarizes the fabrication techniques of $Nb_2Pd_xS_y$ superconductor. Detailed structural and superconducting properties of this superconductor are discussed. The effect of various factors on superconducting properties is also represented in detail. Here, the normal-state properties of this superconductor are also examined. Finally, the conclusion of this chapter is presented at the end.

10.2 Preparation Method

10.2.1 $Nb_2Pd_xS_y$ Superconductor

A standard solid-state reaction route is a simple technique used to prepare materials. In this reaction, starting elements react chemically without any presence of solvent and then heated at high temperature to complete the reaction. Generally, to synthesize superconductors in different morphology, this method is used. $Nb_2Pd_xS_y$ superconductor has also been synthesized using a standard solid-state reaction route inside the glove box using starting elements like Nb, Pd, and S, respectively. The pellet of the mixed elements was sealed in a quartz tube to avoid attacking of air. Finally, the sealed quartz tube was kept in a furnace for the heating sample at temperature 800–900 °C to complete the reaction. Here, normal cooling followed to synthesize polycrystals [22] and the quenching or slow cooling were followed for the growth of single crystals and single crystal fibers [52, 54]. The main advantage of this technique is simplicity in handling and cost-effective.

10.3 Structural and Superconducting Properties

10.3.1 Crystal Structure and Morphology of $Nb_2Pd_xS_y$ Superconductor

Nb_2PdS_5 superconductor has a monoclinic structure with C2/m space group and its schematic crystal structure is exemplified in Fig. 10.1(a).

Its structure comprises 1D chains along the b direction, which are formed by the sheets of a single Nb_4PdS_{10} basic unit on sharing their face and edges [36]. Room temperature powder XRD pattern of Nb_2PdS_5 sample fitted using Rietveld refinements is shown in Fig. 10.1(b). Scanning electron microscopy (SEM) of $Nb_2Pd_xS_y$ fibers showed that the fibers were formed in a cylindrical form, see

Fig. 10.1(c). From XRD pattern, we can clearly observe that there is no extra impurities present in this sample. The lattice parameters were extracted for the single crystal and polycrystal $Nb_2Pd_xS_y$ sample, given in Table 10.1.

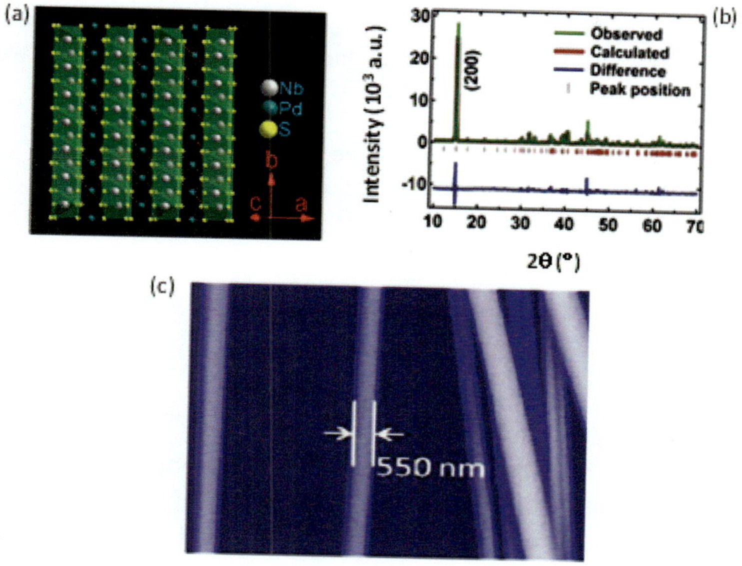

Figure 10.1 (a) Unit cell for $Nb_2Pd_xS_5$ sample [36] (b) Room temperature powder XRD pattern with Rietveld for $Nb_2Pd_xS_5$ sample [52] (C) SEM image of $Nb_2Pd_xS_5$ sample [54].

Table 10.1 Lattice parameters extracted for $Nb_2Pd_xS_y$ single crystal and polycrystal

Sample	Morphology	a (Å)	b (Å)	c (Å)	Ref.
$Nb_2Pd_xS_y$	Single crystal	12.134	3.283	15.09	[52]
$Nb_2Pd_xS_y$	Polycrystal	12.134	3.277	15.023	[22]

10.3.2 Superconducting Properties

10.3.2.1 Temperature-dependent electrical resistivity r (T)

The most important property of the superconductor is the electrical resistivity, which vanishes completely below the critical transition temperature. $Nb_2Pd_xS_y$ superconductor is a low-temperature superconductor and its application may depend on the high value of an upper critical field and low-dimensional behavior. The temperature-dependent electrical resistivity of $Nb_2Pd_xS_y$ sample was measured by several groups to find the superconducting transition temperature. For $Nb_2Pd_{0.81}S_5$ single crystal, the value of critical transition temperature at which resistivity drops to zero was estimated to be equal to 6.5 K [57].

Table 10.2 Critical transition temperature for $(Nb/Ta)_2Pd_x(S/Se)_y$ samples

Compound	T_c (K)	Reference
$Ta_2Pd_{0.92}S_5$	6	[31]
$Ta_2Pd_{0.97}S_6$	3.96	[47]
Ta_2PdSe_5	2.5	[56]
Nb_2PdSe_5	5.9	[24]

Jha et al. have further confirmed the superconductivity in bulk polycrystal Nb_2PdS_5 sample with T_c around 5.9 K [22]. Yu et al. have synthesized series of samples belongs to $Nb_2Pd_xS_{5-y}$ with varying x and y composition. The maximum onset transition at 7.43 K was observed for $Nb_2Pd_{0.963}S_{4.967}$ sample. Also, in the wide range ($0.6 < x < 1$, $y < 0.61$), they found transition temperature in the range of 4.67 to 7.43 K, which suggests that the superconductivity is strongly robust in $Nb_2Pd_xS_{5-y}$ fibers [54]. Later, Yadav et al. have observed the superconductivity with onset transition temperature (the temperature at which resistivity drops to 90% from normal-state resistivity) at ~7.8 K and the temperature at which resistivity drops to zero is 6.5 K. The narrow superconducting transition width is 1.3 K (less than 2K), which indicates the high quality of the single crystal [52]. The observation of low transition temperature from temperature-dependent electrical resistivity by different groups categorized this superconductor as a low-temperature superconductor. We present a list of transition temperature value for $(Nb/Ta)_2Pd_x(S/Se)_y$, new Pd-based ternary superconductors isovalent to $Nb_2Pd_xS_y$ compound, in Table 10.2. Clearly, all the discovered superconductors similar to $Nb_2Pd_xS_y$ superconductor showed transition temperature T_c nearly equal to 7 K and hence these superconductors are also considered as low-temperature superconductors. The variation of resistivity with temperature in Nb_2PdS_5 bulk polycrystal is shown in Fig. 10.2, which gives an idea how its resistivity changes with temperature.

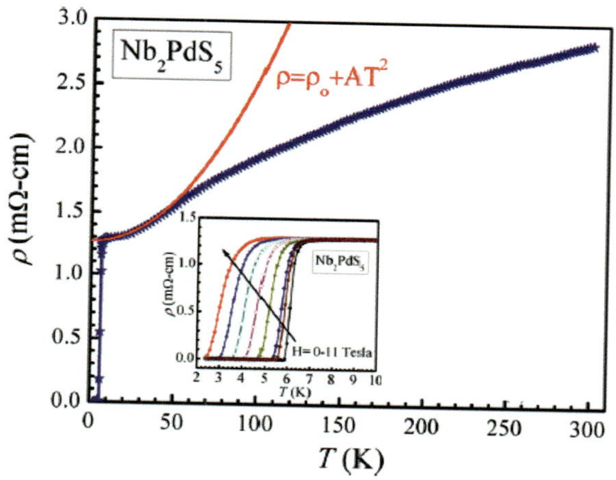

Figure 10.2 Temperature-dependent resistivity in temperature range 2–300 K, inset shows the temperature-dependent electric field under several applied field [22].

Further, to see the effect of the magnetic field on superconductivity, the temperature-dependent electrical resistivity was carried out in superconducting region at several magnetic fields (magneto transport measurement). It was observed that the value of T_c was monotonically decreased with increasing external applied magnetic field. The decrement in T_c with an applied magnetic field (dT_c/dH) determines the value of the upper critical field of superconductor. It has confirmed that the value of T_c shifts toward the lower temperature on increasing applied magnetic field. It was found that for bulk polycrystal, T_c decreases from 5.9 K (zero field) to 2.4 K with an applied magnetic field of 110 kOe [22]. Also, they have observed that the relative decrement of T_c with the applied field, i.e., dT_c/dH is around 0.3 K/10 kOe. A similar study on single crystal was carried by Yadav et al. and they found that the value of T_c decreases with the rate of 0.05 K/kOe and 0.02 K/kOe along H ⊥ b-axis and H || b-axis, respectively [52]. Conclusively, these studies showed that its T_c

value decreases with the slow rate on applying an external magnetic field, which claimed that this compound is superconductivity slowly decreases with the increasing magnetic field. On comparing the value of dT_c/dH obtained for Nb_2PdS_5 bulk polycrystal with other famous superconductors as presented in Table 10.3, it has been claimed that this superconductor is strongly robust against the magnetic field. The value of the upper critical field is evaluated from magneto-transport measurement using Ginzburg–Landau (GL) formula, $H_{c2}(0) = H_{c2}(T)(1 - T/T_c)^2$, where $H_{c2}(0)$ is the upper critical field at zero temperature.

Table 10.3 Comparison of dT_c/dH for Nb_2PdS_5 sample with other different superconductors

Compounds	dT_c/dH	Reference
Nb_2PdS_5	0.3 K/1 T (see Fig. 10.2, inset)	[22]
HTSC	3 K/10 kOe	[30]
Fe pnictides	1 K/1 T	[19]
MgB_2	1.5 K/1 T	[49]
$PbMo_6S_8$	0.45 K/1 T	[38]

The obtained value of the upper critical field for this compound from different studies is exemplified in Table 10.4. Clearly, $Nb_2Pd_xS_y$ system displays a remarkably high value of the upper critical field which crosses the Pauli paramagnetic limit given by $H_{c2} = 1.84T_c$ [10].

Table 10.4 Upper critical field value for $Nb_2Pd_xS_y$ samples

	Morphology	Condition	H_{c2} (T)	Ref.
	Single crystal	Field along b-axis	37	[57]
	Single crystal fibers	For d = 500 nm	42	[54]
$Nb_2Pd_xS_y$	Bulk polycrystal	—	24	[22]
	Single crystal fibers	Field along b axis	39	[52]
	Single crystal fibers	For d = 340 nm	43.2	[20]
	Bulk polycrystal	—	19.1	[12]

Also, the ratio of the upper critical field to transition temperature obtained for $Nb_2Pd_xS_5$ sample is compared with the values extracted for other superconductors, see Table 10.5. Nb_3Sn is popularly used in superconducting applications, such as for the fabrication of superconducting magnet and medical MRI machines [1,25,40]. The ratio of the upper critical field and transition temperature is larger than Nb_3Sn superconductor also, which is technically very important. From generalized Ginzburg–Landau theory equation, $\xi = [\Phi_0/2\pi\mu_0 H_{c2}(0)]^{1/2}$, the value of coherence length (ξ) was also determined using upper critical field value obtained at zero temperature.

Table 10.5 Comparison of ratio of upper critical field and transition temperature of different superconductors

Compounds	H_{c2}/T_c	Reference
$Nb_2Pd_xS_5$	37 T/6.5 K	[57]
Nb_3Sn	30 T/18.3 K	[33]
β-FeSe	20 T/8.7 K	[26]
$Ba_{1-x}K_xFe_2As_2$	70 T/28 K	[55]
$PbMo_6S_8$	60 T/13.3 K	[38]

Here, in this case, the large value of upper critical field implied small coherence length. The value of coherence length for $Nb_2Pd_{0.81}S_5$ single crystal was found to be 3 nm [57].

10.3.2.2 Magnetic measurement

The magnetic measurement needed to confirm the superconducting property of $Nb_2Pd_xS_y$ system as only zero resistance may not mean the bulk superconducting characteristics. Figure 10.3 shows the temperature-dependent real (χ') and imaginary (χ'') part of ac susceptibility for Nb_2PdS_5 polycrystal measured at a frequency of 777 Hz and an ac field amplitude of 10 Oe. The real part shows diamagnetic transition at 5.8 K and the imaginary part has a sharp peak at the same temperature and the saturation is observed at around 3 K. The diamagnetic transition temperature was found to be equal to the transition temperature reported from temperature-dependent electrical resistivity measurement, which confirmed the bulk superconductivity in this system [22]. From resistivity and magnetic measurements, $Nb_2Pd_xS_y$ sample is categorized as low-temperature bulk superconductor.

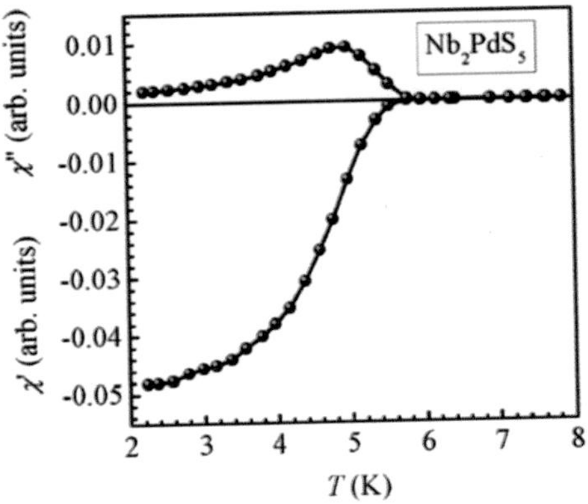

Figure 10.3 Real and imaginary part of ac susceptibility for Nb_2PdS_5 sample in temperature range of 2–8 K field [22].

10.3.2.3 Anisotropy in upper critical field

Anisotropy is observed in single crystals and with the samples in which all grains exhibit common crystallographic orientation. Single crystals are mostly preferred to calculate the anisotropy as it provides the most reliable value. $Nb_2Pd_xS_y$ single crystal shows large direction-dependent upper critical, which indicates anisotropy in this sample [57]. In general, for conventional superconductors, the anisotropy (Γ) is calculated using the ratio of two superconducting parameters (Penetration depth, upper critical field, band dependent effective mass, etc.) in two orientations with respect to the crystallographic axes and applied field. They followed the conventional approach to calculate the anisotropy and its value found to be equal to 2.2. Later, Yadav et al. have followed the conventional and scaling approach to limit the uncertainty in determination γ of as compared to conventional approach [52]. The anisotropy parameter Γ of the $Nb_2Pd_xS_y$ superconducting single crystal is examined by measuring the angular dependence of resistivity under various magnetic fields at different temperatures. From this measurement, the reduced magnetic field was calculated using equation

$$H_{red} = H\sqrt{\sin^2(\theta) + \Gamma^2\cos^2(\theta)}$$

According to the Ginzburg-Landau theory [5], the curves of resistivity versus reduced magnetic field under different magnetic fields should collapse on to each other. In Fig. 10.4(a), we can see that all the curves are merging onto each other and the adjusted parameter known as anisotropy found to be equal to 2.55 ± 0.5. Also, the anisotropy for $Nb_2Pd_xS_y$ sample is changing with temperature dependent for $Nb_2Pd_xS_y$ sample, see Fig. 10.4(b) [52, 57]. The value of anisotropy decreases as the temperature of the sample goes down in the superconducting region. These studies suggest that the temperature-dependent behavior of anisotropy may be due to the multiband nature of this superconductor as the similar behavior reported in multiband superconductors like $BaFe_{1.9}Co_{0.8}As_2$ and $Fe_{1.04}Te_{0.6}Se_{0.4}$ single crystals [43, 44].

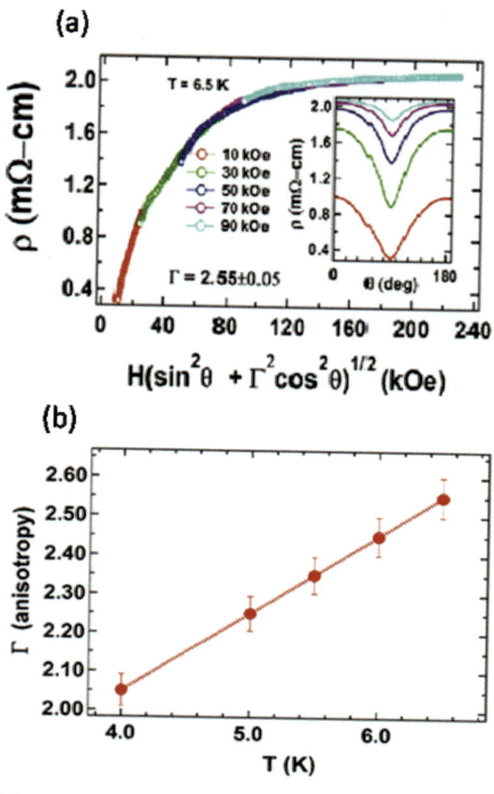

Figure 10.4 (a) Inset shows the angle-dependent resistivity where angle is between b-axis and magnetic field and main panel of figure shows the resistivity plot as a function of reduced magnetic field and (b) temperature-dependent anisotropy curve [52].

10.3.2.4 Specific heat

Specific heat measurement for $Nb_2Pd_xS_5$ sample showed the thermodynamic nature of the superconducting transition. The value of specific heat was measured in the temperature range of 2 to 300 K. The data reached a value of ~250 J/mol-K at 300 K [12]. The temperature-dependent specific heat curve in a range of 2 to 10 K follows

$$C = \gamma T + \beta T^3$$

The first and second terms represent the electronic and phononic-specific heat. The value of Sommerfeld coefficient (γ) and Debye constant (β) was found to be equal to 41.75 mJ/mol-K^2 and 1.14 mJ/mol-K^4, respectively. The value of Debye temperature was calculated to be equal to 251.7 K and this is higher than the superconducting transition temperature [12]. The temperature-dependent electronic specific heat which was estimated after subtraction of phononic specific heat from the total heat capacity, showed an anomaly at the superconducting transition, confirmed bulk superconductivity in this system [12,22,57]. A shallow jump was also observed, which further claimed multiband superconducting nature in Nb_2PdS_5 sample [12,22], as discussed in section 10.3.2.5.

10.3.2.5 Superconducting gap

It is essential to explore the existence of an energy gap in a new low-dimensional superconductor. There are some discrete reports available in open literature regarding the energy gap of the $Nb_2Pd_xS_y$ superconductor. Few reports support a single energy gap and few are favor of two superconducting gaps in this system. Energy gap values have been deduced by using upper critical field [57], anisotropy in upper critical field [52,57], specific heat studies [12,22,57], Muon spin relaxation and rotation measurements [4], penetration depth [13], and spectroscopic technique [39]. The spectroscopic technique is considered a powerful technique to evaluate the energy gap of any superconductor [6]. The temperature-dependent upper critical field increases with decreasing temperature [57] and also a linear dependency of anisotropy on temperature indicate multiband superconductivity in this system, see Section 10.3.2.3 [52,57]. Muon-spin relaxation and rotation measurements and penetration depth measurement, however, claimed a single s-wave energy gap in this system [4,12]. Later, specific heat measurement shows that it is necessary to involve two gap superconductivity to describe the specific heat data. The electronic specific heat (C_{es}) as a function of T/T_c followed exponential function $C_{es} = \exp(-\Delta/k_B T)$, K_B is the Boltzmann constant. This curve was fitted using the two-band model, i.e., $C_{es} = n_1 C_1 + n_2 C_2$, which revealed multiband superconductivity in Nb_2PdS_5 sample [12]. The estimated large and small ratio of the superconducting gap to transition temperature obtained from specific measurements was 6.4 and 1.9, respectively [12]. The direct evidence for two-band superconductivity was first observed by the point contact spectroscopic measurements [39]. The multiband nature of the superconducting gap in this system was supported from the dependency of differential conductance on temperature and magnetic field. For this system, the large and small superconducting gaps values calculated from this measurement were $\Delta_1 = 0.61$ meV and $\Delta_2 = 1.20$ meV, respectively [39]. The corresponding ratio of the large and small superconducting gap to T_c was 4.2 and 2.1, different from the values obtained from specific heat measurements. This difference may be due to the limited temperature range over which specific heat measurement was carried out. From this, it was concluded that specific heat measurements need to be verified again with further low-temperature measurement.

10.4 Factors Affecting Critical Parameters (T_c and H_{C2})

10.4.1 Effect of Doping Elements

The substitutions of elements in the parent compound are important in various aspects. It may enhance the critical temperature and critical field of one compound. Also, one can find the existence of related high transition temperature superconductor. After finding new superconductor, researchers always try to look for suitable doping elements to understand the effect on superconducting properties. Lots of studies have been conducted by number of groups to find the effect of doping elements on the critical parameters of Nb_2PdS_y

system. Various elements like Se, Ni, Pt, Ir, Ag and Ru have been substituted in $Nb_2Pd_xS_5$ superconductor [7,37,45,58]. All the samples were studied structurally using XRD to find the lattice parameters. The change in the value of lattice parameters of $Nb_2Pd_xS_5$ sample was noted with respect to the concentration of doping elements.

The effect of these doping elements on lattice parameters of the parent $Nb_2Pd_xS_y$ sample is given in Table 10.6.

Table 10.6 Effect of doping on lattice parameters of $Nb_2Pd_xS_5$ superconductor

Doping element	Sample	a (Å)	b (Å)	c (Å)	Reference
Se	$Nb_2Pd(S_{1-x}Se_x)_5$	Increasing	No change	Increasing	[37]
Ni	$Nb_2Pd_{1-x}Ni_xS_5$	Increasing	No change	Decreasing	[58]
Pt	$Nb_2Pd_{1-x}Pt_xS_5$	Increasing	No change	Decreasing	[58]
Ir	$Nb_2(Pd_{1-x}Ir_x)_{0.76}S_5$	No change	No change	No change	[45]
Ag	$Nb_2(Pd_{1-x}Ag_x)_{0.76}S_5$	Increasing	No change	Increasing	[45]
Ru	$Nb_2Pd_{1-x}Ru_xS_5$	No change	No change	Decreasing	[7]

Now, the effect of doping elements on the superconducting properties of the $Nb_2Pd_xS_y$ sample is discussed. For $N_2Pd(S_{1-x}Se_x)_5$ polycrystalline samples experimentally, it was observed that superconducting transition temperature in this system was suppressed with increasing Se doping and completely vanished for $x \geq 0.5$. Also, the ratio of the large upper critical field to the transition temperature was calculated for this system and found no change in the ratio with respect to doping [37], see Fig. 10.5. They claimed that spin orbit coupling could play an important role in large upper critical field value in this system.

Later, to confirm the role of spin orbit coupling for the large upper critical field, Zhou et al. have doped lighter heavier elements (Ni and Pt) at Pd site in the Nb_2PdS_5 system [58]. For $Nb_2Pd_{1-x}Ni_xS_5$ polycrystal sample, the superconductivity appeared in system till $x = 0.2$

and for $Nb_2Pd_{1-x}Pt_xS_5$ polycrystal sample, the superconductivity appeared till $x = 0.4$, respectively. The magnetization measurements were also performed to confirm the bulk nature of superconductivity in these systems. Later, the strength of the upper critical field was measured by the ratio of the upper critical field to critical transition temperature (H_{c2}/T_c), which was initially enhanced with the Pt doping until $x = 0.1$ and then decreases with further doping of Pt until $x = 0.4$, see Fig. 10.5. In the case of Ni doped samples, the ratio of H_{c2}/T_c was considerably suppressed with increasing doping concentration, Fig. 10.5.

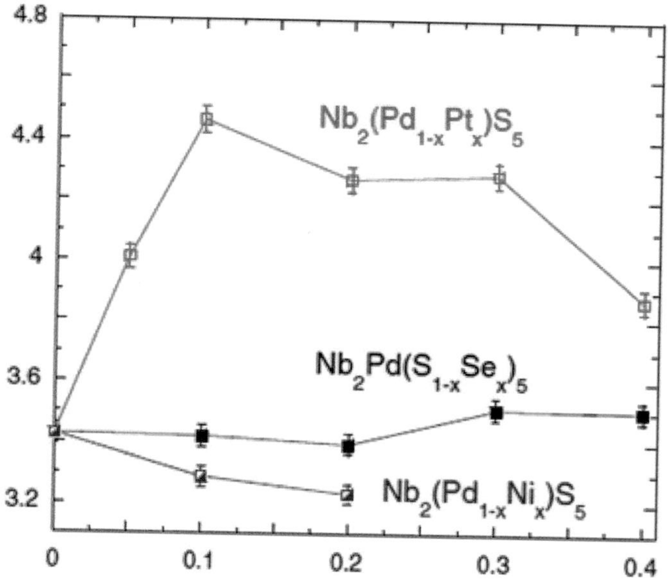

Figure 10.5 H_{c2}/T_c with respect to Se, Ni and Pt doping in Nb_2PdS_5 system [58].

Interestingly, the ratio in Pt doped sample was superior to the samples doped with the same amount of Se or Ni at all levels of doping, see Fig. 10.5. This study supported that large upper critical field in Nb_2PdS_5 system was due to the spin orbit coupling associated with the heavy Pd elements [24,31]. Summarily, when the heavier Pt ions were doped at the Pd site, the spin orbit coupling (SOC) was increased and hence enhanced the ratio of H_{c2}/T_c in the Pt doped system [58].

In 2016, Shen et al. synthesized $Nb_2(Pd_{1-x}Ir_x)_{0.76}S_5$ and found that T_c was first increased to the maximum value at 8 K for $x = 0.4$ and with the further doping T_c drops below 0.5 K for complete replacement of Pd by Ir [45]. While for $Nb_2(Pd_{1-x}Ag_x)_{0.76}S_5$, T_c was suppressed quickly up to $x = 0.4$ and suddenly vanished completely with further increasing doping concentration. They found that the electron doping (Ag) suppressed superconductivity quickly in comparison to hole doping (Ir). Hall Effect measurement revealed that the charge density of this system varied with Ir and Ag doping. Therefore, this study claimed the charge career density could play an important role to tune the superconducting order of this system.

Later, the hole doping with Ru at Pd site in Nb_2PdS_5 system was experimentally performed to compare with the Ir doped samples [7]. The effect of charge carriers and the SOC on the upper critical field was clearly distinguished by comparing Ru doping with Ir doping at Pd site in Nb_2PdS_5 system as the Ru doping hardly change the SOC (directly proportional to Z^4) while Ir doping could enhance the SOC associated with the Pd site. It was demonstrated that superconductivity in Nb_2PdS_5 system remains up to 80% of Ru doping and 100% of Ir doping, which suggest the robust behavior of superconductivity against hole doping in comparison to other Fe-based superconductors [27,50]. The ratio H_{c2}/T_c was greatly enhanced in case of Ir doping as compared to Ru doping and hence suggested that the upper critical field in this system was increased due to the SOC. Conclusively, these studies claimed that the charge carrier and the spin orbit coupling have a significant role in the superconductivity of the $Nb_2Pd_xS_y$ system.

10.4.2 Effect of Diameter of Fibers

As discussed in above section, the upper critical field of $Nb_2Pd_xS_y$ superconductor is high, which crosses Pauli paramagnetic limit. The upper critical field of this system is high in case of fiber morphology as compared to bulk polycrystalline form, see Table 10.4. Yu et al. have observed the strong dependency of upper critical field on the diameter of fibers [54]. Form Fig. 10.6, it is clear that the fibers with small diameter exhibited large upper critical field in comparison to the fibers with small diameters, suggested upper critical field can be enhanced by making the fibers thinner.

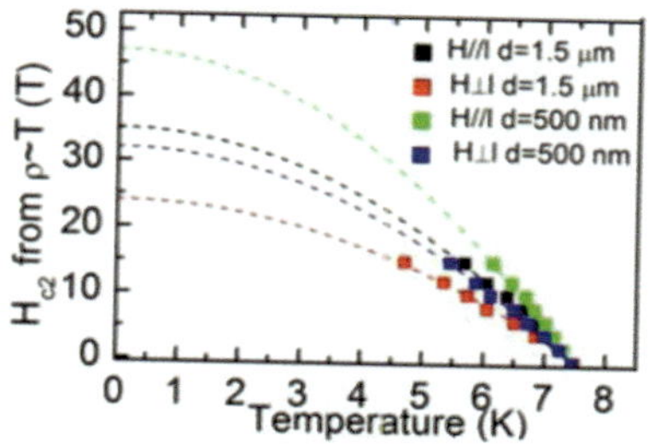

Figure 10.6 Variation of upper critical field with temperature at different diameters [54].

Later, to keep this in mind, Huang et al. experimentally determined the upper critical field and transition temperature dependency on $Nb_2Pd_{0.77}S_5$ fibers diameter and hence conducted a series of experiments [20]. In the experiment outcomes, they observed the marginal change in transition temperature with respect to the diameter of fibers, while the upper critical field was inversely proportional to the diameter of fibers. They also noted that the onset transition temperature corresponds to fibers with different diameters decreases with the applied external field, i.e., followed the Meissner effect. At zero field, there was less change observed in onset transition temperature with decreasing diameter of fibers from 3.26 μm to 0.34 μm, suggest robust behavior of superconductivity with diameter. The value of the upper critical field was calculated using GL formula, described in Section 10.3.2.1. It was observed that the value of upper critical field was increased from 24.3 T to 43.2 T when the diameter of fibers reduced from 3.26 μm to 0.34 μm, respectively. They also claimed that when the sample is very thin, the value of the upper critical field can be extremely large. This experimental study successfully demonstrated that the large vale of upper critical field was strongly dependent on the diameter of $Nb_2Pd_xS_5$ fibers. In the case of Niobium also, the value of the upper critical field was found to be approximately equal to 0.1 T and it could be about 1.2 T due to suitable filament [17]. The current density was also enhanced with decreasing the diameter of fiber [54].

10.5 Hall Effect

Jha et al. performed Hall resistivity and Hall coefficient measurements to understand the contribution of charge carriers in Nb_2PdS_5 system [22]. They found that the Hall resistivity at different temperatures was positive below $T = 100$ K, indicating that the dominant charge carriers are hole-type and above 100 K the Hall resistivity was negative above 100 K, indicating that the dominant charge carriers are electron-type. From this study, it was concluded that the Nb_2PdS_5 sample exhibits a change in type of dominating charge carriers above 100 K. To further confirm this behavior, they have measured Hall coefficient (R_H) and found that R_H changes its sign from negative to positive around 190 K, i.e., from electron type to hole type, which means it is likely a multiband superconductor [41], also supported by other measurements on same system, see Section 10.3.2.4. Later, Chen et al. have also measured the Hall coefficient and supported the multiband nature in $Nb_2Pd_xS_y$ system [7].

10.6 Normal-State Temperature-Dependent Electrical Resistivity

This section provides the normal-state properties of the Nb_2PdS_5 compound from which the sample goes into the superconducting state. Generally, the variation of electrical resistivity in a temperature range from room temperature to the temperature just above the superconducting transition temperature provides normal-state properties of any superconductor. Some representative examples of the measurements for the $Nb_2Pd_xS_y$ sample are offered here to give an idea of the common behavior. In the normal state, the variation of resistivity with temperature showed metallic behavior of the $Nb_2Pd_xS_y$ system [12,22,52,54]. The residual resistivity ratio (RRR) is generally defined as the ratio of resistivity value at 300 K to the resistivity value at $T = 0$ K (RRR = $\rho_{300\text{ K}}/\rho_{Tc}$), where ρ_{Tc} is the value of resistivity at onset transition temperature if the sample is superconducting. The value of RRR indicates the quality of the sample since the impurity presents in the sample increases the residual resistivity and also shows the metallicity of a material. Experimentally, Tomy et al. observed that the value of RRR

(ρ_{300K}/ρ_{8K}) for $Nb_2Pd_xS_y$ single crystal fiber was approximately equal to 3.4, which was less than the corresponding value of good conductors and hence it was categorized as bad metal [52]. Figure 10.2 shows that the temperature-dependent resistivity for Nb_2PdS_5 bulk polycrystal sample in the temperature range of $T_c < T < 50$ K followed the following relation:

$$\rho = \rho_0 + AT^n$$

The exponent "n" was found to be equal to 2, suggested bulk polycrystal followed Fermi liquid behavior [12,22]. The residual resistivity ρ_0 found to be equal to 1.27 mΩ-cm and A= 1.126×10^{-4} mΩ-cmK^{-2}, respectively [22].

10.7 Conclusion

The discovery of superconductivity in quasi 1-D Pd-based ternary-layered compound $Nb_2Pd_xS_y$ brings in a new $(Nb/Ta)_2Pd_x(S/Se)_y$ superconducting family. Here, the crystal structure and various superconducting properties of the $Nb_2Pd_xS_y$ sample are reviewed in detail. Electrical resistivity and magnetic measurements confirmed the bulk superconductivity in these compounds. The large value of the upper critical field suggests that the $Nb_2Pd_xS_y$ compound is very robust against the magnetic field, which is good for practical applications like Nb_3Sn superconductor, which is utilized as a superconducting magnet. Also, flexible superconducting fibers can be utilized for various electronic applications. The spectroscopic technique, specific heat, Hall effect, temperature-dependent upper critical field, and anisotropy in the upper critical field suggest that $Nb_2Pd_xS_y$ is probably a multiband superconductor. In contrast to this, few measurements like Muon-spin relaxation and rotation and temperature-dependent penetration depth measurements support a single s-wave energy gap in this system. The large upper critical field in $Nb_2Pd_xS_y$ system is inversely proportion to the diameter of superconducting fibers. The doping in Nb_2PdS_5 sample via various elements suggests that the upper critical field in this system depends on the charge carrier as well as on the spin orbit coupling associated with its Pd site. According to our understanding, the large upper critical field and its highly flexible nature of $Nb_2Pd_xS_y$ superconducting material could be applicable in various practical applications like high-power transmission cables, superconducting magnets, and electronic devices.

Acknowledgment

One of the authors (Dr. Reena Goyal) is thankful to Shibaura Institute of Technology (SIT) for providing financial support as postdoctoral candidate. We would like to thank Director SIT for allowing us to write a contributing chapter in Jenny Stanford Publishing.

References

1. Baig T, Yao Z, Doll D et al. (2014) Conduction cooled magnet design for 1.5 T, 3.0 T and 7.0 T MRI systems, *Supercond. Sci. Tech.*, **27**, pp. 125012.
2. Bardeen J, Cooper LN, Schrieffer JR (1957) Theory of superconductivity, *Phys. Rev.*, **108**, pp. 1175.
3. Bednorz JG, Müller KA (1986) Possible high T_c superconductivity in the Ba–La–Cu–O system, *Z. Phys. B*, **64**, pp. 189–193.
4. Biswas P, Luetkens H, Xu X et al. (2015) Nodeless superconductivity in quasi-one-dimensional Nb_2PdS_5: A μ SR study, *Phys. Rev. B*, **91**, pp. 100504.
5. Blatter G, Geshkenbein VB, Larkin A (1992) From isotropic to anisotropic superconductors: A scaling approach, *Phys. Rev. Lett.*, **68**, pp. 875.
6. Buzea C, Yamashita T (2001) Review of the superconducting properties of MgB_2, *Superond. Sci. Tech.*, **14**, pp. 115.
7. Chen Q, Yang X, Yang X et al. (2017) Enhanced superconductivity in hole-doped Nb_2PdS_5, *Front. Phys.*, **12**, pp. 127402.
8. Chen X, Wu T, Wu G et al. (2008) Superconductivity at 43 K in $SmFeAsO_{1-x}F_x$, *Nature*, **453**, pp. 761–762.
9. Chu C, Gao L, Chen F et al. (1993) Superconductivity above 150 K in $HgBa_2\ Ca_2Cu_3O_{8+\delta}$ at high pressures, *Nature*, **365**, pp. 323–325.
10. Clogston AM (1962) Upper limit for the critical field in hard superconductors, *Phys. Rev. Lett.*, **9**, pp. 266.
11. Doan P, Gooch M, Tang Z et al. (2012) $Ba_{1-x}Na_xTi_2Sb_2O$ ($0.0 \leq x \leq 0.33$): A layered titanium-based pnictide oxide superconductor, *J. Am. Chem. Soc.*, **134**, pp. 16520–16523.

12. Goyal R, Tiwari B, Jha R et al. (2015) Specific heat of robust Nb_2PdS_5 superconductor, *J. Supercond. Nov. Mag.*, **28,** pp. 1427–1432.

13. Goyal R, Awana V, Patnaik S (2016) Single gap s-wave superconductivity in Nb_2PdS_5, *Physica C*, **524,** pp. 24–27.

14. Goyal R, Gupta G, Srivastava A et al. (2016) Superconductivity at 5.5 K in Nb_2PdSe_5 compound, *J. Supercond. Nov. Mag.*, **29,** pp. 2705–2710.

15. Guo J, Qi Y, Matsuishi S et al. (2012) T_c maximum in solid solution of pyrite $IrSe_2$–$RhSe_2$ induced by destabilization of anion dimers, *J. Am. Chem. Soc.*, **134,** pp. 20001–20004.

16. Guo Q, Zhang P, Bo L et al. (2017) An application of high-temperature superconductors YBCO to magnetic separation, *Int. J. Mod. Phys. B*, **31,** pp. 1745001.

17. Hauser J, Helfand E (1962) Critical field of thin superconducting shapes, *Phys. Rev.*, **127,** pp. 386.

18. Hillier AD, Quintanilla J, Mazidian B et al. (2012) Nonunitary triplet pairing in the centrosymmetric superconductor $LaNiGa_2$, *Phys. Rev. Lett.*, **109,** pp. 097001.

19. Hsu F-C, Luo J-Y, Yeh K-W et al. (2008) Superconductivity in the PbO-type structure α-FeSe, *Proc. Natl. Acad. Sci.*, **105,** pp. 14262–14264.

20. Huang H, Tan M, Ji P et al. (2016) Interesting diameter-dependent upper critical field in $Nb_2Pd_{0.77}S_5$ superconductor, *J. Supercond. Nov. Magn.*, **29,** pp. 309–313.

21. Ishida K, Mukuda H, Kitaoka Y et al. (1998) Spin-triplet superconductivity in $Sr_2\ RuO_4$ identified by ^{17}O Knight shift, *Nature*, **396,** pp. 658–660.

22. Jha R, Tiwari B, Rani P et al. (2014) Robust superconductivity with large upper critical field in Nb_2PdS_5, *J. App. phys.*, **115,** pp. 213903.

23. Kamihara Y, Hiramatsu H, Hirano M et al. (2006) Iron-based layered superconductor: LaOFeP. *J. Am. Chem. Soc.*, **128,** pp. 10012–10013.

24. Khim S, Lee B, Choi K-Y et al. (2013) Enhanced upper critical fields in a new quasi-one-dimensional superconductor $Nb_2Pd_xSe_5$, *New J. Phys.*, **15,** pp. 123031.

25. Lamm MJ (2003) Nb/sub 3/Sn accelerator magnet development around the world, *IEEE Trans. Appl. Superconduct.*, **13,** pp. 1278–1283.

26. Lei H, Graf D, Hu R et al. (2012) Multiband effects on β-FeSe single crystals, *Phys. Rev. B*, **85,** pp. 094515.

27. Li L, Luo Y, Wang Q et al. (2009) Superconductivity induced by Ni doping in $BaFe_2As_2$ single crystals, *New J. Phys.*, **11,** pp. 025008.

28. Liu R, Wu G, Wu T et al. (2008) Anomalous transport properties and phase diagram of the FeAs-based $SmFeAsO_{1-x}F_x$ superconductors, *Phys. Rev. Lett.*, **101**, pp. 087001.

29. Liu Z, Suo H, Ma L et al. (2011) Preparation and superconducting properties of F-doped $SmO_{1-x}F_xFeAs$, *J. Mater Sci. & Technol.*, **27**, pp. 227–232.

30. Liyanawaduge N, Singh SK, Kumar A et al. (2012) Magnetization and magneto-resistance in $Y(Ba_{1-x}Sr_x)_2Cu_3O_{7-\delta}$ (x = 0.00–0.50) superconductors, *Supercond. Sci. Tech.*, **25**, pp. 035017.

31. Lu Y, Takayama T, Bangura AF et al. (2014) Superconductivity at 6 K and the violation of Pauli limit in $Ta_2Pd_xS_5$, *J. Phys. Soc. Jpn.*, **83**, pp. 023702.

32. Luke GM, Fudamoto Y, Kojima K et al. (1998) Time-reversal symmetry-breaking superconductivity in Sr_2RuO_4, *Nature*, **394**, pp. 558–561.

33. Matthias B, Geballe T, Geller S et al. (1954) Superconductivity of Nb_3Sn, *Phys. Rev.*, **95**, pp. 1435.

34. Meissner W, Ochsenfeld R (1933) Ein neuer effekt bei eintritt der supraleitfähigkeit, *Naturwissenschaften*, **21**, pp. 787–788.

35. Nagamatsu J, Nakagawa N, Muranaka T et al. (2001) Superconductivity at 39 K in magnesium diboride, *Nature*, **410**, pp. 63–64.

36. Ning W, Yu H, Wang N et al. (2014) Nonlinear transport in quasi-one-dimensional Nb_2PdS_5 nanowires, *App. Phys. Lett.*, **105**, pp. 172603.

37. Niu C, Yang J, Li Y et al. (2013) Effect of selenium doping on the superconductivity of $Nb_2Pd(S_{1-x}Se_x)_5$, *Phys. Rev. B*, **88**, pp. 104507.

38. Okuda K, Kitagawa M, Sakakibara T et al. (1980) Upper critical field measurements up to 600 kG in $PbMo_6S_8$, *J. Phys. Soc. Jpn.*, **48**, pp. 2157–2158.

39. Park E, Lee S, Ronning F et al. (2018) Spectroscopic evidence for two-gap superconductivity in the quasi-1D chalcogenide $Nb_2Pd_{0.81}S_5$, *J. Phys. Condens. Mat.*, **30**, pp. 165401.

40. Rosner H, Former C (2012) Intermagnetics remembered: From superconductor-based GE spin-off to billion dollar valuation. In: IEEECSC ESAS Eur. Supercond. News Forum.

41. Rullier-Albenque F, Colson D, Forget A et al. (2012) Multiorbital effects on the transport and the superconducting fluctuations in LiFeAs, *Phys. Rev Lett.*, **109**, pp. 187005.

42. Schilling A, Cantoni M, Guo J et al. (1993) Superconductivity above 130 K in the Hg–Ba–Ca–Cu–O system, *Nature*, **363**, pp. 56–58.

43. Shahbazi M, Wang X, Ghorbani SR et al. (2012) Angular dependence of pinning potential, upper critical field, and irreversibility field in underdoped $BaFe_{1.9}Co_{0.1}As_2$ single crystal, *App. Phys. Lett.*, **100,** pp. 102601.

44. Shahbazi M, Wang X, Dou S et al. (2013) The flux pinning mechanism, and electrical and magnetic anisotropy in $Fe_{1.04}Te_{0.6}Se_{0.4}$ superconducting single crystal, *J. App. Phys.*, **113,** pp. 17E115.

45. Shen C, Si B, Bai H et al. (2016) Pd site doping effect on superconductivity in $Nb_2Pd_{0.76}S_5$, *EPL*, **113,** pp. 37006.

46. Stewart G (2011) Superconductivity in iron compounds, *Rev. Mod. Phys.*, **83,** pp. 1589.

47. Tiwari B, Prasad BB, Jha R et al. (2014) Superconductivity at 4 K in Pd-deficient layered $Ta_2Pd_xS_6$, *J. Supercond. Nov. Mag.*, **27,** pp. 2181–2183.

48. Togano K, Badica P, Nakamori Y et al. (2004) Superconductivity in the metal rich Li-Pd-B ternary boride, *Phys. Rev. Lett.*, **93,** pp. 247004.

49. Vajpayee A, Jha R, Srivastava A et al. (2011) The effect of synthesis temperature on the superconducting properties of n-SiC added bulk MgB_2 superconductor, *Supercond. Sci. Tech.*, **24,** pp. 045013.

50. Wang C, Li Y, Zhu Z et al. (2009) Effects of cobalt doping and phase diagrams of $LFe_{1-x}Co_xAsO$ (L = La and Sm), *Phys. Rev. B*, **79,** pp. 054521.

51. Wu M-K, Ashburn JR, Torng C et al. (1987) Superconductivity at 93 K in a new mixed-phase Y-Ba-Cu-O compound system at ambient pressure, *Phys. Rev. Lett.*, **58,** pp. 908.

52. Yadav AK, Sharma H, Thakur AD et al. (2015) Growth and angular dependent resistivity of $Nb_2Pd_{0.73}S_{5.7}$ superconducting single crystals fiber, *Mater. Chem. Phys.*, **164**, pp. 46–50.

53. Yang J, Liang J, Jin D et al. (1997) The influence of fluorine on the structures and properties of $Pr_{2-x}Sr_xCuO_{4-y}$, *J. Phys. Condens. Mat.*, **9,** pp. 1249.

54. Yu H, Zuo M, Zhang L et al. (2013) Superconducting fiber with transition temperature up to 7.43 K in $Nb_2Pd_xS_{5-\delta}$ (0.6< x< 1), *J. Am. Chem. Soc.*, **135,** pp. 12987–12989.

55. Yuan H, Singleton J, Balakirev FF et al. (2009) Nearly isotropic superconductivity in (Ba, K) Fe 2 As 2, *Nature*, **457,** pp. 565–568.

56. Zhang J, Dong J, Xu Y et al. (2015) Superconductivity at 2.5 K in the new transition-metal chalcogenide Ta_2PdSe_5, *Supercond. Sci. Tech.*, **28,** pp. 115015.

57. Zhang Q, Li G, Rhodes D et al. (2013) Superconductivity with extremely large upper critical fields in $Nb_2Pd_{0.81}S_5$, *Sci. Rep.*, **3**, pp. 1–7.

58. Zhou N, Xu X, Wang J et al. (2014) Controllable spin-orbit coupling and its influence on the upper critical field in the chemically doped quasi-one-dimensional Nb_2PdS_5 superconductor, *Phys. Rev. B*, **90**, pp. 094520.

Chapter 11

High-Temperature Superconducting Cable Application to Ship Magnetic Deperming and Its Contribution toward SDG

Megumi Hirota
Representative, Naval Ship M&UEP R. C., 1-3-15, Tokyo 1130023, Japan
megumi.hirota@nmurc.com

Naval ship magnetic deperming is a method to keep ship's safety against weapons equipped with magnetic sensors, such as sea-mines, and has been operated by several navies. Deperming of ship requires a few milli Tesla of magnetic field to saturate the magnetization of steel, which is used as a material in the construction of ships. Recent advancements of high-temperature superconducting (HTS) cable technologies are realizing a deperming coil set on seabed enabling stable operation of the ship. Due to the large size of a ship, which is several tens of meters wide, and high decreasing ratio of magnetic field, depending on a distance from the source of magnetic field, large electric current of more than 100 kA is required to pass

High-T_c Superconducting Technology: Towards Sustainable Development Goals
Edited by Muralidhar Miryala

ISBN 978-981-4877-65-7 (Hardcover), 978-1-003-16468-5 (eBook)
www.jennystanford.com

through one cable. Zero electric resistivity of HTS conductor realizes this system. To achieve a complete coil with MgB_2 conductor cooled by liquid hydrogen supported well-designed radiation shield, R&D steps with conventional HTS conductor and helium gas cooling are to be conducted. The complete system is a low-energy-loss yet an effective deperming system, and furthermore, its technologies with these derivatives contribute to low-loss and large-current electric power line transmissions in field and underwater which support the SDG.

11.1 Introduction

Boats were built with wood when people tried for the first time to go over water in any place in the world. In the fifteenth century, large ships to cruise across oceans were built also by wood with advanced design of keel, mast, and deck. Modern ships with propulsion sources as steam engine and with metal hulls were developed after the Industrial Revolution, which began in the eighteenth century. From the end of the nineteenth century, naval ships were mostly built with metal hulls, especially with steel hull, due to its high stiffness allowing the ship with smaller weight of hull. This is more prominent for larger ships which require to be more resilient to larger forces of bending and twisting of the hull.

Most naval ships are constructed with ferrous steel with some exceptions. Ships commissioned for minesweeping are built mainly by wood [1], glass fiber reinforced plastic (GRP) [2], and magnetic steel [3]. Torpedo boats, which are required to reach high cruising speed and be lightweight, and LNG container ships, which need high strength at low temperature, use aluminum alloy in their major structures [4]. The submarines of the Alpha class of the Russian navy are constructed from titanium alloys for the hull to dive into deep water [5]. The hull of German Navy submarine is constructed from high-tensile nonmagnetic steel [6], then the magnetization is restricted to ferrous machines onboard [7].

Steel has many favorable properties, such as high strength, low expense of sourcing, and mature manufacturing technologies to control its physical properties. Mild steel with low carbon content has been used for shipbuilding and recently developed high-tensile steel is used for large-scale ships to reduce the weight of ship structure.

Steel consists of high amounts of iron atoms which have magnetic moments in its natural state, and therefore most steels have magnetization except for stainless steels and some nonmagnetic steels. A great amount of steel is used for a ship, such as for ship-hull, deck, electric machines, propulsion mechanics, weapons and weapon's launcher, and has magnetization depending on various shipbuilding procedure and ultimately generate magnetic field around the finished ship. The intensity of this magnetic field increases steeply with shorter the distance from the ship, and thus the approach of a ship can be detected by observing the anomalies in magnetic field. Sea-mines equipped with magnetic sensors are destructive weapons against ships [8]. Magnetic sensors laid on seabed are a silent yet reliable defense to detect underwater invasion passing over them. Magnetic sensors onboard antisubmarine aircraft are effective to detect submerged submarines. To counter the threat of magnetic detection, navies have put efforts to reduce the magnetization of their ships since the Second World War [9].

British navy is the first to treat magnetic silencing of their ships against magnetic mines in 1940. They set coils outside of ship-hull to cancel induced magnetization by the Earth's magnetic field [10]. This coil is made from copper wire which was important material in wartime. Because of the urgent threat during wartime, and the lack of power of coils for small ships, the process of wiping magnetization was conducted by using high-powered cable outside of the hull [11].

Magnetization of ship is that induced by Earth's magnetic field and remnant one which is independent to ship motion. They are both magnetization due to the alignment of microscopic magnetization in material, but are separately treated in magnetic silencing of ship.

The treatment to reduce this remnant magnetization of ship is called ship deperming. As for deperming methods of magnetic material, one is to heat the material over Curie temperature and then to cool in the zero magnetic field environment. Another method is to impose AC (alternative current) magnetic field with the maximum intensity to saturate the magnetization of the material, which is several thousands of A/m for ferrous steel, then to reduce intensity gradually to zero. The latter method is applicable to ship deperming.

Complete deperming of ship needs a magnetic field to saturate magnetization of all magnetic components of ship, such as ship hull, support structures, electric generator, motor, propulsion mechanics, and weapon launchers. Since equipment inside the ship is magnetically shielded by hull, deperming of hull structure must be considered carefully. On the contrary, for ships with nonmagnetic hull structures these equipment and machines can be treated separately. However, since most naval ships have hull structure with magnetic steel, deperming of hull is to be considered first.

Even with the magnetic field of medium intensity of a few milli Tesla (a few thousand A/m) on ship to deperm, large volume of ship requires the coil cable to pass large electric current through. Recent development of superconducting material with high critical transition temperature (HTC) and the production technologies for conductors of the cables, design of deperming coil set on seabed can be realized. In comparison with the deperming by conventional conductor, the energy can be saved by this HTC system and contributes to sustainable world development.

11.2 Magnetic Silencing of Ship

Silencing of ship magnetization $\vec{M}$ is implemented by degaussing and by deperming treatment. They are expressed schematically by the following equation (11.1):

$$\vec{M} = \tilde{\chi}\tilde{R}\overrightarrow{H_e} + \sum_i \tilde{c}_i \overrightarrow{L_i} I_i + \overrightarrow{M_0} \tag{11.1}$$

Degaussing is to decrease induced magnetization of the ship by Earth's magnetic field $\overrightarrow{H_e}$ using cables equipped on the ship. Earth's magnetic field is fixed on the Earth, while total magnetic moment vector $\vec{M}$, permanent component of the moment of ship $\overrightarrow{M_0}$, and degaussing coil vector are fixed on ship, rotation matrix of axes $\tilde{R}$ is imposed on $\overrightarrow{H_e}$. $\tilde{\chi}$ is the permeability matrix decided by the magnetic property and the shape of the material. $\overrightarrow{L_i}$ is i-th degaussing coil vector decided by coil shape. I_i is the current passed through the i-th coil and $\tilde{c}_i$ is the boundary matrix decided by the magnetic materials' boundary.

When the ship is cruising, $\tilde{R}$ in the first term of Eq. 11.1 is changing with respect to time. In the second term, $\tilde{c}_i$ and $\overrightarrow{L_i}$ are fixed by the coil shape and its position in the ship. $\overrightarrow{L_i}$ is to change to cancel out the first term. This is the theory of degaussing.

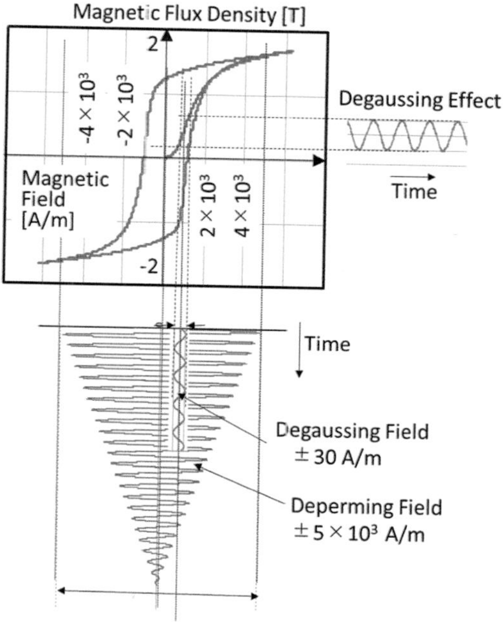

Figure 11.1 Degaussing and deperming magnetic field shown on B-H curve of typical high-tensile steel.

Deperming is a treatment to reduce remnant magnetization of ship $\overrightarrow{M_0}$ by cable temporally set outside of ship during the treatment. The magnetic field imposed to this ship by the coil is of alternately changing signs and of gradual reducing maximum intensity from saturation field. Therefore, the magnetic field required to impose on ship for degaussing is the Earth's field level of several tens of micro Tesla. On the other hand, the magnetic field for deperming ship is a few milli Tesla to saturate magnetization of ship. Degaussing and deperming fields on magnetic property of B-H curve of typical high-tensile steel are shown schematically in Fig. 11.1. The time dependence of deperming field is shown with linearly decreasing peak intensity as a schematic figure. The step to reduce the intensity and the duration of each step is decided mainly by experience. Starting from urgent treatment and experiences in war time, scale model experiments had been conducted and concluded by Baynes [12,13]. Theoretical study on deperming signature using Preizach model calculation and scale model experiment is reported to find the efficient deperming process [14].

11.2.1 Degaussing of Ship

Degaussing of the ship is implemented by cables installed inside of ship hull and electric current passing through these cables controlled by the information of ship motion. The rotary motion of ship, expressed by $\tilde{R}$ with regard to the Earth's magnetic field $\overrightarrow{H_e}$, such as heading, rolling and pitching of the ship are informed by the magnetic sensor output which is fixed onto the ship. The information of vertical and horizontal motion of the ship is from an acceleration sensor or a GPS, and the Earth's magnetic field $\overrightarrow{H_e}$ may be corrected referring the Earth's field map.

The ship hull is made from combination of steel plates with various shapes, thicknesses, and histories. Design of cable installed on the hull and the current to supply is for compensating magnetization induced by surrounding magnetic field based on mathematical simulation which the parameters are obtained from experimental data.

Degaussing cables set on naval ship using HTS technology are reported by Fitzpatrick et al. [15]. This degaussing system was set on the cruiser of the US Navy and confirmed of its degaussing effect [16]. Because of zero electric resistance of the HTS cables, thinner degaussing cable is sufficient to pass the required current in comparison of that of a conventional conductor. Electric power and weight of the cable are reduced whilst compensating the additional power and weight of refrigerator for the superconductor. Compared to the serial connection with controlling turns of each coil in conventional degaussing coil system, short-divided coils with superconducting cable together by controlling current to each coil enable continuous control of signature divided by each coil set. Small power supply to each coil is sufficient because of the zero resistivity of superconducting cable.

11.3 Magnetic Deperming of Ship

Deperming is to reduce remnant magnetization of ship, which is expressed by permanent magnetization of ship. This permanent magnetization is stable while the ship stays in same area of the Earth's magnetic field but once in different field area [17] or suffering mechanical stresses such as by diving deep under the water [18], instability may happen because of steel's magnetic hysteretic property.

Since degaussing is to impose magnetic field for cancelling the magnetization of ship, permanent magnetization can be also cancelled out by degaussing systems imposing DC current through degaussing coils. Deperming works to minimize the electric power consumption. This is especially important for the boats with restricted electric power, such as submarine ships with conventional power source. Another point of importance of deperming is the difficulty of monitoring the magnetization of self-ship. As the intensity of the magnetic field generated by magnetic material decreases steeply with respect to the distance from the material, measured data are affected by the closest magnetic material. Due to this and considering that a ship is a complex structure of various magnetic materials, any measurements attempt of total ship magnetization onboard ship are hardly successful despite numerous research trials conducted. Instead of monitoring ship magnetization inside of the ship, measurement by portable sensors set temporally on seabed is successfully conducted [19]. Compensating the operation of the deperming by DC current of the degaussing system may even provoke some risk to increase the magnetization of the ship.

11.3.1 Deperming Field for Ship

Deperming has been conducted by navies around the world and magnetic field to impose on ship is decided by each navy from the data obtained on their ships and is kept confidential. Here, magnetic field for deperming is derived from the field to saturate magnetization of ship. Ships are built with mostly low carbon steels and recent larger ship designs with thin hull structure to keep larger payload requires high-tensile steel to be used for ship-hull. Figure 11.2 shows the yield strength of shipbuilding steel written by each specification [20–25].

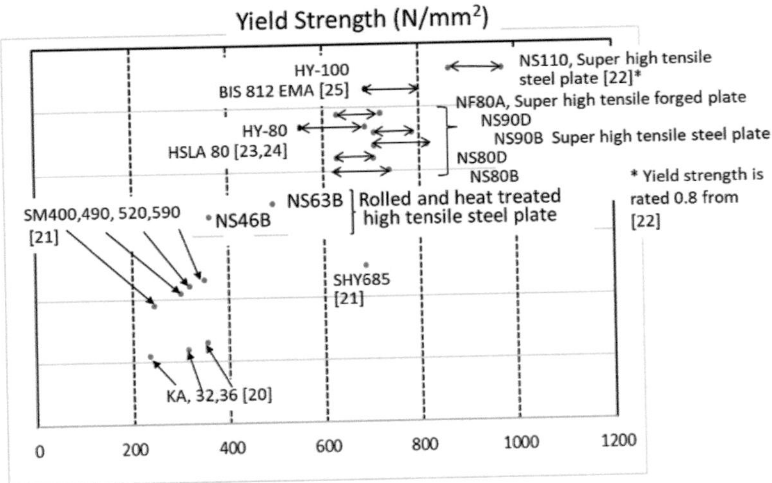

Figure 11.2 Yield strength of steels for shipbuilding.

Yield strength is the maximum stress that the material can endure before it begins to deform plastically, and it is approximately 20% less of its tensile strength which is the limit of fracture. Steels for shipbuilding need high yield-strength and must be easy to bend and weld in the building process. KA designation, under NK standards [20], are mild steels and the number following indicates high-tensile property with the strength written by kg weight/mm^2. NS designation, under National Defense Standards of Japan [22], is the steel for mainly naval-submarine shipbuilding. High-yield steels from standard number 46 to 110 are developed intermittently.

Yield-strength of steel relies on magnetic cohesive force through magnetic domain wall motion [26]. The higher magnetic cohesive force is expected by a steel with higher yield strength. The letters B and D in Fig. 11.2 indicate the thickness of the plate to be 6 to 42 mm and 42 to 65 mm, respectively. The lower maximum yield strength of D-type means the degradation of yield strength for thicker plate.

Magnetic field criteria for deperming are estimated from cohesive force on B-H property of shipbuilding steel. Steel samples of ships for civil use can be obtained in the open market but steel for naval ships such as NS standards are classified and thus cannot be obtained freely. However, tensile strength and component ratio are provided by the standard, from which the information of a similar steel can be found.

Figure 11.3 shows the measured B-H property of shipbuilding steel of SM490 (Steel Marine) and high tensile steel for automobile of SPFC980 (Steel Plate Formability Cold) [27]. The number 980 indicates the minimum tensile strength of 980 MPa from which the 20% reduction results in 800 MPa an estimation of its yield strength, a value similar to that of NS80. Table 11.1 lists their cohesive force H_c and magnetic field H_s to saturate 96% of full magnetization.

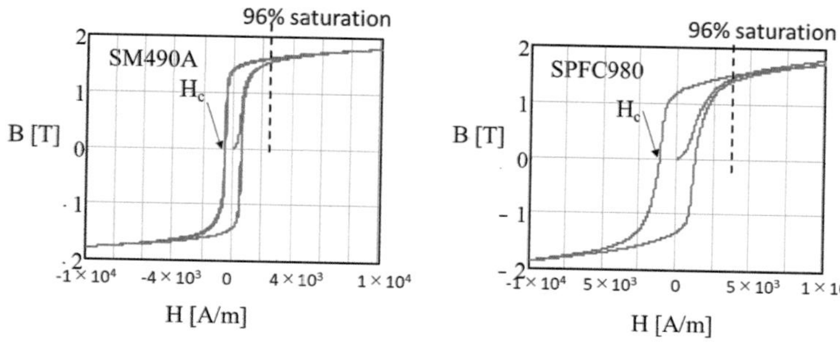

Figure 11.3 B-H properties of high-tensile steels.

Table 11.1 Magnetic cohesive force and 96% magnetization saturation fields of high-tensile steels for clarity

Type	Thickness (mm)	H_c (A/m)	B_s (T)	μ_0	H_s (A/m) 96% saturated
SM490A	4.5	580	1.8	1.18×10^{-3}	2,350
SPFC980A	1	1182	1.8	3.56×10^{-4}	3,700

Considering the thickness of 1 mm for SPFC plate, the H_s of 3,700 A/m to saturate 96% of magnetization of steel is to be reduced to accommodate the thicker plate used for shipbuilding.

From these measurements and estimation, maximum magnetic field to impose the target ship is set to be 3,700 A/m to find a design of seabed deperming coil.

11.3.2 Conventional Deperming Methods

Deperming of ship was first conducted in the Second World War. Since then, there have been developments in various methods to implement deperming magnetic field on ships.

11.3.2.1 Wound cable on ship-hull

The most common and widespread method of ship deperming since the Second World War has been to wind a cable onto the hull of the target ship, as shown in Fig. 11.4(a), and passing intermittent electric current through the cable with reducing peak intensity. When the cable is wound around the hull of the ship with n turns per meter measuring along the ship keel and the current to pass is I Ampere, magnetic field at the center of the cable wrapping is approximately $n \times I$ A/m. This means that required 3,700 A/m of magnetic field is generated by 3,700 A of current with one turn per meter of winding. Using a copper cable with 40 mm of diameter, a ship with 165 m length needs nearly 2 MW of power source for deperming.

By this method, magnetic field is effectively imposed on ship because of the small degradation of internal field factor in the longitudinally elongated geometry. Because of this reason, large ships such as cruisers and aircraft carriers are treated by this method although winding cable onto ship needs heavy manual loads.

The internal field factors calculated for a few ship types [28] are shown in Table 11.2. The calculation on the ship types DDH (Hull classification symbol of Helicopter Destroyers) and SS (*ibid*, Submarine Ship) are done by the spheroidal shell model [29] using the size and weight information of ships, and magnetic permeability of HY-80 [30], which is a commonly used steel for shipbuilding.

Table 11.2 Ship parameters of type of ships and their internal field factors calculated by spheroidal shell model

Type	Length (m)	Width (m)	Draft (m)	Displacement (ton)	Shell thickness [28] (m)	Internal field factor: Longitudinal	Internal field factor: Transverse
DDH	248	38	7	19,950	0.119	0.96	0.81
DDG	165	21	6	7,750	0.128	0.95	0.68
SS	84	9.7	8.5	2,950	0.212	0.85	0.37

In the longitudinal orientation of imposing magnetic field, the internal field factor decreases slightly from DDH ship to SS type, whereas in transverse orientation, there is a sharp decrease calculated in the SS type. This means that imposing transverse magnetic field inside of SS ship needs higher magnetic field than that for DDH ship. The reason of smaller transverse internal field factor is due to the thin and elongated shape of the ship where induced demagnetization appears in short range and wide hull area.

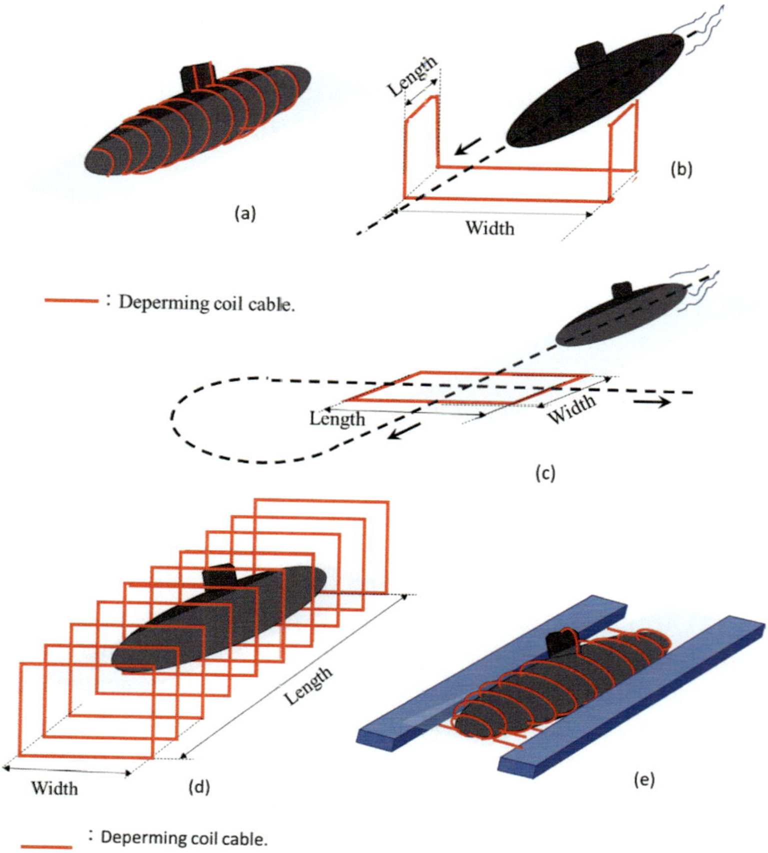

Figure 11.4 Conventional deperming system with (a) coil wound on hull, (b) running through saddle-shaped coil, (c) running over seabed coil and (d) cage type and (e) coil wound-on-hull type supported by piers. Length and width directions are indicated for each coil system.

11.3.2.2 Running through the coil

One of the methods is to set cable on sea surface in a shape of a saddle and to pass the current through the cable shown in Fig. 11.4(b). A target ship goes through two parallel lines of the cable on sea surface and magnetic fields from the surface lines are imposed on the ship. An example in a space with 20 m parallel lines, 955 A/m of magnetic field is created at the center between the lines with a current of 2,000 A on 15 turns of each line. This method restricts the width of the ship applicable due to the spacing of the parallel lines widthwise. Considering internal field factor in this transverse alignment, the intensity of magnetic field is not enough to deperm steel of ship when one considers the criterion obtained in the previous discussion, yet 1 MW of power source is still needed [31]. Intermittent current is imposed on coil and thus the ship must maneuver slow speed through the lines synchronous to the change of the electric current.

A variation of this method, similar in its principle, is to set coil on seabed and the ship runs over this coil, shown in Fig. 11.4(c) [32]. The intensity of magnetic field generated by the seabed coil decreases along the depth of water. Bigger ship with larger dwelling needs larger depth of water and larger space to imposing magnetic field, and this larger depth of water needs higher current through the cable. The cable with high electric current is necessary for the deperming of large ships by this method.

11.3.2.3 Cage-type coil

Square cage-type coil placed along the keel of ship is shown in Fig. 11.4(d) [33]. The magnetic field is imposed along the keel of ship, internal field factor degradation is low and magnetic field is effectively imposed onto the ship. Vertical coil to compensate environmental vertical field is attached easily because thinner cable is sufficient to generate the Earth's magnetic field. This type of coil is suitable for submarines because of its nearly standardized short width and height of 10 meters compared to ones having varying lengths. On the other hand, for surface ships, the size of the ship is restricted by the cage size. This method has benefits of easy operation for ship and does not involve hard manual work for coil wrapping and provides effective magnetic field for better deperming of the ship.

11.3.2.4 Variations of wound-on-hull

There are variations of each type of deperming system. A variation of wound-on-hull method is that supported by floating or landing pier [34]. One end of the cable to be wrapped on the hull is fixed onto the pier as shown in Fig. 11.4(e). Thus, the cable passing underneath the bottom of the ship can be prepared without the aid of divers, a requirement for wound-on-hull method. Restricted power source and hard steel hull of ship to require high magnetic field to impose, deperming cable close to the hull, and longitudinal magnetic field to impose on the ship are absolutely necessary in conventional methods.

11.3.3 Electric Current and Power for Each Type of Deperming Coil

Required electric current and power supply to impose 3,700 A/m of magnetic field onto the target ship by the different methods of deperming are calculated and summarized in Table 11.3. The target ship assumed here is the SS. The cable consists of copper wire with 40 mm of diameter. Parameters required to calculate the current and the power of each system, such as length, width and others, are shown in the table referring to Fig. 11.4.

The difference of current and power between saddle coil and seabed coil is because of the geometry of coil lines. Saddle coil imposes double of magnetic field created by each cable line on ship at the center between the two cable lines on surface. On the other hand, seabed coil imposes 50% of magnetic field by each cable line at the center of the space with double the depth.

Furthermore, magnetic field by coils is imposed transversely on ship for these two cases, and considering inner field factor of 0.37 in Table 11.2, required current is divided by this factor and power may be increasing inverse square of this factor. Decreasing the length of coil shown in the Fig. 11.4(b, c) may decrease the resistance of coil by the impact of rigorous operation for ship to pass through or over the coil.

Comparatively low power for wound-on-hull type is because of the short distance of cable and hull. However, parallel cable lines wound-on-hull makes low magnetic field area on hull just between the cable lines shown in Fig. 11.5. Cable lines are wound over cylindrical hull of 84 m length with 5 m of radius and the space of the cable line is 1 m. (a) is the magnetic field dependence on radial direction from center of the ship to the surface of the hull at under cable line, at one-fourth and at a half of space between cable lines. Magnetic field from center of the ship to 90% of r/r_0 is equal to required field. However, in the area of remaining 10% from the surface, magnetic field between the cable lines is lower than the required level. Figure 11.5(b) shows the magnetic field in surface area along the keel. In this surface area, the imposed magnetic field is attenuated in the area between the cables, although raw magnetic field is immune of inner field factor.

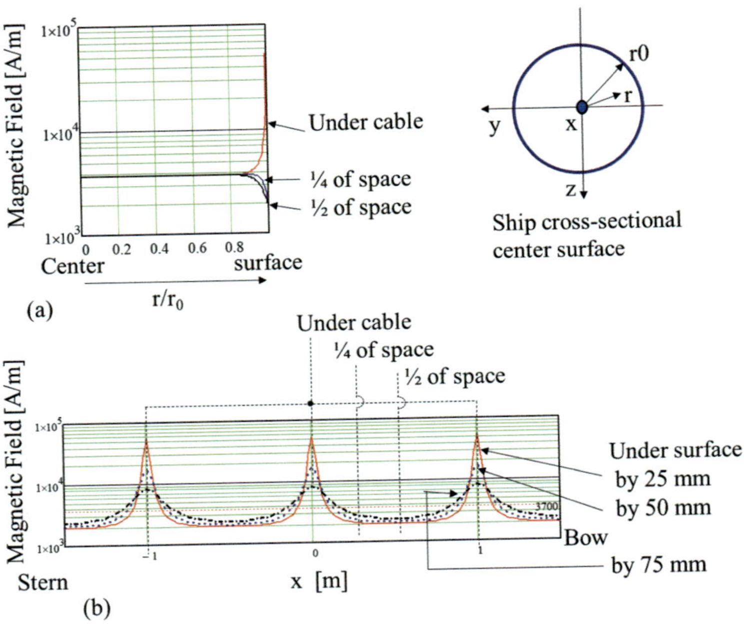

Figure 11.5 Magnetic field generated by cable wound-on-hull along hull surface. (a) Radial position dependence; (b) longitudinal position dependence based on the center of ship.

Cage-type coil generates comparatively homogeneous magnetic field within the volume of the ship in the cage because of the distance between the cable lines and the ship hull and immune from the steep field change close to the cable lines. The ship is required to be maneuvered into the cage and to be kept its position during the deperming treatment. There are a few sets of this cage system constructed and operated for submarines by US Navy [35]. The solid coil structure restricts the size of ship to set in the cage, and large power source is needed to operate.

Table 11.3 Electric current and power for different types of conventional deperming coil system with parameters for calculation

Name	Current (A)	Power (W)	Coil Shape			Refer to Fig. 11.4
			Turns	Length	Other	
Wound on hull	3,700	5.2E+05	1 t/m along keel	—	—	(a)
Going through saddle coil	3,900	1.5E+06	Total 30 t	84 m along keel	20 m width across keel	(b)
Going over seabed coil	7,800	5.1E+06	Total 30 t	84 m across keel	20 m width along keel	(c)
Cage-type coil	3,800	1.9E+06	1 t/m along keel	Total = Ship length + 2 × Coil width	20 m × 20 m each coil	(d)

11.4 HTS Superconducting Deperming

An ideal system of ship deperming is considered as a system with stable and short time of operation for ship, small power source, and free from hard manual work. Flat seabed coil with race-track shaped along a target ship is a candidate of this system [36]. High temperature superconducting (HTS) cable realizes the requirement of extensive high electric current with small power source for this system by its zero-electric resistivity.

11.4.1 Magnetic Field by Deperming Coil

Required magnetic field for deperming is set as 3,700 A/m. Ship type to deperm is assumed as both DDH and SS in Table 11.2. The depth of seabed coil is set as 12 m to allow 2° of pitching of DDH by the following calculation:

$$\text{Draft} + \sin\{(\text{Length}/2)\times(\pi/180)\times 2\} = 11.1$$

The addition of the term Draft in the equation, is to take into account for clearances, to avoid accidental collision at pitching between the seabed and the parts of the bottom of the hull, such as screws, rudder, and sensors.

As a standard unit of the cable, 1,200 m of total length and 120 kA of current to pass a unit seabed coil is designed. The length of the straight part of the seabed coil is to cover the length of DDH with additional length at the bow and stern of the ship by the width of the coil to impose uniform field all over the ship. The lead cable to and from the power source on land is designed to be 400 m. The electric current is decided by the estimation from the data of the developed DC electric power transmission line as around 10 kA. The magnetic fields of the central vertical cross-section of each target ship imposed by three sets of coil on seabed are calculated. The results of both with and without the effect of inner field factors are shown in Fig. 11.6(a). The views of target ships in relation to seabed coils are shown in Fig. 11.6(b).

Despite the large size of the DDH compared to the SS, the required field by DDH is rather less due to the effects of the inner field factor, as shown in Fig. 11.6(a). On the contrary, high magnetic field is required to deperm the inner structure of SS ship by low inner field factor still compensated by low vertical position and small width of SS ship. Deperming is to reduce magnetization of the ship hull. Because inside magnetization is expectedly shielded by the magnetic hull structure, required magnetic field is set at the water surface level to be 3,700 A/m which is generated by 120 kA of passing current through each cable.

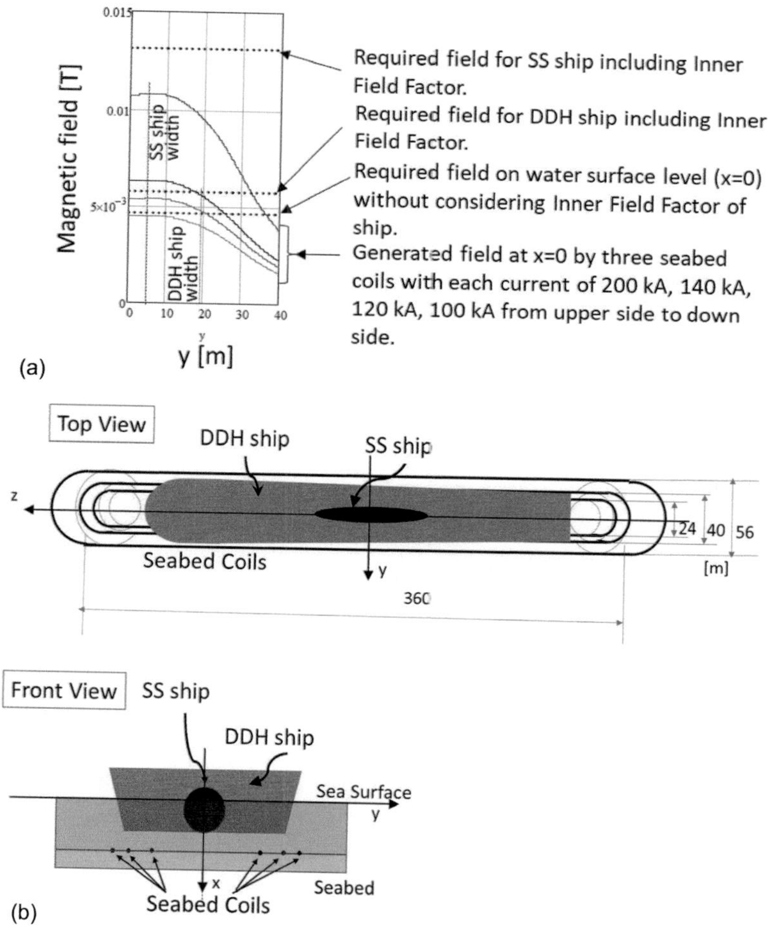

Figure 11.6 (a) Required magnetic field on central cross-sectional surface of target ships. (b) Views of target ships in relation to seabed HTS coils.

11.4.2 Superconducting Cable for Seabed Deperming Coil

Candidates of superconducting material for the cable of seabed deperming coil are discussed from those shown in Fig. 11.7(a), in which respective year of discovery for each superconducting material is shown. As schematically shown in Fig. 11.7(b), the critical temperature, the current and the magnetic field are related to each other, and a much lower operational temperature allows for a higher electric current under higher magnetic field than that at critical temperature.

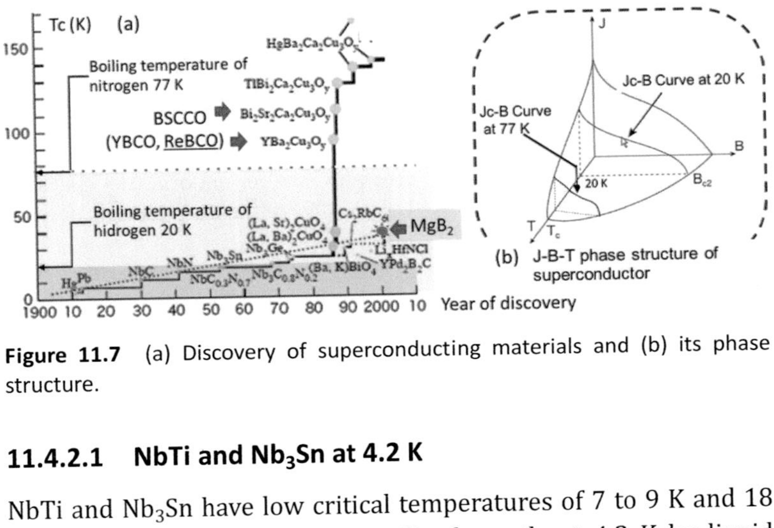

Figure 11.7 (a) Discovery of superconducting materials and (b) its phase structure.

11.4.2.1 NbTi and Nb_3Sn at 4.2 K

NbTi and Nb_3Sn have low critical temperatures of 7 to 9 K and 18 K, respectively and have been utilized mostly at 4.2 K by liquid helium refrigeration. Design technology and fabrication procedure of these superconductors have been matured by the experiences of applying them to high field magnets for nuclear magnetic resonance (NMR), for nuclear fusion furnaces and various experimental magnets. In the 1970s, research and development of application to power transmission line was advanced in several countries from the purposes of low power loss by superconductor [37]. Short cables with mostly 10 kA of current, extending to several tens of meters were fabricated and their expected performance was confirmed [38]. As a whole system of the power transmission based on niobium conductor, the cable with 2,000 MVA of power rating may overcome the economic position over the conventional conductor [39].

Super critical helium gas has single phase at 5.2 K and is better to refrigerate long cables. Extremely low viscosity of liquid helium II under 2.17 K improves the cooling capability by flowing through porous structure of the conductor.

However, the problem with helium is the low cooling efficiency at this temperature range. In addition, helium is a resource developed in restricted areas in the world such as Qatar, the United States, and a few other countries. Furthermore, it is appointed as a strategic material by the United States. Using strategic material may not be a viable solution for cooling down the underwater long cable conductors to extremely low temperatures, used in the deperming coil, especially for which will be required in effective and operationally urgent manner in times when political tension arises.

11.4.2.2 BSCCO at 70 K

BSCCO ($Bs_2Sr_2Ca_2Cu_3O_y$) has critical temperature of 110 K which is much above the boiling point of liquid nitrogen at 77 K. Subcooled liquid nitrogen to nearly 65 K is obtained by evacuating liquid nitrogen or by refrigerator and increases the critical current under self-field by 1.7 times improved from that of 77 K. Conductor of BSCCO is commercialized as tape conductor in 2004 [40] by basically Powder-In-Tube (PIT) method where thin needle-shaped single crystal is packed into the pipe and aligned to make superconducting current flow along these needles. The remaining bubbles and defects in superconductor in tube are improved by thermal treatment while evacuation and high-performance tape-shaped conductor are constantly fabricated in long length [41]. Electric power transmission line project, power cable for railway and many other applications confirmed the capability of this cable. Problem is that material of sheath of the conductor is silver because of the transfer property of oxide molecules while in thermal treatment. One of the HTS conductor's biggest problems is its high expense of the cable itself. Price of silver restricts the lower limit of the expense for BSCCO.

Tape-shaped conductor is developed in response to the requirement of strong stiffness to use Hastelloy material as substrate. Tape conductor are accumulated as layers on core structure or pipe, which allows the flow of the refrigerant. Conventional cable fabrication technology is helpful to fabricate cables with this tape-shaped conductor.

11.4.2.3 ReBCO

YBCO ($YBa_2Cu_3O_y$) is the first-discovered HTS material with critical temperature of 90 K which is higher than liquid nitrogen boiling point. In the thin film to make a superconductor cable element, requirement of all the three crystal axes alignment to pass the superconducting current is controlled by IBAD (Ion Beam-Assisted Deposition) process and conductor with long tape structure is produced [42]. The fabrication process consists of a multi-layer deposition of substrate, a cap layer and the YBCO layer. Atomic position of yttrium is replaced by various rare-earth atoms and as a whole called ReBCO (Rare-Earth Barium Copper Oxide) with their improved superconducting property. Lower critical temperature than BSCCO close to liquid nitrogen boiling point makes ReBCO lower performance at liquid nitrogen temperature. However, its performance of critical current and critical magnetic field improves in lower temperature at around 50 K.

To accumulate this tape-shaped conductor to a cable, twisted stacked-tape cable (TSTC) is developed. They are tapes stacked, twisted, and covered with metal sheath [43]. A cable with tape-conductor wound as multi-layer on round copper core, named CORC®, has been developed and successfully tested as a cable with diameter of several milli meters [44]. The tape-shaped conductor is then treated as a common cable element by these developments. The cable with six CORCs® wound on core is tested with 80 kA of critical current under 12 T of magnetic field at 4.2 K [45]. 100 kA of current lead is tested by stacked ReBCO tapes in metal jacket at 20 K for nuclear fusion furnace [46].

A problem of ReBCO is still the high expense of the tape conductor because of the complex fabrication processes and the scale of its mass production. In response to the global requirement of ReBCO tape-supply, manufacturers are expected to expand their factory to increase the production capacity.

11.4.2.4 MgB_2

The superconducting property of MgB_2 is discovered in 2001 at the critical temperature of 39 K which is a higher temperature than liquid hydrogen boiling point and the highest temperature ever discovered as a metallic superconductor. Compared to the cuprate superconductors of ReBCO and BSCCO, its transition temperature is low and critical current under magnetic field in 20 K is lower. However, the crystal structure is simple and thus the cabling technology developed for low-temperature superconductor can be applied. In addition, MgB_2 consists of cheap material such as magnesium and boron, and thus the expense for the complete cable conductor fabricated from MgB_2 is expected to be smaller.

Refrigeration system to cool the conductor to 20 K is made by circulation of liquid hydrogen or cooled helium gas. Long cable for seabed deperming may be supported by radiation shield cooled by liquid nitrogen.

11.4.2.5 Summary

The question of which superconducting material is most suitable to be used as material for seabed deperming cable must take into account from the magnetic field to generate, current to pass through the cable, conductor to pass the current, production and support background and expense to construct, operate and maintain the system, not only considering the present technology but also the future advancements of new technologies.

As an ideal goal of HTS deperming cable system, conductor consists of MgB_2, Rutherford cable technology may be used and refrigeration is by circulation of liquid hydrogen. However, considering the rapid improvement of the property of ReBCO tape, and their developing mass-production potentials, ReBCO tapes accumulated on small diameters unit cable to bundle may be a realistic candidate of the deperming cable conductor. In this case, operational temperature can be more flexible than MgB_2.

As an expected goal within a few years of range, a cable design is discussed with ReBCO tape conductor accumulating on CORC® technology, operational at 20 K.

11.4.3 Expected Goal of Complete System

Required magnetic field of 3,700 A/m to impose on the target ships of DDH and SS is assumed as a predetermined condition. High electric current cable line to generate the magnetic field covering the required volume can be realized by HTS conductor with low-power consumption. Once the cable is set on seabed, its solid structure is required to keep low thermal input from the environment. Depending on the high electric current to pass through the flat coil on seabed, only a cable of HTS conductor is sufficient to create required magnetic field with small power supply, which is completely different design from other conventional deperming coil systems.

Considering the spatial uniformity of the generated magnetic field, required current is divided by the three cables in separate positions and 120 kA of current is to pass each cable. Making this a standard cable system, cable components, power supply, and refrigerator unit can also be standardized.

As a cabling of conductor, ReBCO tape with 4 mm width is considered. Number of tape conductor accumulated by TSTC or CORC® of several millimeters in diameter of thin wire is a unit cable. Taking the parameter of number of tape conductor on core in CORC case, calculated maximum internal magnetic field with respect to the required electric current is shown in Fig. 11.8. Critical current versus imposed magnetic field of ReBCO with artificial pin from recent reports [47,48], are plotted in the figure.

At the operational temperature of 20 K, one bundle of six CORCs with more than 36 tapes in one CORC is sufficient to generate 120 kA of total current. The current source of this cable is 20 kA and 300 nΩ of electric resistivity at cable, assuming typical 100 nΩ at the connection, every 400 m of tape, and parallel connection of tapes in one CORC. Power source of this current is 720 W only considering the superconducting conductor. Small quantity of tape is expected for the 1.200 × 36 × 6 = 259 km for one unit cable system. This causes 12.9 M$ of expense for each unit system, provided 50 $/m of tape expense. Cases with bundles of two and four conductors are shown in Fig. 11.8. Decreasing the required current to each tape allows the increase of operational temperature. In relation to the increasing quantity of superconducting tape and increasing expense for the conductor, current source requires 10 kA and 5 kA of lower current capacity, respectively, and temperature of 50 K is sufficient to operate the system, which is realized by cooled helium gas flow.

Three sets of the unit cable system installed on the seabed generate deperming magnetic field of 3,700 A/m in the water surface area of the target ships.

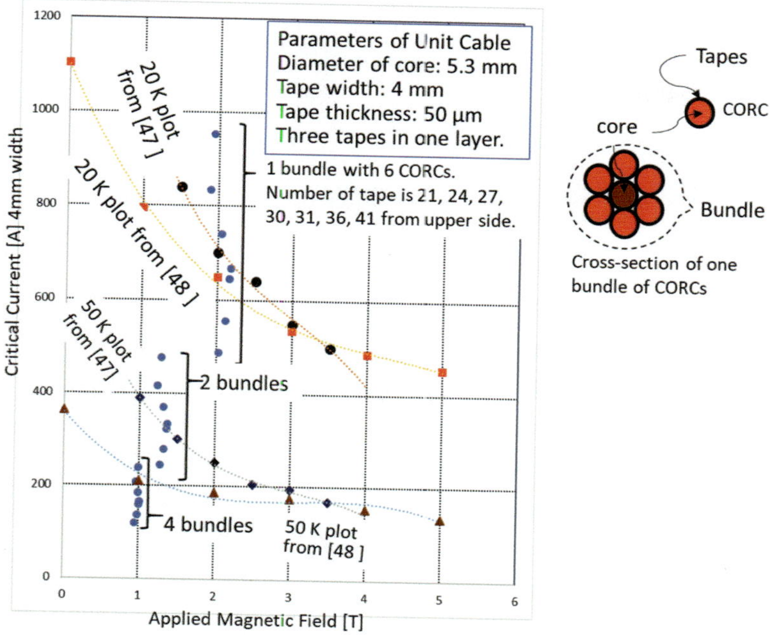

Figure 11.8 Maximum internal magnetic field on tape conductor accumulated on core along the parameters of number of tapes in one CORC and of bundle which is consisting of six CORCs.

11.4.3.1 Refrigeration of cable

The cross-section of deperming cable with one bundle of conductor is schematically shown in Fig. 11.9(a) and its refrigeration system in (b).

The concept of the refrigeration of the radiation shield by liquid nitrogen follows the experimental results of Ishikari Project [49] referenced. Cooling the conductor from room temperature to 20 K by liquid hydrogen needs 9 m^3 and 5 h providing 30 L/min of constant supply. In practice, rapid evaporation of liquid hydrogen at the beginning of cooling will occur, and thus slower refrigeration is recommended. After reaching the equivalent state of the conductor at 20 K, heat input to the conductor is estimated as 10 W for the radiation from the radiation shield, and 500 W for the direct contact and through the cable inlet [49]. This is cooled by liquid hydrogen of

31 L/min supply allowing the temperature rise of the coolant from the input to the output of the cable by 2 K.

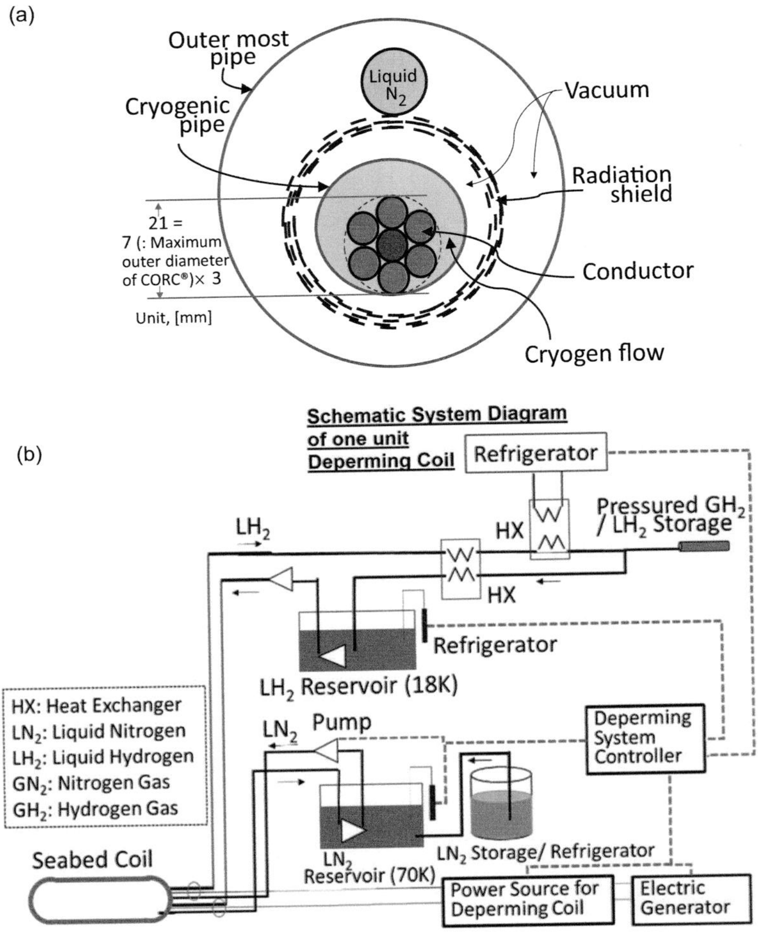

Figure 11.9 (a) Concept of the cross-section of seabed deperming coil cable and (b) its refrigeration system.

In hope of decreasing expense for tape conductor, increasing the quantity of tape is acceptable which decreases electric current through one tape and decreases the scale of current source and the expected heat input through lead.

11.4.3.2 Electromagnetic force

Race-track shape of coil with the current of I = 120 kA in one line generates magnetic field around the cable and imposes field to other sets of cable. Long parallel lines in three sets of race-track coil are under the magnetic field of other lines and electromagnetic force of the sum of $\mu_0 I^2/2\pi\times$(distance between two lines) is imposed. The widths of three race-track coils is 24 m, 40 m, and 56 m, to cover the water surface area of the target ships. The electromagnetic force on the straight part of each coil is 822 N/m, 222 N/m, and −357 N/m from inside to outside coil lines. This means the conductor must be supported in the cryogenic pipe and every intermediate structure in the outer most pipe resisting against this force. As an example of SUS304, yield strength is 427 MPa at 77 K and less than 2 mm^2 of contact area is enough to support 822 N of force. Thermal conductivity of SUS304 is 7 W/m/K at 77 K and 2 W/m/K at 20 K [50]. Heat input of contact area S and length L between T_1 and T_2 is $\rho\times S\times(T_2-T_1)/L$ to be less than heat input through radiation.

From the requirement of setting the cable on seabed, the cable needs to have flexibility in bending and every pipe structure of the cable is to be corrugated. Research work is needed to complete this cable design.

11.4.4 Research Step toward the Complete System

Design of cable cross-sectional structure is to be confirmed by a short test cable cooled to the operational temperature under the operational electromagnetic circumstance. The test cable has to pass the same current as the operational cable to confirm the electromagnetic property of the operational cable. This requires the similar cross-section as that of the operational cable, but shorter cable is allowed.

The superconducting material for the conductor of complete cable must be selected considering the expense of HTS unit cable because it is the highest among the other expenses. However, considering rapidly advancing HTS material property and cabling technology from which lower expense of the material is expected in near future, testing with short cable in the present time may allow the expense of HTS material.

ReBCO tape conductor is considered as a unit to accumulate on a unit cable with a diameter of several mm of TSTC or CORC® structure. The tape conductors in a unit cable are connected parallelly and the passing current I_u is the product of the tape number n_t in each unit cable and the current i_e on each tape conductor, so that $I_u = n_t \times i_e$. The magnetic field H generated by straight line is proportional to the total current of the coil I_t, which is augmented by serial connection n_s of each unit cable and is $H_t \propto I_t = n_s \times n_t \times i_e$. Here, i_e is a function of temperature and surrounding magnetic field and thus it is the function of n_s and n_t through the geometrical design. While the larger i_e causes the smaller n_s and n_t, and reduces the quantity of HTS material.

On the other hand, the current source of the coil needs to supply $I_u = n_t \times i_e$ of current. The power of current source is expected from the connection between the superconducting tape of less than μΩ, but resistivity of high current lead with normal conductor has not a negligible effect. Only for the test and evaluation, a few kA of current source is available and the conductor with the same total current is designed by increased n_s.

The length of the test cable is better to be shorter considering the expense for HTS conductor material and smaller load for refrigeration. However, serial connection of unit cable obliges a loop structure of the cable and electromagnetic force imposing on each part of the line must be concerned. The support of the conductor in the cable has to resist 822 N/m, the separation of cable parallelly goes and back is more than 3.5 m. The length of the test cable is set as 40 m and parallel placing is to be planned for test and evaluation. A specification of the test cable system is shown in Table 11.4.

The heat input to the radiation shield at 77 K is expected as 580 W (=500+2×40), and for the cryogenic pipe at 20 K to be 202 W, provided that the heat input through the support is the same as that by radiation and through direct contact is 500 W and 200 W, respectively. Designing the cable to achieve the smaller heat input is one of the most important research items of this development.

Table 11.4 A specification of a test cable to evaluate the maximum electric current to pass

			Temperature		Conductor						
Term Unit	Length (m)	Current (A)	Conductor (K)	Radiation shield (K)	Unit tape material	Current width (A, mm)	Tapes numbers	Unit cable numbers	Support force (N/m)	Current Source (kA)	Term Unit
Specification	40	120,000	20	77	ReBCO	23i, 4	21	24	822	5	Specification

11.5 Contribution to Sustainable Development Goal

After the end of the Second World War, a large number of large ships suffered severe damage from weapons, mostly by underwater mines [51,52]. Since then, deperming of ship became an indispensable operation for a naval ship for safety from destructive underwater weapons. The magnetic field to deperm ship is almost 100 times of the intensity of the Earth's field which is smaller than the magnetic field generated by a tiny rare-earth magnet. However, in order to impose on a large volume of ship, high electric current to be passed through the cable, and large power supply, in levels of MW, is to be provided in conventional methods. Furthermore, various treatment and manual work must be expensed for ship deperming operations.

HTS deperming of ship by seabed coil needs power supply in the levels of tens of kW class and some extra power for refrigeration by liquid nitrogen and liquid hydrogen. This smaller electric power consumption, comparing to the deperming by other methods, contributes to save electric energy and also expense for the construction of power supply systems. Once the cable system set on the seabed, ships ranging from small tugboats to large cruisers and hard ships such as submarines are only to stay over the coil while deperming, which not only cuts shorter the time for the operation but also decreases the heavy manual work to prepare for the operation.

Acknowledgments

The application of HTS cable set on seabed for ship deperming has not been realized yet, and it was hard to find any related information. We appreciate the information to design long HTS cable from Professor Yamaguchi S., Chubu University, and cable unit fabrication from Dr. Van der Laan and Dr. Takayasu M. This work was conducted under the framework of original research project by Naval Ship N&UEP R. C. We also appreciate the information provided by Mr. Mukoyama, Furukawa Electric Co., Dr. Yoshida S., then Taiyo Nissan Co. Ltd, and Mr. Aoki I., Jekk Torisha Co. to advance the project.

References

1. Sanders, S. (2011). *JANE's Fighting Ships 2011-2012* (Janes Information Group) p. 435.
2. *ibid*, p. 889.
3. *ibid*, p. 289.
4. Naito, R. (1982). *Journal of the M. E. S. J.*, **17**, pp. 130–137 (in Japanese).
5. Polmar, N. and Moore K. J. (2005). *Cold War Submarines: The Design and Construction of U. S. and Soviet Submarines, 1945-2001* (Potomac Book Inc., USA) pp. 137–138, p. 289.
6. Moore, J. (1986). *JANE's Fighting Ships 1986-1987* (Janes Information Group) p. 206.
7. WTD71 (2019). Magnetic Ship and Item Ranging, *Poster presented at the 11th Conference of Ship Signature Management*, Kiel, Sept. 24–26.
8. Hartmann, G. K. and Truver, S. C. (1991). *Weapons That Wait, Annapolis* (Naval Institute Press, MD) p. 90.
9. Holmes J. J. (2006). *Exploitation of a Ship's Magnetic Field Signatures* (Morgan & Claypool Publishers) pp. 1–3.
10. Potts, W. C. (1946). The Magnetic Field of a Ship and Its Neutralization by Coil Degaussing, *J. Institute Electrical Engineers*, **93**, Issue 71, Part 1, pp. 488–495.
11. Ayliffe, S. H. (1946). Process Applied to a Ship to Alter Its State of Magnetization, *ibid*, pp. 508–517.
12. Baynes, T. M., Russell, G. J. and Bailey, A. (2002). Laboratory Simulation of the Deperm Process, *DSTO-TR-1340*.
13. Baynes T. M., Russell G. J. and Bailey A. (2002). Comparison of Stepwise Demagnetization Techniques, *IEEE Trans. Magn.*, **38** (4) Part II pp. 1753–1758.
14. Ju, H. S., Chung, H. J., Im, S. H. et al. (2015). Efficient Deperming Protocols Based on the Magnetic Properties in Demagnetization Process, *IEEE Trans. Magn.*, **51** (11) 7301704.
15. Fitzpatrick, B. K., Kephartl, J. T. and Golda, E. M. (2007). Characterization of Gaseous Helium Flow Cryogen in a Flexible Cryostat for Naval Applications of High Temperature Superconductors, *IEEE Trans. Appl. Supercon.* **17** (2) 1752–1755.
16. Gubser, D. U. (2011). U. S. Navy's Superconductivity Programs Scientific Curiosity to Fleet Utility, *IEEE Trans. Appl. Supercon.* **21** (3) 931–935.

17. Gilroy, L. (2019). MEASURE Signature Management Trial (RIMPASSE Lite), *Proc. the 11th Conference of Ship Signature Management*, Kiel, Sept. 24-26.

18. Robertson, I. M. (1991). Magneto-Elastic Behaviour of Steels for Naval Applications, *MIRL Technical Report*, NRL-TR-90-27.

19. Davidson, S. J. et al. (2009). Electromagnetic Underwater Signature Management: Signature Design Techniques and Redeployable Measurement Systems, *Proc. Marine Electromagnetics.* Stockholm, July.

20. ClassNK Standard, https://www.classnk.or.jp

21. Japanese Industrial Standards, https://www.jis.go.jp

22. National Defense Standards, https://mod.go.jp>atla>nds. (in Japanese).

23. HY-80 Steel, http://www.matweb.com/search/datasheet_print.aspx?matguid.

24. *Steel Plate, Alloy, Structural, High Yield Strength (HY-80 and HY-100)* (1987). Military Specification MIL-S-16216K(SH).

25. Holmquist, T. J. (1987). Strength and Fracture Characteristics of HY-80, HY-100, and HY-130 Steels Subjected to Various Strains, Strain Rates, Temperature, and Pressures, *AD-A233 061*.

26. Ebine, N., Kitsuta, K, Attila, G., Ara, K., Kamisaka, M. and Miya, K. (1996). Correlation between the Magnetic and Mechanical Properties of Ferromagnetic Structural Steels, A533B and SUS410, *J. Magnetics Soc. Japan*, **20,** pp. 665–668. (in Japanese).

27. Hirota, M. (2019). Magnetic Property of Ship Building Steel, *The 80th JSAP Autumn Meeting*, Sapporo, Sept., 21a-PA1-13. (in Japanese).

28. Hirota, M. (2017). A Study on Ship Deperming Coil System as an Application of HTS Power Cable Technology, *Proc. Ocean Eng. Symposium* (Tokyo), OES26-0006A.

29. Frumkis, L. and Kaplan, B. Z. (1999). Spherical and Spheroidal Shells as Models in Magnetic Detection, *IEEE Trans. Magn.*, **35** (5), Sept. pp. 4151–4158.

30. Holmes, J. J. (2008). *Reduction of a Ship's Magnetic Field Signatures* (Morgan & Claypool Publishers). p. 10.

31. Yoshii, T. (1969). *Jikikougaku* (Kaibundo, Tokyo), pp. 248–249 (in Japanese).

32. Kama, S. (2019). STL Overview of Deperming Ranges, Presentation and catalogue at *UDT2019* (Stockholm), May 14th.

33. Mayo, B. (2019). Magnetic Silencing Facility Provides Unique Service, *Navy News Service*, https://www.navy.mil/submit/display.asp?story_id=51513.

34. Baynes T. M. (2002). Simulation and Analysis of the Magnetic Treatment of Naval Vessels, *Presentation at Australian Defence Forum Academy*, Sept.

35. Shang, P. (2017). Where the Fleet Begins, Ship Signature Department (Code 70) Overview, May. 01.

36. Hirota, M. (2017). A Study on Ship Deperming Coil System Using High Temperature Superconducting Cable Technology, *J. Ship. Ocean Eng.* **7**, pp. 93–99.

37. Rogalla, H. and Kes, P. H. (2012). *100 Years of Superconductivity* (CRC Press, Florida). p. 788.

38. Furuto, Y., Miura T. and Ikeda, M. (1974). Electrification Test of Superconducting Model Cables, *Furukawa Electric Review*, March.

39. Appleton, A. D. (2013). British Developments in Superconductivity Applications, *Proc. 1970 Cryogenic Engineering Conference* (Colorado), June. Advances in Cryogenic Eng. (Springer Science & Business Media.) p. 11.

40. Kalsi S. S. (2011). Applications of High Temperature Superconductors to Electric Power Equipment, *IEEE PRESS* (John Wiley & Sons, New Jersey), p. 13.

41. Ayai, N., Kikuchi M. et al. (2006). Achievement of High-Temperature Superconducting Wire with Critical Current Exceeding 200 A, *SEI Technical Review* (Japan), December.

42. Iijima, Y., Kakimoto, K. and Takeda, K. (2001). Ion Beam Assisted Growth of Fluorite Type Oxide Template Films for Biaxially Textured HTST Coated Conductors, *IEEE Trans. Appl. Supercond.* **11**, pp. 3457–3460.

43. Takayasu M., Chiesa L. and Minervini J. V. (2014). Investigation of REBCO Twisted Stacked-Tape Cable Conductor Performance, *J. Phys. Conf. Series,* **507**, 022040.

44. Van der Laan D. C., Weiss J. D. and McRae D. M. (2019). Status of CORC® Cables and Wires for Use in High-Field Magnets and Power Systems a Decade after Their Introduction, *Surpercond. Sci. Technol.*, **32**, 033001.

45. Mulder, T., Dudarev, A. and ten Kate, H. (2017). REBCO CORC Wire and Cable-In-Conduit Conductor Development at CERN, *Presentation at WAMHTS-4* (Barcelona), Feb.

46. Terazaki, T., Yanagi, N. et al. (2015). Measurement and Analysis of Critical Current of 100-kA Class Simply-Stacked HTS Conductors, *IEEE Trans. Appl. Supercond.* **25**, June, 4602905.

47. Nakasaki, R. et al. (2015). Continuous Improvements in Performance and Quality of 2G HTS Wires Produced by IBAD-MOCVD for Coil Applications, *Presentation at Int. Conf. Magnet Technology 24*, Oct. 18-23, Seoul.

48. Yoshida, Y. et al. (2017). Yttrium-Based Coated Conductors Doped Artificial Pinning Centers, *Fujikura Technical Report* **1** (130) pp. 22-28.

49. Watanabe, H. et al. (2017). Cooling and Liquid Nitrogen Circulation of the 1000 m Class Superconducting DC Power Transmission System in Ishikari, *IEEE Trans. Appl. Supercond.* **27** (4) June, 5400205.

50. Touloukian, Y. S. (1995). *Thermophysical Properties of Matter* (IFI/ Plenum 1970).

51. Avrey J. (1998). The Naval Mine Threat to US Surface Forces, *Surface Warfare*, May/June, pp. 4–9.

52. Finlan, A. (2004). *The Royal Navy in the Falklands Conflict and the Gulf War: Culture and Strategy* (Routledge, Frank Cass, London).

Chapter 12

High-T_c Superconducting Bearings Design: Towards High-Performance Machines

I. Valiente-Blanco and D. López-Pascual
Área de Ingeniería Mecánica, Universidad de Alcalá,
Ctra. Madrid-Barcelona km, 33.600, 28805 Alcalá de Henares, Spain
i.valiente@uah.es

Superconducting magnetic bearings (SMBs) are finding new opportunities high-tech applications because of their very attractive characteristics. They can provide passive stable levitation without contact, eliminating tribological problems and lubrication requirements. In addition, they can operate in vacuum, clean, and cryogenic environments with minimum rotational energy losses.

In this chapter, a review of properties of materials used in superconducting bearings design is provided. A set of design rules is discussed for various types of superconducting bearings including journal, thrust rotational bearings and linear bearings in the Meissner and mixed state. Special attention is paid to the dimensioning and improvement of load capacity and stiffness, force

High-T_c Superconducting Technology: Towards Sustainable Development Goals
Edited by Muralidhar Miryala

ISBN 978-981-4877-65-7 (Hardcover), 978-1-003-16468-5 (eBook)
www.jennystanford.com

relaxation, hysteresis, coefficient of friction, temperature influence, and their behaviour under harmonic vibrations.

Finally, recent advances in some selected applications of SMBs are reviewed.

12.1 Introduction to Bulk Superconducting Levitation

Since the discovery of high temperature superconductivity of $YBa_2Cu_3O_{7-\delta}$ in 1987 [1], applications using high-temperature superconductors have experienced a great boost. YBaCuO becomes superconducting at 93 K, so it can be easily cooled-down using relatively inexpensive liquid nitrogen (LN_2). This fact has allowed researchers all over the word to investigate the fascinating properties of such material and has open new possibilities for cutting-edge applications. This chapter aims to provide electromechanical and cryogenic engineers with the tools for understanding and designing high-T_c superconducting magnetic bearings. Special attention has been paid to passive SMBs using bulk YBaCuO superconductors.

12.2 Bearing Materials and Cryogenics

Passive bulk SMBs are mainly composed by bulk superconductors and permanent magnets operating in a cryogenic environment. One critical aspect of a SMB design is an optimal selection of the materials and a proper understanding of temperature effects in their electromagnetic, mechanical, and thermal properties. In this section, main properties of different relevant materials used in SMBs are reviewed and discussed.

12.2.1 Permanent Magnet Materials

Permanent magnets are also referred as hard magnetic materials due to their tendency to oppose to demagnetization. This capacity is defined by the material coercivity (H_C) and it represents the opposing magnetic field that is required to fully eliminate the remnant magnetic field (B_r) inside a material. On the contrary, soft magnetic materials (such as pure iron) present a much lower coercivity than permanent magnets. Therefore, they are much more easily demagnetized. Figure 12.1 compares the intrinsic magnetic hysteresis curve of a hard magnetic material (NdFe N50) and a soft magnetic steel. In the family of hard magnetic materials, rare-earth permanent magnets present the highest magnetic performance. In particular, $Nd_2Fe_{14}B$ and Sm_2Co_{17} alloys present the highest magnetic product remanence, and coercivity.

Corrosion resistance should be evaluated in a case-by-case basis for the environment to which the magnets are exposed [2,3]. Qualitatively, SmCo are less prompt to corrosion (especially $SmCo_5$) and behave better at extreme temperatures. On the contrary, NdFe magnets, due to their high Fe content, need to be coated to avoid corrosion.

NdFe magnets present a spin-reorientation transition (SRT) at 135 K [4]. This SRT causes the magnetization of a NdFe magnets to decrease from this point. Despite the loss of magnetization, the axis of magnetization remains mostly invariant for macroscopic permanent magnets, as demonstrated in [5]. Replacement of Nd by Pr eliminates such spin transition, providing a more stable cryogenic behaviour [6,7]. Coercivity of rare earth magnets increases even at lower temperatures than the spin-reorientation temperature [8].

From a structural point of view, NdFe magnets show better mechanical properties and are easier to manipulate. They show lower tendency to chip than SmCo magnets. From the thermal point of view, cooling NdFe magnets takes more energy, and thermal conductivity is lower than in SmCo magnets. In addition, NdFe magnets present lower coefficient of thermal expansion. Finally, NdFe magnets present higher electric resistivity.

Other permanent magnet materials, very common in room temperature applications, are AlNiCo and hard ferrite magnets. However, their use in cryogenics is very unusual. Previous experience has demonstrated that both magnet materials present a permanent loss of magnetization when exposed to cryogenic temperatures [9,10]. For practical applications, a reasonable minimum operational temperature of about −75°C is recommended for AlNiCo magnets and −60°C for ferrite magnets.

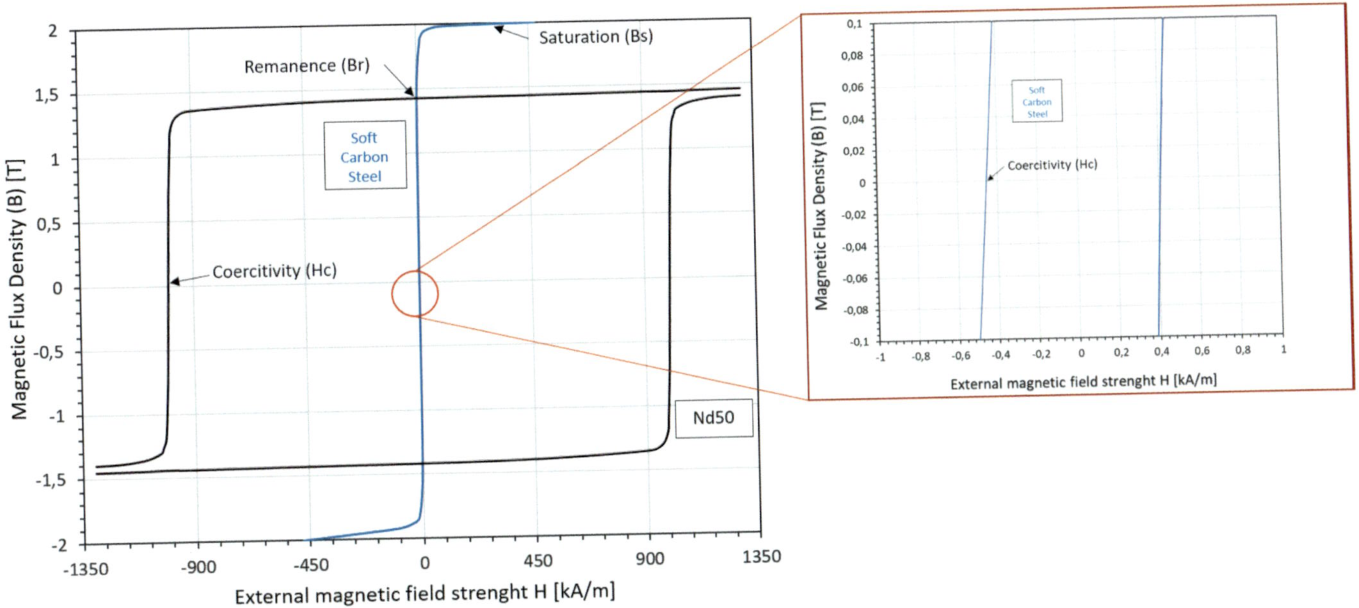

Figure 12.1 B-H intrinsic curves of NdFe N50 and a soft carbon steel.

Table 12.1 summarizes the main properties, at room temperature, of the most relevant permanent magnet materials. Obtained from specifications of various commercial manufacturers and papers.

Table 12.1 Properties of most relevant SMB permanent magnet materials

Material	$Nd_2Fe_{14}B$ N55	Sm_2Co_{17} N35	$SmCo_5$ N25	Units
Density	7500	8300	8400	kg/m^3
Electric resistivity	180	90	55	μΩ·m
Remanence, (Br)	1.49	1.19	1	T
Br, Temp. coefficient (αBr)	−0.12	−0.035	−0.05	%/°C
Coercivity, (Hc)	935	880	775	kA/m
Hc, Temp. coefficient (αHc)	−0.79	−0.25	−0.24	%/°C
Maximum energy product	55	35	25	MGOe
Curie temperature	310	820	725	°C
Coefficient of thermal expansion, (α)	//7.5; ⊥−0.1	//11; ⊥13	//7; ⊥15	$\times 10^{-5}$
Thermal conductivity, (k)	7.6	10	11	W/mK
Specific heat, (Cp)	460	350	370	J/kgK
Flexural strength	285	120	120	MPa
Compressive strength			1000	MPa
Young modulus, (E)			140	MPa

Note: For directions parallel and perpendicular to the magnetization direction.

12.2.2 Superconducting Materials

Superconductivity is a phenomenon that some materials exhibit, exclusively for the moment, at very low temperatures. It is characterized by two very particular properties. First, a total loss of resistivity, a fact which was discovered by H. Kamerly Onnes in 1911 [11]. Second, a total exclusion of the magnetic field from inside the superconductor, a fact discovered by W. Meissner and R. Ochsenfeld in 1933 [12]. Despite presenting this common features, superconducting materials can be classified in different ways according to the differences in their behaviour:

1. According to the magnetic behaviour:

a. Type I superconductors: External magnetic fields are fully expelled of the superconductor until the field (H_{c1}) is high enough to break superconductivity. Coherent length is larger than the characteristic magnetic field variation length ($\xi > \sqrt{2}\lambda$). Pure metals typically belong to this group.
b. Type II superconductors: They exhibit a mixed state where magnetic field and superconductivity can coexist inside the superconductors ($H_{c1}<H<H_{c2}$). They are typically made of metal alloys or oxidized ceramics as YBaCuO. These superconductors satisfy that $\xi < \sqrt{2}\lambda$.
c. Type "1.5 superconductors": They are multicomponent superconductors with more than one coherent length that satisfy: $\xi_1 > \sqrt{2}\lambda$ and $\xi_2 < \sqrt{2}\lambda$, $MgBr_2$ belongs to type 1.5 superconductor [13].

2. According to their critical temperature:
   a. Low-temperature superconductors (LTS) typically refer to superconductors with a critical temperature below 30 K. Most of them are type I superconductors.
   b. High-temperature superconductors (HTS) for superconductors with transition temperatures above 30 K. Since the discovery of BaLaCuO in 1986 [14], the race for increasing the transition temperature of superconductors is a hot topic in physics. Lanthanum hydride (LaH_{10}) under very high pressure, currently holds the temperature record in just 250 K (−23 °C) [15]. The dream of a room temperature superconductor seems to be, now, closer than ever.

3. The theory that explains their behaviour:
   a. Conventional superconductors, which can be explained by the BCS theory. Type I superconductors usually belong to this group.
   b. Exotic superconductors which behaviour cannot be explained by BCS theory. Phenomenological theories, like Landau theory, and the Beam model are used to predict their behaviour. Most type II and HTS superconductors belong to this group [16].

12.2.3 Bulk YBaCuO Properties

$YBa_2Cu_3O_{7-\delta}$ (YBaCuO) is a ceramic compound that exhibits a transition temperature (T_c) at 93 K [1]. It can be classified as an exotic, type II, HTS. It is, probably, the most popular superconducting material and its discovery represented a major breakthrough for SMBs technology. It can be cooled down using LN_2 at room pressure, opening the gate to fast and inexpensive experiments in laboratories worldwide. The properties of YBaCuO superconductors strongly depend upon the manufacturing process [17]. In addition, gathering all properties required in a detail design involves a significant effort combining information from different sources. Table 12.2 summarizes typical magnetic properties of thin film and bulk melt textured superconductors at 77 K [18–22]:

Table 12.2 Typical magnetic properties of YBaCuO superconductors at 77 K

YBaCuO	(T_c) [K]	Current density (Jc) [A/cm²] $B_{applied}$<1 T.	Meissner transition field (B_{c1}) [mT][a] // ab	// c	Upper transition field (B_{c2}) [T] // ab	// c
Thin film	93	Up to 6×10^7			From 35 to 200	
Bulk melt textured	93	From 6×10^3 to 4×10^4	From 4 to 25	From 9 to 55	18 to 25	≈50

[a]Value of magnetic flux density calculated in a vacuum $\mu_0 = 4\pi \times 10^{-7}$H/m.

The critical magnetic field at any temperature below the critical temperature is given by the relationship [23]:

$$B_{c\_(T)} = B_{c\_(ref)} \cdot \left(1 - \left(\frac{T}{T_{ref}}\right)^2\right) \tag{12.1}$$

Understanding of the SMB will require knowledge on mechanical and thermal properties of YBaCuO as well. A very important parameter that describes the mechanical behaviour is the Young modulus. It is usually indirectly obtained from macro and micro-indentation surface measurements, which is an excessively ideal approach for porous materials such as bulk YBaCuO. A large dispersion can be found in the measured value of the Elastic modulus in the literature, maybe caused by this fact. In [24], a study of Young modulus, tensile and compressive strength of single-domain melt-textured YBaCuO is developed for bulks at room temperature and 77 K in presence of no magnetic field. In addition, a technique to improve such values up to 45% by making small drills in the superconductors is demonstrated. Such technique has also been reported to significantly improve load capacity of SMBs [25]. Table 12.3 summarizes the main mechanical properties of YBaCuO single domain melt-textured superconductors at 77 K [25]. Values can vary for different manufacturing techniques.

Table 12.3 Typical mechanical properties of melt-textured YBaCuO

YBaCuO	Density [kg/m^3]	$\Delta L/L \times 10^{-3}$ at 50 K [26]		Young modulus [GPa]	Compressive strength [MPa]		Flexural strength [MPa] [27]		Vickers hardness (kgf/mm^2)
		//ab	//c		//ab	//c	//ab	//c	
Bulk melt-quenched	6300	−1.7	−2.6	52		390	80	70	540

12.2.4 Performance of Materials in Cryogenics

Complex machines using SMB technology require a detailed design that considers not only the magnetic properties of the SMB, but also mechanical and thermal aspects. Behaviour of materials in cryogenics and their physical properties may differ significantly from what we experience at room temperature. In addition, inherent difficulties of testing in the low-temperature range are a potential source of discrepancies in materials properties obtained from experimental data. In an extraordinary effort for homogenization, the American National Institute of Standards and Technology (NIST) has published and updated a reference database for materials properties in cryogenics. Another very complete guide for cryostat and cryogenic machinery design is provided in [28].

NIST approach is to provide a fit curve for different materials properties as a polynomial function of temperature (T):

$$\log_{10\_(\text{PROPERTY})} = a + b\cdot\log(T) + c\cdot(\log(T))^2 + d\cdot(\log(T))^3 + e\cdot(\log(T))^4 + f\cdot(\log(T))^5 + g\cdot(\log(T))^6 + h\cdot(\log(T))^7 + i\cdot(\log(T))^8 \tag{12.2}$$

In this section, main thermal and mechanical properties of relevant materials in a SMB design will be reviewed. Unless otherwise specified, all material properties are obtained from NIST updated material database.

12.2.4.1 Mechanical properties

When a superconducting bearing is composed of different materials in mechanical contact, and it is cooled down or warmed up, it may face high differential contractions. Such effect can potentially result in high stress or system misalignments. In order to have a control on such contractions, the thermal coefficient of expansion (α) of the materials shall be considered. It can be calculated as:

$$\alpha = \frac{1}{L}\cdot\frac{dL}{dT} \tag{12.3}$$

where L is the characteristic length of the inspected sample and T its temperature.

Expansion coefficient is temperature dependent. This dependency becomes very relevant in cryogenic applications. Therefore, it is much more common for cryogenic engineers to express the thermal contraction/expansion as a total accumulated thermal linear expansion with respect to a reference temperature. This reference temperature is typically 293 K:

$$\frac{\Delta L}{L} = \frac{L_T - L_{293}}{L_{293}} \tag{12.4}$$

Figure 12.2 shows the total thermal linear expansion for different relevant materials. In almost every case, the thermal expansion coefficient becomes temperature independent at low temperatures below 50 K, and differential thermal contractions becomes asymptotic.

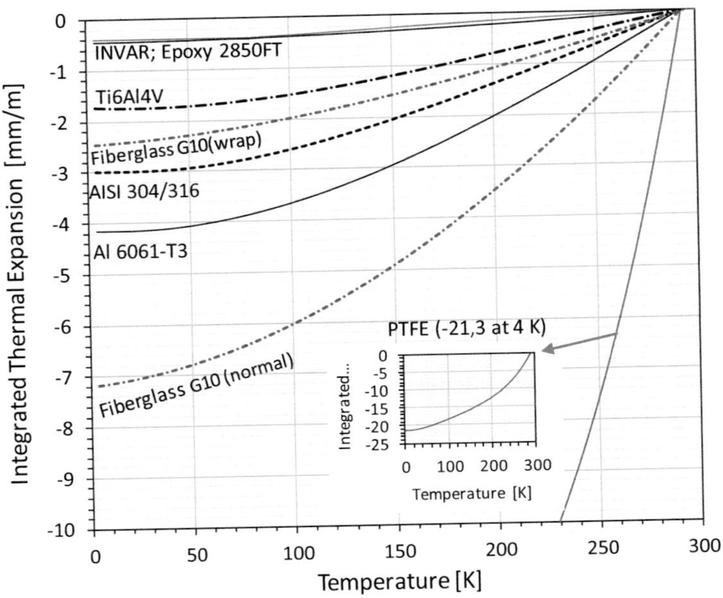

Figure 12.2 Integrated thermal expansion vs. temperature of various materials.

The elastic (or Young) modulus of most metals suffers little variation from room temperature to 4 K. Yield strength and ultimate strength of most metal alloys, however, frequently increase at low temperatures. Finally, fracture toughness can be significantly reduced for alloys as Ti_6Al_4V (ELI) or some nickel-steel alloys. Figure 12.3 shows the Yield strength vs. temperature of various metals and metal alloys typically used in cryogenics.

12.2.4.2 Thermal properties

Heat transportation mechanisms in solid materials differ from metals (electron dominated), thermal insulators (lattice phonon dominated), or semiconductors (mixture of phonons and electron). In addition, pure metals' thermal conductivity (k) highly depends on the presence of impurities. Contrary to room temperature applications, thermal conductivity and specific heat (Cp) cannot be considered as a constant value for the cryogenic temperature range. Table 12.4 summarizes the coefficients of polynomial Eq. 12.2 for calculation of thermal conductivity. Ti_6Al_4V data has been obtained from [29].

Table 12.4 *k* coefficients for various representative materials

k coefficients	Al 6061-T6	SS AISI 304/316	Copper RRR = 50	INVAR FeNi36	Ti6Al4V	PTFE	Fiberglass epoxy G10-CR (normal)
A	0.07918	−1.4087	1.8743	−2.7064	−587.282	2.738	−4.1236
B	1.0957	1.3982	−0.41538	8.5191	1924.845	−30.677	13.788
C	−0.07277	0.2543	−0.6018	−15.923	−2608.943	89.43	−26.068
D	0.08084	−0.626	0.13294	18.276	1870.452	−136.99	26.272
E	0.02803	0.2334	0.26426	−11.9116	−747.609	124.69	−14.663
F	−0.09464	0.4256	−0.0219	4.40318	157.928	−69.556	4.4954
G	0.04179	−0.4658	−0.051276	−0.86018	−13.775	23.32	−0.6905
H	−0.00571	0.165	0.0014871	0.068508	0	−4.3135	0.0397
I	0	−0.0199	0.003723	0	0	0.33829	0
Temperature range	4–300 K	4–300 K	4–300 K	4–300 K	4–300 K	4–300 K	4–300 K
Fit error	0.5%	2%			2%		5%

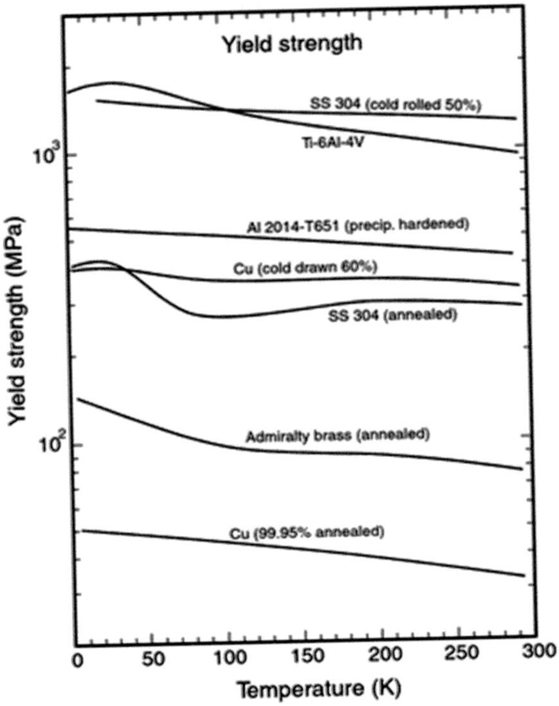

Figure 12.3 Yield strength of various materials. With permission [28].

Equation 12.2 requires a little modification to represent the thermal conductivity of oxygen free cooper (OFHC) versus temperature. Only for such material, the following equation applies:

$$\log_{(k_{\mathrm{OFHC}})} = \frac{a + c\cdot T^{0.5} + e\cdot T + g\cdot T^{1.5} + i\cdot T^{2}}{1 + b\cdot T^{0.5} + d\cdot T + f\cdot T^{1.5} + h\cdot T^{2}} \tag{12.5}$$

Thermal conductivity of OFHC is highly dependent on the presence of impurities. It is usually provided for different values of residual resistance ratio (RRR) from 20 to 1000. See Section 12.2.4.3 for further details on the meaning of RRR. Figure 12.4 shows the thermal conductivity vs. temperature from 4 K to 300 K for some relevant cryogenic materials used in manufacturing of superconducting magnetic bearings:

When assemblies of different parts are considered in a vacuum environment, the real area of contact between parts is significantly smaller than the macroscopic contact area. The result is that a contact thermal resistance appears that reduces overall heat transfer capacity by conduction. This contact resistance can be very significant and it depends on the materials properties, contact pressure distribution, and surface roughness [30]. A simple way, and a very common practice, to reduced contact resistance is to use interstitial filling materials such as Apiezon® grease or Indium foils.

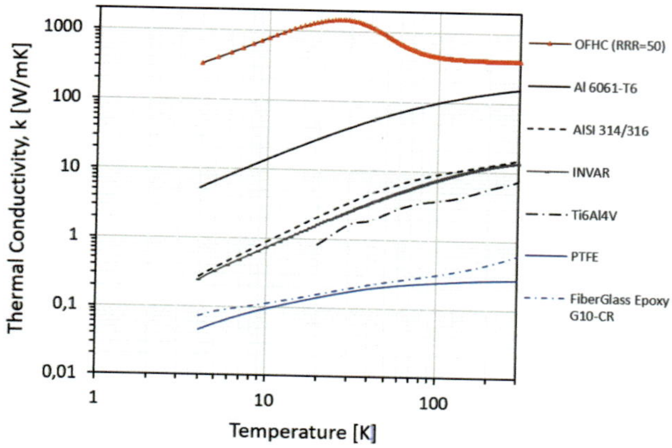

Figure 12.4 Thermal conductivity vs. temperature of various materials.

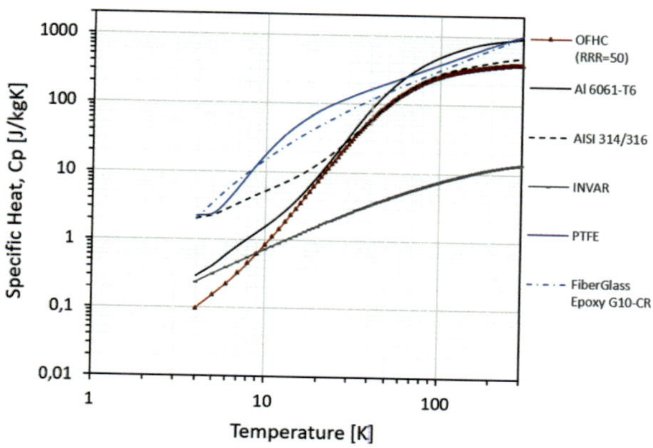

Figure 12.5 Specific heat vs. temperature of various materials.

Specific heat is a very relevant thermal property in cryogenic engineering too. It represents the amount of energy per unit mass required to increase (or decrease) one degree a material temperature. Table 12.5 represents the coefficients for some relevant cryogenic materials:

Table 12.5 *Cp* coefficients for various representative materials

Cp, coefficients	Al 6061-T6	SS AISI 304/316	Copper RRR= 50	INVAR FeNi36	PTFE	Fiberglass epoxy G10-CR
A	46.6467	22.0061	−1.91844	28.08	31.88256	−2.4083
B	−314.292	−127.5528	−0.15973	−228.23	−166.51949	7.6006
C	866.662	303.647	8.61013	777.587	352.01879	−8.2982
D	−1298.3	−381.0098	−18.996	−1448.423	−393.44232	7.3301
E	1162.27	274.0328	21.9661	1596.567	259.98072	−4.2386
F	−637.795	−112.9212	−12.7328	−1040.294	−104.61429	1.4294
G	210.351	24.7593	3.54322	371.2125	24.99276	−0.24396
H	−38.3094	−2.239153	−0.3797	−56.004	−3.20792	0.015236
I	2.96344	0	0	0	0.16503	0
Temp. range	4–300 K	4–300 K	4–300 K	4–27 K	4–300 K	4–300 K
Fit error	0.5%	2%	2%	2%	1.5%	2%

Figure 12.5 depicts the specific heat for various relevant materials.

12.2.4.3 Electric resistivity and magnetic susceptibility

Electric resistivity of pure metals and metal alloys is reduced at low temperatures. The influence of impurities and material imperfections has a great impact at temperatures below 40 K. In order to evaluate the impact of such imperfections, RRR is defined by Eq. 12.6 [31]:

$$RRR = \frac{\rho_{ref}}{\rho_{0K}} \tag{12.6}$$

where ρ_{ref} is the resistivity at the reference room temperature and ρ_{0K} is the theoretical resistivity at absolute zero.

Figure 12.6 shows the electrical resistivity for various RRR pure aluminium materials.

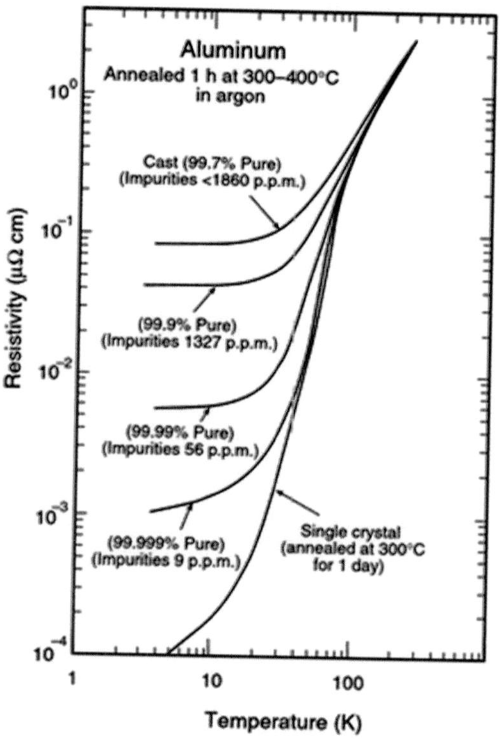

Figure 12.6 Electric resistivity vs. temperature of RRR aluminium. With permission [28].

Austenitic steel alloys (AISI 300 series) are non-magnetic at room temperature. However, magnetic susceptibility can be temperature dependent. A very complete description and experimental characterization of various relevant stainless-steel alloys is provided in [6] and [7]. Soft ferromagnetic materials change their magnetic properties at low temperatures too. Valuable references including experimental characterization of B-H and core-loss curves of some ferromagnetic materials are presented in [8] and [9].

12.3 Superconducting Bearings Classification

12.3.1 According to Their Motion Degree of Freedom

Degree of freedom (DoF) is a term used by mechanical engineers to describe a preferred direction of space for the motion of a mechanism. A rigid solid in free space presents three translational and three rotational independent DoF. SMBs will present preferred DoF in the directions where magnetic fields remain invariant for a spatial transformation.

Most common SMBs are free to rotate with very low resistance. Therefore, they present a rotational DoF between the moving and the stationary element. Other SMBs establish a linear kinematic joint and they are free to slide in one direction of space with little resistance while they remain stiff in the others [32]. SMBs with multiple degrees of freedom are possible, like for example a magnet floating over a superconducting torus at the Meissner state [33]. However, such configurations are much less common in practical applications. Figure 12.7 shows a typical rotational SMB and a linear SMB together with the magnetic field that a NdFe N35 magnet (magnetized in the Z direction) generates on the top and lateral surfaces of the superconductors at the normal state (before cooling).

In some cases, SMBs are designed to provide a known stiffness in one or more DoF rather than a free motion capability. This is the case of SMBs designed to provide vibration isolation and they will be discussed more in depth in Section 12.4.9.

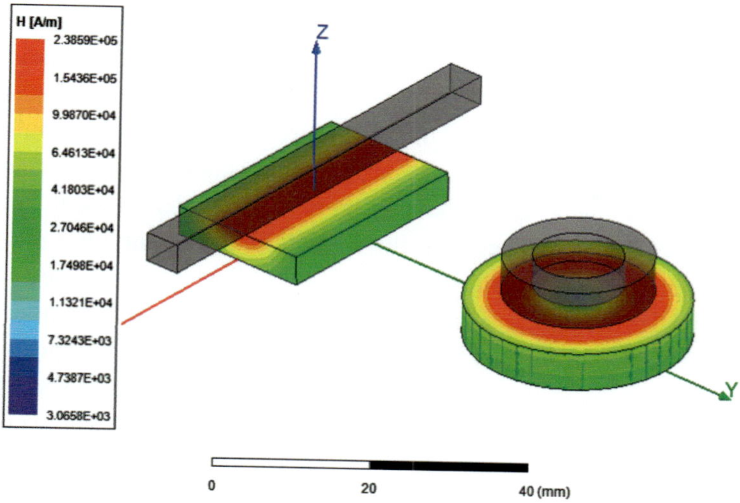

Figure 12.7 Surface magnetic field in a linear and rotational SMB.

12.3.2 Meissner and Mixed State Bearings

Type II superconducting magnetic bearings can either exist at the Meissner state (where applied magnetic field is fully expelled from inside the superconductor) or in a mixed state where magnetic flux is trapped inside the superconductor. A SMB will remain in the Meissner state always that the first critical field (B_{C1}) is not overpassed.

Meissner state bearings show a perfect non-hysterical behaviour, very attractive for mechanisms in cryogenics [34]. However, their load capacity is low. In contrast, mixed state bearings present a larger load capacity but a strong hysterical behaviour which depends on temperature [35]. In both cases, stable levitation is possible; however, mixed state bearings are much common in practical applications.

12.3.3 According to Their Load Bearing Configuration

SMBs can be classified based on how they bear the loads. Thrust SMBs are a particular type of rotational bearings which are designed to bear a predominantly axial load. They are typically composed of one or more superconducting bulks and one or more floating permanent magnets.

Journal SMBs present one or more superconducting bulks coaxially arranged with a set of permanent magnets. These bearings generally show a better balance between radial and axial load capacity. The airgap between the rotor in journal bearings can be minimized and the main load (typically the weight) can be easily arranged in the rotational axis of symmetry of the bearing. Frequently, magnets in journal bearings are arranged with facing opposing polarities to increase load capacity and stiffness [36]. Furthermore, soft ferromagnetic materials are used as flux concentrators between the magnets to increase the load capacity and stiffness of the bearings [37].

Journal and thrust configurations can be combined in a single bearing to generate hybrid bearings with characteristics from both of them. Finally, SMBs that present permanent magnets in both the stator and the rotor are widely described in the literature. In these configurations, magnets are typically arranged with facing opposing polarities and they exert most of the levitation force. According to the Earnshaw theorem, such interaction is unstable. Superconducting bulks are used to stabilize the overall bearing and to further increase axial and radial load capacity and stiffness.

Figure 12.8 shows a cut section of three bearing configurations (in the normal state). The Z axis is coincident with revolution symmetry axis of the bearings. Magnets are made of NdFe N35 axially magnetized as indicated by the arrows.

12.3.4 Active and Passive Superconducting Bearings

Passive magnetic bearings do not present any additional mean to control stability during levitation. On the contrary, active SMBs contain an active system to control levitation. This additional mean can be a set of actuation coils or a position-adjusting device. Active SMBs can present improved stiffness in a "weak" direction (for example radial direction in thrust bearings)[38–40] or a better

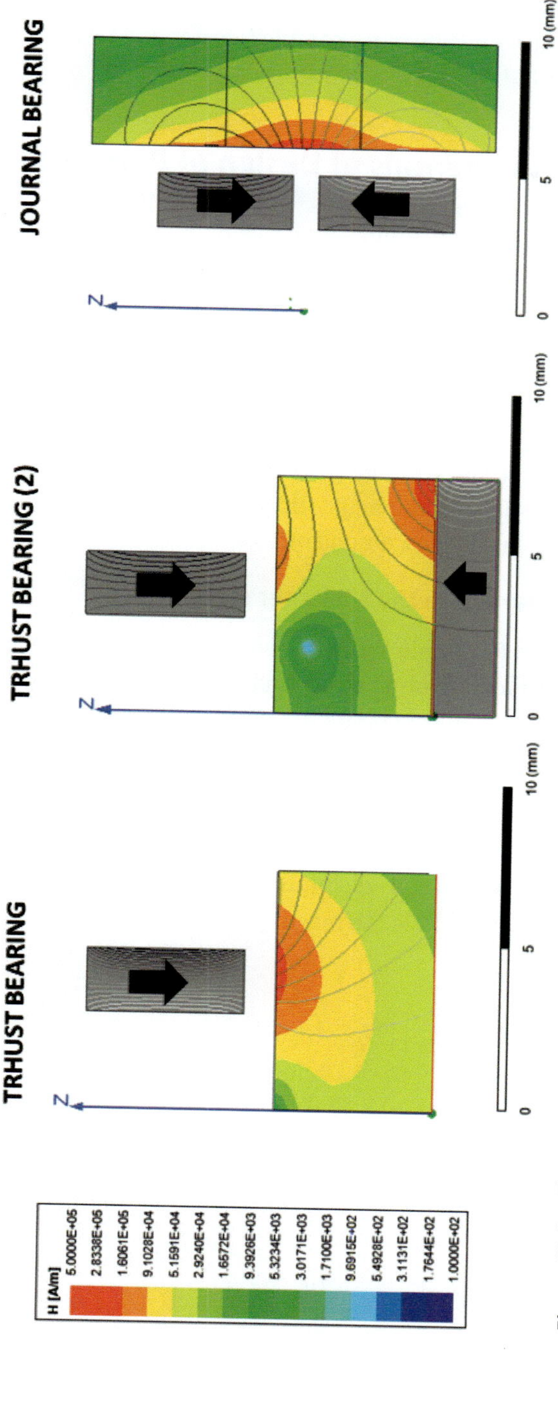

Figure 12.8 Three SMB configurations.

vibration response and stability [41], specially at low frequencies. Active SMBs are typically more complex, costly, and weighty because of the power and control electronics.

12.4 Fundamentals of Design of Passive SMB

In this section, the behaviour of passive SMBs mainly composed by permanent magnets and bulk YBaCuo superconductors is described. In addition, some basic rules for design are proposed and discussed.

12.4.1 State of the Superconducting Bearing

Load capacity of SMBs depends on several factors such as temperature, cooling strategy (field cooled or zero field cooled), the intensity and shape of the magnetic field, the bearing configuration and materials properties, among others.

12.4.1.1 Bearings in the Meissner state

Load capacity of a simple thrust bearing can be estimated assuming the superconductor to be a perfect image of the levitating permanent magnet. For any superconductor shape, the force ($\vec{F}$) that the permanent magnet is experiencing is [19]:

$$\vec{F} = -2\mu_o \iint_{SC} \left(\overrightarrow{n_s} \times \vec{H}^{app}\right) \times \vec{H}^{app} dS \qquad (12.7)$$

where $\overrightarrow{n_s}$ is the normal vector to the surface (S), μ_o is vacuum permeability and $\vec{H}^{app}$ is the magnetic field applied.

And the torque:

$$\vec{M} = -2\mu_o \iint_{SC} \vec{r}_{PM} \times \left(\overrightarrow{n_s} \times \vec{H}^{app}\right) \times \vec{H}^{app} dS \qquad (12.8)$$

where $\vec{r}_{PM}$ is the position vector of each differential element of the surface.

For a simple geometry of a cylindrical permanent magnet (PM) floating over an identical superconductor, the following equation can be derived [17]:

$$F_{ZC} = -\frac{3\mu_0}{2\pi} \frac{M^2 A^2 L^2}{r^4} \qquad (12.9)$$

where M is the magnetization density, A is the area of the magnet, L its length, and r the distance between centres of mass of PM and superconductor.

Maximum force is clearly obtained when the distance $r = L$, and the maximum force per unit area of the permanent magnet can be estimated as:

$$F_{ZC} = -\frac{3\mu_0}{8} M^2 \left(\frac{D}{L}\right)^2 \tag{12.10}$$

The magnetic moment of a real permanent magnet can be calculated from the field it generates in the direction of the magnetization axis $B_{(r)}$ as:

$$B_{(r)} = 2\frac{\mu_0 m}{4\pi \cdot r^3} \tag{12.11}$$

Measurement distance shall be far enough from the magnet so higher order components of the magnetic fields can be neglected. A start measurement distance of 10 times the characteristic length of the magnet should provide accurate results in the calculation of m. Then, the total magnetization of the magnet can be calculated as:

$$M = \frac{m}{V} \tag{12.12}$$

where m is the magnetic moment of the magnet, and V its volume.

For typical geometries of permanent magnets in a SMB (disks and cylinders) made of sintered NdFe, a good approximation for the maximum magnetization at room temperature of the material is $M = 6 \times 10^5$ A/m. Therefore, a maximum theoretical load capacity per unit area of about 17 Ncm^{-2} can be estimated using Eq. 12.10. However, the assumptions in the previous calculation are only valid when the superconductor remain at the Meissner state. This assumption will only be valid for certain values of and applied magnetic fields ($B < B_{c1}$). This condition is usually satisfied for relatively large airgaps between magnet and superconductor, drastically limiting the force density of real applications based in superconducting bearings at the Meissner state.

12.4.1.2 Mixed state: Field cooled bearings

Levitation of superconductors at the mixed state is highly enhanced by the phenomena of flux pinning [42]. Probably the most extended calculation method for mixed state superconductors is the critical state or Bean's model [16,43]. Superconductors at the mixed state show a strong hysterical behaviour.

Figure 12.9 shows a typical magnetization B-H curve (at 77 K) of a bulk YBaCuO superconductor of 16.5 mm diameter and 6.32 mm height and a critical current density of $J_c = 1.4 \times 10^4$ A/cm^2 at 1 T [44].

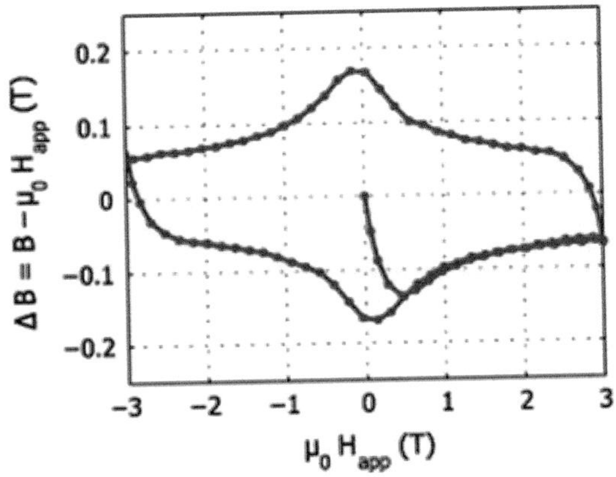

Figure 12.9 Typical YBaCuO magnetization curve. With permission [44].

For applications where hysteresis is not important, a combination of a "frozen" mirror image with a diamagnetic image can be used to properly describe some properties of field cooled (FC) superconductors at the mixed state [22]. In this book, the behaviour of field cooled YBaCuO bulk superconducting bearings will be described by analysing a set of representative configurations.

In order to describe the levitation performance (load capacity) of a superconducting bearing, two parameters are defined. One of them is load surface density: the load capacity (F_{max}) of the bearing compared to the PM or the superconductor surface characteristic area (A), as shown in Eq. 12.13.

$$F_{density\_surface} = \frac{F_{max}}{A} \qquad (12.13)$$

In other situations, is more practical to compare the load capacity with the volume of the envelope of the bearing (V). This approach is useful for evaluation of the cost and other technical requirements such as cooling needs or weight of the bearing. Then, a parameter of force density per unit volume can be defined as:

$$F_{density\_volume} = \frac{F_{max}}{V} \quad (12.14)$$

12.4.2 Thrust Bearings: Force and Stiffness

Axial force and stiffness will depend on a wide set of parameters such as the materials used, the operational temperature, geometry of the bearing, or the magnetic arrangement. Typical bulk YBaCuO bearings operating at 77 K can provide a force density per unit surface between 1 and 5 N/cm^2. In this section, the typical behaviour of thrust bearing will be discussed.

Figure 12.10 shows the levitation force vs. axial displacement on a bearing composed by multiple hexagonal YBaCuO superconductors and a NdFe N48 magnet ring axially magnetized. Results are presented for different heights of field cooling (HFC) and displacements at 77 K and ambient pressure [36].

Levitation force (Fz) depends on the displacement from the field cooling position and from HFC itself in a non-linear way. The maximum bearing load capacity will occur when Z = HFC = 10.5 mm. However, some "airgap" shall be provided for the bearing to operate safely. For a minimum airgap of 1.5 mm, a maximum load capacity of about 3.5 N/cm^2 is calculated.

Stiffness of the SMB can be evaluated as:

$$K = \frac{\partial F}{\partial Z} \quad (12.15)$$

Stiffness is a critical parameter of the performance of a superconducting bearing, especially in dynamic applications. Maximum axial stiffness of bearing in Fig. 12.10 arrangement is about 17 N/mm. Since the previous bearing is highly hysteretic when operating at LN_2 temperature, one must consider that the value of the stiffness highly depends on the displacement amplitude. Then, stiffness should be measured based on the characteristic displacement of each particular application with an adequate testing method.

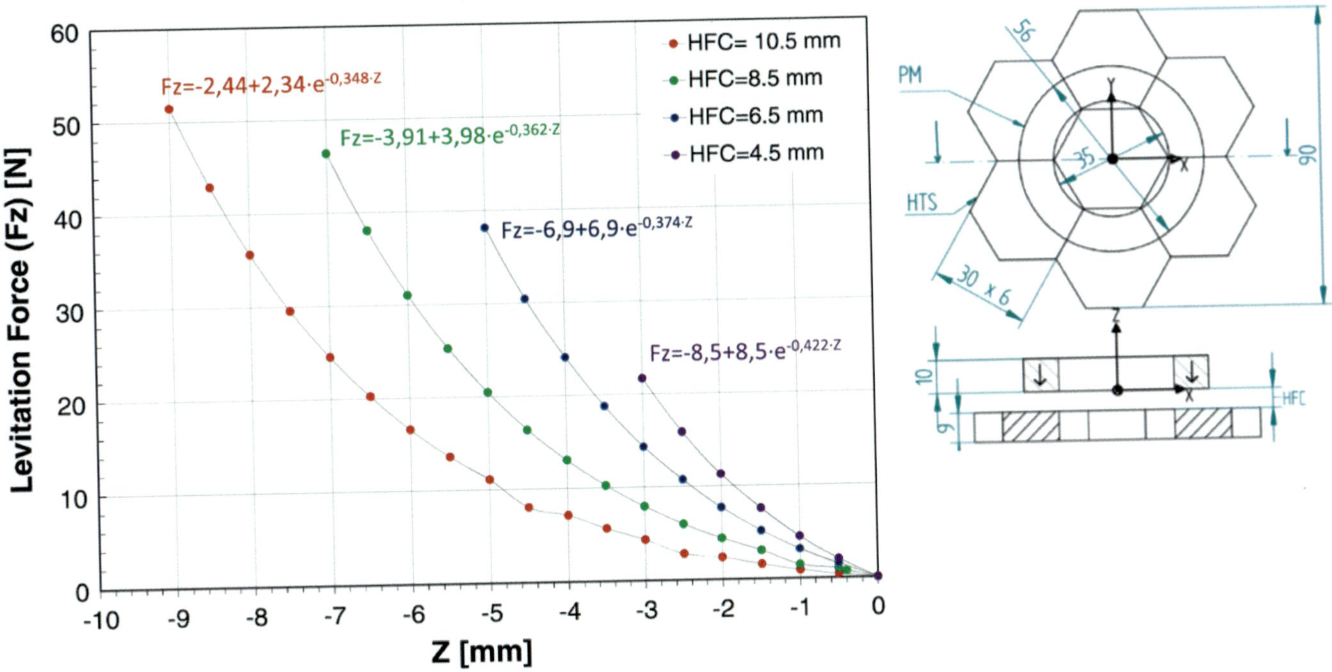

Figure 12.10 Thrust bearing dimensions and axial force vs. axial displacement (Z) at various HFC. With permission [36].

Radial force of previous thrust bearing is dependent on the radial displacement (*X*) and on the HFC. It can be also observed that due to the low hysteresis of the radial force (for small radial displacements), it is a good approach to define the radial stiffness as a single value. Values of radial load capacity and stiffness are 7 times less than in the axial direction.

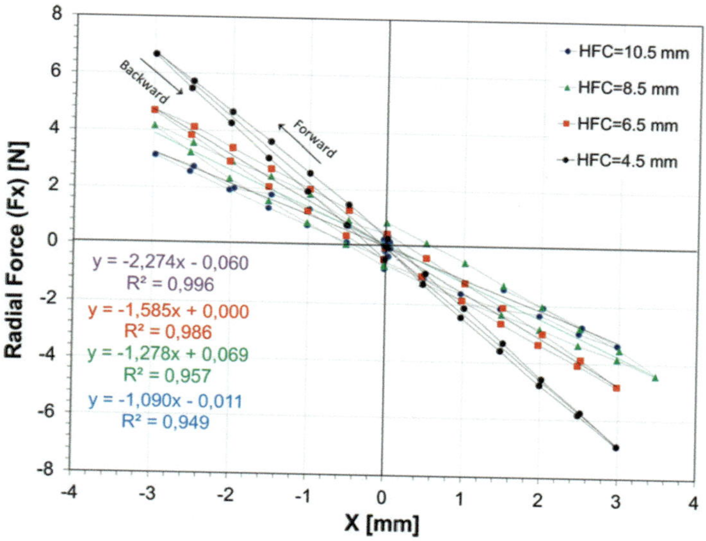

Figure 12.11 Radial force vs. radial displacement (*X*) for various HFC. With permission [36].

Table 12.6 summarizes the performance of the bearing tested at 77 K and HFC = 10.5 mm. Area of the PM is considered to calculate the load surface density:

Table 12.6 Summary of properties of thrust bearing in [36]

Direction	Load capacity [N]	Load density_ surface [N/cm^2]	Load density_ volume [N/cm^3]	Max. stiffness [N/mm]
Axial ($Z = 9$ mm)	52	3.5	0.24	17
Radial ($X = 3$ mm)	6.4	0.43	0.03	2.1

12.4.3 Journal Bearings

Figure 12.12 shows a typical journal bearing and the axial force vs. axial displacement after field cooling. Levitation and radial force were measured for an environmental temperature of 77 K [37]. Note that axial force reaches a maximum when the travel distance of the magnet slightly exceeds half of its length and then it starts to decay. Definition of a force density per unit area of PM is probably less clear than in the case of thrust bearings. However, in order to compare values of both configurations of SMB, the lateral (radial) area of the ring PM will be considered:

$$A_{PM} = \pi Dh \tag{12.16}$$

where D is the outer diameter of the magnet and h its height.

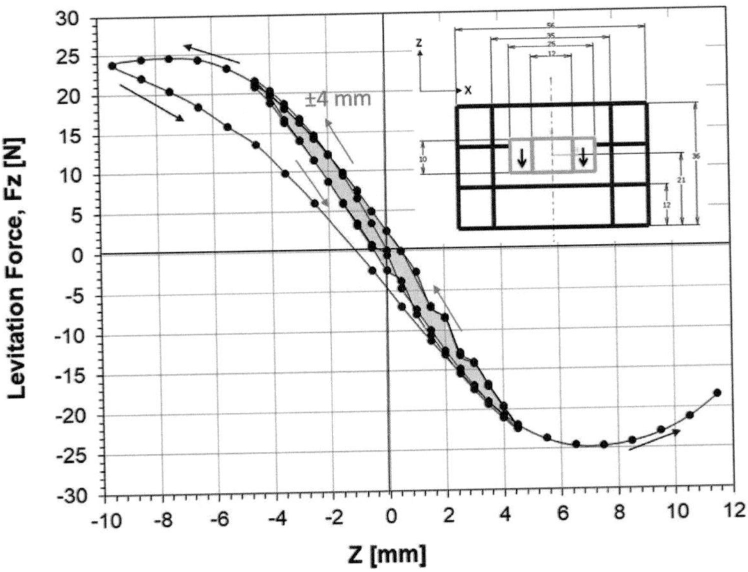

Figure 12.12 Levitation force (Fz) vs. axial displacement (Z) of a typical journal bearing.

Radial stiffness in a journal bearings is frequently independent of the radial displacement due to the small airgaps. A sufficiently large "airgap" should be provided to guarantee a safe use of the bearing during every operational condition. The definition of this "airgap" will significantly affect the size of the magnet and the load capacity and stiffness of the bearing. For this particular case, a safety airgap of 2 mm was considered. Radial force vs. displacement of such bearing is depicted in Fig. 12.13.

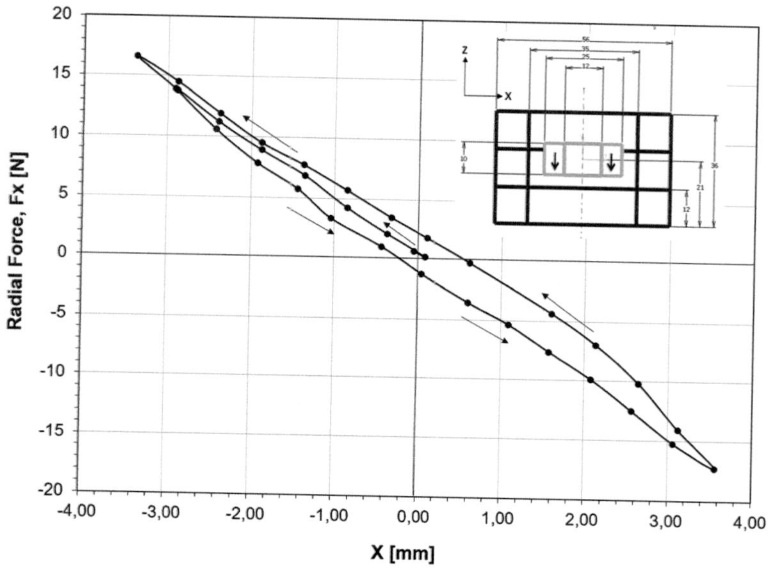

Figure 12.13 Radial force (*Fz*) vs. radial displacement (*X*) of a typical journal bearing.

Radial stiffness of a journal bearing is very similar to its maximum axial stiffness. Its radial load capacity, however, is limited by the allowable radial travel distance. In this case, radial load capacity is about 25% of the axial one.

12.4.4 Improved Magnetic Arrangement

In order to improve the load capacity and stiffness of the bearing, a magnetic arrangement that enhances magnetic flux density and gradient on the superconductor is a technique that provides very good results. Among this magnet arrangements, Halbach array [45] is one of the most common.

Figure 12.14 shows a classic thrust bearing compared to an alternative magnet arrangement. Note that, by changing the orientation of one of the magnets, the levitation force, stiffness and load capacity density can be very significantly improved.

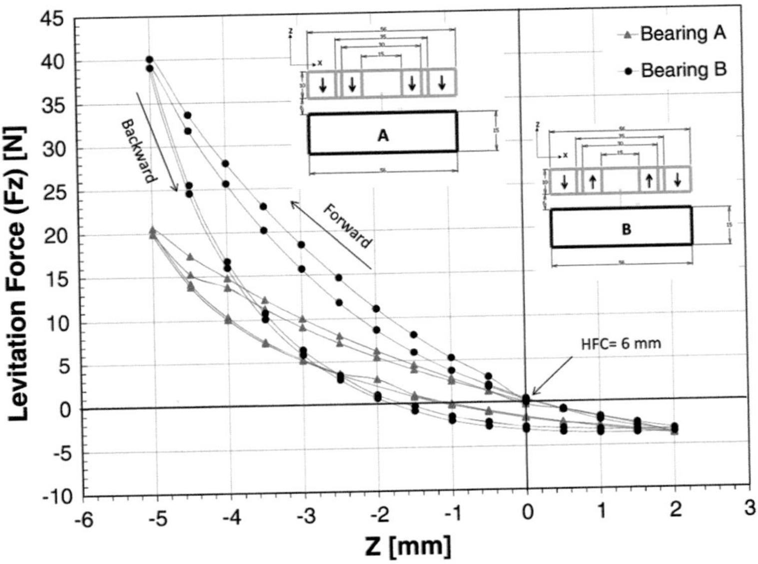

Figure 12.14 Dimensions and levitation force (*Fz*) vs. axial displacement of two thrust bearing. With permission [36].

Radial stiffness of bearing B in Fig. 12.14 is also enhanced by a factor of 4 with regard to radial stiffness in bearing A [36]. Table 12.7 compares the performance of journal and thrust bearings presented in Sections 12.4.3 and 12.4.4.

Table 12.7 summarizes the maximum load capacity and stiffness (between $Z = -5$ and $Z = -4$ mm) for the thrust bearings in this section, measured at 77 K. A HFC = 6 mm was set. The front area of the permanent magnet arrangement (outer diameter = 5.6 cm and inner diameter = 1.5 cm) is considered for calculation of load density per unit surface.

Table 12.7 Comparison of performance of thrust and journal bearings

SMB configuration	Direction	Magnetic arrangement	Load capacity [N]	Load density surface [N/cm^2]	Load density volume [N/cm^3]	Max. stiffness [N/mm]
Thrust	Axial (Z = 5 mm)	Normal (A)	20	0.87	0.25	5
		Improved (B)	40	1.75	0.5	12
	Radial (X = 1 mm)	Normal (A)	1.8	0.08	0.023	1.8
		Improved (B)	8	0.35	0.1	8
Journal	Axial (Z = 5 mm)	Normal (1 PM)	21	3.2	0.28	6
		Improved (3xPM)	77	2.6	0.87	22
	Radial (Z = 1 mm)	Normal (1 PM)	5	0.95	0.08	5
		Improved (3xPM)	14	0.72	0.24	14

In journal bearing, a simple way to enhance load capacity and stiffness is to increase the number of permanent magnets in the bearing. These techniques are particularly effective when poles are faced with opposing magnetic polarity, as indicated in Fig. 12.15.

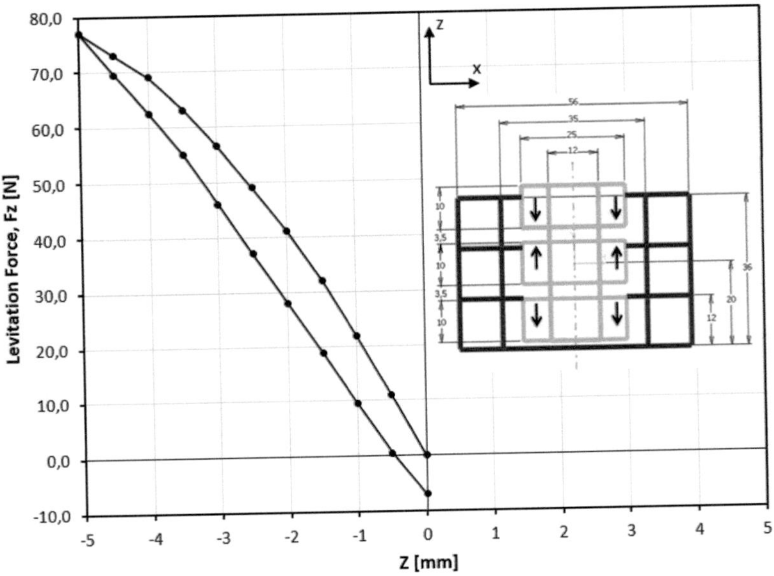

Figure 12.15 Levitation force (*Fz*) vs. axial displacement (*Z*) of a journal bearing with improved magnetic arrangement.

Improvement of the load capacity and stiffness of the bearing are proportional to the number of poles covered by the superconductor. Figure 12.16 shows the axial and radial load capacity vs. the number of poles in the journal SMB. A maximum axial displacement $Z = 5$ mm and radial displacement $X = 1$ mm were defined for a straightforward comparison.

Selection of the ideal number of poles and magnet thickness is a beautiful problem of optimization that shall consider load capacity, stiffness and travel stroke.

Table 12.7 compares the performance of journal and thrust bearings presented in Sections 12.4.3 and 12.4.4.

From the previous results, it can be stated that the journal bearing present a better balance between the axial and radial load and stiffness. On the contrary, radial stiffness of thrust bearings is just between 60% and 30% of their axial stiffness. In addition, the journal bearing provide better load density per unit surface and unit volume. Therefore, a journal bearing should be considered when the most relevant parameter in the design is the load density and stiffness of the bearing.

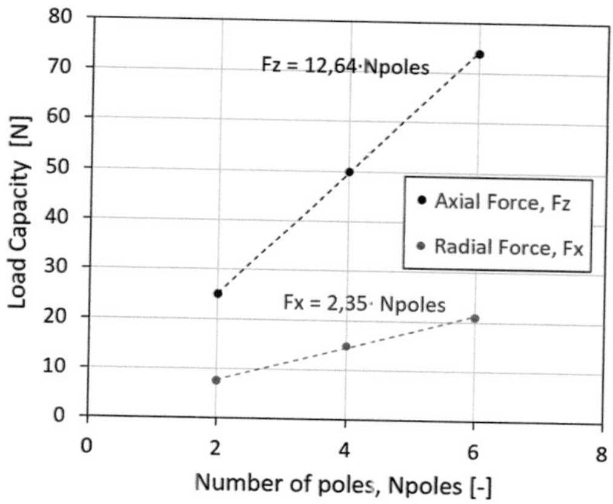

Figure 12.16 Load capacity (*Fz*) vs. number of poles in a journal bearing

12.4.5 Linear Bearings

In a linear SMB, magnetic fields applied on the superconductors present a translational symmetry in a preferred DoF. The idea of using such bearings is very attractive to allow relative liner motion in cryogenic and vacuum environments. In this line, there are plenty of examples of SMB systems proposed for high-accuracy linear positioning and conveyors. Iizuka and Fujita [46] published a conveyor with 2.8 mm stroke and 40 μm accuracy. Lin et al. [47] controlled the height of levitation of a PM over a field cooled superconductor bulk by changing the operating current in a Helmholtz coil. Maximum precision obtained by the authors was in the range of μm for a total stroke around 140 μm.

One particular configuration demonstrated a positioning resolution of 70 Nm in a stroke of ±7.5 mm with very high stability in other DoF [48]. The device is proposed for far infrared interferometry applications in space [49,50]. It consists of a linear slider (1), mainly composed of a 160 mm long NdFe N42 permanent magnet, field cooled over a set of melt-seeded YBaCuO superconducting disks (2) of 45 mm in diameter. A set of two coils (3), placed at each end of the stroke of the mechanism, are used to actively control the position of the slider. Figure 12.17 shows a picture of the prototype.

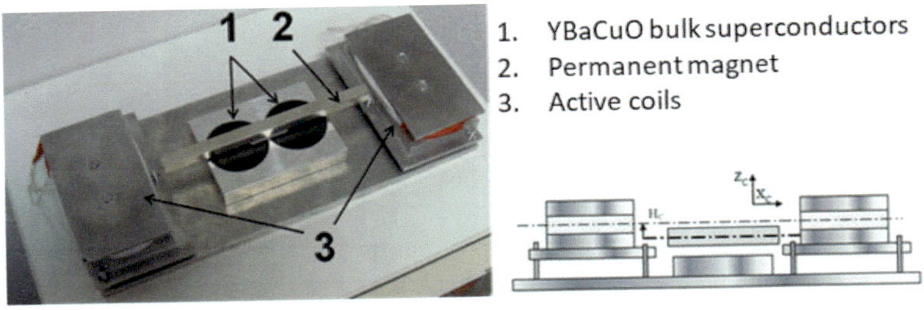

Figure 12.17 Active linear superconducting positioner.

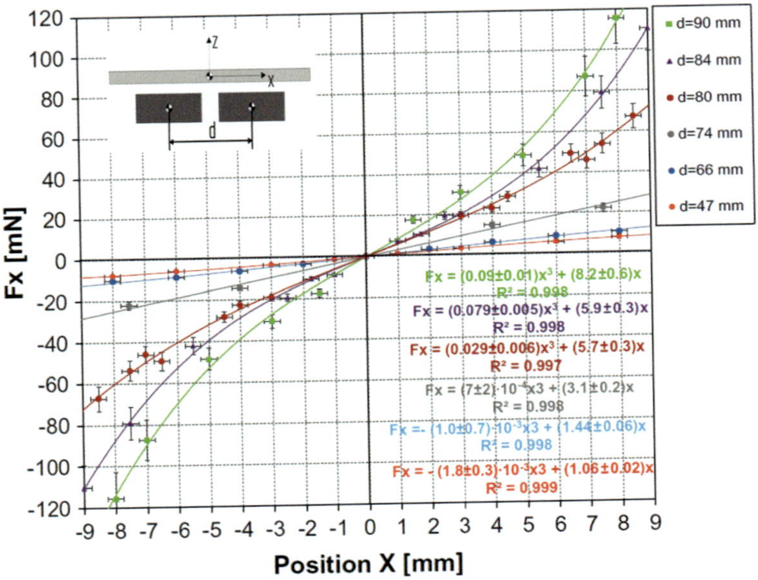

Figure 12.18 Drag force (*Fx*) vs. linear displacement (*x*) from the field cooling position. Measured at 77 K. With permission [32].

Due to variations of the applied field on the superconductors for linear displacements (*X*) of the slider, small drag forces appear. However, the system remains much stiffer in the other DoF. This "friction force" depends cubically on the position X of the slider and quadratically on the distance between the centre of the superconducting disks (*d*). Figure 12.18 shows the results of friction force vs. displacement for values of *d*.

12.4.6 Force Relaxation

Force relaxation refers to the reduction of the levitation force with time that is observed in high-temperature superconductors in the mixed state. Force relaxation follows an exponential decay function [37,51]. The equation that describes the levitation force vs. time is:

$$F_{(t)} = F_{(0)} \cdot e^{-\beta t} \tag{12.17}$$

where $F_{(0)}$ is the levitation force at the initial time, β is the time constant and t is time.

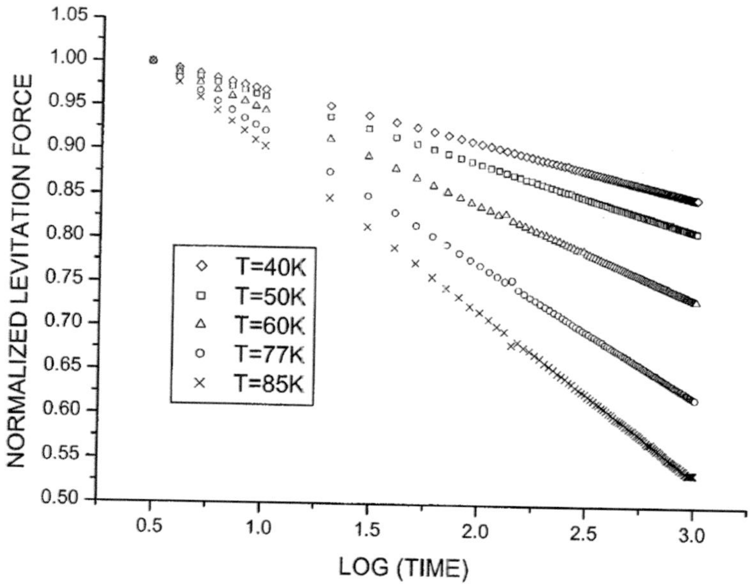

Figure 12.19 Normalized levitation force vs. time (*h*) for a thin YBaCuO SMB at various temperatures. With permission [52].

The higher the superconductor temperature the faster is the force decay. For relatively small samples, the force decay can be as much as 40% in a relatively short time of thousands of seconds [52].

For relatively large samples in a SMB exposed to fields below 1 T at 77 K, a fist order approximation is to expect an asymptotic relaxation with time of about 10–15% [51]. Figure 12.19 shows the normalized force ($F(t)/F(0)$) vs. time for different environmental temperatures of a thin YBaCuO SMB [52].

12.4.7 Hysteresis and Damping

A superconductor at the mixed state will present a predominantly elastic behaviour at low-displacement amplitudes and a very pronounced hysteretic behaviour at larger amplitudes. Hysteresis of a typical force vs. displacement loop of a field cooled YBaCuO SMB can be seen in Fig. 12.20 [22].

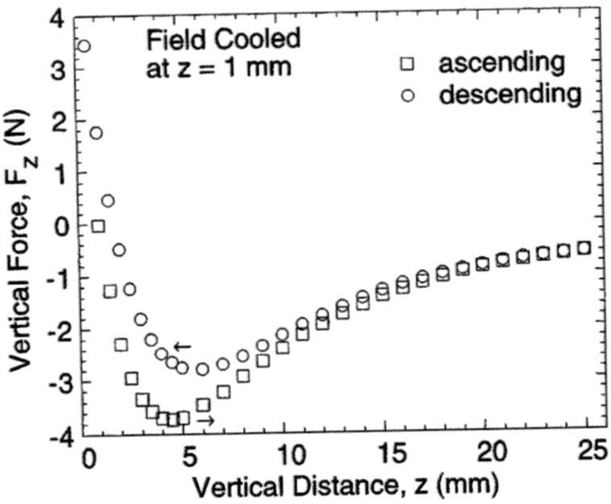

Figure 12.20 Typical hysteresis curve of a thrust superconducting magnetic bearing. With permission [22].

Hysteresis will cause a drag force and an energy dissipation which is proportional to the motion displacement. For reciprocating motion systems (like vibrations), this hysteresis is frequently defined as a hysteresis loss factor (η) [53]:

$$\eta = \frac{D}{2\pi W} \tag{12.18}$$

where D is the dissipated energy and W is the storage energy in the cycle.

The storage energy can be defined, for a linear spring system as:

$$W = \frac{1}{2}kx^2 \tag{12.19}$$

where k is the stiffness of the motion and x is the displacement amplitude.

For large displacements in a SMB, the stiffness cannot be considered constant and the previous equations must be reconsidered.

The loss factor can be related to the damping ratio (ζ) in low damped hysteretic systems to determine resonance amplification and vibration isolation as [53]:

$$\eta = 2\zeta \quad \text{if } \eta \leq 0.28 \tag{12.20}$$

Dependency of the hysteresis loss factor on temperature is reviewed in the next section.

12.4.8 Temperature Influence

Temperature has a major impact on the bearing performance. This section is focused in the evaluation of the impact on temperature on a journal bearing performance. Performance is studied in a high-vacuum from 12 to 90 K [35]. The representative journal bearing is composed of 3 $SmCo_{32}$ magnets and a stack of 6 YBaCuO melt-growth superconducting rings as shown in cross section in Fig. 12.21.

The bearing is designed with a maximum displacement of ± 3.5 mm from the field cooling position. Figure 12.21 depicts the axial force vs. displacement from the field cooling position from 12 to 90 K.

The load capacity (Lc) of the bearing is defined as:

$$\text{Lc} = \frac{F_{max} - F_{min}}{2} \tag{12.21}$$

where F_{max} is the maximum force and F_{min} the minimum force in the stroke.

In order to provide an estimation of the bearing stiffness, the following linearization equation is proposed:

$$K = \frac{F_{max} - F_{min}}{Z_{max} - Z_{min}} \tag{12.22}$$

where Z_{max} and Z_{min} are the ends of the stroke.

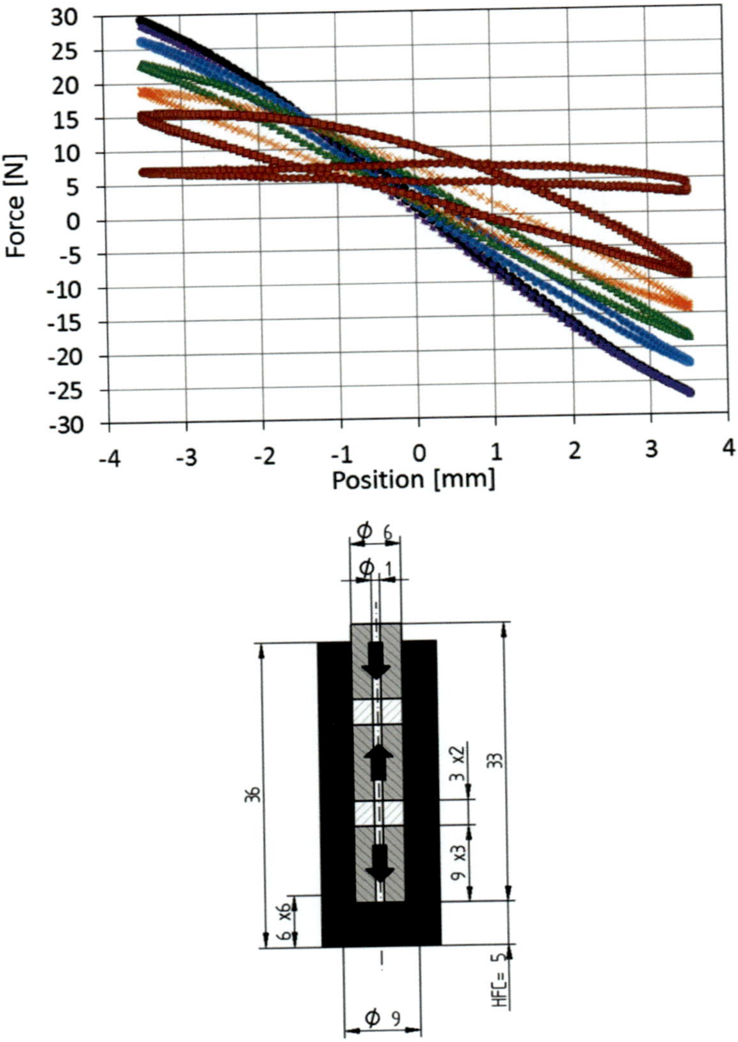

Figure 12.21 Dimensions and axial force vs. axial displacement from the field cooling position of a typical journal bearing. With permission [35].

The previous equation does not exactly meet the stiffness of the bearing, which is a complex function that depends on the displacement amplitude and position. However, it is very useful for definition of an average stiffness.

Figure 12.22 shows the temperature influence on the bearing load capacity and stiffness. Both load capacity and stiffness depend quadratically with temperature. An asymptotic behaviour is found at lower temperatures than 30 K.

Results in Fig. 12.22 demonstrate that the loss factor is exponentially reduced at low temperatures.

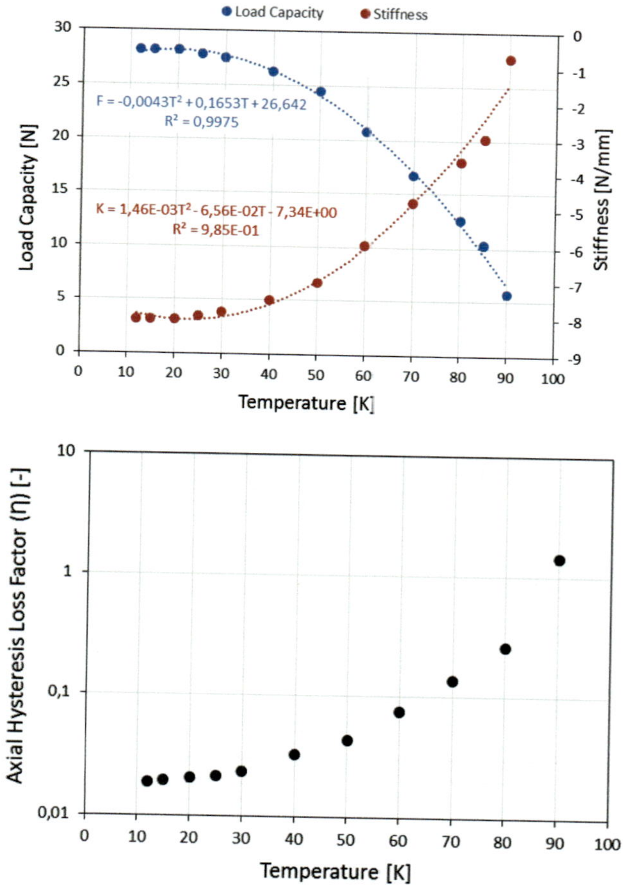

Figure 12.22 Load capacity, stiffness and hysteresis loss factor vs. temperature. With permission [35].

The value measured at 14 K is just about 0.02. This very low value of hysteresis is of the order of typical values of hysteresis factors associated to structural damping in solid metals.

Finally, force and stiffness were measured in the radial direction. Radial stiffness and load capacity are less sensitive to temperature than the axial ones. Radial hysteresis loss factors decrease from 0.08 at 60 K to 0.03 at 15 K. Figure 12.23 depicts load capacity and stiffness measured in the radial direction at different temperatures.

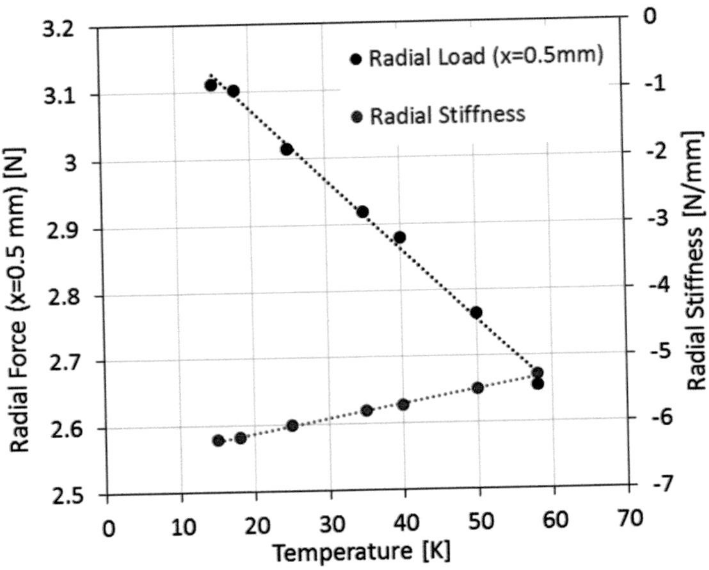

Figure 12.23 Radial load capacity and stiffness vs. temperature. With permission [35].

12.4.9 Vibration Isolation

A vibrating system with one DoF is represented in Fig. 12.24 for the journal bearing investigated in the previous section.

The proposed motion equation for a single DoF SMB under harmonic vibrations is:

$$m\ddot{x}+2\zeta\omega_0\dot{x}+\omega_0 x=\vec{F} \tag{12.23}$$

where m is the moving mass, $\ddot{x}$ the acceleration, ζ the damping ratio and ω_0 is the natural frequency of the bearing. $\vec{F}$ is the summatory of all external forces.

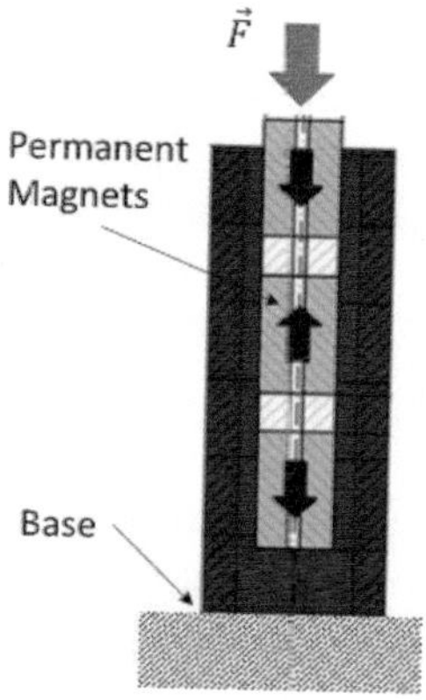

Figure 12.24 Diagram of a journal bearing under harmonic forced excitation.

Damping ratio can be estimated as:

$$\zeta = \zeta_e + \frac{\eta}{2} \tag{12.24}$$

where ζ_e is the damping ratio associated to the eddy current power losses and η the hysteresis loss factor.

This model can be applied with little error for a linear system (small displacement amplitudes and $k \approx$ constant) under harmonic excitations when hysteresis loss factor is low. A threshold value of $\eta \leq 0.28$ is proposed in [53].

When previous conditions are satisfied, Eq. 12.23 could be used with little error. Then, two possibilities can be considered to evaluate the performance of the system under harmonic vibrations. One is to consider an infinitely rigid base (1) and evaluate the relationship between the moving element response (x_Ω) compared to the static deflection (x_{st}):

$$\left|\frac{x_\Omega}{x_{st}}\right| = \frac{1}{\sqrt{\left(1-\Omega^2\right)^2 + \left(2\zeta\Omega\right)^2}} \tag{12.25}$$

where $\Omega = \dfrac{\omega}{\omega_o}$, x_{st} is the static displacement amplitude, ω the excitation angular frequency and

$$\omega_o = \sqrt{\frac{k}{m}}. \tag{12.26}$$

The second possibility is to consider a case of base motion. Then, the moving mass is vibrating with an amplitude of amplitude (x_Ω) which is a function of the base motion (z_Ω). In that case, a transmissibility modulus of vibrations can be defined as:

$$|T| = \left|\frac{x_\Omega}{z_\Omega}\right| = \sqrt{\frac{1+(2\zeta\Omega)^2}{\left(1-\Omega^2\right)^2+(2\zeta\Omega)^2}} \tag{12.27}$$

Figure 12.25 shows the predicted transmissibility modulus vs. frequency for various damping ratios for the bearing model in Fig. 12.24. Such damping ratios were defined using Eq. 12.23 from the hysteresis loss factors obtained for a journal bearing at temperatures between 85 and 14 K (see Section 12.4.8):

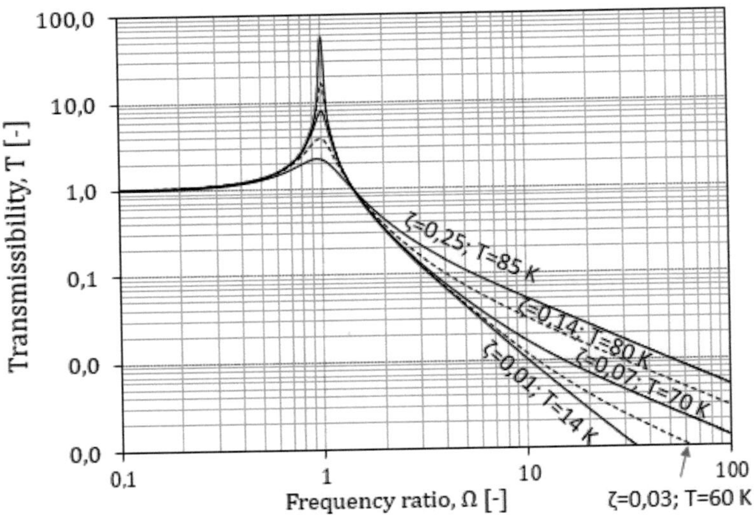

Figure 12.25 Transmissibility vs. frequency ratio estimated at various temperatures.

From the previous figure, a set of relevant conclusions can be extracted:

- No vibration isolation ($T < 1$) is obtained for frequencies below $\sqrt{2} \cdot \Omega$.
- Vibration isolation at high frequencies is lower for higher values of the damping ratio/temperature.

- Amplification of vibrations at the resonance increase for lower values of the damping ratio/temperature.

Therefore, during a detailed design of a passive SMB a trade-off must be met between the isolation of vibrations at high frequencies and amplification of force/motion at the resonance.

12.4.10 Coefficient of Friction

An ideal rotational SMB in a high vacuum would present no drag torque if the magnetic field is perfectly axisymmetric. However, practical developments showed, from the very begging of the technology, an equivalent drag torque due to small asymmetries of the generated magnetic field of coils or permanent magnets. Energy losses due to magnetic drag for a superconductor at the critical state is given by Bean's model [16]:

$$E_{mag} = \frac{A\left(\Delta B_{pp}\right)^3}{J_c} \tag{12.28}$$

where A is a geometric factor, ΔB_{pp} is the peak to peak change of magnetic field and J_c is the critical current of the superconductor at the operative temperature.

Note that energy losses are proportional to $(\Delta B_{pp})^3$ and inversely proportional to the critical current (J_c). Therefore, a main goal on the design of a SMB should be to provide a very homogeneous magnetic field source. Three conditions shall be satisfied by the bearing permanent magnet to be compliant with such requirement:

- To present very homogeneous material properties and magnetization.
- The magnets shall present a very regular shape and tight tolerances.
- To be very well assembled whit its magnetization direction very well aligned with the rotational axis of the bearing.

Fabrication aspects such as tolerances, surface roughness, thickness of the coating, porosities and material inhomogeneities can play a very relevant role too. Techniques focused on maximization of the magnetic homogeneity are sometimes referred as "magnetic balancing" [54,55]. In addition, operating at lower temperatures will reduce the energy losses in the bearing.

Other magnetic power losses can include hysteresis losses and eddy current dissipation in the constructive materials of the bearing and shall be very carefully considered when designing the SMB. Finally, the rotor of a SMB must be statically and dynamically balanced to minimize this friction losses generated by vibrations.

Friction coefficient in a SMB is related to the previously discussed energy losses. A coefficient of friction for SMBs was first introduced by J. Hull related to the observed drag force as[56]:

$$\mu = F_D/F_L \tag{12.29}$$

where F_D is the drag force and F_L is the levitation force.

On one hand, the levitation force is:

$$F_L = m \cdot g \tag{12.30}$$

where m is the mass of the flywheel and g the acceleration of gravity.

On the other hand, in a rotational system, we can correlate the drag force with the drag torque (T_D) of the bearing using Eq. 12.31.

$$T_D = F_D \cdot R_b \tag{12.31}$$

where R_b is the pitch radius of the bearing.

This drag torque, will cause a deceleration of the flywheel as:

$$T_D = -I \cdot \frac{d\omega}{dt} \tag{12.32}$$

where I is the moment of inertia referred to the rotational axis of the flywheel, ω the angular speed of the flywheel and t is time.

By combination of Eqs. 12.30 to 12.32, it can be demonstrated that:

$$\mu = -\frac{I}{mgR_b} \cdot \frac{d\omega}{dt} \tag{12.33}$$

Therefore, the friction coefficient of a SMB can be easily determined by measuring the angular deceleration rate of the bearing.

For a flywheel that can be approximated to a thin solid rim, the previous equation can be further simplified to:

$$\mu = -\frac{R_f^2}{gR_b} \cdot \frac{d\omega}{dt} \tag{12.34}$$

where R_f is the radius of the flywheel rim.

It is also very useful for flywheel applications to correlate the friction coefficient with the ratio of energy loss. For that, Eq. 12.35 can be used:

$$\frac{E}{E_o} = \left(1 - \frac{\mu g R_b}{\omega_o R_f^{\,2}} \cdot t\right)^2 \tag{12.35}$$

SMBs present the lowest coefficient of friction of all bearing technologies up to date with experiments showing coefficients of friction typically ranging from 10^{-5} to values as low as 10^{-8} [57,58]. Figure 12.26 shows the friction coefficient of a set of low-speed superconducting flywheels. Reading of [22] is recommended for further discussion on the impact of vibration modes and rotor dynamics in the CoF of superconducting bearings.

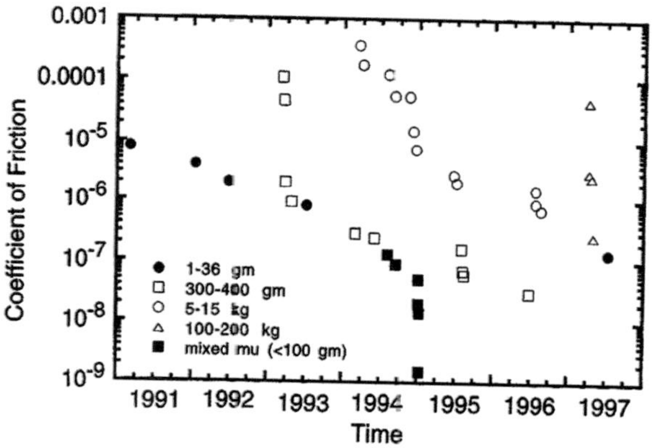

Figure 12.26 CoF of various low mass flywheel development. With permission [22].

12.5 Applications of Superconducting Bearings

12.5.1 Introduction

Classical mechanical bearings play a key role in our modern-day world. They are a fundamental technology that allows relative motion between moving parts, making operation of machinery more efficient and sustainable. Bearings can be found everywhere: from bicycle wheels, to wind turbines in the North Sea or even in space telescopes. Bearings come in all forms and sizes, adapted to whatever task they must perform. This performance is frequently limited by a diverse set of factors, such as vibrations, wear, lubricant conditions, operation temperatures, and so on.

SMBs by being non-contact solutions, are claimed to solve many of the limitations of their traditional counterparts. SMB can be currently found as active or fully passive solutions, and they have been used in more applications than what an inexperienced reader may initially think. Active SMBs have already been implemented in numerous systems [59] and are capable of working under high loads in vehicles, pumps, and flywheels. Either passive or active SMBs typically require cooling system which should be justified by the advantages that the technology provides. Active superconducting bearings require input power to operate [60]. On the other hand, they potentially provide improved quasi-static (stiffness) and dynamic performance (e.g., vibration isolation at low frequencies).

In both cases, active and passive SMB are in an early stage of development when real-life applications are considered. Despite this drawback, they present very promising capabilities that are meant to solve modern technology problems related to energy storage, space applications, cryogenic machinery, vibration isolation, transportation, etc. Since the development of new superconducting materials in the 1980s [1,61–63], many of these potential solutions have become possible and some are still being studied. This section presents a review of some advanced applications of SMBs.

12.5.2 Cryogenic Machinery

Cooling SMBs is a necessity, so it is natural that they were first applied in machinery operating in cryogenic environments. In such environments, conventional mechanical bearings present severe tribological problems associated with lubrication and wear. Some of these limitations are extensive to other vacuum, clean and cold environments, not necessarily cryogenic, such as space or manufacturing environments. In these cases, outgassing generated by lubricants can represent a major drawback and a limitation for conventional bearings that may be overcome by SMBs.

Efficiency of SMBs is also a main advantage of this technology. As discussed in Section 12.4.10, the CoF of superconducting bearings is extremely small, resulting in very low power losses and improved systems efficiency. SMB have been proposed and evaluated for: cryogenic flowmeters [64], advanced cryocooler designs, and cryogenic turbopumps [65]. In a recent design study, a HTS bearing was used to design and fabricate a helium cold compressor [66] that achieved a design speed of 50,000 rpm and a levitation force of 77 N.

12.5.3 Aerospace Applications

There are numerous applications that could benefit from superconducting levitation in space. Cryogenic missions are numerous: Planck and Astri-H Athena, James Web or LISA missions are some examples [67–69]. In such space telescopes, cryogenic environments are in fact available and SMB may be used without imposing significant additional requirements to the cooling needs of the telescope.

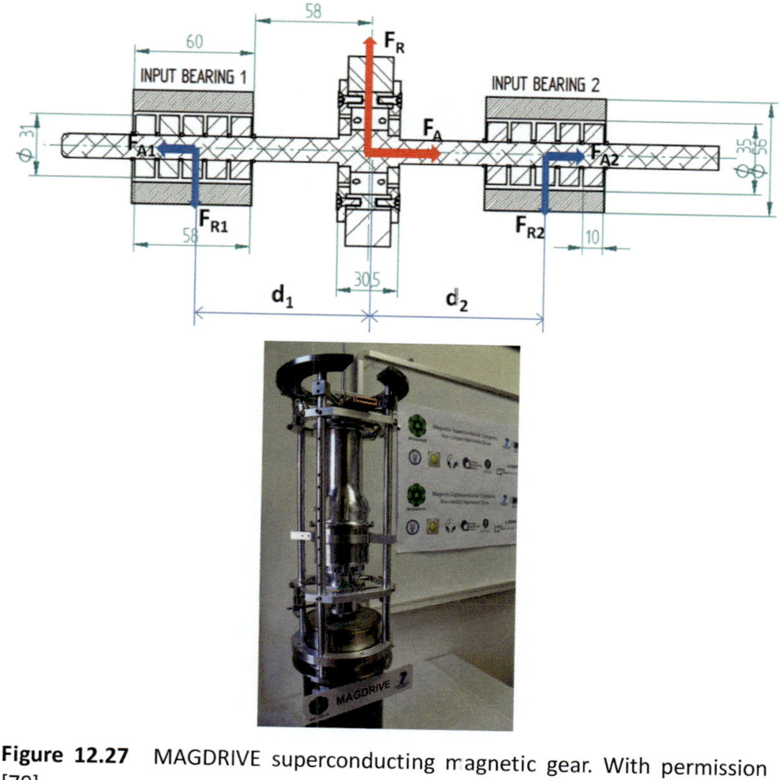

Figure 12.27 MAGDRIVE superconducting magnetic gear. With permission [70].

In this context, a superconducting magnetic gear has been demonstrated for torque multiplication in cryogenic conditions [70]. In the FP7 Space MAGDRIVE project, a prototype of a magnetic gear supported by passive SMBs was demonstrated at an operation temperature of 40 K. The gear is supported by two bulk journal SMBs that provide a radial stiffness of 220 N/mm at 77 K. Figure 12.27 shows a diagram for the tested prototype.

SMBs were used for demonstration of a nano-positioning device for far infrared interferometry in space applications too [71]. The prototype, tested at 77 K and 15 K, demonstrated a 70 nm resolution in a stroke of ±7.5 mm with an estimated power consumption of just 6 mW. A lateral run out of just 2 μm and angular run outs of the order of tens to hundreds of μrad were measured [48]. Figure 12.28 shows a picture of the prototype during testing.

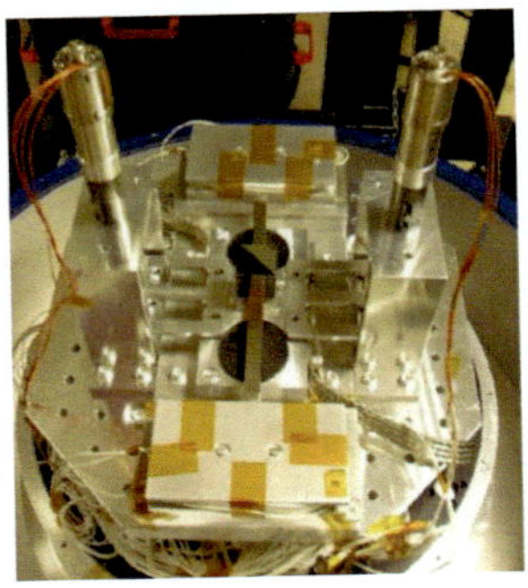

Figure 12.28 Prototype of a superconducting nano-positioner.

Finally, superconducting magnetic levitation has been evaluated for its potential use in the ESA ATHENA mission as vibration isolation and thermal disconnect mechanism. ATHENA is a large-class cryogenic mission for observation of the high energy universe. In that mission, vibrations generated by the cryocoolers are a main concern. These vibrations reduce focusing capability and generate heat with reduces the instruments performance. The use of superconducting levitation for vibration isolation and "perfect" thermal disconnect has been evaluated for its potential use in the supports of the focal plane array (FPA) of the X-IFU instrument [72]. The FPA weight is about 12 kg and operates at extremely low-cryogenic temperatures [73]. The proposed vibration isolator is mainly composed of a set of YBaCuO bulks and SmCo magnets.

The technology, aims to a reduction of vibrations by a factor of 10 at 30 Hz and to reduce thermal leakages below 5 mW, highly reducing the cooling power requirements.

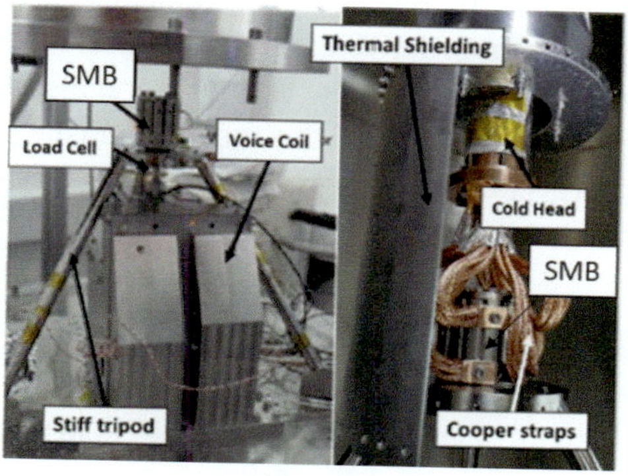

Figure 12.29 SMB manufactured by MAG SOAR during testing. Adapted from [73].

Cryocoolers are used in other space missions to cool sensors to the proper operating temperature to allow precise measurements. Infrared sensors are operated at cryogenic temperatures of around 2 K [74] in order to extract the low level signal from the background noise. Cryocoolers are also used to refrigerate X-ray detectors to Sub-Kelvin temperatures [75]. The application of SMB to these systems has already been subject of study [76], and with NASA's plan to use high-temperature superconducting generators and engines on the recent future [77], their spread is a firm possibility. Cryopumps are critical components of a launching systems since space shuttles use cryogenic fuels, and liquefied oxygen (LOX) is used to provide breathable air to astronauts. HTS bearings have been proved to be more efficient for this application than previous self-acting gas bearings [78].

Finally, SMBs have been also explored for use in spacecraft attitude control and energy storage systems (ACESF)[41].

12.5.4 Energy Storage

Storage of energy is a key element in power availability. A classical and widely spread example are fossil fuels that storage a great amount of chemical energy that is then transformed into power on internal combustion engines (in a rather inefficient manner). For storage of electrical energy, different types of batteries are used, the most common being Lead-Acid and Lithium-ion. The production of these conventional energy storage systems has a serious impact in the environment. In order to continue the sustainable transition process to renewable energy, new energy storage systems must be developed and applied. One of the most promising solutions for sustainable energy storage is flywheel systems (store energy in the form of kinetic energy).

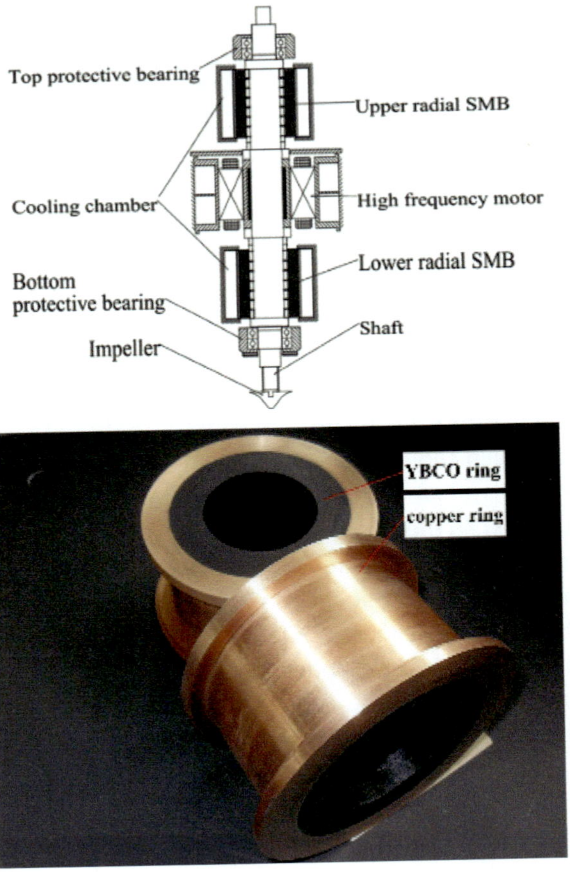

Figure 12.30 Schematic of helium cold compressor with SMBs. With permission [66].

In the short term, electrical power fluctuation is one of the main concerns surrounding the application of renewable energy to country grids as the main source of power generation. Large-scale flywheels provide a natural storage system for energy and angular momentum, which can be significant in the solution of this issue. Traditionally, one of the limitations in the design of these high-speed rotating bodies has been friction losses in the bearings.

Flywheel Energy Storage Systems (FESS) using SMB present lower power losses due to friction than conventional FESSs because they have the lowest CoF of all bearing technologies, as discussed in section 12.4.10. Attempts have been carried out to apply SMBs to support FESSs. Most of those attempts used a combination of SMB and other types of bearings, either magnetic or conventional [79–81].

First large-scale SMB flywheel (5 kWh) was developed by Boeing in 1995 [82]. Since then, several developments have demonstrated the potential of this technology, including a prototype from NASA presented in 2006 intended to operate in the International Space Station [83].

A flywheel that uses superconducting bulks and superconducting coils was installed in the Yamanashi photovoltaic power plant [84]. It achieved a maximum speed of 6000 rpm, being the current most powerful SMB flywheel with a 100 kWh energy storage capacity [85,86].

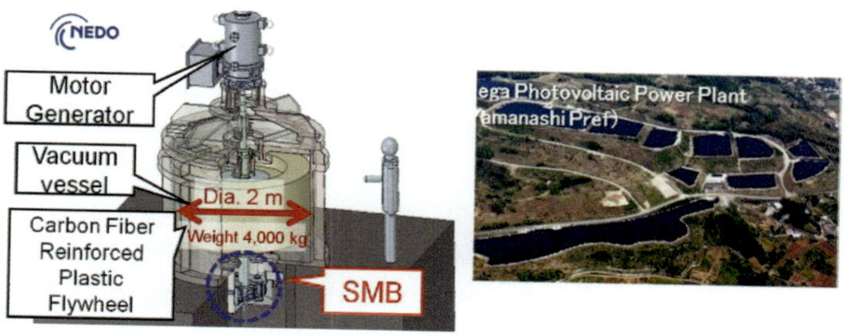

Figure 12.31 Diagram of the flywheel with SMB technology installed in the Yamanashi Power Plant. With permission [95].

FESSs can also act as energy storage for pulsed power applications, like high-power lasers and fusion reactors, due to their quick rate of charge and discharge [87]. Other applications of conventional magnetic bearings in flywheels that could benefit from the use of SMBs are Rotary Uninterruptible Power Supply (UPS) units [88].

12.5.5 Transportation

Flywheel using SMBs has been also proposed for storage systems in transportation vehicles. The objective of such flywheels is to provide a primary or parallel energy source in electric vehicles, minimizing battery needs. Flywheels present a very quick rate of charge, which is one of the main drawbacks of conventional batteries. Nowadays, the main use of conventional flywheel in transportation is Kinetic Energy Recovery Systems (highly developed in motorsport) to store breaking energy that is then used to power an electric motor and reduce the energy consumption of the vehicle.

Volvo presented in 2013 a kinetic energy recovery system, which flywheel (weighting just 6 kg) spinning at up to 60,000 rpm, with a power of 60 kW [89]. Volvo claimed to achieve a 25% reduction in energy consumption when the mentioned KERS system was attached to an S60 model. Also a 100 kW FESS for a hybrid bus was developed in 2013 [90].

Modern Light-Rail trains also use this technology. Trains equipped with regenerative braking systems dump the energy recovered in the breaking back into the catenary. FESS help control the variation in voltage in the line. This technology is especially useful in Subway systems, where trains accelerate and decelerate constantly [91]. One example of this application can be seen in Fig. 12.32, where a 35 kWh FESS system for electric power stability of subway stations is shown.

There are several examples of "mini-trains" using superconducting magnetic levitation that have been developed recurrently for demonstration and didactic proposes [92]. However, the potential of this technology goes much further. Probably, the best example is the superconducting MAGLEV LO Series train. This train connects Tokyo and Osaka and currently holds the world speed record with 603 km/h. In such train, superconducting active magnetic bearings are used for stable levitation of the train.

Finally, another widely studied application is the use of superconducting motors for electrical vehicles propulsion. Such motors, present a potential high-power density and efficiency. Superconducting motor are also under development for efficient aircraft propulsion. A 1 MW motor with 20 W/kg power density is currently under development [93]. Superconducting motors have also been proposed for spacecraft magnetic propulsion engines [94].

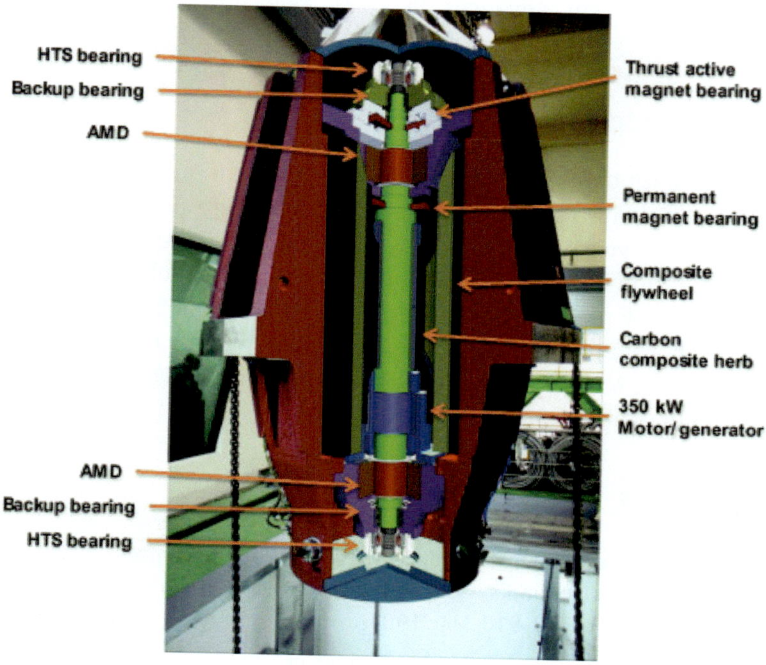

Figure 12.32 Sectional view projected on the photo of 35 kWh SFES installed in KEPRI. With permission [96].

12.5.6 Conclusions

Superconducting Magnetic Bearings present a huge potential for improving the performance of advanced machinery: they can operate passively, without contact between moving parts; they eliminate all the tribological issues such as wear, generation of debris or need of lubrication and moreover, they provide very little energy losses, resulting in improved machine efficiency and sustainability.

In this chapter, we gather relevant properties of very common materials used in SMBs. Fundamental rules of design of different types of SMBs are provided, including: load capacity and stiffness, force relaxation, coefficient of friction, temperature influence, hysteresis and behaviour under harmonic vibrations. Such design rules will provide the mechanical and cryogenic engineer with the required tools for optimization of SMBs from a multidisciplinary point of view.

Finally, recent advances in some relevant applications are discussed.

Acknowledgment

The authors thank Professor J. L. Perez-Diaz and all MAGSOAR SL team for their contributions to SMB technology development to though the years. We would like to thank Alba Martinez for her collaboration in generation of some figures in this book.

The work presented in this chapter has been partially funded by the "Subprograma Estatal de Incorporación Ramón y Cajal 2017" of the Spanish Ministry of Science, Innovation and Universities under grant number RYC-2017-23684.

References

1. M. K. Wu et al., "Superconductivity at 93 K in a new mixed-phase Yb-Ba-Cu-O compound system at ambient pressure," *Phys. Rev. Lett.*, vol. 58, no. 9, pp. 908–910, Mar. 1987.
2. I. Gurrappa, "The corrosion behaviour of SmCo5 permanent magnets in different environments," *Mater. Chem. Phys.*, vol. 78, no. 3, pp. 719–725, Feb. 2003.
3. I. Gurappa, "Corrosion characteristics of permanent magnets in acidic environments," *J. Alloys Compd.*, vol. 360, no. 1–2, pp. 236–242, Oct. 2003.
4. K. P. Belov, A. K. Zvezdin, A. M. Kadomtseva, and R. Z. Levitin, "Spin-reorientation transitions in rare-earth magnets," *Sov. Phys. - Uspekhi*, vol. 19, no. 7, pp. 574–596, 1976.
5. E. Diez-Jimenez, J. L. Perez-Diaz, F. Canepa, and C. Ferdeghini, "Invariance of the magnetization axis under spin reorientation transitions in polycrystalline magnets of Nd 2Fe 14B," *J. Appl. Phys.*, vol. 112, no. 6, p. 063918, Sep. 2012.
6. Y. He, X. Bao, and Q. Zhou, "Magnetic field stability of PrFeB magnets developed by GBD for cryogenic permanent magnet undulators," *J. Rare Earths*, vol. 36, no. 4, pp. 385–389, 2018.

7. A. Bollero, O. Gutfleisch, K. H. Müller, L. Schultz, and G. Drazic, "High-performance nanocrystalline PrFeB-based magnets produced by intensive milling," *J. Appl. Phys.*, vol. 91, no. 10, pp. 8159–8161, 2002.

8. C. Benabderrahmane et al., "Nd 2Fe 14B and Pr 2Fe 14B magnets characterisation and modelling for cryogenic permanent magnet undulator applications," *Nucl. Instruments Methods Phys. Res. Sect. A Accel. Spectrometers, Detect. Assoc. Equip.*, vol. 669, pp. 1–6, 2012.

9. R. J. Parker and R. J. Studders, *Permanent Magnets and Their Application*. J. Wiley and Sons, 1962.

10. R. Poggiani, "Materials and components of possible interest for cryogenic operation of Einstein Telescope," 2009.

11. H. K. Onnes, "Further experiments with Liquid Helium. G. On the Electrical Resistance of Pure Metals, etc. VI. On the Sudden Change in the Rate at which the Resistance of Mercury Disappears.," in *Through Measurement to Knowledge, Boston Studies in the Philosophy of Science*, Springer, Ed. 1971, pp. 267–272.

12. W. Meissner and R. Ochsenfeld, "Ein neuer Effekt bei Eintritt der Supraleitfähigkeit," *Naturwissenschaften*, vol. 21, pp. 787–788, 1933.

13. V. Moshchalkov et al., "Type-1.5 superconductivity," *Phys. Rev. Lett.*, vol. 102, p. 117001, 2009.

14. J. G. Bednorz and K. A. Müller, "Possible high T_c superconductivity in the Ba-La-Cu-O system," *Zeitschrift für Phys. B Condens. Matter*, vol. 64, pp. 189–193, 1986.

15. A. P. Drozdov et al., "Superconductivity at 250 K in lanthanum hydride under high pressures," *Nature*, vol. 569, no. 7757. pp. 528–531, 2019.

16. C. P. Bean, "Magnetization of hard superconductors," *Phys. Rev. Lett.*, vol. 8, no. 6, pp. 250–253, 1962.

17. F. C. Moon, *Superconducting Levitation: Applications to Bearing and Magnetic Transportation*. 2007.

18. Y. L. Jiao, L. Xiao, H. T. Ren, M. H. Zheng, and Y. X. Chen, "Jc-B characteristics for bulk single domain YBCO superconductors," *Phys. C Supercond. Appl.*, vol. 386, pp. 266–270, 2003.

19. E. Díez Jiménez, J. L. Pérez Díaz, and I. Valiente, *Máquinas y mecanismos compuestos por imanes y superconductores Transmisión de fuerzas y pares sin contacto entre los partes móviles*. Publicia, 2014.

20. J. L. Smith et al., "Low-temperature critical field of YBCO," *J. Supercond.*, vol. 7, no. 2, pp. 269–270, Apr. 1994.

21. N. Miura, H. Nakagawa, T. Sekitani, M. Naito, H. Sato, and Y. Enomoto, "High-magnetic-field study of high-T_c cuprates," *Phys. B Condens. Matter*, vol. 319, no. 1–4, pp. 310–320, Jul. 2002.
22. J. R. Hull, "Superconducting bearings," *Supercond. Sci. Technol.*, vol. 13, no. 2, pp. 1–15, 2000.
23. J. W. Rohlf, *Modern Physics from A to Z*, 1st ed., vol. 47, no. 12. Willey, 1994.
24. T. Hlasek et al., "Enhanced mechanical properties of single-domain YBCO bulk superconductors processed with artificial holes," *IEEE Trans. Appl. Supercond.*, vol. 29, no. 5, pp. 1–4, 2019.
25. F. Antončik et al., "Radial and axial stiffness of superconducting bearings based on YBCO single-domain bulks processed with artificial holes," *Supercond. Sci. Technol.*, vol. 33, no. 4, 2020.
26. M. Zeisberger, I. Latka, W. Ecke, T. Habisreuther, D. Litzkendorf, and W. Gawalek, "Measurement of the thermal expansion of melt-textured YBCO using optical fibre grating sensors," *Supercond. Sci. Technol.*, vol. 18, no. 2, p. S202, 2005.
27. K. Konstantopoulou, Y. H. Shi, A. R. Dennis, J. H. Durrell, J. Y. Pastor, and D. A. Cardwell, "Mechanical characterization of GdBCO/Ag and YBCO single grains fabricated by top-seeded melt growth at 77 and 300 K," *Supercond. Sci. Technol.*, vol. 27, no. 11, 2014.
28. J. Ekin, *Experimental Techniques for Low-Temperature Measurements*. Oxford Publishing, 2006.
29. W. T. Ziegler, J. C. Mullins, and S. C. P. Hwa, "Specific heat and thermal conductivity of four commercial titanium alloys from 20° to 300°K," *Adv. Cryog. Eng.*, vol. 8, pp. 268–277, 1963.
30. M. Bahrami, J. R. Culham, and M. M. Yovanovich, "Modeling thermal contact resistance: A scale analysis approach," *J. Heat Transfer*, vol. 126, no. 6, pp. 896–905, 2004.
31. J. E. Campbell, EldridgeE.A., and J. K. Thompson, "Handbook on Materials for Superconducting Machinery," 1974.
32. I. Valiente-Blanco, E. Diez-Jimenez, and J. L. Perez-Diaz, "Engineering and performance of a contactless linear slider based on superconducting magnetic levitation for precision positioning," *Mechatronics*, vol. 23, no. 8, pp. 1051–1060, 2013.
33. J. L. Perez-Diaz, E. Diez-Jimenez, I. Valiente-Blanco, and J. Herrero-De-Vicente, "Stable thrust on a finite-sized magnet above a Meissner superconducting torus," *J. Appl. Phys.*, vol. 133, no. 6, p. 063907, 2013.

34. E. Diez-Jimenez, I. Valiente-Blanco, V. Castro-Fernandez, and J. L. Perez-Diaz, "Design and analysis of a non-hysteretic passive magnetic linear bearing for cryogenic environments," *Proc. Inst. Mech. Eng. Part J J. Eng. Tribol.*, vol. 228, no. 10, pp. 1071–1079, 2014.
35. I. Valiente-Blanco, J. L. Perez-Diaz, C. Cristache, J. Sanchez-Garcia-Casarrubios, and F. Rodriguez-Celis, "Experimental characterization of the temperature influence on the stiffness and damping of a field-cooled superconducting magnetic suspension," *IEEE Trans. Appl. Supercond.*, vol. 29, no. 8, pp. 1–5, Dec. 2019.
36. I. Valiente-Blanco, E. Diez-Jimenez, C. Cristache, M. A. Alvarez-Valenzuela, and J. L. Perez-Diaz, "Characterization and improvement of axial and radial stiffness of contactless thrust superconducting magnetic bearings," *Tribol. Lett.*, vol. 54, no. 3, pp. 213–220, Jun. 2014.
37. C. Cristache, I. Valiente-Blanco, E. Diez-Jimenez, M. A. Alvarez-Valenzuela, N. Pato, and J. L. Perez-Diaz, "Mechanical characterization of journal superconducting magnetic bearings: stiffness, hysteresis and force relaxation," *J. Phys. Conf. Ser.*, vol. 507, 2014.
38. S. Sivrioglu and K. Nonami, "Active permanent magnet support for a superconducting magnetic-bearing flywheel rotor," *IEEE Trans. Appl. Supercond.*, vol. 10, no. 4, pp. 1673–1677, Dec. 2000.
39. T. A. Coombs, A. M. Campbell, and D. A. Cardwell, "Development of an Active Superconducting Magnetic Bearing," *IEEE Trans. Appl. Supercond.*, vol. 5, no. 2, pp. 630–633, 1995.
40. J. R. Fang, L. Z. Lin, L. G. Yan, and L. Y. Xiao, "A new flywheel energy storage system using hybrid superconducting magnetic bearings," in *IEEE Trans. Appl. Supercond.*, vol. 11, no. 1 II, pp. 1657–1660, 2001.
41. J. Tang, J. Fang, and S. S. Ge, "Roles of superconducting magnetic bearings and active magnetic bearings in attitude control and energy storage flywheel," *Phys. C Supercond. Appl.*, vol. 483, pp. 178–185, 2012.
42. T. Matsushita, *Flux pinning in superconductors*. Springer, 2007.
43. C. P. Bean, "Magnetization of high-field superconductors," *Rev. Mod. Phys.*, 1964.
44. M. P. Philippe et al., "Magnetic characterisation of large grain, bulk Y-Ba-Cu-O superconductor-soft ferromagnetic alloy hybrid structures," *Phys. C Supercond. Appl.*, vol. 505, pp. 20–30, 2014.
45. G. G. Sotelo, A. C. Ferreira, and R. De Andrade, "Halbach array superconducting magnetic bearing for a flywheel energy storage system," *IEEE Trans. Appl. Supercond.*, vol. 15, no. 2, 2005.

46. T. Iizuka and H. Fujita, "Precise positioning of a micro conveyor based on superconducting magnetic levitation," in *MHS 1997 - Proceedings of 1997 International Symposium on Micromechatronics and Human Science*, 1997.
47. P. P. Lin, T. Y. Yang, S. Y. Chen, and I. G. Chen, "Study of micro/nano scale displacement by electro-maglev system using bulk superconductors and permanent magnet," *IEEE Trans. Appl. Supercond.*, vol. 17, no. 2, pp. 2063–2066, 2007.
48. I. Valiente-Blanco, E. Diez-Jimenez, J. Sanchez-Garcia-Casarrubios, and J. L. Perez-Diaz, "Improving resolution and run outs of a superconducting noncontact device for precision positioning," *IEEE/ASME Trans. Mechatronics*, vol. 20, no. 4, pp. 1992–1996, 2015.
49. J. L. Perez-Diaz, I. Valiente-Blanco, E. Diez-Jimenez, and J. Sanchez-Garcia-Casarrubios, "Superconducting non-contact device for precision positioning in cryogenic environments," *IEEE/ASME Trans. Mechatronics*, vol. 19, no. 2, pp. 1–8, 2014.
50. J. Serrano-Tellez et al., "Experience on a cryogenic linear mechanism based on superconducting levitation," *Proc. SPIE*, vol. 8450, p. 84501Y, 2012.
51. C. Cristache, E. Diez-Jimenez, I. Valiente-Blanco, M. A. Alvarez-Valenzuela, J. Sanchez-Garcia-Casarrubios, and J. L. Perez-Diaz, "Force relaxation and hysteresis in a frictionless superconducting magnetic bearing," *Int. J. Surf. Sci. Eng.*, vol. 8, no. 4, pp. 318–326, 2014.
52. Y. S. Tseng, C. H. Chiang, and W. C. Chan, "Levitation force relaxation in YBCO superconductors," *Phys. C Supercond. Appl.*, vol. 411, no. 1–2, pp. 32–34, 2004.
53. M. Carfagni, E. Lenzi, and M. Pierini, "Loss factor as a measure of mechanical damping," in *Proceedings of the International Modal Analysis Conference - IMAC*, 1998.
54. C. Liu and G. Liu, "Field Dynamic Balancing for Rigid Rotor-AMB System in a Magnetically Suspended Flywheel," *IEEE/ASME Trans. Mechatronics*, vol. 21, no. 2, pp. 1140–1150, 2016.
55. J. Fang, Y. Wang, B. Han, and S. Zheng, "Field balancing of magnetically levitated rotors without trial weights," *Sensors*, vol. 13, no. 12, pp. 16000–16022, Nov. 2013.
56. J. R. Hull, T. M. Mulcahy, K. L. Uherka, and R. G. Abboud, "Low rotational drag in high-temperature superconducting bearings," *IEEE Trans. Appl. Supercond.*, vol. 5, no. 2, pp. 626–629, 1995.

57. J. P. Lee, S. C. Han, and B. C. Park, "Experimental estimation on magnetic friction of superconductor flywheel energy storage system," *J. Magn.*, 2011, Vol. 16, no. 2, 124–128.

58. B. Bolund, H. Bernhoff, and M. Leijon, "Flywheel energy and power storage systems," *Renew. Sustain. Energy Rev.*, vol. 11, no. 2, pp. 235–258, 2007.

59. A. Chiba, T. Fukao, O. Ichikawa, M. Oshima, M. Takemoto, and D. G. Dorrell, "Magnetic bearings and bearingless drives," Elsevier/Newnes, 2005, p. 381.

60. E. Rodriguez et al., "Passive magnetic bearing system," in *22nd International Conference on Magnetically Levitated Systems and Linear Drives*, 2014.

61. R. J. Cava et al., "Bulk superconductivity at 91 K in single-phase oxygen-deficient perovskite $Ba_2YCu_3O_9$," *Phys. Rev. Lett.*, vol. 58, no. 16, pp. 1676–1679, 1987.

62. R. M. Hazen et al., "Superconductivity in the high-T_c Bi-Ca-Sr-Cu-O system: Phase identification," *Phys. Rev. Lett.*, vol. 60, no. 12, pp. 1174–1177, 1988.

63. R. J. Cava, R. B. Van Dover, B. Batlogg, and E. A. Rietman, "Bulk superconductivity at 36 K in $La_{1.8}Sr_{0.2}CuO_4$," *Phys. Rev. Lett.*, vol. 58, no. 4, pp. 408–410, 1987.

64. S. Pietropinto et al., "Superconducting instrumentation for high Reynolds turbulence experiments with low temperature gaseous helium," in *Phys. C Supercond. Appl.*, 2003, vol. 386, pp. 512–516.

65. Q. Lin et al., "Research of radial high temperature superconducting magnetic bearings for cryogenic liquid pumps," *IEEE Trans. Appl. Supercond.*, vol. 22, no. 3, 2012.

66. Y. Zou, J. Shang, X. Guan, X. Bian, J. Wu, and Q. Li, "Design and experimental research of superconducting magnetic bearing for helium cold compressor," *Cryogenics (Guildf).*, vol. 106, p. 103047, Mar. 2020.

67. D. K. and et al., "LISA proposal to ESA call 3 for L3 mission concepts," 2017.

68. P. A. R. Ade et al., "Planck early results. II. The thermal performance of Planck," *Astron. Astrophys.*, vol. 536, p. A2, Dec. 2011.

69. ESA, "L2-ATHENA-cryogenic vibration isolator and thermal disconnect: Statement of Work. Call Reference: AO8201-14.164.16." 2016.

70. J. Perez-Diaz et al., "Performance of magnetic-superconductor non-contact harmonic drive for cryogenic space applications," *Machines*, vol. 3, no. 3, pp. 138–156, Jul. 2015.

71. J. L. Perez-Diaz, I. Valiente-Blanco, E. Diez-Jimenez, and J. Sanchez-Garcia-Casarrubios, "Superconducting noncontact device for precision positioning in cryogenic environments," *IEEE/ASME Trans. Mechatronics*, vol. 19, no. 2, pp. 598–605, 2014.

72. C. Liu, X. Jing, S. Daley, and F. Li, "Recent advances in micro-vibration isolation," *Mechanical Systems and Signal Processing*, vol. 56. Academic Press, pp. 55–80, 01-May-2015.

73. J. Navarro-Zafra et al., "Contactless suspension for cryogenic applications," in *8th Eyropean Conference for Aeronautics and Space Science (EUCASS)*, 2019, pp. 1–9.

74. J. H. Bhatt and J. J. Barve, "Control of spaceborne linear cryocoolers: A review," *Progress in Aerospace Sciences*, vol. 109. Elsevier Ltd, p. 100544, August 2019.

75. Y. Ogawara, "The Astro-E mission," *Symp. - Int. Astron. Union*, vol. 188, pp. 75–78, 1998.

76. T. Kuriyama et al., "Cryocooler directly cooled 6 T NbTi superconducting magnet system with 180 mmroom temperature bore," *Cryogenics (Guildf).*, vol. 34, no. SUPPL. 1, pp. 643–646, Jan. 1994.

77. R. Radebaugh, "Cryocoolers for aircraft superconducting generators and motors," in *AIP Conference Proceedings*, 2012, vol. 1434, no. 57, pp. 171–182.

78. V. Iannello, J. S. Marshal, and W. Dood, "Superconducting Meissner Effect Bearings for Cryogenic Turbomachines," 1989.

79. F. N. Werfel et al., "Progress toward 500 kg HTS bearings," in *IEEE Trans. Appl. Supercond.*, 2003, vol. 13, no. 2 II, pp. 2173–2178.

80. U. Floegel-Delor, R. Rothfeld, D. Wippich, B. Goebel, T. Riedel, and F. N. Werfel, "Fabrication of HTS bearings with ton load performance," in *IEEE Trans. Appl. Supercond.*, 2007, vol. 17, no. 2, pp. 2142–2145.

81. H. Mitsuda, A. Inoue, B. Nakaya, and M. Komori, "Improvement of energy storage flywheel system with SMB and PMB and its performances," in *IEEE Trans. Appl. Supercond.*, 2009, vol. 19, no. 3, pp. 2091–2094.

82. M. Strasik et al., "An overview of Boeing flywheel energy storage systems with high-temperature superconducting bearings," *Supercond. Sci. Technol.*, vol. 23, no. 3, 2010.

83. R. H. Jansen and T. P. Dever, "G2 Flywheel Module Design," in *2nd International Energy Conversion Engineering Conference*, 2004, pp. 1–20.

84. Y. Miyazaki et al., "Development of superconducting magnetic bearing for flywheel energy storage system," *Cryogenics (Guildf).*, vol. 80, pp. 234–237, Dec. 2016.

85. S. Mukoyama et al., "Development of REBCO HTS magnet of magnetic bearing for large capacity flywheel energy storage system," *Phys. Procedia*, vol. 65, pp. 253–256, 2015.

86. S. Mukoyama et al., "Test of REBCO HTS magnet of magnetic bearing for flywheel storage system in solar power system," *IEEE Trans. Appl. Supercond.*, vol. 25, no. 3, pp. 1–4, 2015.

87. H. Wang, K. Liu, B. Zhu, J. Feng, P. Ao, and Z. Zhang, "Analytical investigation and scaled prototype tests of a novel permanent magnet compulsator," *IEEE Trans. Magn.*, vol. 51, no. 8, Aug. 2015.

88. "Flywheel UPS Systems, 50–1000 kVA using TLE or SG Series UPS GE Industrial Solutions Incorporating Flywheel Technology."

89. T. Mathews and D. Nisanth, "Flywheel based kinetic energy recovery systems (KERS) integrated in vehicles," *Int. J. Eng. Sci. Technol.*, vol. 5, no. 9, pp. 1694–1699, 2013.

90. X. Lu, K. L. V. Iyer, K. Mukherjee, and N. C. Kar, "Study of permanent magnet machine based flywheel energy storage system for peaking power series hybrid vehicle control strategy," in *IEEE Transportation Electrification Conference and Expo*, 2013, pp. 1–7.

91. H. Lee, S. Jung, Y. Cho, D. Yoon, and G. Jang, "Peak power reduction and energy efficiency improvement with the superconducting flywheel energy storage in electric railway system," *Phys. C Supercond. Appl.*, vol. 494, pp. 246–249, 2013.

92. S. Rosenzweig et al., "A superconducting levitation transport model system for dynamical and didactical studies," *Phys. Procedia*, vol. 36, pp. 1037–1042, 2012.

93. H. Oyama, T. Shinzato, K. Hayashi, K. Kitajima, T. Ariyoshi, and T. Sawai, "Application of superconductors for automobiles," *SEI Tech. Rev.*, vol. 77, no. 8, 2008.

94. F. R. C. Diaz et al., "The Development of VASIMR engine," *Int. Conf. Electromagn. Adv. Sp. Appl.*, no. July, p. 6, 1999.

95. Y. Miyazaki et al., "Development of superconducting magnetic bearing for flywheel energy storage system," *Cryogenics (Guildf).*, vol. 80, pp. 234–237, 2016.

96. Y. H. Han, B. J. Park, S. Y. Jung, S. C. Han, W. R. Lee, and Y. C. Bae, "The improved damping of superconductor bearings for 35 kWh superconductor flywheel energy storage system," *Phys. C Supercond. Appl.*, vol. 485, pp. 102–106, 2013.

Chapter 13

Low-Frequency Rotational Loss in an HTS Bearing and Its Application in Sensitive Devices

Wenjiang Yang, Long Yu, and Yu Ji
School of Astronautics, Beihang University,
37 Xueyuan Road Beijing, 100191, People's Republic of China
yangwjbuaa@buaa.edu.cn

13.1 A Brief Overview of HTS Bearings

It has been more than 30 years since the high-temperature superconductor (HTS) was discovered. Most types of HTS have a critical temperature above the boiling temperature of liquid nitrogen, which is 77 K. An HTS tries to minimize the magnetic flux density when is placed in a magnetic field and this kind of property is known as diamagnetic. An HTS also tries to keep the flux within itself constant when it experiences a magnetic field change. This property is known as a flux pinning effect. High-temperature superconductors passively achieve levitation stability by coupling of diamagnetism with flux pinning [1].

High-T_c Superconducting Technology: Towards Sustainable Development Goals
Edited by Muralidhar Miryala

ISBN 978-981-4877-65-7 (Hardcover), 978-1-003-16468-5 (eBook)
www.jennystanford.com

Taking advantage of the levitation stability achieved by HTS, many researchers have made efforts to develop levitated magnetic bearings to support rotating structures in the last 30 years. Such levitated bearings are free from friction, mechanical wear, mechanically induced torque ripple, and energy dissipation, thus result in long-life operation with high efficiency and less maintenance. Advantages of HTS bearings include passive levitational stability, and very low rotational drag torque. These qualities have suggested the use of HTS bearings in various applications from high-efficiency flywheel energy systems to low-frequency sensitive instruments.

13.1.1 Brief Introduction of HTS Bearing

In 1987, F. Hellman et al. of Bell Labs found that a 7-mm-cubic NdFeB permanent magnet can be suspended above a YBCO high-temperature superconducting circle with a diameter of 25 mm and a thickness of 6 mm [2]. After only 5 years, high-temperature superconducting magnetic levitation axial bearings have achieved 520,000 rpm. As shown in Table 13.1, compared with other types of bearings, HTS bearing is particularly suitable for high-speed and medium-load applications.

Table 13.1 Comparison of different types of bearing

	Working conditions	Loading capacity	Stability	Applications
Mechanical bearings	Lubricating	High	Usually stable, short life at high speeds	Wide applications
Air bearing	Compressed air	Medium	Affected by jet pressure	Floating platform
Ultrasonic bearing	Highly intensive sound source	Low	Affected by temperature	Material processing and physical experiments
Active magnetic bearing	Precision sensor feedback system	High	High	Generators and motor

	Working conditions	Loading capacity	Stability	Applications
Passive magnetic bearing	Magnetic circuit structure	Medium	medium	Uncommon
HTS bearing	Mixed-state flux pinning	Medium	High	Flywheel energy storages

In an HTS bearing system, axisymmetric magnetic field of permanent magnet is locked by superconductor depending on flux pinning effect, and the movement in the axial direction will cause the change of the current and magnetic field states of the superconductor thus a levitation force would be generated. However, when the permanent magnet rotates around the axial direction, the magnetic field locked by the superconductor changes a little, thereby freedom of rotation is provided.

13.1.2 Classification of HTS Bearing

As shown in Fig. 13.1, HTS bearings are generally classified into axial and radial types. The axial HTS bearing is composed of a HTS stator placed under a permanent magnet rotor. The axial and radial stiffness of axial HTS bearings with a permanent magnet rotor diameter of about 220 mm is only 144 N/mm and 69 N/mm [3].

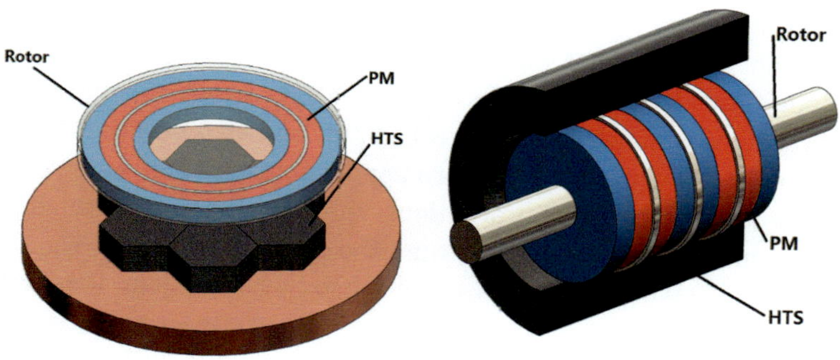

Figure 13.1 Axial and radial types of HTS bearings.

Radial HTS bearings are divided into two types: inner permanent magnet rotor type and outer permanent magnet rotor type. The superconducting stator is generally a cylindrical structure. Compared with the axial HTS bearing, the radial HTS bearing has a more compact structure, a larger axial offset distance and better stiffness characteristics. At present, the maximum axial bearing capacity and stiffness of radial HTS bearings have reached 10.08 kN and 4.5 kN/mm [4], while the maximum radial bearing capacity and stiffness have reached 6.9 kN and 5.1 kN/mm (28 K) [5].

Besides increasing the levitation force and stiffness, extensive advances have also been made in reducing the rotational losses of HTS bearings [1].

13.1.3 Application of HTS Bearing in Flywheel

Flywheels have long been used to store energy in the form of rotational kinetic energy. A basic flywheel is a device that has a large moment of inertia and spins around only one axis to store rotational energy.

While past applications of the flywheel usually used conventional mechanical bearings that had relatively higher losses due to friction, high-speed idling (that is, no power input or output) losses of which is about the order of ≈1%/h. With the development of magnetic bearings using HTS, the idling loss has been greatly decreased to as little as 0.1%/h. Experimental versions of flywheels that employ HTS bearings have already stored >2 kWh.

The energy capacity of such a storage device relies on the moment of inertia of the wheel as well as the angular velocity. Modern flywheel applications utilizing HTC bearings and operating in vacuum can reach rpms between 23,000 and 40,000 with a usable storage energy [6]. These modern applications face the same safety issues of the previous incarnations, though the improvements of mechanical bearings improve the efficiencies immensely. The G2 flywheel shown in Fig. 13.2 was designed by NASA Glenn Research Center to test integrated power and attitude control system in low earth orbit working cycle with available energy of 525 Wh, charge and discharge power of 1 kW, diameter of 12 in, length of 30 in, and total weight of 218 lb [7,8].

Figure 13.2 NASA Glenn research center G2 flywheel.

Superconducting flywheel is often considered as one of the major uses of bulk HTSs, thus considerable development effort of this technology has occurred since the discovery of HTSs [9]. As shown in Table 13.2, recent superconducting flywheel has achieved 100 kWh stored energy.

Table 13.2 Superconducting flywheel systems situation and parameters

Company	Bearing	Stored energy (kWh)	Diameter (mm)	Weight (Kg)	Angular speed (rpm)	Year
Gebze Technical University	SMB	—	—	—	750	2017
NEDO	SMB+AMB	100 kWh	2000	4000	6000	2016
Shanghai University	SMB		160	25	15000	2016
Boeing	SMB	5 kWh		164	15000	2010
NEDO	HSMB	5 kWh		125	11000	2010
ATZ and L-3 Magnet-Motor	SMB	10 kWh	1000	600	10000	2010
NEDO	SMB	10 kWh	—	425	5000	2006
Dynastore	SMB	11 kWh	—	1200	10000	2006
NASA Glenn	SMB	0.5 kWh	—	—	60000	2002

13.2 Rotational Loss of HTS Bearings

13.2.1 Rotational Loss Phenomenon and Coefficient of Friction

13.2.1.1 Rotational loss phenomenon

When a permanent magnet (PM) disc is levitated and spinned stabilization over an HTS, the absence of direct physical contact between the PM and the HTS produces the illusion that the rotation is frictionless. However, previous studies of the high-speed rotation in PM/HTS bearings display surprisingly residual drag and loss [10,11]. As was mentioned in 13.1.3, idling loss HTS bearings is less than 0.1%/h. To realize energy efficiency maximization, a number of researchers have investigated the origin of this friction and developed multiple means to reduce it.

To study this rotation loss, the simplest way is to record the rotational frequency decay of the levitated PM. As shown in Fig. 13.3, a rotor consisting of a PM attached to an adjusting level is positioned to set up the field cooling height. Vacuum pumps including a molecular pump start to lower the pressure of the vacuum house. When the pressure is sufficiently low (below 10^{-3} Pa), an GM cryocooler works to refrigerate the cold head and the attached HTS bulk.

When the HTS reach a stable temperature, the height adjusting level is lowered and moved away. At this moment, the rotor usually sinks and then levitates stably. Spin-up of the PM rotor is accomplished by injecting the gas helium to the rotator along the tangential direction. The rotation rate is measured with an optical transducer, and a CCD camera is installed above the top window to observe the rotating state of the rotor.

By this way, the PM rotor is spun up to a given rotation speed, and then losses gradually slow the rotation. The slope of the measured speed curve would be the same order with the friction torque. Figure 13.4 shows the measuring results of the rotational decaying speed.

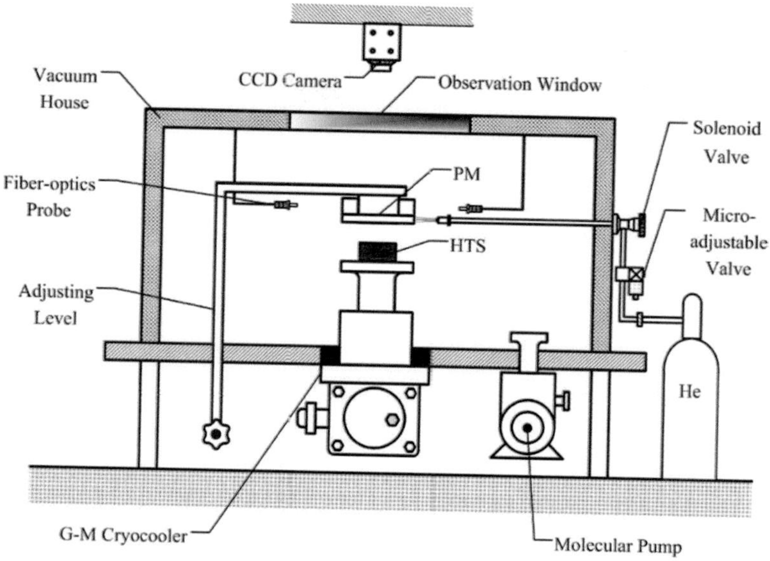

Figure 13.3 A rotational loss measuring device.

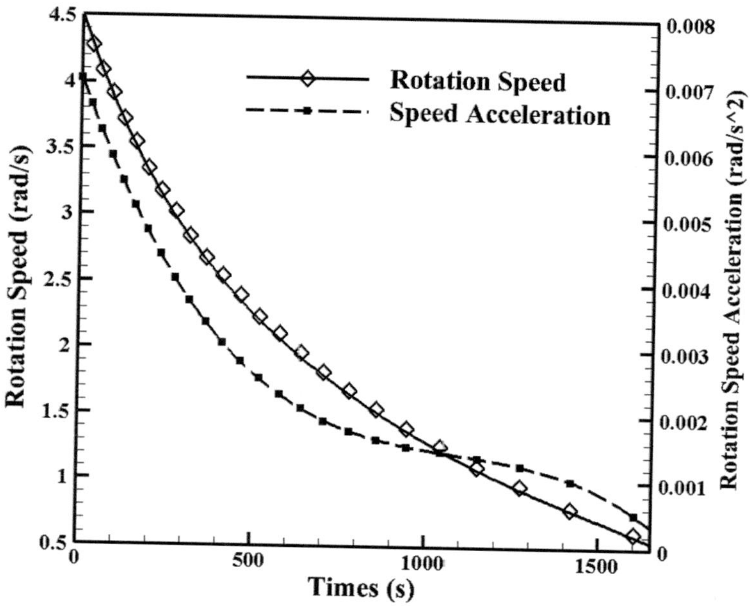

Figure 13.4 Measuring results of rotational decaying speed.

13.2.1.2 Coefficient of friction

J. R. Hull advised the use of an effective coefficient of friction, μ, as a dimensionless parameter to compare the rotational loss for different bearing designs [12].

$$\mu = F_D/F_L \tag{13.1}$$

where F_D is the drag force and F_L is the lift force. The lift force is the weight to be levitated

$$F_L = Mg \tag{13.2}$$

where M is the mass and g is the acceleration of gravity.

Drag force is calculated from the drag torque

$$\tau_D = F_D R_D = -I\alpha \tag{13.3}$$

where R_D is the mean radius at which the drag force acts, I is the total moment of inertia of the rotating object, and α is the angular acceleration,

$$\alpha = 2\pi(df/dt) \tag{13.4}$$

where f is the frequency of rotation. It is convenient to describe the moment of inertia in terms of the gyration radius R_γ as follows:

$$I = MR_\gamma^2 \tag{13.5}$$

Combining Eqs. (13.1)–(13.5),

$$\mu = -(2\pi R_r^2 df/dt)/(gR_D) \tag{13.6}$$

In Eq. (13.6), all terms are easily measured or calculated except R_D. Usually we take R_D to be the outer radius of the bearing part of the rotor.

The coefficient of friction is useful when comparing the performance of different bearing designs because it helps to normalize the effects of bearing size and differences in moment of inertia ($R\gamma$ term) among experimental devices. By this analysis of coefficient of friction (*COF*), in 1990, F. C. Moon et al. explored the high-speed rotation drag of a magnet on a HTS at both atmospheric pressure and a partial vacuum [11]. The result indicated that the flux drag torques were constant and independent of speed. It was believed that drag torques on rotating magnets were related to asymmetry in the flux density of the magnet. In 1991, B. R. Weinberger et al. studied effects of magnet asymmetry, atmospheric resistance, and eddy currents on rotation loss of HTS bearing consist of a magnet and an HTS bulk [10].

If we think the desired idle loss is 0.1%/h for flywheel energy storage, this requires coefficient of friction μ is less than 1.4×10^{-5} [13]. However, if the cryogenic power for the cooling of the HTS material is considered, the required μ should be an order of magnitude with 10^{-6}.

The *COF* can be affected by multiple factors and the research of these factors is vital to the reduction of the rotation loss. The *COF* for a mechanical roller bearing is of the order of 10^{-3} and the *COF* for an active magnetic bearing is $\approx 10^{-4}$. Measured *COF*s for simple HTS bearings are as low as 10^{-7}, which are 3~5 orders of magnitude lower than the best mechanical bearings [14].

13.2.2 Loss Sources Consideration and Theory

Rotation loss sources causing the rotation drag are not obvious. Efforts to identify and minimize the loss sources in HTS bearings were made in many research groups. By conclusion, the rotational losses in the HTS bearing mainly consist of windage loss caused by air drag, hysteresis loss on superconductors, and eddy current loss in adjacent metal structures.

13.2.2.1 Air drag loss

Air drag loss ΔE_a increases synchronously as ω^3 (ω is rotating angular speed), which makes it particularly prominent at high rotating speed. It can only be ignored in a vacuum chamber under the pressure lower than 10^{-6} torr according to the study of Xia et al. [15].

$$\Delta Ea \propto C\rho S\omega^3 \tag{13.7}$$

where ρ is air mass density and S and C are aerodynamic drag area and aerodynamic coefficient, respectively.

The chamber pressure can have a major influence on windage losses. Study of the loss characteristics at different pressures is of great significance for the application of energy storage flywheel. Despite the low background pressure under which these bearings were deployed, a measurable dependence of the rotational frequency decay rate upon the pressure could still be observed.

Hull studied the influence of chamber pressure on the *COF* at the high velocities in 1996 [16]. In his work, a diffusion pump was used to adjust the chamber pressure. The dependence of the *COF* on pressure was nearly linear. By extrapolating to zero pressure, they concluded that the contribution of chamber pressure to the *COF*, under the usual operating conditions (0.0002 Pa), is about 2×10^{-10} at the velocity around 0.1 m/s, whereas the lowest *COF* measured in this condition was about 6×10^{-8}.

Weinberger [10] also measured the decay rates for the PM rotor in the frequency range 750–950 Hz. Even though the pressure was less than 10^{-4} Torr, the mean free path of the gas molecules in the chamber is orders of magnitude larger than the magnet superconductor gap. Therefore, the frictional effect of collisions between gas molecules and the high-speed rotor were still detectable. For low-frequency consideration, the air friction loss can be ignored.

13.2.2.2 Hysteresis loss

Time varying magnetic field in a type II superconductor creates hysteresis loss. According to the critical-state model of superconductors [17], the energy loss ΔE_h due to the hysteresis loss is given by,

$$\Delta E_h \propto \frac{(\Delta B)^3}{J_C} \tag{13.8}$$

where ΔB is the peak-to-peak variation of the time varying magnetic field in superconductors, and J_c is the critical current density of the superconductor.

In the HTS bearing, ΔB arises from azimuthal inhomogeneity of the magnetic field about the rotation axis due to imperfection in the fabrication process of the ring-shape rotor. Most magnet rotors mentioned in the existing researches were made by joining small segments together, which lead to an unavoidable circumferential variation of the magnetic field, especially across the joints. However, these joint defects could be moderated by appropriate insertion of shimming, as demonstrated by Higasa et al. [18]. Still, the best way to reduce this field variations was to use monolithic magnets which can obtain a typical value less than 3% in the circumferential change of the magnetic flux density [19].

During the rotation of the rotor with the increasing frequency, the magnet vibration in different modes also introduces time varying magnetic field to the superconductors.

The hysteresis loss is inversely proportional to the critical current J_c. The superconductor has higher critical current as the temperature decreases. Zeiberger et al. reported that the critical current of a bulk YBCO sample increases by factor of 20~30 from 77 K to 4 K when the applied magnetic field is in the range of 0–1 T [20]. Therefore, the *COF* is expected to decrease as temperature of the HTS bulks decreases.

13.2.2.3 Eddy current loss

Any time varying magnetic field in electrically conductive materials induces an EMF and therefore eddy current. This eddy current dissipates as Joule heat and contributes as friction of the rotating magnet. The energy loss ΔE_e due to the eddy current is

$$\Delta Ee \propto \sigma (\Delta B)^2 \omega^2 \tag{13.9}$$

Eddy current loss roughly increases as ω^2. It means that the eddy current loss would dominate at high frequencies. This presents a great challenge to the levitation support subsystems, especially the cryosystem for the cooling of bulk superconductors. To avoid using eddy-current metal structures around the bearing, one of the alternative materials is G-10, a fiber glass epoxy composite with high strength and zero conductivity, which can greatly reduce the eddy current loss.

ΔB sources of eddy current loss are similar as the case discussed in the hysteresis loss. Because the stator HTS is usually not a continuous ring but an array of HTS tiles, those tiles finally get separate trapped field distribution in the magnetization of the PM rotor field. The field that is trapped in each tile has higher concentration of magnetic flux at the center of the tile and lower of that as approaching to the edge of the HTS tile. When the PM rotor rotates above the array of HTS tiles, time-varying field from the HTS tiles creates friction to the rotor due to the eddy current interaction.

13.2.3 Effects of Bearing Structures and Scales

Bearing type, structure, and scale will affect the rotation drag characteristics. Figure 13.5 shows the historical records in attainable coefficient of rotation friction for various types of HTS bearing in the world. Depending on bearing size, loading capacity and applications, we just divide them into three categories: Small-scale bearings, medium-scale bearings, and large-scale bearings. The historical data demonstrate that low rotation losses can be achieved in a wide variety of bearing prototypes if optimized designs are considered, and size does not seem to be a limit on attaining low *COF*s.

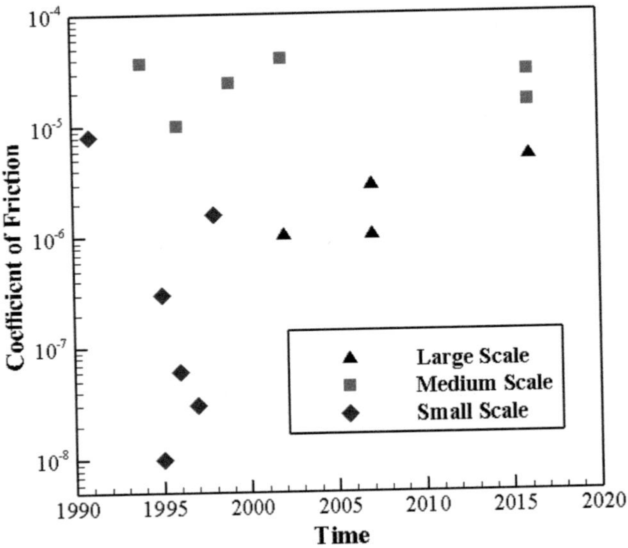

Figure 13.5 Historical records in attainable *COF* of various types of HTS bearing in the world.

13.2.3.1 Small-scale HTS bearing

Small-scale category usually represents a single PM disc levitated over a single HTS or a single-piece PM ring levitated over an array of HTSs [10]. By the existing researches, the mass of the magnetic rotor is mostly less than 1 kg. This category pays more attention to the primary and basic understanding of HTS bearing principle and rotation mechanism. By the rotational drag study, one of HTS bearings achieved the rotational loss with an extremely low level, and the low-speed *COF* of 3×10^{-8} was measured.

As shown in Fig. 13.6(a), it is a very traditional bearing model that a single PM disc is levitated over a single HTS. The HTS is field cooled in the presence of the PM field to trap magnetic flux to make the levitation stable. But limited by both consideration of HTS for levitation force and stability of PM, this configuration usually shows with the low levitational loading capacity. By the investigations of the rotation friction on the traditional HTS bearing model, the rotation loss sources, related parameters, and general laws were discussed in detail [10,12,16].

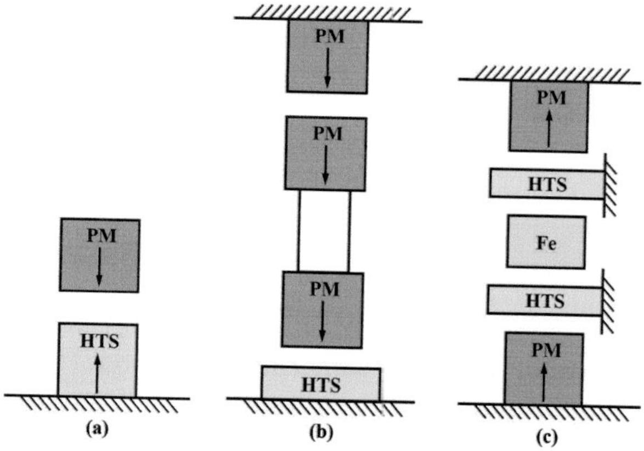

Figure 13.6 Small-scale HTS bearing types. (a) Traditional type bearing, (b) Evershed-type bearing, (c) mixed-μ type bearing.

In this small-scale category, special bearing configurations such as an Evershed-type bearing design [21] and a mixed-μ bearing design (where μ is the relative magnetic permeability) [22] are considered to augment the levitation loading and reduce the levitation frictions. Figure 13.6(b) shows the typical form of the Evershed-type bearing. The top PM of the rotor is attracted toward the stator PM to support less than 100% of the rotor weight, and the remaining weight and stability of the rotor is provided by the stabilizing PM interacting with the HTS. Besides increasing vertical loading, this design supplies additional stiffness for both vertical and radial stability. Figure 13.6(c) shows the typical form of the mixed-μ bearing. A soft Fe cylinder of the rotor is in attractive levitation between two stator

PMs and stabilized by two stator HTSs. A magnetization is induced in the Fe by the PM pair, and then the magnetization interacts with the PM field to obtain an upward levitation force, while the magnetization provided trapped flux in the HTSs to stabilize the rotor. Without rotating PMs or HTSs in the rotor, this design benefits the rotor with high tensile material to provide larger rim velocity. The mixed-μ design displayed the lowest rotational loss in small-scale HTS bearings, and but it has weaker stability and difficulty in larger size applications.

In these researches on small-scale bearings, the velocity-dependent loss was believed to include eddy current loss induced on cryochamber surfaces by the asymmetry of the PM field and eddy current loss in PMs by the inhomogeneity of the trapped field of the magnetized HTSs, respectively. The velocity-independent loss is the hysteresis loss in the HTSs due to the azimuthal inhomogeneity of the PM field.

Progresses in reducing *COF* here mostly benefit from improvement in increasing J_c of bulk superconductors [23], optimizing the HTS array geometry [24], and improving the azimuthal homogeneity of the PM magnetic field [25].

13.2.3.2 Medium-scale HTS bearing

Medium-scale category represents laboratory prototypes and engineering consideration of HTS bearings, and the bearing started to be used for the conceptual design of flywheel energy storage system (FESS). About the bearing structural configuration, it is not only a PM ring configuration levitated over an HTS array (named as axial or thrust type HTS bearing and shown in Fig. 13.7(a) [26]), but also a PM rotor configuration levitated inside an HTS hole (named as radial or journal type HTS bearing and shown in Fig. 13.7(b) [27]). Sometimes, it is a combination of the two types using more surfaces of the superconductor ring [28,29], shown in Fig. 13.7(c). In the medium-scale category, the rotor mass is usually from several kilograms to dozens of kilograms.

Progress in reducing the *COF* in medium-scale category is mostly due to the better construction techniques for the PM rings [30], using nonmetal cryochamber designs such as G-10 fiberglass-epoxy structures [16], and improving the dynamics of the bearing and flywheel structures [31]. Due to more complex structures and working states, compared to the small-scale bearing, the medium-scale bearing or flywheel showed with the increasing *COF*.

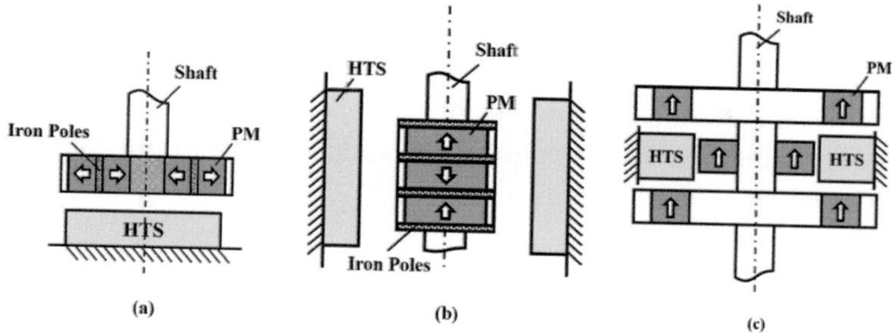

Figure 13.7 Medium-scale HTS bearings: (a) Thrust type bearing; (b) journal type bearing; (c) multi-surface type bearing.

13.2.3.3 Large-scale HTS bearing

Large-scale category represents engineering applications of HTS bearings in flywheel energy storage systems, usually with the combination of other bearing types such as active magnetic bearings (AMB). The flywheel mass is from several hundred kilograms to tons. In the last two decades several large FESS projects have been performed, which are a 5 kWh/100 kW Boeing FESS [13], a 10 kWh/100 kW NEDO FESS project [32], a 5 kWh/250 kW ATZ FESS [4], and a 100 kWh/300 kW Yamanashi FESS [33], respectively.

All of these FESS configurations use a PM clutch in the attractive approach to support the weight of the rotor and lower the burden on the HTS bearing. While the first FESS project (Boeing) uses an axial HTS bearing at the bottom, the second (NEDO), and third (ATZ) FESSs are equipped with journal HTS bearings to stabilize the rotors. The NEDO flywheel is additionally restrained by two radial AMBs in the upper and lower positions to stabilize the 380 kg rotor and keep the resonance small (<1 mm). The rotational loss was 68 W (<0.1%) at 77 K, mostly caused by the AMBs.

As the world's largest-class flywheel energy storage system, the 100 kWh/300 kW Yamanashi FESS uses a full superconducting bearing at the bottom consisting of five stacked double pancake REBCO coils and liquid nitrogen cooled GdBCO bulks [34], as shown in Fig. 13.8. The full superconducting bearing gives an adjustable and higher levitation loading because the magnetic field of the stacked pancake coils can easily be 5 T. Another advantage is that azimuthal field homogeneity of the pancake coils is much higher than that obtained in the PM structure, so there is the potential to achieve lower hysteresis and eddy current losses. Compared to the PM rotor, the main disadvantage of this superconducting rotor is that the levitated HTS bulks require to be cooled by the rarefied helium gas cooling system.

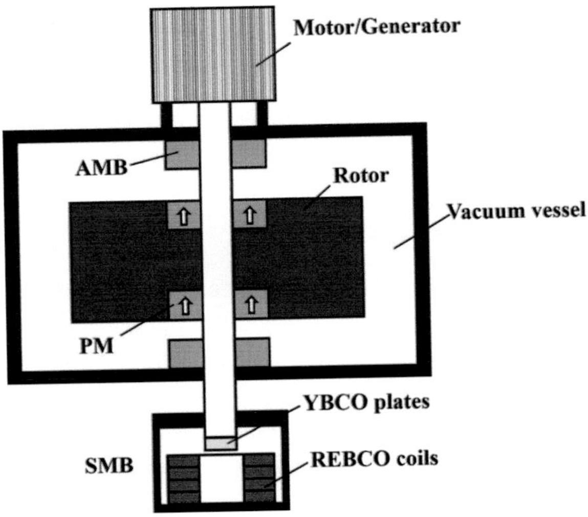

Figure 13.8 Full superconducting FESS with REBCO coils and GdBCO bulks.

13.2.4 Effects of Mechanics and Dynamic Behavior with Rotational Frequency

Besides the bearing type and scale, the motion characteristics of the rotor or flywheel will make essential effect on the rotation loss in the fixed bearing design. The effect of rotational frequency on loss in the HTS bearing is multifaceted. After using the extremely low background pressure to reduce the air drag loss, the major effect of the frequency change is still on the eddy current loss and the hysteresis loss caused by the rotational vibration. The reason for effect of increased frequency on the rotation loss is recognized in many researches [14,20,21].

By analyzing the motion of a spinning PM disc levitated over a single YBCO bulk in a vacuum chamber, Hull and Cansiz studied dynamics of the spinning PM and showed the *COF* as a function of rotational frequency [35]. Our research also showed the increase of *COF* with the rotation frequency is not obvious except near the frequency about 10 Hz [36], showing the direct effect of vibration and even resonance on the spinning speed decay, as shown in Fig. 13.9. During the spinning down of the PM, its rotational dynamic behavior can be divided into three regions: Above the resonance (region I), in the resonance (region II), and below the resonance (region III).

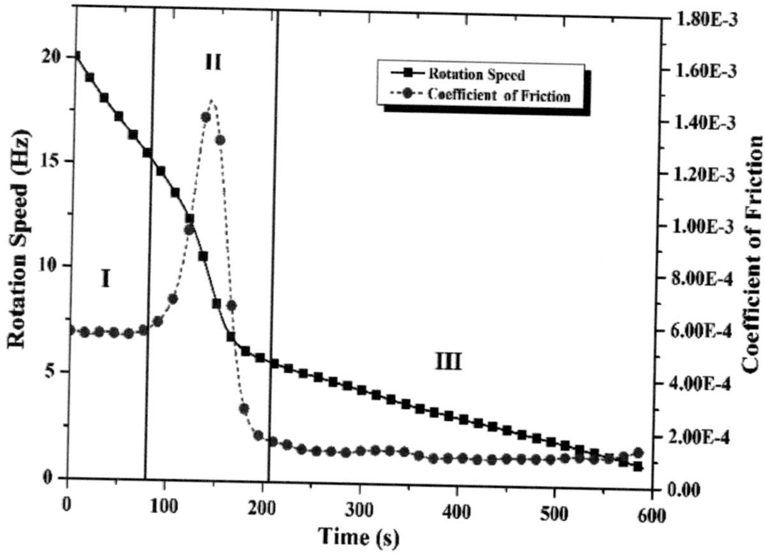

Figure 13.9 Rotation speed decay and *COF* measuring results.

Above the resonance (region I >15 Hz), the *COF* remains constant and shows a small dependency on the rotation frequency, but the *COF* above the resonance is larger than that below the resonance. It is believed that this transition of rotation center makes the PM rotate about its center of mass rather than its center of magnetism. If the rotor is not in perfect design consideration, there would be an offset between the center of mass and the center of magnetism. The offset makes the rotation about the center of mass produce a whirling motion of the center of magnetism, and then additional hysteresis loss by the whirling field happens in the HTS. So, the rotation above the resonance leads to more loss than that below the resonance.

In the resonance region II (15→5 Hz), the *COF* value was observed to peak near the resonance frequency about 10 Hz, possibly due to the large variation of axisymmetric magnetic field benefited from the large-amplitude radial oscillations. We hope the transition region is as small as possible. However, because the rotor is not well designed, the coupling of multiple vibration modes often results in a broad transition area. Further study showed that the effect of frequency increase on rotational loss can be reduced by adjusting the center of magnetism closer to the center of mass using proper counterweights (or balance) [16].

Below the resonance (region III <5 Hz), the *COF* of the spinning PM first is constant and the *COF* at a low speed (less than 1 Hz) increases as the frequency continue to decrease. It is believed that the PM rotor rotates about the center of magnetism before the resonance frequency, and the rotation loss is mainly caused by the hysteresis effect in the HTS sensing the inhomogeneity of the PM field.

Although mostly the axial bearing supports the flywheel and keep the vertical direction well, there is still some tilting angle between vertical axis and rotor axis. Existing tilting angle leads to coupling of mechanics in several directions and make the vibration complex. Cansiz gave a more complete dynamical analysis by studying both translational and angular vibrational characteristics [35]. More resonance peaks were observed. He found the strong cross-couplings among vertical, radial, tilt and rotation directions, and explained the macroscopic vibration modes of the PM rotor were actually combinations of pure vibration frequency modes in these directions.

With the increase of the rotation frequency, the vibration and even resonance of the rotor give the dramatic adding of the rotation loss in the bearing. In the engineering applications of the bearing and flywheel, the radial vibration of the rotor often dominates the dynamic behavior. In addition to the understanding of the natural frequency characteristics of the rotor, the improvement of the radial stiffness is often an important approach to suppress vibration and resonance. For instance, PM/PM bearing is often used to improve the radial stiffness in the small-scale FESS, and AMB bearing often increases the radial stiffness by more than 10 times in the large-scale FESS. Although these methods introduce more bearing structures and then lead to higher idling loss, they can effectively suppress vibration and reduce the possibility of rotor instability.

13.2.5 Effects of Superconducting Material Properties

The research on the influence of superconducting materials on the rotation loss was carried out as early as the 1990s. Experiments compared sintered and melt processed YBaCuO bulks, and showed that the HTS bearing with higher J_c superconductor can obtain smaller rotation loss because the hysteresis loss was effectively resisted [10, 23]. SmBaCuO showed higher J_c values, lower creep and ac loss than YBaCuO [37]. Increasing the critical current becomes one of the important approaches to improve the rotation loss.

Because YBCO thin films have the obvious higher J_c value of >1 MA/cm^2 at 77 K than bulk superconductors, Cansiz thought the thin film between the PM and bulk HTS would shield the bulk HTS from the varying applied field, and the use of the thin film in the HTS bearing has the potential to reduce rotational drag [38]. However, the primary experiments showed the superconducting film did not take the desired reduction of the rotation decay. The *COF* of the combined bulk and film was approximately equal to the sum of the *COFs* for the bulk alone and the film alone. The possible reason was the thickness of the superconducting film is only 1 μm and easy in saturation of magnetization.

In the general construction of HTS bearings, multiple cylindrical, hexagonal, or rectangular superconducting bulks are used to build the planar HTS stator for the thrust type bearing [3], and YBCO arc blocks are joined into a circle to make the annulus HTS stator for the journal type bearing [32]. Because the trapped magnetic field fluctuation in the arrayed HTS bulks will produce the eddy current loss in the rotor magnet and the metal surfaces around, it is also an important way to provide the reduction of rotation loss by increasing the size and circumferential uniformity of the single domain bulk superconductor. For instance, a dense MgB_2 bulk cylinder is up to 65 mm in diameter and 100 mm in height, which has fewer mechanical problems, even if working below the temperature 30 K. The lack of superconductive connections in the MgB_2 bulk cylinder reduces rotational loss in the bearing [39].

On the other hand, several new stator concepts were investigated by stacking superconducting tape and pancake coils as bulk superconductors for both trapped field magnets and for magnetic bearing [40]. An HTS annuli stack consisted of 294 layers was made from 46 mm wide tape to form a composite bulk hollow stator with an inner bore of 26 mm, and stable levitation force up to 550 N was measured to demonstrate this kind of stack bearing [41]. The tape stack bearing has the advantages in more uniform magnetic field distribution and better area expansibility in superconducting region, which will benefit the reduction of eddy current loss of the PM rotor.

13.2.6 Effects of Magnetic Rotor Structures

The inhomogeneity of the PM rotor's circumferential magnetic field causes hysteresis loss and eddy current loss. Even for a PM with the best homogeneities, the inhomogeneity (1%) is still sufficient enough to cause a detectable decay in rotational frequency [19]. Currently, most PM rotor use multiple magnet segments to form a PM ring [30]. As shown in Fig. 13.10, this kind of rotor can provide a strong magnetic field gradient to achieve large levitation stiffness, but its circumferential magnetic field must be uniform enough to reduce levitation loss. A variety of permanent magnet structure solutions have been proposed to meet this demand.

As shown in Fig. 13.11, Sotelo [42] published a Halbach array PM rotor to increase the levitation force and reduce the rotational loss of superconducting bearings. In order to improve the uniformity of the circumferential magnetic field, the performance difference between materials must be as small as possible, and more attention are also placed on the assembly process.

In addition, some researchers have further improved the uniformity of the circumferential magnetic field of the rotor by applying magnetic thin film sheets. Hull suggested the using of soft-ferromagnetic shims to reduce the azimuthal inhomogeneity of the PM rotor and studied the *COF* results with and without this method. The spin-down results show that this method could reduce low-velocity hysteresis losses but does not change velocity dependent [16].

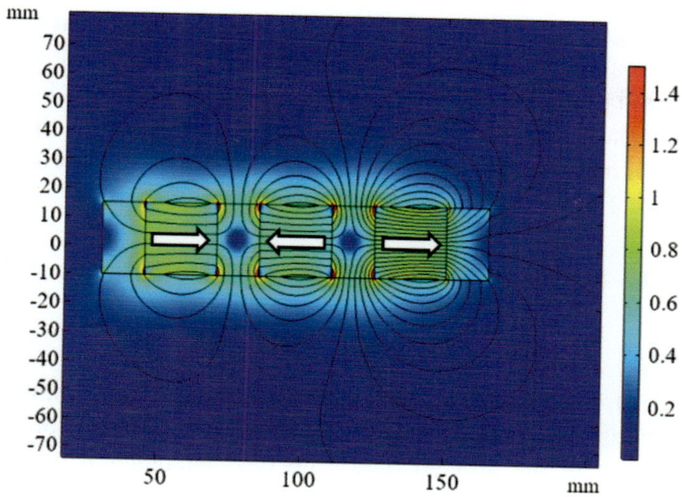

Figure 13.10 Flux density for flux shaper configuration.

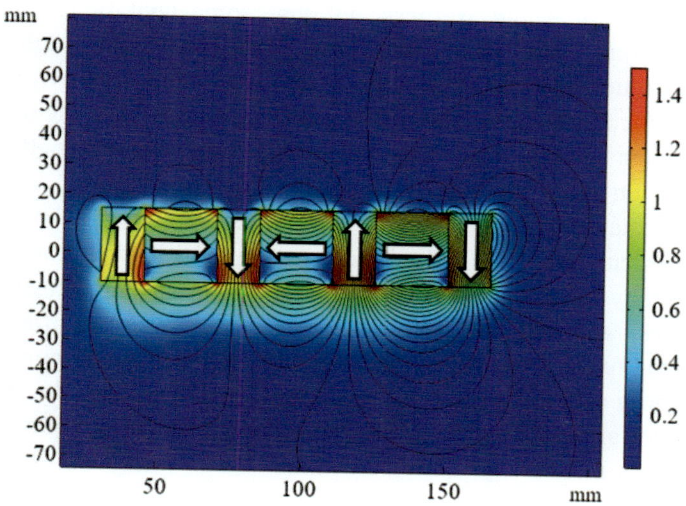

Figure 13.11 Flux density for Halbach configuration.

The magnetic field of the PM rotor was measured using a three-axis Hall probe to establish the functional dependence between magnetic field and the azimuthal angle. Three pairs of free spin-down experiments were performed, each including an original rotor and a shimmed rotor which levitated at about the same height. The measuring results indicated that the frequency dependence of the *COF* was approximately the same for all six spin-down experiments, while the *COF* for the shimmed PM rotors was always smaller than that for the original PM rotors in low frequencies. In this method, the figure of merit for the inhomogeneity of the PM rotor reduced by a factor of 2~3.

13.2.7 Effects of Magnetization and Working Conditions

13.2.7.1 Magnetization and levitation heights

Besides applied field flux density and distribution, the magnetization conditions of superconductors still include field cooling (FC) positions or heights in the applied field. The field cooling position often determines the trapped field and diamagnetism of superconductors, and then decides the levitation gap between applied field magnets and superconductors. Increasing the FC height will improve the levitation force and stiffness, but lose some levitation stability. Furthermore, increasing FC height raises the levitation height and then reduces the amplitude effect of azimuthal inhomogeneity of the magnetic field on the rotation loss in the bearing [12]. Significant levitation height was possibly at 23 mm, and a low-speed coefficient of friction of 3×10^{-8} was measured [21].

In the design of superconducting bearing, the FC height is first calculated to be suitable for the required loading condition, especially when superconductors have to provide both levitation stiffness and stability. With engineering applications of the HTS bearing to flywheel storage system, hybrid bearing configurations are more considered, and the HTS bearing just provides the lateral stiffness and stability. In this situation, PM or electromagnetic clutch supports most of flywheel weight, and the flywheel and the HTS bearing can work at larger FC heights, and then added levitation heights lead to the reduction of the rotation loss in the whole system.

13.2.7.2 Low T_c Cooling Conditions

Low T_c cooling condition is the key factor in the design and operation of the superconducting bearing. Because bulk superconductors generally have the small thermal conductivity (less than 20 $W \cdot m^{-1} \cdot K^{-1}$), the temperature uniformity of superconductors in cooling is poor. In order to provide perfect thermal conduction and structural strength for the cooling of superconductors, good thermal conductive materials are required for refrigeration structures. Copper structures are usually the preferred selection, but the interaction between the copper surfaces and inhomogeneous applied field often leads to serious eddy current loss. At present, AlN and Al_2O_3 are considered to be materials with potential to solve the above contradiction between heat transfer and eddy current. They have both good thermal conductivity (about 150 $W \cdot m^{-1} \cdot K^{-1}$) and weak electrical conductivity, but material brittleness and poor structure machinability also affect the structural reliability of bearings.

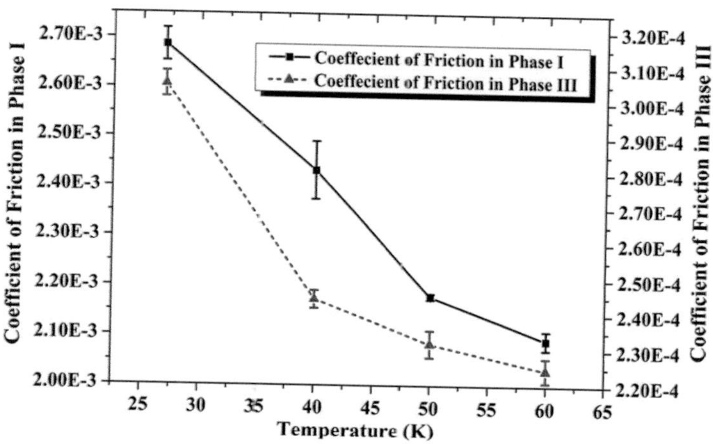

Figure 13.12 *COF* versus different cooling temperatures.

Lowering cooling temperature is often incompatible in affecting the rotation loss of the HTS bearing. One hand, the critical current density of superconductors will be increased in several orders of magnitude with the decrease of the cooling temperature, and then hysteresis loss of superconductors will be resisted. On the other hand, the lower temperature will aggravate the eddy current effect in refrigeration structures. Ye et al. studied effect of different temperatures of HTS on the rotation loss [36], as shown in Fig. 13.12, which are 27, 40, 50, and 60 K at FC height of 5 mm, respectively. In the Phase I and III of the rotor dynamics, the total loss decreased with the cooling temperature rise, indicating the decrement of the eddy current loss is larger than the increase amount of the hysteresis loss.

13.2.8 Rotation Properties at Extreme Low Frequencies

At extremely low rotational speeds (below 1 Hz), the drag and dynamic behaviors of the HTS bearing were complex. The deceleration of the PM rotor becomes rapid and there is a phenomenon of acceleration and deceleration around a position [43]. An oscillating rotation with a small angle is noticed by eye, which is similar to the oscillation of a sphere in a concave shown in Fig. 13.13.

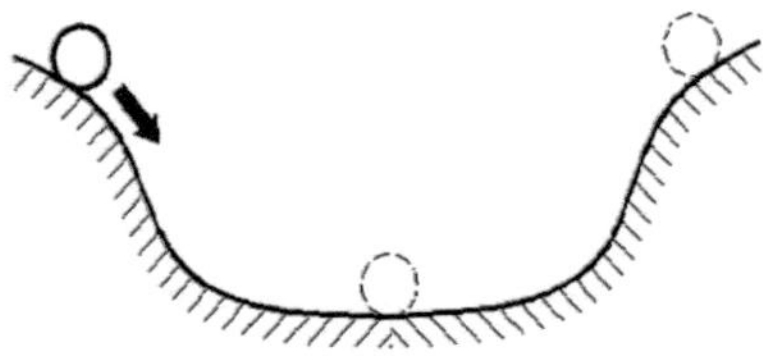

Figure 13.13 Oscillation of a sphere in a concave.

Figure 13.14 shows a basic oscillating angle curve when the PM is released at the FC height of 10 mm. It can be seen that angle amplitudes in each cycle have a good symmetry. The amplitudes decay very slowly with the increase of the time, and the decaying rate is only 0.07865 deg/s, indicating the rotation friction loss existing at the low speed is very small and has little effect on the angular amplitude in a short period.

Figure 13.15 gives curves of oscillating angle corresponding to different FC heights, which are 6, 8, and 10 mm respectively. Obviously, with the increase of the FC height, the decaying rate of the angle amplitude becomes smaller, and the time period of the oscillating rotation also decreases. The negative correlation between the decaying rate and the FC height may be explained by that, the suspension height drops with the reducing FC height, which causes a more serious inhomogeneous field distribution at the smaller gap and then leads to bigger friction loss due to the increasing hysteresis loss and eddy current loss.

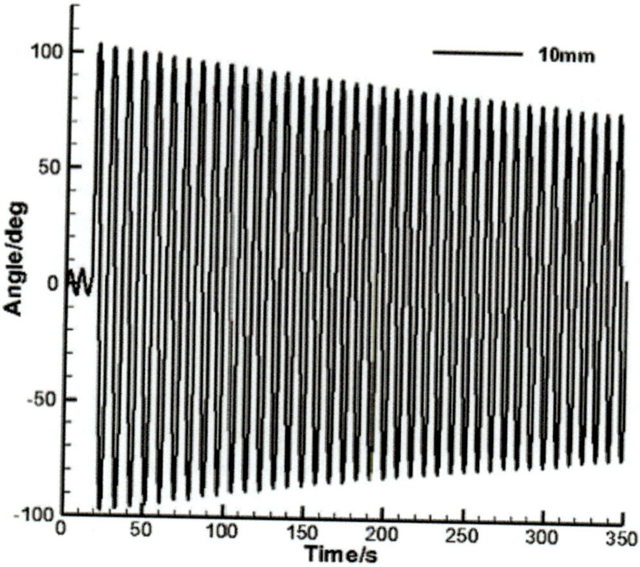

Figure 13.14 Oscillation angle of PM rotor vs. time corresponding to the FC height of 10 mm.

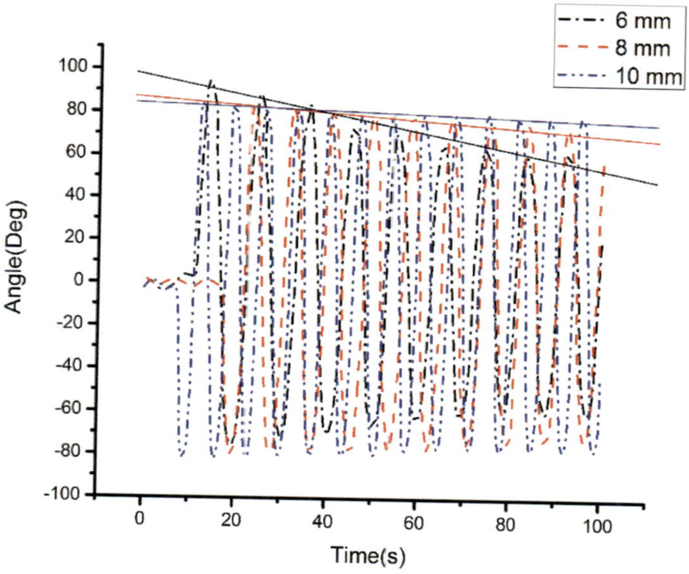

Figure 13.15 Oscillating angle versus time corresponding to different FC heights.

Cansiz first discussed the oscillating rotation at the extreme low frequencies and named it as "cogging torque effect" in his thesis [38]. Two possible reasons are used to understand the oscillating rotation. First, they considered the phenomenon was possibly due to unbalanced permanent magnet around its center of mass. The inhomogeneities of magnetic field from the magnet and superconductor cause the eccentricity between the center of magnetic and the center of mass. Sometimes tilting of the levitated magnet happens and causes precession and nonlinear drag on the spinning PM, which in turn result in oscillating rotation and drag torques associated with a very small friction loss.

Second, the above mechanism of the vibrating PM at very low speed is also affected by the thermodynamic changes in the environment. A thermodynamically phenomenon related to bearing dynamics creates the possibility of self-oscillation that may lead to rotation of a PM disc initially levitated without motion. The cause of the phenomenon appears to be a temperature gradient across the PM in which the temperature dependence of the magnetization causes the center of levitation force to reside below the center of gravity, a situation that is mechanically unstable. If the oscillation frequency and the thermal diffusion times in the PM are appropriately matched, small oscillations can grow to larger amplitudes and eventually become a self-sustaining rotation. Therefore, the small variations in the temperature of the levitation PM also affect the magnetic potential which is almost impossible to recover. Therefore, the low speed response obtained by experiments was incorporated in the magnetomechanical and thermodynamic analysis.

13.3 Low-Frequency Applications of HTS Bearings

The developments based on the rotation of the levitated PM over the HTS have the potential to provide various applications. Besides high-frequency applications such as flywheel storage systems, this chapter mainly concludes the low-frequency applications of the HTS bearing.

As different records of low rotational loss, a low-speed coefficient of friction of 3×10^{-7} in the bearing consisting of an

89-mm-diameter PM ring was achieved in the levitation over an array of HTS cylinders, and extremely low friction coefficients $<10^{-8}$ were observed for rotors with mixed-μ HTS bearings. Characterized by low rotational drag, HTS bearings have expanded their special applications in a lunar telescope [44], sensitive gravimeters [45], polarimeters [46], and micro-force sensing devices [47]. The low-frequency applications of the HTS bearing have been paid close attention and become important for high-precision instruments. As follows, detailed description about these applications are given.

13.3.1 Lunar Telescopes

Since 1990s, development in superconducting magnetic bearings can be beneficial for lunar telescopes in the space program [48,49]. A telescope on the moon is considered to be an ideal observatory with the astronomers. It must be lightweight, yet maintain high-tracking resolution and repeatability, and must work in a cold and dusty vacuum environment for long periods without on-site human maintenance.

The harsh environment on the lunar surface does require special consideration in the design of the telescope bearing. Conventional contact bearings and gear drives have numerous failure modes under such a restrictive and harsh environment. However, the superconducting magnet bearing has the unique capability to adapt to the low temperature and vacuum environments in space or on the Moon. These bearings are stable, light, passive, and essentially frictionless, allowing high-precision electronic positioning control. By passive levitation, the superconducting bearing does not wear out and requires neither maintenance nor power.

K. Ma et al. in University of Houston developed a prototype hybrid superconductor magnet bearing system for the lunar telescope azimuth mount [44,50]. The hybrid HTS-magnet bearing system consists of three parts: Inner support tube, outer support structure, and rotor. The rotor has two sets of two concentric ring magnets attached to the cylindrical rotating structure at one quarter of the height from the top. They are assembled by fitting one into the other with opposite polarity and its outer diameter is 10 cm. The inner support tube is located inside the rotor structure and its diameter is about 4 cm. The two stator magnets for each bearing are contained within thin-walled stainless steel tube. Eight HTSs are located 45 degrees apart around the circle of 7.62 cm. Each HTS is 2.54 cm in diameter and 0.6 cm in height. HTS is made of melt-textured YBaCuO and the HTSs are housed within two cold stages which are fixed separately to the outer support structure at one quarter of the height from the top and one sixth of the height from the base.

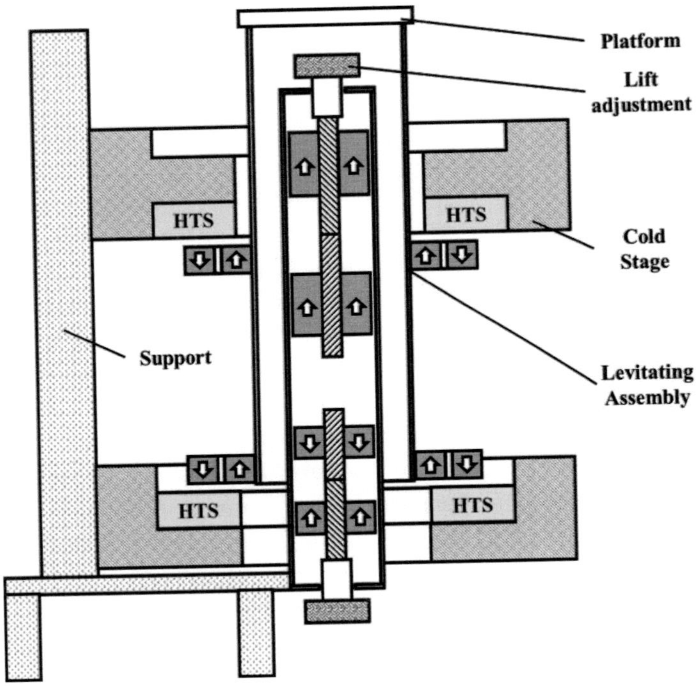

Figure 13.16 Schematic of the hybrid HTS-magnet bearings for the lunar telescope.

The entire structure is about 1 m tall and its mass is about 7 kg. The bearing assembly can support a load of about 4.5 kg, while its own weight is about 6.8 kg. The assembled bearing structure makes two hybrid HTS- magnet bearings. The total axial stiffness of the bearing is about 10~20 kN/m when the HTS gap was set at 8 mm and 6 mm. The smaller the gap, the higher the stiffness.

Characterization experiments have been conducted to understand its dynamic behavior and to estimate the required motor torque for its driving system design [1]. The experiments show the drag torque measured at three different rotational velocities from 0.02 rad/sec to 1 rad/sec. These results demonstrate that the torques oscillate with periodicity. The data show that all of these match well with a sinusoidal friction having an amplitude of about 1 mN•m. Eight bumps were shown in per cycle superimposed on the sinusoidal function. These oscillations could be identified with elastic restoring forces associated with a slightly uneven distribution of trapped magnetic flux around the ring of superconductor disks.

The oscillation of torque with periodicity seems to be caused by two facts. First, the magnets do not have uniform magnetic field density and the rotor magnets are not perfectly symmetric about the axis of rotation. Second, the superconducting ring is not one continuous annulus but instead eight separate disks or pieces. This oscillation will make its fine control difficult especially at low velocity. However, the magnitude of drag torque is velocity independent and very small. It makes the bearing attractive for high speed application.

Based on experimental results, a few design guidelines can be suggested. To achieve high system performance, it should be designed using permanent magnets with uniform magnetic field density, and the HTS ring should be made of one piece. Static and dynamic mechanical balance needs to be secured including the uniform gap size between HTS and the magnets. Future work involves the study of creep, velocity dependence of energy loss, and hysteresis energy loss from magnet asymmetry. Due to hysteresis drag torques, active control may be needed for high system performance. The relationship between rotational velocity and oscillation period needs to be further studied.

13.3.2 Polarimeter

The polarization of the cosmic microwave background (CMB) radiation field encodes new and valuable information about the early universe, and measurements of the polarization of the CMB radiation are expected to significantly increase our understanding of the early universe and the "big bang." Normally the polarization is measured by a modulator named polarimeter which is a scientific instrument used to measure the angle of rotation caused by passing polarized light through an optically active substance [51].

As shown in Fig. 13.17, the polarimeter generally consists of an optical element, a half-wave plate (HWP), and a mechanism to continuously rotate it at a few Hz [52]. While for the measurement of polarization of the incident radiation, the optimal technique currently available is by turning the HWP. However, over a broad range of wavelengths the challenge in using the technique to CMB polarimetry is coupling the rotation mechanism to bolometric detectors. Mechanical rotation mechanisms, such as gears and bearings, have stick slip friction which induce vibrations, which deposit mechanical energy in the detectors and cause excess noise.

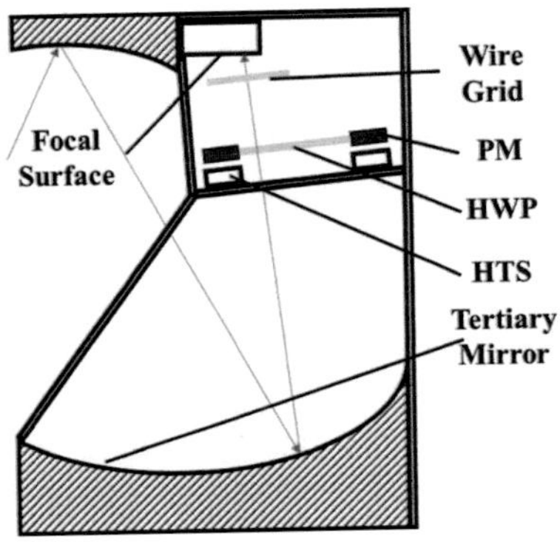

Figure 13.17 HWP and the magnetic bearing.

In 2003, S. Hanany et al. from University of Minnesota reported a prototype design for the CMB polarimeter in which a cryogenically cooled HWP rotates by means of an HTS bearing [51]. As shown in Fig. 13.17, the HWP and its rotation mechanism are located at the aperture stop of the optical system, which is between the tertiary mirror and the focal surface of the system. A polarizing wire grid mounted at the entrance to the photometers transmits only one component of the incident polarization vector. The HWP consists of a 3.2 mm thick z-cut sapphire disk with a radius of 2.54 cm which is mounted inside a permanent ring-shaped magnet. The sintered NdFeB magnet has an inside radius of 2.54 cm, an outside radius of 3.56 cm, thickness of 1.2 cm, and mass of 0.2 kg. The magnet and the HWP are the rotor of the magnetic bearing that is levitated above a ring of YBCO HTS material. The magnet is held at an appropriate distance above the ring of HTS material which consists of 12 pieces of melt-textured YBCO. The distance is typically between 4 and 10 mm.

As shown in Fig. 13.18, two clamps, each resembling a pair of pliers, hold the rotor in place during the cool-down of the system. A cryogenic stepping motor that drives a shaft and a pair of cams is used to open and close the clamps. The HTS material and the clamps are mounted on a G-10 board. Rotation of the rotor is achieved by means of a half gear that is driven by a second stepping motor that is mounted outside of the cryostat. During cool-down the half gear and the gear at the outer circumference of the rotor are engaged. Once the system has been cooled to 4 K, the clamps are opened and the half gear is turned with a preprogrammed acceleration. The axial spring constant at 77 K is measured to be 256 and 136 N/m for levitation distances of 8.6 and 10.7 mm, respectively. The corresponding resonant frequencies are about 5.5 and 4 Hz.

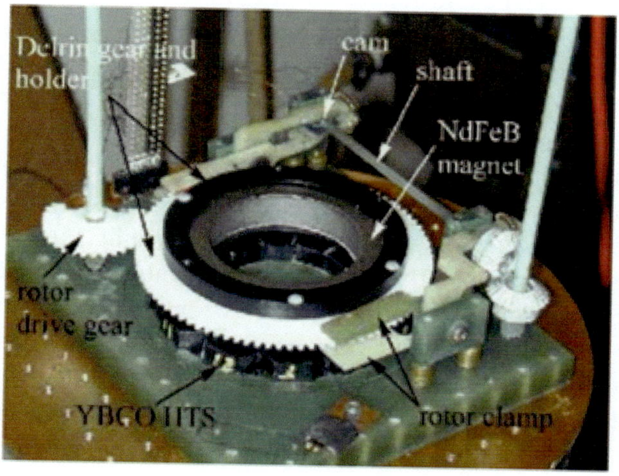

Figure 13.18 Magnetic bearing apparatus during the testing in a liquid helium cryostat [51].

The coefficient of friction as a function of several parameters including temperature between 15 and 83 K, rotation frequency between 0.3 and 3.5 Hz, levitation distance between 6 and 10 mm, and ambient pressure between 10^{-7} and 1 torr were measured [9]. The *COF* value is not less than 5×10^{-6}, and is mostly independent of rotational speed, except at temperatures near 77 K. This suggests that eddy current effects are not significant in the experiments. The general trend is for lower *COF*s to occur at lower temperatures and is consistent with J_c increasing with decreasing temperature.

It is generally understood that hysteresis loss is the dominant deceleration mechanism for most HTS bearings, and according to the critical-state model the hysteresis loss is inversely proportional to J_c. The low rotational drag of the HTS bearing allows rotations for long periods of time with minimal input power and negligible wear and tear thus making this technology suitable for a future satellite mission.

The polarimeter for the CMB polarization experiment has been fabricated and investigated in recent years. In 2011, a long duration balloon-borne CMB experiment from Antarctica, *EBEX*, employs an SMB system for the polarization modulator successfully [53]. From 2015, Tomotake Matsumura et al. continues to improve the SMB system design for use in a space-borne CMB polarization experiment.

In order to estimate the heat dissipation and the rotor temperature of superconducting magnetic bearing below 10 K, a typical experiment was carried out by conducting the spin-down measurements of 400 mm diameter of HWP rotation. By taking in account the mass for the HWP on the bearing, the corresponding expected heat dissipation is 2~20 mW from the hysteresis contribution depending on the levitation height [46,54]. They discussed several potentials for the improvements of heat dissipation. One is to minimize the source of eddy current loss by replacing the aluminum material of the prototype with the nonmetallic one. The other possibility to lower heat dissipation is to achieve the higher critical current of YBCO. Another possibility is to minimize effect of the amplitude of the magnetic field nonuniformity on the HTSs by constructing the magnetic rotor with lower axial nonuniformity and increasing the levitation height between the rotor bottom and the top surface of the HTS.

In order to understand the potential interference to the cosmological observations due to the vibration of the HWP, it is essential to characterize the vibrational properties of the levitating rotor of the SMB. In 2017, Tomotake Matsumura et al. presented the vibration characteristics of the superconducting magnetic bearing system with the diameter of 400 mm operating at below 10 K [55]. The spring constant of the rotor is in the order of 10^4~10^5 N/m and the damping coefficient is 15~45 depending on levitation heights. These vibration parameters are inversely proportional to the levitation height.

Thermal characteristics of the levitating rotor in the SMB are important to reduce the detector noise for the CMB polarization experiment. In 2018, the thermal characteristics of the superconducting magnetic bearing operating below at 10 K were investigated [56]. It is obtained that the requirement of heat loads from the SMB system is to be less than 1 mW. The rotation mechanism requires achieving a continuous rotation with a minimal heat dissipation at cryogenic temperature. Rotation stability is less than 1% with the rotation frequency between 0.5 Hz and 3.0 Hz. A thermal performance test was preformed, and the obtained heat dissipation at the rotor is 9.0 mW [57].

13.3.3 Micro-Thrust Measurement Devices

13.3.3.1 Traditional micro-thrust measurement methods

Recent years have witnessed many breakthroughs in research on microsatellite technology. Thanks to its large specific impulse and high efficiency, microsatellites have been widely used in the field of attitude adjustment and orbit control. Microsatellite ground thrust measurement is a key step in its development which can provide a necessary reference for the design, parameters selection and performance evaluation of the micro-thrusters.

Various micro-thrust measurements have been developed and widely used including inverted pendulum, hanging pendulum and torsional pendulum [58]. Hanging pendulum's construction is the simplest and the most stable while its sensitivity is normally unsatisfied and lower than others. Sasoh Akihiro [59] developed a hanging pendulum thrust stand with a resolution of 0.5 mN, which was pretty high for micro-thrust. It was primarily applied to thrusters with high thrust-to-weight ratio.

Inverted pendulums have a better sensitivity but worse stability [60]. Therefore, it is typically used when high sensitivity is needed but the space is limited. A typical inverted pendulum measurement designed by Cassady [61] had a sensitivity of 3.00±0.32 mV/mN, mainly used for high-power steady-state plasma thrusters. One problem of the inverted pendulum is that its stability highly relies on the flexure stiffness, which can be easily affected by many external factors like temperature and vibrations. Xu [62] developed an inverted pendulum stand for micro-thrusters with a thrust between 1 mN and 5 N, using a damper to compensate for vibrations and a multi-loop cooling system to minimize thermal effects.

The restoring force of torsional pendulum can be produced independent of thrusters' weight, which allows it has the highest sensitivity. Evidence shows that its resolution can be as low as 20 nN [63]. Haag [64] had developed a torsional type thrust stand to measure impulse of 300 μNs for pulsed plasma thrusters (PPTs).

13.3.3.2 Micro-thrust stand using HTS bearing

In the application of HTS magnetic levitation to measure micro-thrust, two different measurement principles are used, namely by swing method and angular momentum method. The schematic graphics of both methods are displayed in Fig. 13.19. Thrusters to be tested were fixed on the top of rotor where the direction of the thrust was parallel to the tangent direction of the rotor.

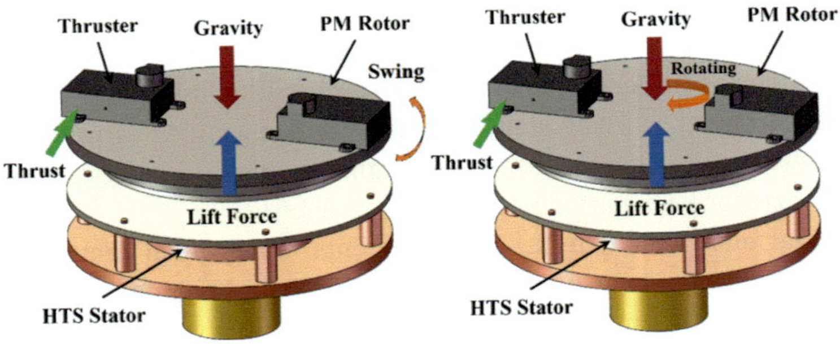

Figure 13.19 Schematics of both methods; (a) Swing method; (b) Angular momentum method.

The principle of the swing method is depending on *Hooke's law*,

$$F = k_s \cdot x \tag{13.10}$$

where F is the restoring force produced by the PM rotor; k_s is the stiffness factor and x is the angular displacement of the PM rotor. A gravity well is constructed by making the rotor slope slightly with a offset load fixed on the rotor. Obviously, the offset load is at the lowest point of the rotor in a static state, while the interference of external forces will cause it to swing from the lowest position around vertical axis and therefore generate a restorative force under the action of gravity.

The rotor will produce an angular displacement under the action of tangential thrust, and therefore the restoring force increases with the displacement until it is equal to the thrust. Consequently, the rotor will remain in a specific position by the primary suspension. After the calculation of the stiffness factor by proper calibration, the thrust can be obtained according to the measuring angular displacement of the rotor.

The principle of the angular momentum method is angular momentum theorem,

$$F_T R_T - F_R L_R = I\alpha \tag{13.11}$$

where F_T is the thrust of the micro-thruster; R_T is the moment radius of the thrust; I is the rotational inertia of the entire rotor; α is the angular acceleration of the rotor (of which the unit is rad/s^2); F_R is the circumferential resistence force and L_R is the moment radius of the resistance. Parameters such as angular velocity and angular acceleration are collected by speed measurement of the PM rotor.

Obviously, the resistance force of the rotor is a crucial factor affecting the range and accuracy of the thrust measurement. The relation between F_T and COF can be described as

$$F_T = M \cdot (R_g{}^2 d\omega/dt + COF \cdot gL_R)\,/\,R_T \tag{13.12}$$

where ω is the rotation speed of the rotor (of which the unit is rad/s), R_a is the radius of gyration, M is the entire rotor mass and g is the acceleration of gravity.

The measurement accuracy of F_T mainly depend on the measurement accuracy of the rotation speed while the measurement range can be roughly determined by the mass of rotor and COF. To ensure the high accuracy of the measurement, the resistance force should be at least an order of magnitude lower than thrust. Most existing HTS bearings can reach a COF of 10^{-6} even 10^{-7} and the corresponding loading weight for different thrust level is shown in Table 13.3.

Table 13.3 Rotor mass level consideration corresponding to resistance level

Required thrust level	Resistance level	Rotor mass level (kg)		
		$COF = 10^{-6}$	$COF = 10^{-7}$	$COF = 10^{-8}$
1 mN	0.1 mN	10	100	1000
0.1 mN	10 μN	1	10	100
10 μN	1 μN	0.1	1	10
1 μN	0.1 μN	0.01	0.1	1

13.3.3.3 Prototype design

W. J. Yang has been studying an applicational design of HTS bearing for micro-thrust measurement for space propulsion since 2013 [65,66]. During this time, the friction loss and cogging torque characteristics of the superconducting levitation under low-frequency rotation were studied. The overall system designed for milli-newton measurement, shown in Fig. 13.20, consists of an HTS bearing, azimuth and rotation speed measurement, cooling devices, and lifting structures. The system is deployed in a vacuum chamber where the pressure was lower than 0.1 Pa. The load capacity of the HTS bearing was up to 20 kg, and the entire propulsion system was placed on the levitation system independently to simulate the real operation of the thruster working in space environment. Two different measuring methods were designed to fitting for several operation modes of the micro-thruster, and the measured impulse level was from 1 to 100 mNs.

The thrust type HTS bearing used in the micro-thrust measuring stand, composing of an PM rotor and an HTS stator, can sustain more than 20 kg effective load which is good enough for deploying the entire propulsion system, including thruster, power supply, and control units.

Hence, lowering the friction loss was the primary design consideration. Two key indexes are chosen as the validation factors which are the average hysteresis loss of HTS bulks and variation of the magnetic field at certain circumferential.

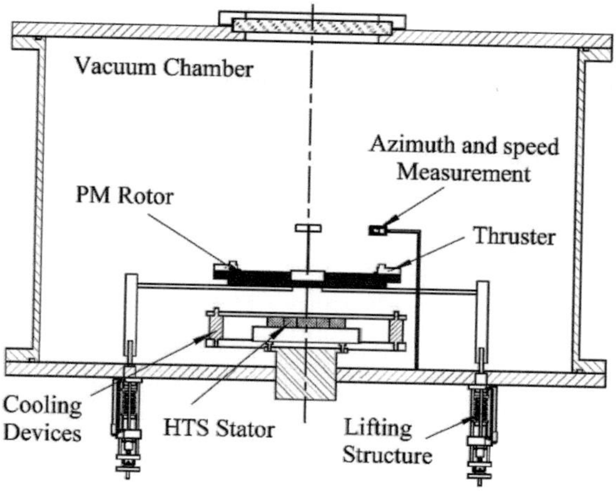

Figure 13.20 Overall design of milli-newton measurement system.

A total of 8 YBCO bulks, diameter of 30 mm and 18 mm in height, are arranged in a ring right under the PM rotor. All these bulks are selected with max levitation force within 70 N to 75 N at zero field cooling (ZFC) condition, which was tested by the commonly used PM moving method, to gain a levitation state closer to horizontal level to reduce the friction. Meanwhile, the average hysteresis loss of all this bulks was 70.15 mJ within one levitation force test circle. The diameter of the superconducting ring is 65 mm where the magnetic flux density of the magnetic rotor structure is the highest.

The PM rings are made of radially polarized NdFeB materials, separated and fixed by three steel rings. The variation of the magnetic field at certain circumferential, mainly caused by segments of PM ring, was mitigated by this steel rings. The magnetic flux in each PM ring is in opposite direction, gathered and directed outward by the middle steel ring, which leads to a high gradient within the HTS stator area, therefore, high levitation stiffness could be achieved. The magnetic flux at radial boundary are also constrained by the external and internal steel rings so that the magnetic flux density within the adjacent metal parts would be very small and the eddy current loss could be greatly reduced. Furthermore, a shielding steel plate is employed to prevent the electronics above the rotor from being interfered by the magnetic flux of the PM rings.

13.3.3.4 Use for EMDrive and Mach-effect thruster

In 2018, to investigate the EMDrive and Mach-effect thruster concepts, amongst others at the Institute of Aerospace Engineering at Technische Universität Dresden (TU Dresden) designed a rotational thrust balance using an HTS bearing [67]. By measuring the change in angular velocity of the magnetically levitated system inside a vacuum chamber onto which the thrusters apply a torque, thrusts in the range of ~1 μN was possible to be detected. As shown in Fig. 13.21, the balance is based on the magnetic levitation bearing utilizing HTSs and permanent magnets to provide a frictionless rotational degree of freedom.

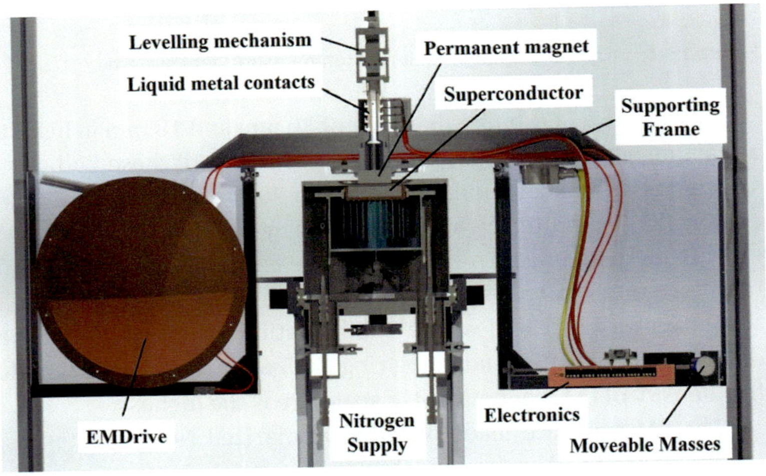

Figure 13.21 Sectional view of the magnetically levitated structure and the cryostat [68].

In comparison to electromagnetically levitated magnets, this HTS bearing offers an increased axial load capacity on a smaller construction volume, while not depending on closed-loop control of the electromagnetic coil. Bearing system is a combination of a cylindrical NeFeB permanent magnet and an YBCO HTS disc. The permanent magnet has an outer diameter of 50 mm and a height of 20 mm with a quality specification of N50. Two different sizes of YBCO HTSs are available for the balance. Both of which with a height of 16 mm and outer diameters of 54 mm/80 mm.

Measurements of the levitation force resulted in a maximum axial load capability of 22 kg for the HTS bearing, which is an order of magnitude above electromagnetically levitated thrust balances. Insufficient alignments of the center of mass of the levitating structure lead to complications in thrust measurements by introducing a scaling factor of the gravitational acceleration within the rotational plane. The measured friction torque reached a minimum value of 2.4×10^{-6} Nm for an angular velocity of 0.02 rad/s [67].

13.4 Summary

This chapter provides preliminary information about rotational loss properties in the HTS bearings. In the definition of the coefficient of friction and analysis of rotational loss sources in the HTS bearings, we concluded that the rotational losses in the HTS bearing mainly consist of windage loss caused by air drag, hysteresis loss on superconductors and eddy current loss in adjacent metal structures.

Several main factors including bearing structure types, dynamic behaviors, superconducting material properties, magnetic rotor structures, and working conditions were investigated to understand their effects on the rotational losses. Progress in reducing the rotation loss and *COF* can be attributed to methods for improving the azimuthal homogeneity of the magnetic field of the magnetic rotor structure as well as continual improvements in J_c of superconductors, optimizing the HTS array geometry and the low eddy-current cryochamber, suppressing the dynamic behaviors of the rotor and thermodynamical properties. For designing a superconducting bearing for different applications, especially for very low rotational frequencies, the geometric and magnetic considerations must be taken into account.

HTS bearing has great applications in a variety of low-frequency cases where the low rotational loss or coefficient of friction is required. Here we just list the analysis of using HTS bearings in lunar telescopes, polarimeters, and micro-thrust measurement devices. Special considerations of HTS bearings in the low-frequency applications demonstrate their unique performance in high stability and sensitivity compared to any other mechanical bearings, which create a special approach to contribute the sustainable development of bulk superconductors and the attended superconducting bearings.

References

1. Lee E., Ma K. B., Wilson T. L. and Chu W.-K., (1999) Superconductor-magnet bearings with inherent stability and velocity-independent drag torque, *1999 IEEE/ASME International Conference on Advanced Intelligent Mechatronics (Cat. No. 99TH8399)*, pp. 806–811.
2. Hellman F., Gyorgy E., Johnson Jr D., O'Bryan H. and Sherwood R., (1988) Levitation of a magnet over a flat type ii superconductor, *Journal of Applied physics*, **63**, pp. 447–450.
3. Day A., Strasik M., McCrary K., Johnson P., Gabrys J., Schindler J., Hawkins R., Carlson D., Higgins M. and Hull J., (2002) Design and testing of the HTS bearing for a 10 kWh flywheel system, *Superconductor Science and Technology*, **15**, pp. 838.
4. Floegel-Delor U., Rothfeld R., Wippich D., Goebel B., Riedel T. and Werfel F. N., (2007) Fabrication of HTS bearings with ton load performance, *IEEE Transactions on Applied Superconductivity*, **17**, pp. 2142–2145.
5. Walter H., Bock J., Frohne C., Schippl K., May H., Canders W., Kummeth P., Nick W. and Neumueller H., (2006) First heavy load bearing for industrial application with shaft loads up to 10 kN, *Journal of Physics: Conference Series*, **43**, pp. 995.
6. Sung T.-H., Han S., Han Y., Lee J., Jeong N., Hwang S. and Choi S., (2002) Designs and analyses of flywheel energy storage systems using high-T_c superconductor bearings, *Cryogenics*, **42**, pp. 357–362.
7. Jansen R. and Dever T., (2004) G2 flywheel module design, *2nd International Energy Conversion Engineering Conference*, pp. 5603.
8. Yang Y. and Cao X., (2006) Design and development of the small satellite attitude control system simulator, *AIAA Modeling and Simulation Technologies Conference and Exhibit*, pp. 6124.
9. Hull J. R., Hanany S., Matsumura T., Johnson B. and Jones T., (2004) Characterization of a high-temperature superconducting bearing for use in a cosmic microwave background polarimeter, *Superconductor Science and Technology*, **18**, pp. S1.
10. Weinberger B., Lynds L., Hull J. and Balachandran U., (1991) Low friction in high temperature superconductor bearings, *Applied Physics Letters*, **59**, pp. 1132–1134.

11. Moon F. and Chang P. Z., (1990) High-speed rotation of magnets on high t c superconducting bearings, *Applied Physics Letters,* **56**, pp. 397–399.
12. Hull J. R., Mulcahy T., Uherka K. and Abboud R., (1995) Low rotational drag in high-temperature superconducting bearings, *IEEE Transactions on Applied Superconductivity,* **5**, pp. 626–629.
13. Strasik M., Johnson P. E., Day A., Mittleider J., Higgins M., Edwards J., Schindler J., McCrary K., McIver C. and Carlson D., (2007) Design, fabrication, and test of a 5-kWh/100-kW flywheel energy storage utilizing a high-temperature superconducting bearing, *IEEE Transactions on Applied Superconductivity,* **17**, pp. 2133–2137.
14. Hull J. R., (2000) Superconducting bearings, *Superconductor Science and Technology,* **13**, pp. R1.
15. Xia Z., Ma K., Cooley R., Fowler P. and Chu W., (1996) Superconducting magnetic bearing and its application in flywheel kinetic energy storage, *Proceedings of the 10th Anniversary Workshop on Physics, Materials and Applications, Houston, TX,* pp.
16. Hull J. R., Labataille J. F., Mulcahy T. M. and Lockwood J. A., (1996) Reduced hysteresis loss in superconducting bearings, *Applied Superconductivity,* **4**, pp. 1–10.
17. Bean C., (1962) Magnetization of hard superconductors, *Bean CP (1964) Rev Mod Phys,* **36**, pp. 31.
18. Higasa H., Ishikawa F., Kawauchi N., Yokoyama S., Nakamura S., Imaizumi H. and Ito N., (1994) Experiment of a 100 Wh-class power storage system using high-T_c superconducting magnetic bearing. (Springer)
19. Ma K., Postrekhin Y. and Chu W., (2003) Superconductor and magnet levitation devices, *Review of Scientific Instruments,* **74**, pp. 4989–5017.
20. Zeisberger M. and Gawalek W., (1998) Losses in magnetic bearings, *Materials Science and Engineering: B,* **53**, pp. 193–197.
21. Hull J. R., Mulcahy T. M. and Labataille J. F., (1997) Velocity dependence of rotational loss in Evershed-type superconducting bearings, *Applied Physics Letters,* **70**, pp. 655–657.
22. Hull J., Hilton E., Mulcahy T., Yang Z., Lockwood A. and Strasik M., (1995) Low friction in mixed-mu superconducting bearings, *Journal of Applied Physics,* **78**, pp. 6833–6838.
23. McMichael C., Ma K., Lin M., Lamb M., Meng R., Xue Y., Hor P. and Chu W., (1991) Effects of material processing in high temperature superconducting magnetic bearings, *Applied physics letters,* **59**, pp. 2442–2444.

24. Ye M., Yang W., Liu Y., Li Y. and Yu L., (2018) A design method of HTS bulks array for decreasing rotation loss in a superconducting maglev microthrust stand, *IEEE Transactions on Applied Superconductivity,* **28**, pp. 1–5.

25. Hull J., Rossing T. D., Mulcahy T. and Uherka K., (1992) Improvement of azimuthal homogeneity in permanent-magnet bearing rotors, *Presented at the International Magnetics Conference,* Stockholm, Sweden, April 13-16, 1993.

26. Sotelo G., Rodriguez E., Costa F., Oliveira J., de Santiago J. and Stephan R., (2016) Tests with a hybrid bearing for a flywheel energy storage system, *Superconductor Science and Technology,* **29**, pp. 095016.

27. Han Y., Hull J., Han S., Jeong N., Oh J. and Sung T.-H., (2004) Losses of superconductor journal bearing, *AIP Conference Proceedings,* **710**, pp. 1899-1905.

28. Xu K.-x., Wu D.-j., Jiao Y. and Zheng M., (2016) A fully superconducting bearing system for flywheel applications, *Superconductor Science and Technology,* **29**, pp. 064001.

29. Basaran S. and Sivrioglu S., (2017) Radial stiffness improvement of a flywheel system using multi-surface superconducting levitation, *Superconductor Science and Technology,* **30**, pp. 035008.

30. Mulcahy T. M., Hull J. R., Uherka K. L., Abboud R. G., Wise J. H., Carnegie D. W., Bakis C. E. and Gabrys C. W., (1996) A permanent-magnet rotor for a high-temperature superconducting bearing, *IEEE Transactions on Magnetics,* **32**, pp. 2609–2612.

31. Komori M. and Uchimura Y., (2005) Improving the dynamics of two types of flywheel energy storage systems with SMBs, *IEEE Transactions on Applied Superconductivity,* **15**, pp. 2261–2264.

32. Matsunaga K., Tomita M., Yamachi N., Iida K., Yoshioka J. and Murakami M., (2002) YBCO bulk of the superconducting bearing for a 10 kWh flywheel, *Superconductor Science and Technology,* **15**, pp. 842.

33. Mukoyama S., Nakao K., Sakamoto H., Matsuoka T., Nagashima K., Ogata M., Yamashita T., Miyazaki Y., Miyazaki K. and Maeda T., (2017) Development of superconducting magnetic bearing for 300 kW flywheel energy storage system, *IEEE Transactions on Applied Superconductivity,* **27**, pp. 1–4.

34. Ogata M., Matsue H., Yamashita T., Hasegawa H., Nagashima K., Maeda T., Matsuoka T., Mukoyama S., Shimizu H. and Horiuchi S., (2016) Test equipment for a flywheel energy storage system using a magnetic

bearing composed of superconducting coils and superconducting bulks, *Superconductor Science and Technology,* **29**, pp. 054002.

35. Cansiz A., Hull J. and Gundogdu Ö., (2005) Translational and rotational dynamic analysis of a superconducting levitation system, *Superconductor Science and Technology,* **18**, pp. 990.

36. Ye M., Yang W., Wang Y., Yu L. and Liu Y., (2015) Effect of temperature on rotation loss in a superconducting device for microthrust measurement of electric propulsion system, *IEEE Transactions on Applied Superconductivity,* **25**, pp. 1–4.

37. Otani T. and Murakami M., (2000) Performance of Sm-Ba-Cu-O bulk superconductors for a magnetic bearing, *Superconductor Science and Technology,* **13**, pp. 866.

38. Cansiz A., (1999) Force, stiffness and hysteresis losses in high temperature superconducting bearings, Thesis (PhD). Illinois Institute of Technology, Source DAI-B 61/01, p. 331, Jul 2000, 135 pages.

39. Perini E. and Giunchi G., (2009) Field cooling of a MgB_2 cylinder around a permanent magnet stack: Prototype for superconductive magnetic bearing, *Superconductor Science and Technology,* **22**, pp. 045021.

40. Patel A., Hopkins S. C., Baskys A., Kalitka V., Molodyk A. and Glowacki B. A., (2015) Magnetic levitation using high temperature superconducting pancake coils as composite bulk cylinders, *Superconductor Science and Technology,* **28**, pp. 115007.

41. Patel A., Hahn S., Voccio J., Baskys A., Hopkins S. C. and Glowacki B. A., (2016) Magnetic levitation using a stack of high temperature superconducting tape annuli, *Superconductor Science and Technology,* **30**, pp. 024007.

42. Sotelo G. G., Ferreira A. C. and De Andrade R., (2005) Halbach array superconducting magnetic bearing for a flywheel energy storage system, *IEEE Transactions on Applied Superconductivity,* **15**, pp. 2253–2256.

43. Yu L., Yang W., Ye M. and Liu Y., (2013) Rotation friction loss in superconducting levitation applied for a micro thrust measuring device, *IEEE Transactions on Applied Superconductivity,* **24**, pp. 1–4.

44. Lee E., Ma K. B., Wilson T. L. and Chu W.-K., (1999) Characterization of superconducting bearings for lunar telescopes, *IEEE Transactions on Applied Superconductivity,* **9**, pp. 911–915.

45. Hull J. R. and Mulcahy T. M., (1999) Gravimeter using high-temperature superconducting bearing, *IEEE Transactions on Applied Superconductivity,* **9**, pp. 390–393.

46. Matsumura T., Kataza H., Utsunomiya S., Yamamoto R., Hazumi M. and Katayama N., (2016) Design and performance of a prototype polarization modulator rotational system for use in space using a superconducting magnetic bearing, *IEEE Transactions on Applied Superconductivity*, **26**, pp. 1–4.

47. Yang W., Ji Y., Ye M. and Tang H., (2019) A micro-force measurement system based on high-temperature superconducting magnetic levitation, *Measurement Science and Technology*, **30**, pp. 125020.

48. McMichael C., Ma K., Lamb M., Lin M., Chow L., Meng R., Hor P. and Chu W., (1992) Practical adaptation in bulk superconducting magnetic bearing applications, *Applied Physics Letters*, **60**, pp. 1893–1895.

49. Ma K., Lamb M., Chen P., Coolep R., Xia H. and Fopler C., (1994) A natural application for high temperature superconductors: A bearing for the azimuth mount of a lunar telescope, *Applied Superconductivity*, **2**, pp. 479–486.

50. Lamb M., Ma K. B., Cooley R., Mackey D., Meng R., Chu C. W., Chu W. K., Chen P. C. and Wilson T., (1995) High temperature superconducting bearings for lunar telescope mounts, *IEEE Transactions on Applied Superconductivity*, **5**, pp. 638–642.

51. Hanany S., Matsumura T., Johnson B., Jones T., Hull J. R. and Ma K. B., (2003) A cosmic microwave background radiation polarimeter using superconducting bearings, *IEEE Transactions on Applied Superconductivity*, **13**, pp. 2128–2133.

52. Sakurai Y., Matsumura T., Katayama N., Imada H., Komatsu K., Kanai H., Takaku R., Sugiyama S., Ghigna T. and Iida T., (2018) Design and development of a polarization modulator unit based on a continuous rotating half-wave plate for LiteBIRD, *Millimeter, Submillimeter, and Far-Infrared Detectors and Instrumentation for Astronomy IX*, **10708**, pp. 107080E.

53. Klein J., Aboobaker A., Ade P., Aubin F., Baccigalupi C., Bao C., Borrill J., Chapman D., Didier J. and Dobbs M., (2011) A cryogenic half-wave plate polarimeter using a superconducting magnetic bearing, *Cryogenic Optical Systems and Instruments XIII*, **8150**, pp. 815004.

54. Sakurai Y., Matsumura T., Kataza H., Utsunomiya S. and Yamamoto R., (2017) Estimation of the heat dissipation and the rotor temperature of superconducting magnetic bearing below 10 k, *IEEE Transactions on Applied Superconductivity*, **27**, pp. 1–4.

55. Sakurai Y., Matsumura T., Sugai H., Katayama N., Ohsaki H., Terao Y., Terachi Y., Kataza H., Utsunomiya S. and Yamamoto R., (2017)

Vibrational characteristics of a superconducting magnetic bearing employed for a prototype polarization modulator, *Journal of Physics: Conference Series,* **871**, pp. 012091.

56. Sakurai Y., Matsumura T., Iida T., Kanai H., Katayama N., Imada H., Ohsaki H., Terao Y., Shimomura T. and Sugai H., (2018) Design and thermal characteristics of a 400 mm diameter levitating rotor in a superconducting magnetic bearing operating below at 10 k for a CMB polarization experiment, *IEEE Transactions on Applied Superconductivity,* **28**, pp. 1–4.
57. Sakurai Y., Matsumura T., Katayama N., Iida T., Komatsu K., Sugai H., Ohsaki H., Terao Y., Hirota Y. and Enokida H., (2019) Development of a contact-less cryogenic rotation mechanism employed for a polarization modulator unit in cosmic microwave background polarization experiments, *Journal of Physics: Conference Series,* **1293**, pp. 012083.
58. Patel A. and Frederick R. A., (2017) Overview of small satellite micro-propulsion devices, ground test capabilities, and performance measurement techniques, *53rd AIAA/SAE/ASEE Joint Propulsion Conference,* pp. 4886.
59. Sasoh A. and Arakawa Y., (1993) A high-resolution thrust stand for ground tests of low-thrust space propulsion devices, *Review of Scientific instruments,* **64**, pp. 719–723.
60. Polzin K. A., Markusic T. E., Stanojev B. J., DeHoyos A. and Spaun B., (2006) Thrust stand for electric propulsion performance evaluation, *Review of Scientific Instruments,* **77**, pp. 105108.
61. Cassady L., Kodys A. and Choueiri E., (2002) A thrust stand for high-power steady-state plasma thrusters, *38th AIAA/ASME/SAE/ASEE Joint Propulsion Conference & Exhibit,* pp. 4118.
62. Xu K. G. and Walker M. L., (2009) High-power, null-type, inverted pendulum thrust stand, *Review of Scientific Instruments,* **80**, pp. 055103.
63. Phipps C. R. and Luke J. R., (2000) Micro laser plasma thrusters for small satellites, *High-Power Laser Ablation III,* **4065**, pp. 801–809.
64. Haag T. W., (1997) Thrust stand for pulsed plasma thrusters, *Review of Scientific Instruments,* **68**, pp. 2060–2067.
65. Yang W., Ji Y., Yu L., Liu X., Long J., Liu Z. and Song D., (2019) Low frequency rotational loss in a high-temperature superconducting bearing and its application in micro-thrust measurement for space propulsion, *Superconductor Science and Technology,* **33**, pp. 014001.

66. Ye M., Yang W., Yu L. and Liu Y., (2013) A novel micro-thrust measuring method based on magnetic levitation using high temperature superconductors, *Proceedings of 2013 2nd International Conference on Measurement, Information and Control,* **1**, pp. 41–44.

67. Kößling M., Monette M., Weikert M. and Tajmar M., (2019) The spacedrive project-thrust balance development and new measurements of the mach-effect and EMDrive thrusters, *Acta Astronautica,* **161**, pp. 139–152.

68. Neunzig O., Kössling M., Weikert M. and Tajmar M., (2019) Characterization of a rotational thrust balance for propellantless propulsion concepts utilizing magnetic levitation with superconductors, *Presented at the 36th International Electric Propulsion Conference University of Vienna*, Vienna, Austria, September 15-20, 2019.

Chapter 14

Superconducting Motor Using HTS Bulk

Alexandre Colle,[a] Thierry Lubin,[b] Sabrina Ayat,[c] and Jean Leveque[b]

[a]*Kyoto University, Graduate School of Engineering, Kyoto 610-1103, Japan*
[b]*GREEN Université de Lorraine, Vandoeuvre, 54506, France*
[c]*SAFRAN Tech, Magny-lès-Hameaux, France*
Jean.leveque@univ-lorraine.fr

14.1 Introduction

14.1.1 Growing Air Transport

The International Civil Aviation Organization (ICAO) has standardized a unit of measurement called the "Revenue Passenger-Kilometer" (RPK), which measures the actual demand for air transport. This indicator is calculated as the number of passengers who purchased a ticket multiplied by the total distance travelled. Following RPK, the air transport has grown by about 9% every year since 1950 and has made the world smaller than ever. However, commercial flights remain expensive and represent 2% of human CO_2 emissions. As a result, airlines and the aviation industry are

High-T_c Superconducting Technology: Towards Sustainable Development Goals
Edited by Muralidhar Miryala

ISBN 978-981-4877-65-7 (Hardcover), 978-1-003-16468-5 (eBook)
www.jennystanford.com

looking to the transportation of tomorrow. This will be achieved by reducing costs, noise, and greenhouse gas emissions. To meet these expectations, electrical energy is chosen for the development of the future aircraft. Thus, several concepts appear as "the more electric aircraft" or "the electrically propelled aircraft."

14.1.2 Electrification of Aircraft

In an aircraft architecture used for civil transport, the energy sources of the on-board equipment can be mechanical, hydraulic, pneumatic or even electric. Power is obtained from the combustion of fuel in the engines. This energy is mainly used to propel the aircraft, and the rest is used for on-board systems, also called non-propulsive energy, existing in four forms:

- Pneumatic energy is obtained from high-pressure compressors. This type of energy is used to pressurize the cabin, to supply air for the crew and passengers or for the ice protection system.
- Mechanical energy is transferred from the main motor to hydraulic pumps but also to electric generators via gearboxes.
- Hydraulic energy is supplied by hydraulic pumps to power most actuation systems. It includes the landing gear deployment system, the brakes, the flap and slat management, ...
- Electric energy is produced with the main generator to supply the cabin, the aircraft lighting, or the on-board entertainment systems. The advantage of this energy is that generally the electrical equipment can be rather compact, is not subject to leaks, and is flexible to use. On the other hand, safety management could be more complex, particularly in the event of a short circuit.

This architecture is the product of decades of development, but systems are becoming more and more complex today. The interaction between the equipment reduces the efficiency of the whole system. A "more electric" aircraft is an airplane where a part of the non-propulsive energy becomes more electric. This change implies an increase of the on-board electrical power. For a "more electric" airplane with 300 passengers, the power required is greater than 1.5 MW [ROS07].

The on-board electrical power is generated by several synchronous generators. These generators are coupled to the turbine via a variable ratio gearbox. The current electrical network is three-phase of 115 V AC at a constant frequency of 400 Hz. Replacing the gearbox with power electronics has been shown to improve efficiency, to reduce cost, and to reduce maintenance. Thus, the airplane's turbojet engines drive the electric generators supplying electricity at variable frequency which is then transformed into a constant frequency by a power converter. An increase of the voltage to 230/400 V AC at 400 Hz is explored, in order to reduce the weight of the converters (lower current)

14.1.3 Electrification of Propeller

The second aspect of aircraft electrification concerns the propulsion. This solution would reduce greenhouse gas emissions, reduce fuel consumption for hybrid propulsion solutions, but also reduce aircraft noise. However, to be used on a commercial aircraft, the electrical technology must be more competitive than current ones. Some architectures are studied:

- The all-electric architecture uses only electrical components.
- Hybrid architecture is a coupling between electric actuators and kerosene turbines. The distribution on the propulsion can be adjusted according to the flight phases.
- The turboelectric architecture: a turbine drives a generator which powers the converters and the electric motors connected to the propellers.

The level of reduction in CO_2 emissions depends on the architecture chosen. For example, for the all-electric solution the only source of power is the batteries, so the direct CO_2 emission is reduced by 100%. While for hybrid structures the reduction is therefore less significant.

Several manufacturers, governments, and universities have started working on hybrid or electric aircraft systems. Some programs are listed in Table 14.1. [NAP16], there we find the type of architecture chosen or the specific power required for batteries or motors.

Table 14.1 Some projects of electric powered aircraft

Project	Aircraft	Electric motors and generators	Batteries
BOEING SUGAR	Long range (+100 seats)	Motor: P: 5.3 MW P_m: 5 kW/kg	E_m : 750 Wh/kg
NASA N3X	Long range (+100 seats)	Generator: P: 30 MW - P_m: >10 kW/kg Motor: P: 4 MW - P_m: >10 kW/kg	
Airbus E-Fan X	SMR (100 seats)	Motor: P: 2.5 MW	I
Zunum AERO	SMR (12–50 seats)	Generator: P : 1 MW	E_m: 400 kWh

One of the most important parameters for aeronautical systems is mass energy (Wh/kg) for storage systems and mass power (kW/kg) for electric actuators or power converters. This is why the data based on these criteria are presented in Table 14.1.

14.1.4 State of the Art of Electrical Motor

The Boeing 787 provides the starting point for the state of the art, as it is the most "electric" aircraft today. The total installed electrical power is 1 MW. This aircraft incorporates electric generators whose power density is 2.2 kW/kg. Optimistic projections estimate that the specific power could reach 9 kW/kg for conventional machines (permanent magnets, wound copper rotor, etc.) of 1–3 MW. This objective could be achieved with the use of composite materials [ZHA16]. However, to reach the target of upper of 10 kW/kg, disruptive technologies must be studied. Even if there are today many obstacles to a possible "cryogenic" aircraft, the use of materials such as superconductors could significantly increase the specific power of engines or generators.

The work presented in this chapter relates to the study of a superconductive electrical machine design for aeronautical use. They are also exploring the gains in mass power possible with the performance of usual superconducting materials for machines up to MW.

14.1.5 Superconducting Motor

In this part, a state of the art of the main superconducting machines manufactured is presented.

14.1.5.1 History of superconducting machine

Research and development around superconducting machines began in the 1960s, shortly after the development of NbTi multi-filament wires with sufficiently stable performance [NIT18]. Interest in superconducting machines may be the result of two factors. The first was the oil shock that occurred in the early 1970s and the second was the belief that electricity demand would double from year to year. We were therefore preparing to have to design increasingly powerful alternators with better efficiency for energy saving. The problem was that the conventional alternators reached their limits beyond 1.5 GW, in terms of transport (tunnel, road, etc.) but also of the dimensions of the machine (peripheral speed limit, boom effect, etc.). After several experimental prototypes of low power around the world (France 1 kW, the United States 50 kW, Russia 20 kVA, etc.), major projects began in the early 1970s [TIX91].

Research on low critical temperature superconducting machines continued until the late 1990s with the culmination of the SuperGM project in Japan. From the 2000s, the use of high-temperature superconductive (HTS) tapes for electrical machines began [BAR05]. In Table 14.2, some important low-temperature superconductors (LTS) superconducting machine projects are listed.

When superconductors are used for the armature and the magnetizing windings, the machine is called fully superconducting. The partially superconducting machine refers to a structure where only the magnetizing winding made with superconductors, and the armature winding is made of conventional conductors.

Fully superconducting machines can reach the highest power densities. Indeed, the capacity for superconductors of carrying a high current enables to significantly increase the magnetizing field and the linear current density of the armature. The thickness of the air gap is fixed by mechanical separation, so it is roughly the same as standard electrical machines. In addition, efficiency is also increased, by eliminating DC losses. However, AC losses in superconducting conductors are still an obstacle to the development of these machines today. They are due to the AC currents in the superconductor of the armature but also to the variable electromagnetic environment in a machine. Partially superconducting machines, on the other hand, also have the potential to increase power density over standard electrical machine by reducing or eliminating the use of ferromagnetic materials. The air gap is more important than a standard or fully superconductive machine, due to the thickness of the thermal insulation between the armature and the magnetizing winding. The superconductors are powered by direct current, and the magnetic AC environment can be easily reduced by electromagnetic screens, placed between the superconductors and the conventional conductors. AC losses are therefore less problematic, which explain why the majority of machines built to date are partially superconducting. In this following part, the magnetizing winding is replaced by superconducting bulks and conducts to superconducting machines with an unconventional topology.

Table 14.2 Some realizations of LTS machines

	Country	**Company**	**Power-Speed**	**Information**	**Ref.**
1970–1980	USA	Westinghouse	5 MVA 3600 rpm	Electric power plant	[BLA12]
	USA	GE	20 MVA 3600 rpm	Electric power plant	[KEI85]
1990–2000	JAP		70 MW 3600 rpm	Electric power plant • Project SuperGM • Efficiency: 98.3 %	[YAM99]

14.1.5.2 Superconducting bulk

According to Lenz's law, any conductor subjected to a variable magnetic field will induced electrical currents. These currents generate a magnetic field whose direction is opposite to the applied field. As a superconductor is characterized by almost zero resistance, large currents can be induced. These currents persist even when the applied field is removed, so a current has been trapped in the material. From these properties, magnets or magnetic screens can be created.

14.1.5.3 Bean's model

This model considers the transition from the superconducting state to the normal state to be infinitely sharp. The change appears when the current density is equal to $\pm J_c$.

$$\begin{cases} |\mathbf{J}| < J_c & E = 0 \\ |\mathbf{J}| > J_c & E \neq 0 \end{cases} \tag{14.1}$$

The current density in the superconducting bulk can have two different values: $\pm J_c$ or 0. So when the current enters the sample, the magnetic field is governed by the equation:

$$\nabla \times \boldsymbol{H} = \pm \boldsymbol{J} \tag{14.2}$$

Although limited to simple geometries, the distribution of the current and the magnetic field inside the superconductor can be obtained analytically with the Bean model. It facilitates the understanding of the behavior of the superconductors under a magnetic field. We will take as an example an infinitely long superconducting cylinder of radius R shown in Fig. 14.1, under an AC external magnetic field $B_a(t)$ along the z-axis.

The cylinder is infinitely long in the z-direction, the magnetic problem is therefore invariant along this direction. If we assume that $\mu = \mu_0$ for the superconducting material (type II), then the Maxwell equations combined with the Bean model are written:

$$\nabla \times \boldsymbol{H} = \pm \boldsymbol{J} \tag{14.3}$$

$$\nabla \times \boldsymbol{E} = -\partial \boldsymbol{B} / \partial t \tag{14.4}$$

$$\nabla \cdot \boldsymbol{H} = 0 \tag{14.5}$$

$$\nabla \cdot \boldsymbol{E} = 0 \tag{14.6}$$

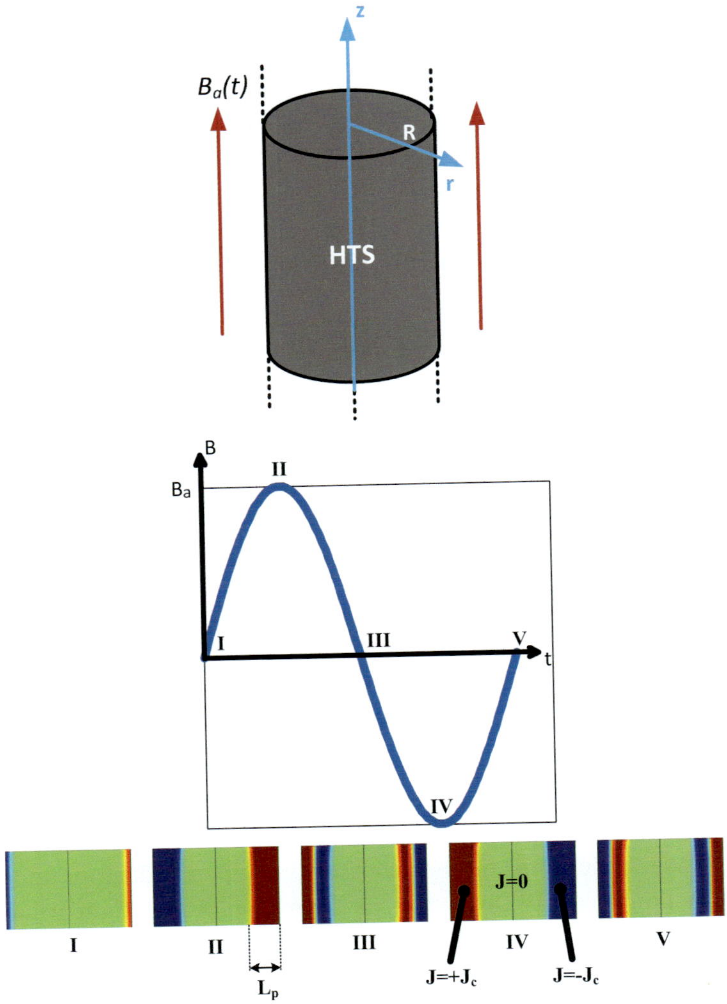

Figure 14.1 Superconducting cylinder subjected to an external field B_a and current distribution in a portion of the cylinder obtained with the Bean model.

Assuming that a sinusoidal is applied field, the current density and the electric field are both oriented along the circumferential direction Θ. The current density (state II in Fig. 14.1) is 0 for $0 < r < L_p$ and $+J_c$ for $R\text{-}L_p < r < R$, where L_p is the penetration length of the current in the cylinder. The problem being one-dimensional, Eqs. 14.3 to 14.6 can be simplified.

$$\frac{1}{r}\frac{\partial rE_\theta}{\partial r} = -\frac{\partial B}{\partial t} \tag{14.7}$$

$$-\frac{\partial B_z}{\partial r} = \pm\mu_0 J \tag{14.8}$$

So, we find:

$$E_\theta(r,t) = -\frac{\partial B}{\partial t}\frac{r^2 - L_p^2(t)}{2r} \tag{14.9}$$

$$B_z(r,t) = -\mu_0 J_c r + B_{a(t)} \tag{14.10}$$

The field in the cylinder decreases linearly until reaching 0 for a depth L_p which is given by:

$$L_p = \frac{B_{max}}{\mu_0 J_c} \tag{14.11}$$

14.1.5.4 Magnetization of superconducting bulk

Permanent magnets have a remnant induction independent of the sample volume. The magnetic field produced by permanent magnets is therefore limited by the properties of the materials. In contrast, superconducting magnets are created from trapped current. The advantage of a superconducting magnet is that the value of the magnetization depends on the amount of current in the material.

From the Bean model presented above, a superconducting magnet is fully magnetized when the penetration length L_p is equal to the radius R of the superconducting bulk. So, the maximum induction that can be trapped in a cylindrical pellet is:

$$B_{trapped} = k\mu_0 J_c R \tag{14.12}$$

where k is a correction coefficient to consider the demagnetizing field due to the finite thickness E_p of the bulk [AIN15]. In the previous example of the infinite cylinder, k was equal to 1, while for a cylinder of finite thickness, we have:

$$k = \frac{E_p}{2R} ln\left(\frac{R}{E_p} + \sqrt{1 + \left(\frac{R}{E_p}\right)^2}\right) \tag{14.13}$$

The quality of the material and the operating temperature determine the critical current density and therefore the maximum field that can be trapped.

A lot of research has been carried out on a possible demagnetization of the superconductor once exposed to an AC magnetic disturbance. This is the case in an electric machine where the superconducting magnets are submitted to the harmonic field produced by the armature windings. In [BER16], a bulk is submitted to a sinusoidal magnetic field of a few hundred mT. Tests conducted for several frequencies show a decrease of less than 1%. However, the magnetic environment of an electric motor is more complex than what is considered in these works. For example, the pulse width modulation (PWM) power supply is composed of a bunch of harmonics with frequencies above kHz. In [SMA19], demagnetization is tested on stacks of magnetized tapes. This time, the magnets are used in a machine environment. The decrease in magnetization at 77 K is greater and reaches almost 30%. The use of superconducting magnets for electric machines is therefore attractive due to the high value of the possible magnetization, if the demagnetization is controlled. There are two methods of magnetization which will not be developed in this chapter, because they are already well developed in different books [AIN15]:

1. Field cooling (FC)
2. Pulse field magnetization (PFM)

14.1.5.5 Superconducting screen

The diamagnetic response of superconductors is a reaction, caused by Lenz's law, to a variable field. The induced currents developed are large and are distributed in order to keep the initial magnetic state in the unpenetrated zone of the superconducting bulk.

Numerical modeling, see Fig. 14.2, shows what happens when the magnetic field applied to a bulk superconductor is increased from a zero value up to 1 T. This current is distributed so as to maintain the initial field inside, here 0. We also note that the magnetic field outside the bulk is influenced by this current. Thus, the magnetic field is reduced when $r < R$ while it is increased up to 2 T for $r > R$.

This property is exploited to create magnetic shields for current limiters [LEV02, LEV09], magnetic lenses [PRI18] or even electric machine as it will be presented. One advantage of using the superconducting screen in electrical machine rather than the superconducting magnet is the reduction of AC losses under harmonic fields because *Ic* is generally greater in this condition (less current induced in the superconducting bulk).

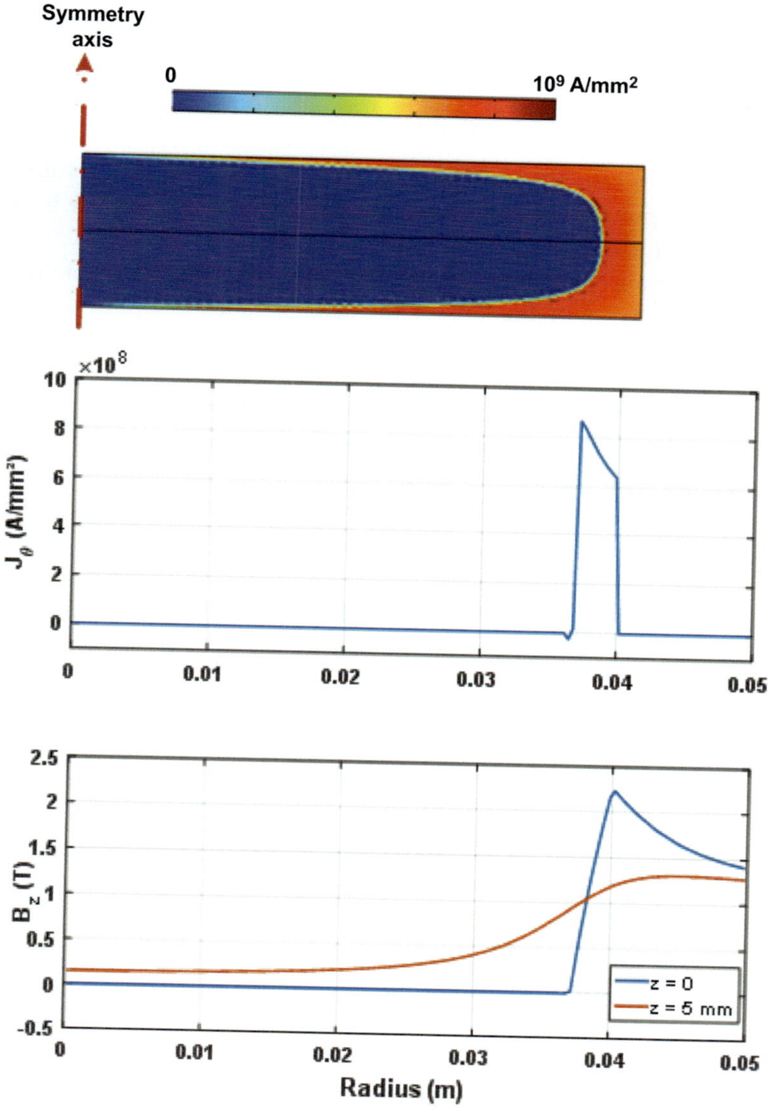

Figure 14.2 Diamagnetic response of a superconducting bulk subjected to a magnetic field ramp. The magnetic flux inside and 5 mm above the superconducting bulk is shown.

14.1.5.6 Topology of superconducting machine using bulk

For many projects, the topology is identical to standard electric machines. In this part, we present the possibilities of realization of machines using superconducting bulk.

Superconducting permanent magnet

This type of structure has a very strong interest for applications with high-power density, by the possibility of magnetizing the bulk superconductors at values well above the permanent magnets.

The main difficulty results in the magnetization of the superconductors in situ. So, for the sake of space, the pulsed field technique is preferred in many cases. However, temperature rises during magnetic field pulses are very detrimental to the properties of superconductors. It is for this reason that today the field trapped with this technique is hardly better than the use of permanent magnets.

Table 14.3 lists some machines with superconductive magnets manufactured and Fig. 14.3 shows some realizations

Table 14.3 Some realizations of machines with superconductive magnets

Country	Institution	Power speed	Superconductor	Ref.
JAP		0.94 kW	YBCO	[ITO95]
JAP	UMST	10 kW 850 rpm 16 kW 760 rpm	GdBCO	[MIK06]
JAP	CJR	1.5 kW 600 rpm	YBCO	[HIR03]
JAP	UMST	3.1 kW 700 rpm	GdBCO	[MAT05] [MIK06]
UK	Cambridge University	7.8 kW 1500 rpm	YBCO	[XIA11]
JAP		30 kW 720 rpm	GdBCO	[MIK10]
JAP	TUMSAT	30 kW 190 rpm	YBCO	[SHA00]

Flux modulation machine

These machines have been developed and realized by the GREEN laboratory in Nancy and used the diamagnetic response behavior.

The inductor of a flux modulation machine is shown in Fig. 14.3. In this figure, two coaxial superconductive solenoids are supplied with opposite currents. Thus, the main path of the magnetic flux density is along the radial component between the solenoids, and superconducting bulks are then placed. The superconducting bulk behavior can be seen as a magnetic screen, where the magnetic flux density is diverted by these screens and distributed between it.

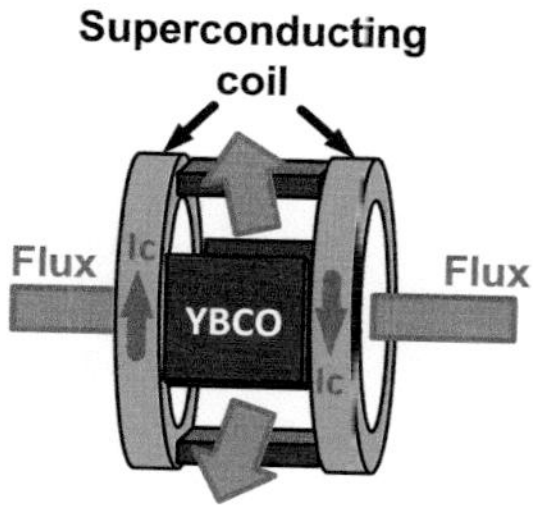

Figure 14.3 Principle of the flux-modulation inductor.

The variation of the magnetic flux density observed in the air gap, is therefore an alternation between a strong magnetic field between the pellets and a weak magnetic field behind a bulk, see Fig. 14.4. This distribution is special because it is not like a standard inductor where the magnetic field changes between a positive and negative value.

Flux barrier machine

The flux barrier machine also uses the diamagnetic response of the superconductors. The inductor is shown in Fig. 14.5. Here, the two superconductive solenoids are fed by currents in the same direction. The variation of the magnetic field in the air gap is created by the insertion of one superconducting bulk between the solenoids, as shown in Fig. 14.5.

The advantage of this structure is the creation of a change the magnetic flux density close to a sinusoid. However, it is difficult to imagine a machine with more than one pair of poles, and high-power machine design requires large screens.

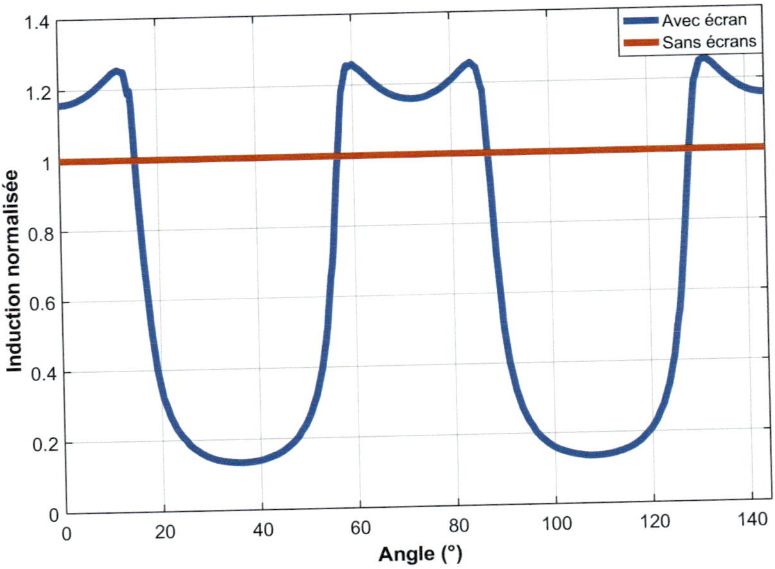

Figure 14.4 Distribution of magnetic flux in the air gap [NET07].

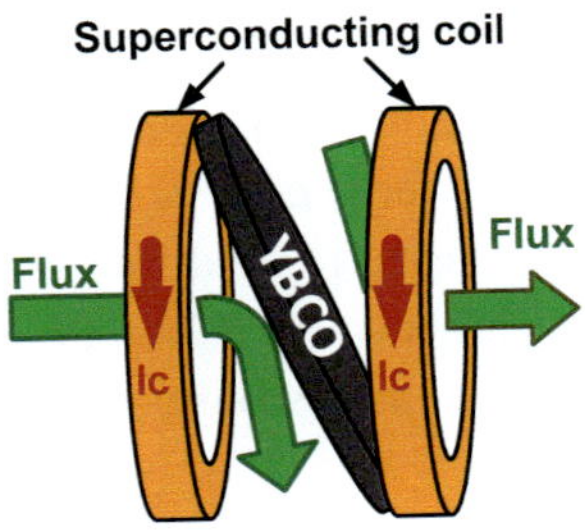

Figure 14.5 Principle of the flux barrier inductor [ALH16].

14.2 Sizing of a Superconducting Motor

The comparison between electrical machines is made using the specific torque or the volume torque. Indeed, the torque is a quantity which is determined by the magnetic interactions in the air gap and the volume of the machine. On the other hand, the power depends on the speed of rotation which is fixed by the supply frequency of the armature.

In the aeronautics field, the competitiveness between technologies is evaluated in terms of specific power. Two applications are studied according to their rotation speed. The first application concerns the distributed propulsion of the aircraft which today seems to be the most promising architecture. Several electric motors are arranged on the wings of the aircraft. The power of these motors is between 1 and 2 MW with a rotation speed of 2,500 to 5,000 rpm. The second application is electrical generation with a power between 5 and 10 MW for rotation speeds of 10,000 to 40,000 rpm. Rotational speeds are higher because these generators are connected to high-speed turbines.

Whatever the application, a power-to-mass ratio of 10 kW/kg is a target announced by NASA for all-electric aircraft. Our study relates to the first category of application, namely a superconducting motor dedicated to propulsion. With the actual performance of superconductors, the cooling temperature is set around 30 K. A lower temperature would improve the electrical properties of the superconductors but also increase the cost of the cooling system.

14.2.1 Specifications

The specifications of our application are given in Table 14.4.

Table 14.4 Specification of the superconducting motor

Power (kW)	50	
Speed (rpm)	5000	
Operating temperature (K)	30	
Current density in armature (A/mm^2)	Liquid: 10 [PYR13]	Air: 5 [PYR13]

The specifications impose also two restrictions on the topology of the machine:

1. **No rotating power supply system** The rotating power system is a device able to transfer the current between a stationary electrical source and several coils which rotate with the rotor of an electrical machine. This system is prohibited in our application for maintenance reasons.

The other solution is electromechanical, it is a rotating transformer. A fixed primary winding induces a voltage in a rotating secondary winding. The induced voltage is then rectified with diodes to supply the inductor. For our project, the rotary transformer is excluded for reasons of weight, because it constitutes to an additional module.

2. **Possibility to disable the excitation** The second criterion is the possibility of cutting the power supply in the event of a short circuit in the machine, i.e., rapid demagnetization

   A superconducting magnet keeps its magnetization as long as it is cooled. There is therefore no way to cancel the field of induction apart from heating the magnets. Unfortunately, the thermal time constant is often long and does not allow for a rapid system reaction. So, we chose to study a flux modulation structure. With this topology, if the supply of the superconducting coils is stopped, the field of induction is cancelled because the pellets are not magnetized. Of course, the superconducting coils need to be fixed in order to avoid the rotating power supply. The structure of such a machine is developed in the next section.

14.2.2 Structure of the Machine

We will recall the components that constitute the inductor.

- Superconducting coil

  These elements generate a large magnetic field. As a reminder, it should not rotate. In order to create an electromagnetic torque, the magnetic flux density in the air-gap has to change; this is why the following components are added.
- Superconducting bulk

  When the magnetic field increases during the supply of the superconducting coil, currents are induced on the periphery of the screens by Lenz's law. These elements rotate and constitute the rotor of the machine.

The two associated components form the superconducting inductor. This behavior can be used in either an axial machine or a radial machine as shown in Fig. 14.6.

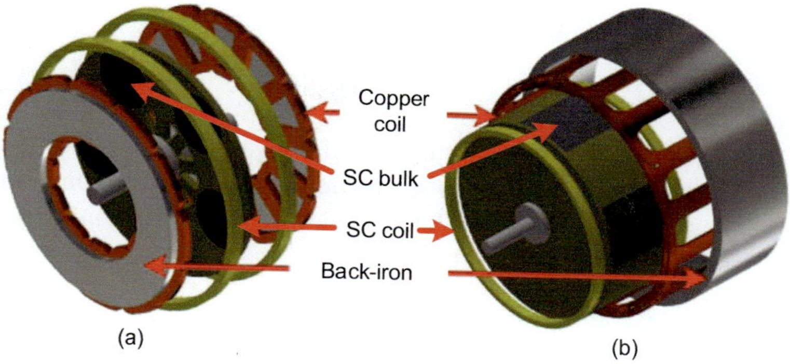

Figure 14.6 Exploded view of the active elements of a flow machine: (a) axial, (b) radial.

14.2.3 Sizing

14.2.3.1 Polarity of the motor

The first step will be to determine the polarity of the electrical machine. The electromagnetic power delivered by a three-phase electric generator is written:

$$P = 3EI \tag{14.13}$$

where E is the electromotive force generated by the rotation of the superconducting bulk and I the current flowing in the coils of the armature. A comparison criterion K based on a simplification of E is introduced below. The electromotive force depends on the speed of rotation Ω, on the number of turns N of the armature, on the polarity p, and on the shape of the flux under a pole. Assuming, a sinusoidal variation of the flux, E is given by:

$$E = \frac{d\Phi}{dt} = pN\Omega\hat{\varphi}\cos p\Theta \tag{14.14}$$

We introduce a criterion K, which does not consider:

- The number of turns which is a data item resulting from the sizing of the armature
- The rotation speeds.

However, the mass of the active elements M of the inductor must appear because the objective is to find the topology with the most significant specific power.

The unit of K is Wb/kg and its maximum value is written:

$$E = \frac{E}{MN\Omega} = \frac{p\widehat{\varphi}}{M} \tag{14.14}$$

A parametric study as a function of the geometry of the motor shows the variation of K as a function of the polarity, and the maximum value is obtained for 10 poles, which corresponds to 5 superconducting bulks.

14.2.3.2 General relationship for design

In this study, a double armature axial machine is chosen. This step is necessary in order to estimate the geometric dimensions of the target machine (50 kW, 100 Nm and 5000 rpm). For this sizing, an analytical formula is firstly used to calculate the electric power of an axial machine [PAR05].

$$P = \frac{4}{3}\frac{\sqrt{2}}{\sqrt{3}}\pi\widehat{B_1}K_1R\Omega \tag{14.15}$$

where B_1 is the maximum value of the fundamental of the axial magnetic induction in the air gap and K_1 is the effective surface current density of the armature both produced in the air gap. Finally, R and Ω represent the external radius of the armature and the speed of rotation of the rotor. In the literature, the external radius of the armature, R is worth approximately 0.15 m for machines with axial permanent magnets with high-power density whose torque value is around 100 Nm [LAM15, EVO, PHI]. This value is kept as a starting point and will be modified later with the performance of the superconducting machine.

14.2.3.3 Calculus of inductor field

It is possible to use the superconducting coil to generate a very large magnetic field in the air gap. It conducts to reduce the volume of the machine but at the expense of the weight of the machine because a large quantity of superconducting wire is necessary for the magnetizing coil. Based on work on superconducting machines for aeronautics, we note that the induction in the air gap is generally less than 1 T [FEL11, TER18]. We choose to carry out a study at constant power P, the value of K_1 in expression (2–8) is adjusted in order to maintain P = 50 kW. Figure 14.7 shows the normalized specific power curve. This figure shows an interesting result, the machine has a better specific power when the variation B_1 is 0.6 T.

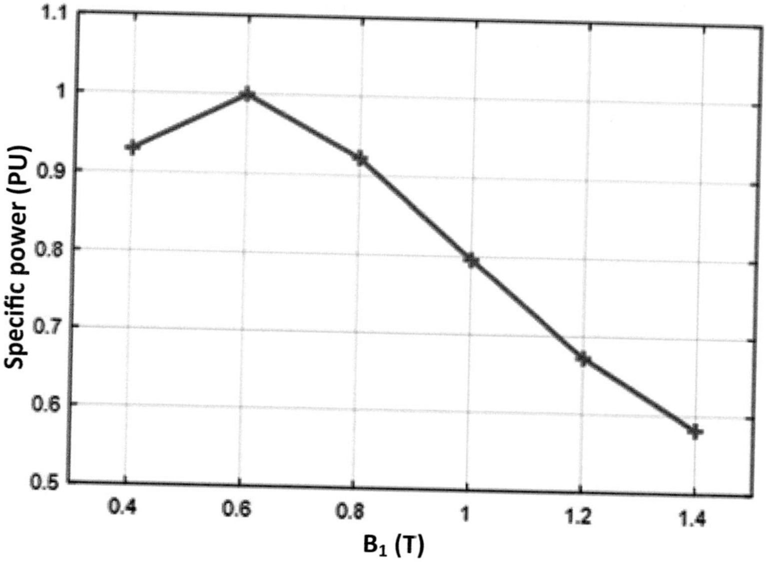

Figure 14.7 Electromagnetic mass power depending on the level of induction.

14.2.3.4 Calculus of armature

In the literature, values of linear current density can be found for standard electrical machines [PYR13, PAR05]. However, these values are only applicable in the presence of a magnetic circuit. For this superconducting machine, the choice was made to remove the ferromagnetic teeth always in the logic of reducing the weight. In order to achieve high-power density, let's set the limit of a superconducting machine without ferromagnetic material. Thus, the winding of the armature is immersed in air.

Considering the whole three-phase winding, the expression of the fundamental of K_1 is developed for each active part of the winding where H_n is the magnetic field produced by one active part and obtained simply with the Ampere law:

$$K_1 = \sum_{n=1}^{6} H_n cos\left(p\frac{2\pi}{X}\left(x-(n-1)\frac{X}{6}\right)\right) \qquad (14.15)$$

And

$$H_{1n} = \int_{(2n-3)x/12p}^{(2n-1)x/12p} H_{\max}\cos\left(\frac{2\pi}{2d}\left(x-(n-1)\frac{x}{6}\right)\right)$$
$$\cos\left(p\frac{2\pi}{X}\left(x-(n-1)\frac{x}{6}\right)\right)dx \qquad (14.16)$$

The sinusoidal distribution can be approached with winding techniques such as the distribution of the coils, polar shortenings.

As the choice of the type of armature is not constrained by the AC losses in the pellets, as we will present later, we will choose the lightest. This is obviously concentric winding due to the shortening of the coil heads. A 45% reduction in copper mass can be obtained between a distributed and concentric topology

As for most of the partially superconducting machines studied, the armature is made with a copper conductor. We will recall several points which were previously presented:

The optimal polarity of the machine is 5 pairs of poles and the concentric-type winding has been chosen. With this polarity, a structure with 12 copper coils is the best choice for a concentric winding, it will reach the best performance.

There are no ferromagnetic teeth. Therefore, a Litz wire must be used and the losses are evaluated with analytical equation. This Litz wire is type 8, a rectangular conductor made with cylindrical filaments. This shape choice allows a more compact winding and therefore more efficient cooling.

The analytical solution of losses AC in a cylindrical filament of radius R, of length l, of resistivity ρ and subjected to a maximum magnetic induction B sinusoidal of frequency f is given by:

$$P_{cf} = \frac{l(\pi f)^2 (2R)^4 B^2}{8\rho} \qquad (14.17)$$

In order to increase the accuracy of the calculation, we will consider the complex magnetic environment of an electric machine. An extension of these results is described in [WAN04]. It is necessary to obtain the Fourier series decomposition of the field seen by the superconducting bulks. Thus, the formula (14.17) becomes:

$$P_{tot} = \frac{l(\pi f_1)^2 R^4}{8\rho s}\sum_{k=1}^{s}\sum_{j=1}^{l} n_j \sum_{i=1}^{m} i^2 \left|B_{ijk}\right|^2 \qquad (14.18)$$

where m is the number of harmonics, l is the number of layers of filaments in the axial direction and s is the number of radial sections to consider the 3D effects. Figure 14.8 shows the eddy current losses as a function of the number of filaments that make up the Litz wire. The number of filaments fixes the radius R in expression (14.18).

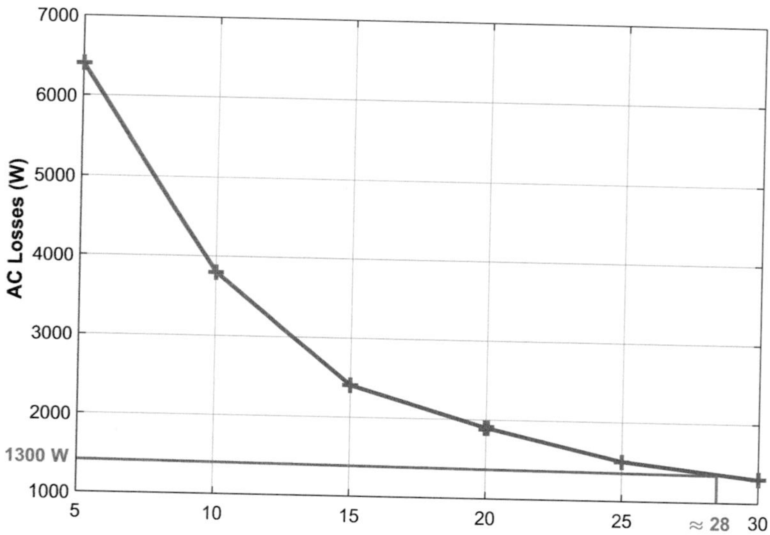

Figure 14.8 Estimation of eddy current losses as a function of the number of filaments.

The Joules losses of the prototype are estimated at 1200 W. To guarantee an efficiency greater than 95%, it is necessary that the AC losses are less than 1300 W. This condition is fulfilled for a number of filaments greater than 28.

Finally, a thin weak magnetic yoke is added and glued on the back of the armature. An alloy of iron and silicon at 3% was chosen for the prototype. The thickness of the cylinder head is 5 mm.

14.2.3.5 AC losses in superconducting bulk

Here we will justify the hypothesis of negligible losses in superconducting bulk. The armature of an electric machine creates rotating field. Ideally, this rotating field has a sinusoidal spatial distribution and therefore no harmonics are present. In reality, this is not the case. The rotating field is made up of different harmonics with different rotating speeds than the rotor speed. The superconducting bulks exhibit AC losses due to this harmonic field.

To model and calculate them, a 2D axisymmetric digital model is developed. The applied field is shown in Fig. 14.9. The rise of the magnetic field by the superconducting coil must be slow in order to minimize the AC losses during the supply. For the simulation, a rise time of 60 s was chosen.

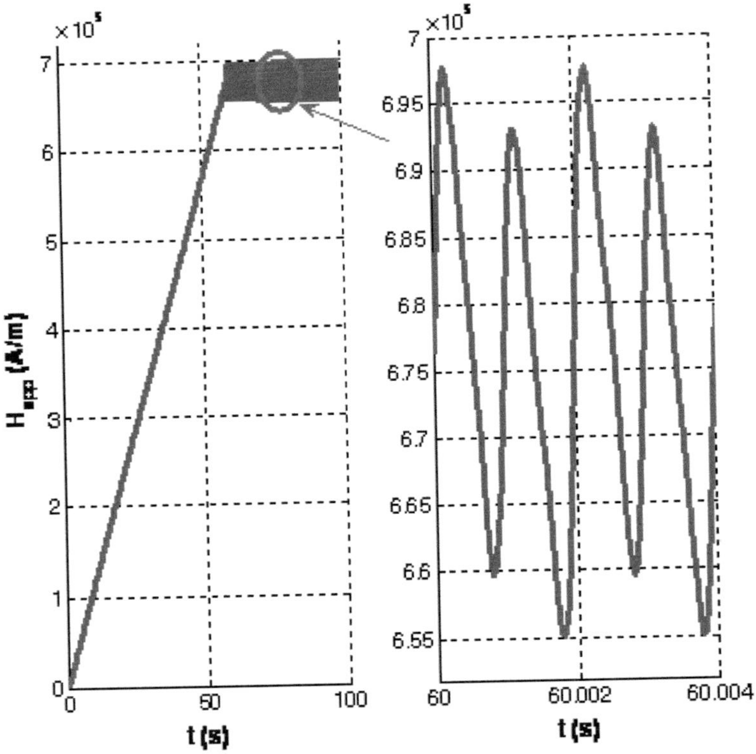

Figure 14.9 Magnetic field applied on the bulk in nominal operation.

The current density and the electric field in the bulk have only an azimuth component. The instantaneous losses throughout the volume of the bulk are given by:

$$P = 2\pi \iint E_\theta(r,z) J_\theta(r,z) r dr dz \quad (14.19)$$

Expression (14.19) must be averaged to obtain the value of the hysteresis losses of the pellet. The value must be calculated after a few periods to avoid transient phenomena as shown in Fig. 14.10.

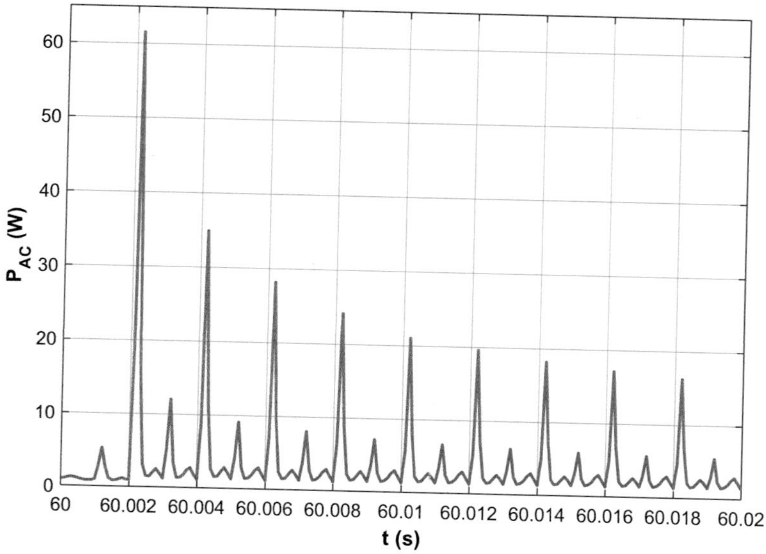

Figure 14.10 Transient phenomenon of AC losses for 10 periods.

From Fig. 14.10, the losses stabilize after 10 periods.

$$P_{\mathrm{AC}} = \frac{1}{T}\int p(t)dt \tag{14.20}$$

The average value over the last period for all the pastilles is 3.3 W. This loss value is low enough to be neglected

14.2.4 Optimization

The aim of this optimization is to extrapolate our results to higher power machines. A total of five different powers (50, 100, 200, 500, and 1000 kW) will be optimized.

14.2.4.1 Considering only active element

The first curves in Fig. 14.11 show the specific power of the machines, considering only the active parts (coil and HTC screen, armature coils) with the two generations of superconductive tape (BSCCO and YBCO). This is not realistic because a superconductive machine necessarily needs cryostats, insulating supports, or thermal conductors. But this curve is intended as a limit toward which we can tend.

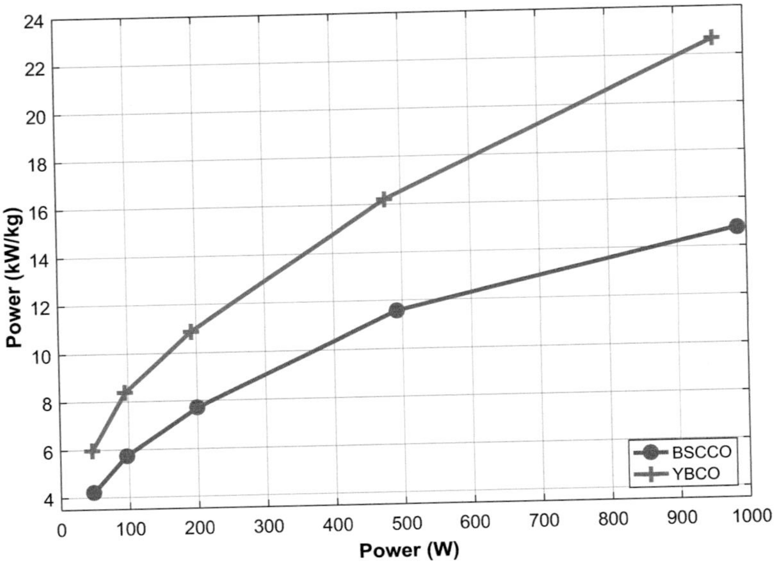

Figure 14.11 Specific power for a flux modulated superconducting machine.

As shown in Fig. 14.11; the specific power is better with a superconducting coil made from a second-generation tape. At 1 MW, the limit of a YBCO machine is more than 23 kW/kg and a BSCCO machine is more than 14.5 kW/kg. Obviously, the proposed results are obtained with the superconducting materials available today. There is significant margin for improvement of these tapes, especially the second generation.

For each study carried out, a trend curve is proposed in order to estimate the specific power of a machine that would not be included in the chosen scan. Thus, the power density for a BSCCO machine is given by the equation:

$$P_{mu} = 0.881\ P^{0.4108} \tag{14.21}$$

And using YBCO tape, we have the following expression:

$$P_{mu} = 1.071\ P^{0.4434} \tag{14.22}$$

where P_{mu} is the specific power in kW/kg and P power in kW.

14.2.4.2 Considering the whole machine

A full mass balance is now used. Figure 14.12 shows the estimated specific power. It can be seen that the weight of the non-active elements has a significant impact on the total mass of the machine. The term "estimated" is used to characterize the specific power because the real mass of the machine can only be given by completely drawing the machine and by considering the weight of the screws, bearings and other elements which cannot be included in this first study. However, the heaviest non-active elements were considered in the mass balance.

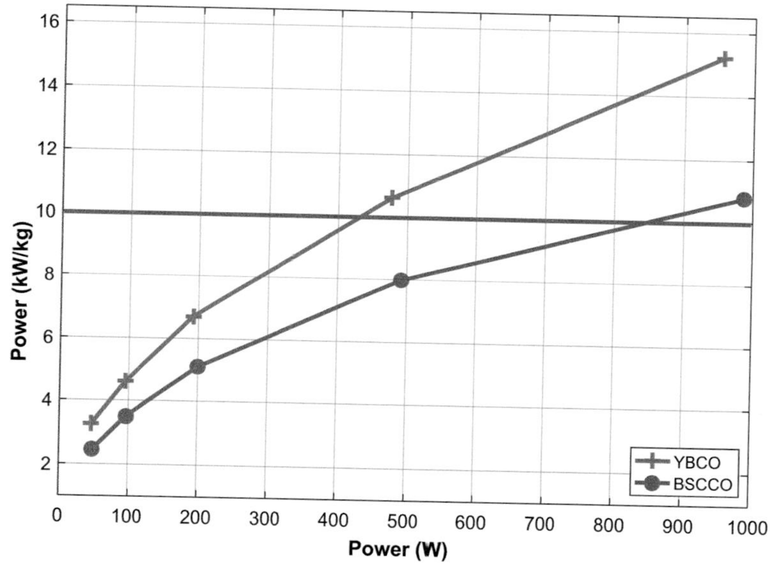

Figure 14.12 Estimated specific power for a superconducting machine with flux modulation.

A green line at 10 kW/kg has been added to the figure to remind the needs of the N3-X project carried out by NASA for electric motors or generators. The YBCO and BSCCO technologies cross this line respectively from 450 kW and 850 kW.

At 1 MW, the estimated specific power of a YBCO machine is around 16 kW/kg and for a BSCCO machine around 11 kW/kg.

The trend curve of the estimated specific power of a BSCCO machine is given by:

$$P_{me} = 0.399\ P^{0.4795} \tag{14.23}$$

Using YBCO tape, we obtain:

$$P_{me} = 0.4425\ P^{0.5154} \tag{14.24}$$

where P_{me} is the specific power in kW/kg and P the power in kW.

14.2.4.3 Superconducting machine and cooling system

For this third scenario, the weight of the cooling system M is integrated into the optimization process from the trend curve (14.25) based on the existing cooling systems [BRA02]:

$$M = 157e^{-0.0533\,T} P_{th}^{0.009\,T+0.1275} \tag{14.25}$$

Where T is the cooling temperature and the P_{th} the heat losses.

This relationship between mass, heat losses, and temperature simplify the optimization problem because we only work with one goal which is the minimization of the total mass.

We will investigate whether there exists an optimal temperature for which the superconducting system is the lightest. For this, two machine configurations are analyzed:

A 50 kW BSCCO machine

A 1 MW YBCO machine

Figure 14.13a shows the mass of the components of the 50 kW BSCCO system. A temperature of 66 K seems to be the optimal temperature. At this temperature the total mass of the superconducting system is 86 kg. Between a 30 K operation and a 66 K operation, the weight of the cooler system is divided by three and the total mass by about two.

Figure 14.13b shows the same curves but for the YBCO system and a 1 MW machine. This time, the optimal temperature is 50 K. At this temperature the total mass of the superconducting system is 166 kg, less than twice as heavy as a 50-kW superconducting system. The more powerful the machine, the lower this temperature will be. Indeed, as the mass of the cooler changes little depending on the application, there will come a point where the mass of the machine will be more substantial and therefore an operation at low temperature will allow to gain the specific power.

These results are once again given from current technologies of superconductors and coolers. They emphasize the importance of considering the cooler. It is obvious that with a lighter cooling as announced by NASA, the operating temperature would be much lower than the results obtained.

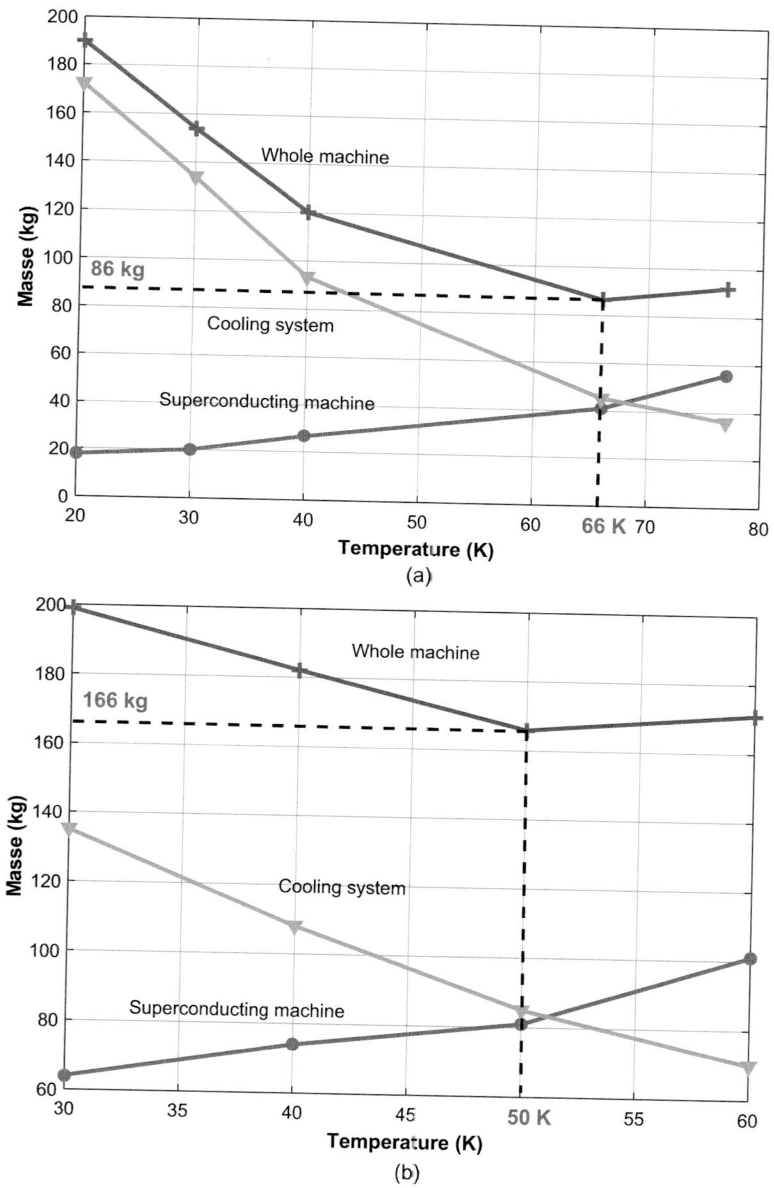

Figure 14.13 Optimal temperature for the machine: (a) BSCCO of 50 kW, (b) YBCO of 1 MW.

14.2.4.4 Improvement margin for high-power machines

It is obvious that a partially superconducting machine can be competitive only if its armature is efficient. This requires real thermal sizing of the armature which was not carried out in this work, making it possible to increase the current densities in the copper. The above results were obtained with a current density in copper of 10 A/mm^2. This value corresponds to a low threshold that can be reached with liquid cooling. In the literature, much higher current density values can be achieved. By opting for cooling by circulation of water in a pipe, the current density varies between 10 and 15 A/mm^2 or even 30 A/mm^2 when the coolant is in direct contact with the active parts [GIE13].

14.2.4.5 Comparison with conventional technology

Generally, electrical machines are not designed to be as light as possible. Industrial motors (pumps, fans, compressor, etc.) are developed to be as reliable as possible and sometimes under severe conditions. The specific powers of these machines are in the range of 0.1–0.5 kW/kg.

More precise sizing techniques are used in the automotive, where the criterion is price reduction. The large-scale production of this type of machine encourages economic growth. Despite this, the mass powers of automobile machines are greater than for the first category, between 1 and 3 kW/kg.

However, the target specific powers for the electric propulsion of aircraft are even higher, as a reminder 10 kW/kg or more. Figure 14.14 compares the YBCO superconducting machine for a current density in the armature of 10 and 15 A/mm^2 compared to standard electrical machines dedicated to aeronautics [ZHA16]. For the comparison to be “fair,” the speed of rotation of the selected machines is between 2500 and 6000 rpm.

One of the benefits of using superconducting materials is the removal of ferromagnetic materials. In the mass distribution of an electric machine in the 2010 Toyota Prius, the contribution of the iron parts amounts to almost 50% of the total weight of the machine. For low powers even in the absence of ferromagnetic material, the weight of cryostats or thermal supports and other parts specific to cryogenic electrical machines are penalizing. On the other hand, from 250 kW, the mass power of the superconducting machine is greater than 33% or 83%, depending on the current density in the armature, compared to standard technologies. The following figure shows a comparison of the different machine types.

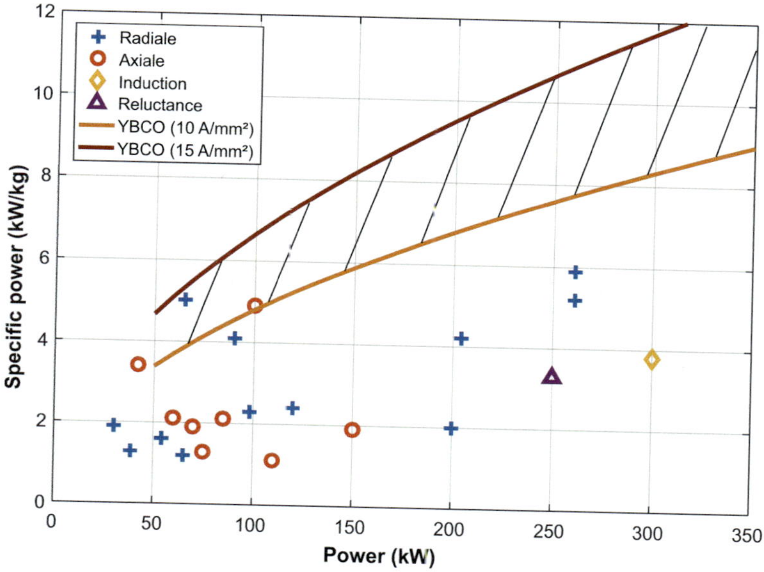

Figure 14.14 Specific power between a superconducting machine with flux modulation and standard electric machines.

14.3 Realization of the Motor

This part is devoted to the construction of the prototype of a superconducting machine with axial flux modulation. The size of the actuator is the following:

One objective was to reduce the length of the electromagnetic air gap. The electromagnetic air gap of a partially superconductive machine is made up of different layers which provide thermal insulation, vacuum, and mechanical air gap. Two types of configuration are shown in Fig. 14.15: the first with a rotating cryostat and the second with a fixed cryostat. In the latter case, the electromagnetic air gap is composed of two layers: a vacuum thickness for thermal insulation and ensuring the rotation of the rotor and a fixed cryostat for maintaining the vacuum.

Table 14.5 Specification of the superconducting motor

	Values
Polarity of the motor	5
Number of coils of the armature	2*12
Current in the superconducting coil	230 A
Linear current density of the armature	2500 A/m
External diameter of the motor	400 mm
Length of the motor	160 mm

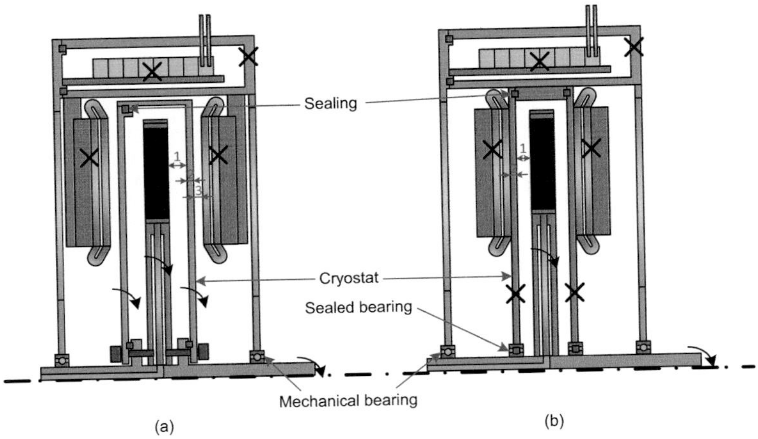

Figure 14.15 Superconductive machine with (a) a rotating cryostat and (b) a fixed cryostat.

The advantage is the reduction of the air gap. This reduction makes it possible to increase the torque produced by the superconductive machine because the modulation of the magnetic field is better near the pellets and decreases rapidly as it moves away.

However, certain precautions must be taken when choosing the material for the cryostat. As this cryostat is fixed, the variable magnetic field seen by the cryostat is produced by the rotation of the screens. The magnetic field has an important value, so the losses and the braking torque can be significant without any precautions. Therefore, we carried out a study on the type of material (insulator or electrical conductor) that could be used for an application with a fixed cryostat.

14.3.1 Cooling System

A helium loop, shown in Fig. 14.16, is used to cool the superconducting machine. We used a Gifford-McMahon AL325-type cold head powered by a CP1110 compressor. The cooling power of this set is more than 100 W @ 30 K.

For this cooling system, the cold head is not directly in contact with the application. There is a helium circuit that circulates through a transfer line using a CP830 compressor. The compressor in this second circuit circulates helium at room temperature while the application needs cold helium, cooled by contact with the cold head. A thermal decoupling is therefore necessary, which is achieved with a copper exchanger.

The cold head and exchanger are placed in a double wall cryostat which is filled with liquid nitrogen in order to reduce radiation losses. The vacuum of the application and the cryostat of the head are not communicating, it is therefore necessary to use 2 vacuum pumps.

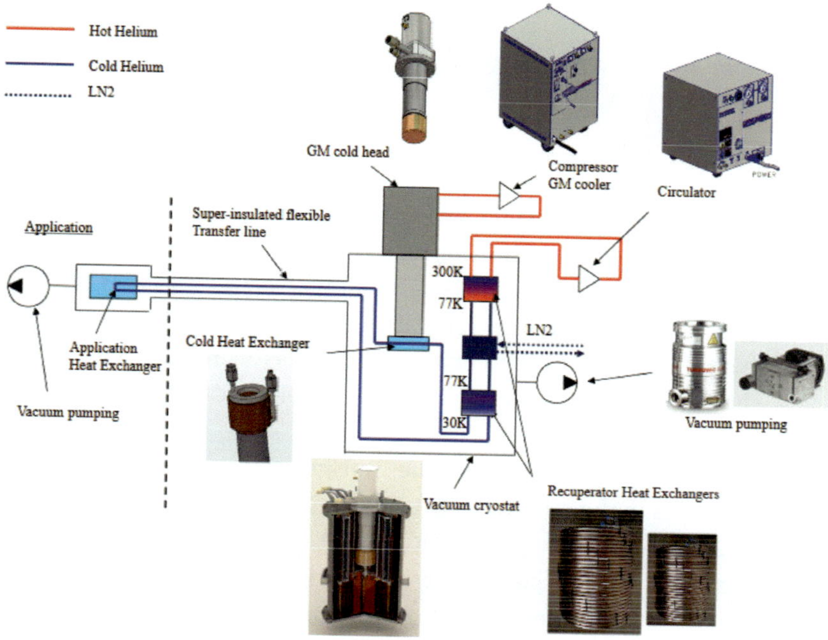

Figure 14.16 Helium loop used for cooling the superconducting machine [TRO14].

The cooling power is measured at the application, after 4 m from the transfer line. This power, logically lower than that which a cold head can have, is shown in Fig. 14.17. With this system, 80 W @ 30 K is available. Using the analytical formulas for calculating heat losses in the designed motor, they are estimated at 30 W for the operating temperature. The system is therefore more than sufficient to bring the machine to a temperature of 30 K.

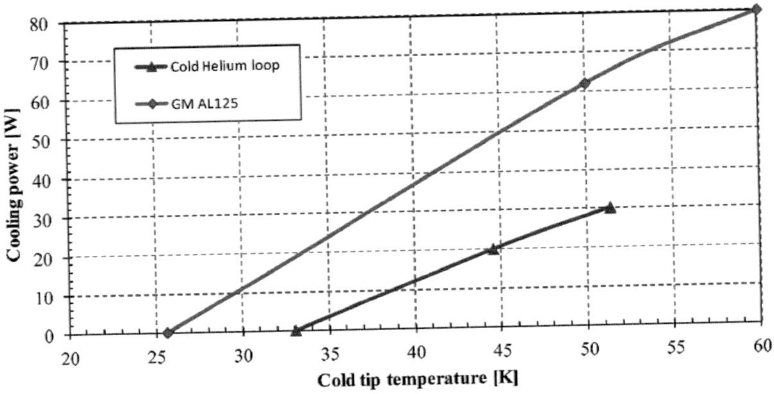

Figure 14.17 Cooling capacity of the cold head only and of the circulator used.

14.3.2 Superconducting Coil

Based on the dimensions given in Table 14.5, the superconducting electromagnet is made. There are 6 double-solenoids of 75 turns made with DI-BSCCO-type H tape manufactured by SUMITOMO. The length of each double-solenoid is 155 m. A 930 m of BSCCO tape was used. All solenoids are connected in series using a copper bridge shown in Fig. 14.18. These junctions are electrically insulated using Kapton-type tape.

Helium gas cooled by the refrigeration system enters and circulates in the cryostat of the HTC coil via a copper pipe, as shown in Fig. 14.18. This pipe is in contact with a thermal bus which is a solid copper piece. The cooling is therefore transmitted by conduction. In order to homogenize the temperature of the coil as much as possible, copper strips are inserted between each double-solenoids. These elements are in contact with the thermal bus. Then, several layers of multi layers insulation (MLI) are added around the whole to reduce radiation losses as we can see in Fig. 14.19.

Figure 14.18 Superconducting coil assembly.

Figure 14.19 PWM layer around the superconducting coil.

Figure 14.20 shows the HTC coil on its support in G10. The cryostat for the superconducting coil, made of aluminum, is shown in Fig. 14.20. The cryostat also constitutes the housing of the superconducting machine, this is the reason why we see flanges to hold and stabilize the rotor of the machine.

The mass of this part is detailed in Table 14.6. Important safety margins have been taken for the cryostat. A consequent reduction in mass can be made on the flanges of the cryostat. The weight of the HTC coil is 12 kg, the weight of its copper cooling circuit is included.

(a) (b)

Figure 14.20 (a) HTC coil on its support. (b) Assembled assembly.

Table 14.6 Balance of masses

	Mass (kg)
Aluminum cryostat	*8*
Two aluminum flanges	*10.8*
Superconducting coil	*12*
Coil support	*1*
Total weight	*32.8*

14.3.3 Rotating Part

The material chosen for the screens is YBCO. The radius of the pellets is 40 mm and thickness 10 mm. These pellets are made from 4 seeds.

The superconducting bulk are inserted in a copper rings shown in Fig. 14.22. The thermal contact is ensured by tightening these rings. Tightening at room temperature should be moderate to avoid damage to the copper rings but enough for the cooling. In addition, during cooling down everything shrinks and the contraction of the copper is greater than that of the YBCO pellets which will increase the tightening.

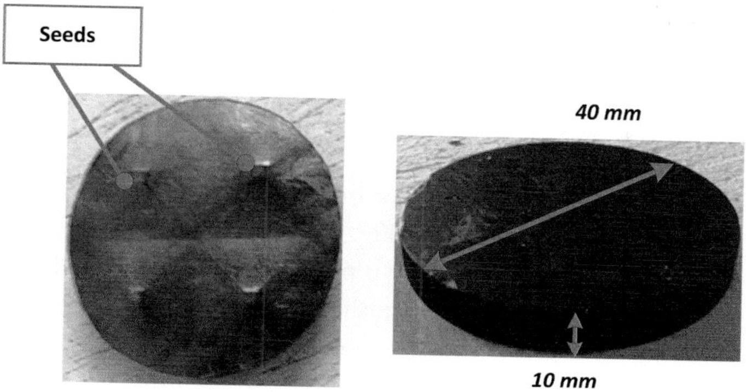

Figure 14.21 Shows the appearance of the bulk before and after polishing.

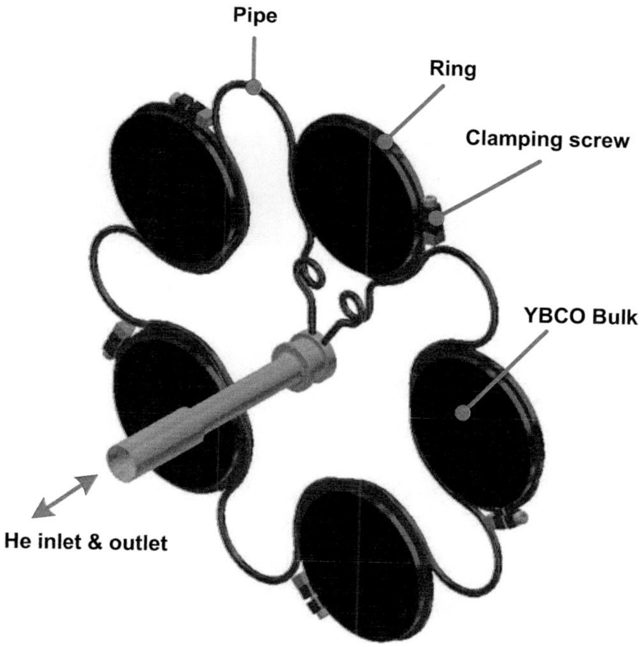

Figure 14.22 Superconducting screen cooling.

A copper pipe in which the cold helium flows is brazed on the different rings. The cooling of the pellets is therefore done in series, which does not guarantee an equivalent temperature of the pellets because the gas heats up along its path. However, the temperature rise is very low so the performance of the different superconducting bulks should not change.

The entry and exit of helium gas in copper pipes is shown in Fig. 14.23. The complexity of cooling is to ensure continuity of circulation between the fixed rod (olive green color) and the rotor (grey color). For this, the helium entering in a pipeline is directed to the end of the fixed rod, see Fig. 14.23a. A Parker seal forces the gas to flow only into the copper pipe. The helium outlet, illustrated in Fig. 14.23b, takes place on another plane. A small ball bearing is added to the end of the fixed rod to guide it. The bearing must be degreased so as not to freeze on contact with helium gas.

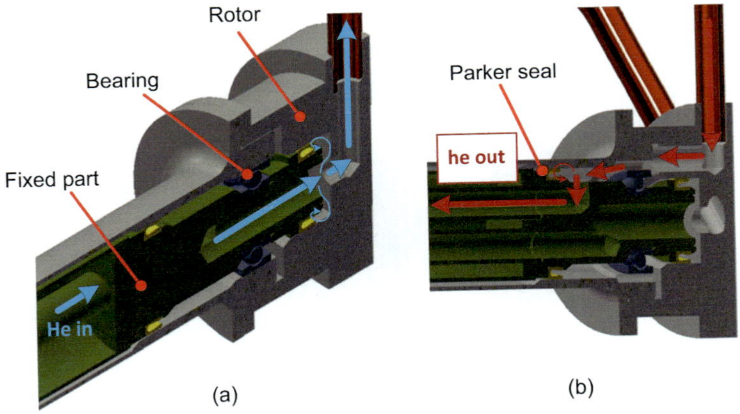

Figure 14.23 Connection between fixed rod and rotor: (a) for entry, (b) for exit.

The support makes the connection between the HTC screens and the mechanical shaft in order to transmit the torque. It is made up of two elements, as we can see on Fig. 14.24:

- A G10 wheel in which the superconducting pellets are housed.
- A reinforcement in Ti6Al4V which is a titanium alloy with a low thermal conductivity for a metal. A complex form "honeycomb" has been used to reduce conduction losses.

The mechanical shaft is finally fixed on the heart of the support, see Fig. 14.25. Two sets of bearings are added:

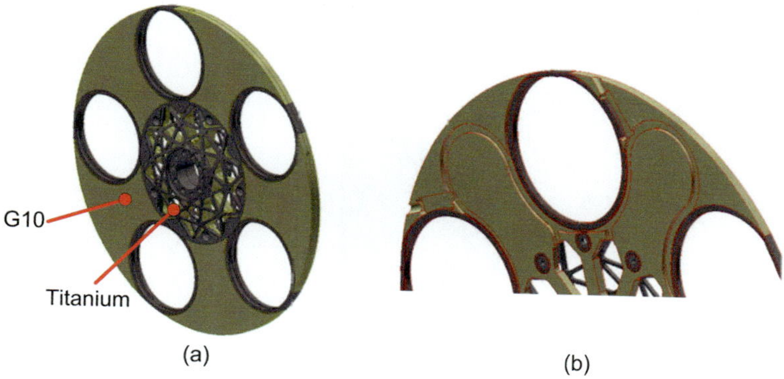

Figure 14.24 Support for superconducting bulk: (a) whole, (b) half.

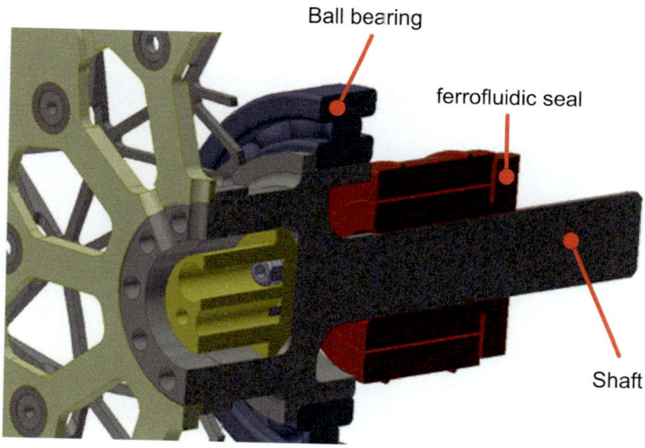

Figure 14.25 Fixing the mechanical shaft.

1. Ball bearings
2. Ferrofluidic seals whose role is to maintain the vacuum between the rotor and the fixed cryostats.

In Fig. 14.26a, we find all the parts that constitute the rotor. The constructed and assembled rotor is shown in Fig. 14.26b. The rotor position is adjusted with several shims which can be seen in Fig. 14.26b.

The detail of the rotor mass is given in Table 14.7. It can be seen that the rotor mass is significantly lower than that of the HTC coil assembly. It would be difficult to drastically reduce the mass of the rotor.

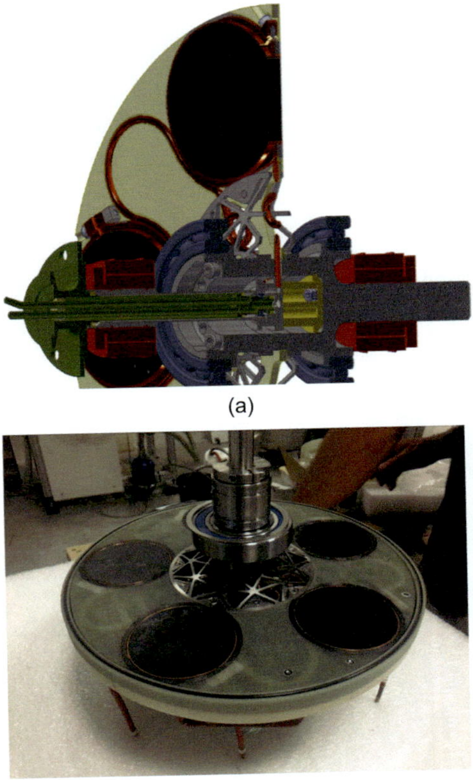

(a)

(b)

Figure 14.26 (a) Technical drawing of the rotor. (b) Rotor built and assembled.

Table 14.7 Rotor mass balance

	Mass (kg)
Bulk mass	*1.5*
Support in G10	*0.87*
Titanium flange	*0.22*
Ferrofluidic seals	*1.07*
Total weight	*5.96*

14.3.4 Armature

Figure 14.27 shows the Litz wire that we used. It is a type 8 (rectangular section) made up of 22 filaments. This number is lower than the result of the study on AC losses (28 filaments), this is due to supply difficulties. The losses will therefore be slightly higher (around 1.8 kW according to Fig. 14.8).

Figure 14.27 Litz wire.

Figure 14.28 Fixed cryostats made in G10.

The fixed cryostat is made with an electrically insulating material and the coils will be glued to these cryostats. Figure 14.28 shows the shape of the cryostat. It contains "pockets" in which the armature coils are placed. It is performed in G10. The material is also an electrical insulator which has good mechanical properties and is flexible. To reduce radiation losses from this part, we have taped a sheet SLI (single-layer insulation) with a low emissivity.

The assembled armature is shown in Fig. 14.29. The winding and the ferromagnetic yoke are bonded to the cryostat. The cryostat is fixed on a steel "bell," and the latter is screwed onto the flanges of the cryostat of the HTC coil.

Figure 14.29 Armature assembled in its cryostat.

Table 14.8 Armature balance mass

	Mass (kg)
Copper coil	*4.55*
Cryostats	*1.24*
Bell	*7.11*
Total weight	*12.9*

The mass detail of the rotor is presented in Table 14.8. Steel "bells" are additional components specific to the use of a stationary cryostat. We note that these parts represent more than 50% of the total mass of the armature. For these elements, significant safety margins have been taken.

14.3.5 Motor and Test Bench

The three parts of the machine are shown in Fig. 14.30 and the recall of their masses is given in Table 14.9. Photos of the manufactured and assembled machine are shown in Fig. 14.31.

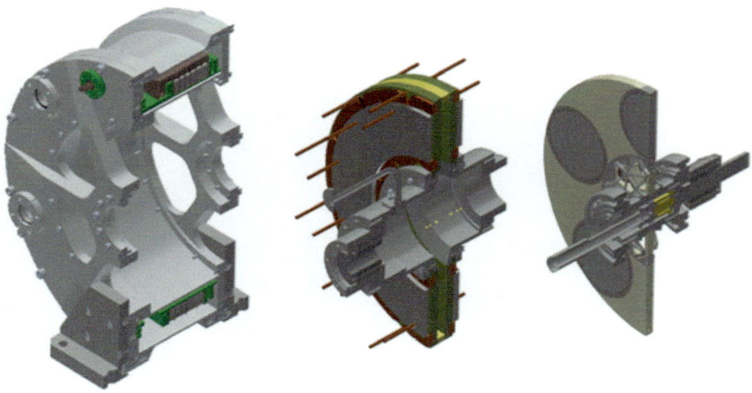

Figure 14.30 The three parts constituting the superconductive machine.

Figure 14.31 Views of the assembled machine.

Table 14.9 Whole machine balance mass

	Mass (kg)
Superconducting coil	32.8
Rotating part	5.96
Armature	12.9
Total weight	51.66

The test bench for the superconducting machine is shown in Fig. 14.32. An asynchronous motor, driven by an inverter, drives the superconducting machine. The electrical energy produced by the superconducting machine will be injected into a resistive load.

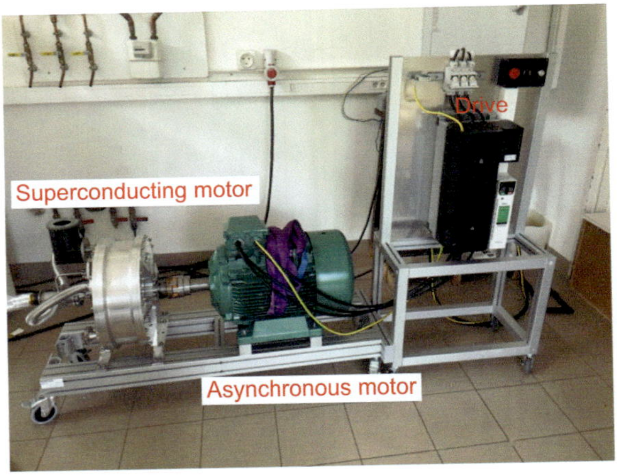

Figure 14.32 Test bench.

14.4 Experimental Results

The measurements will be compared to the theoretical results obtained.

Here is the test plan:

- Characterization of the superconducting coil
- The validation of the modulation of the magnetic flux
- Generator tests.

14.4.1 Characterization of Superconducting Coil

For the first test, the HTC coil was cooled alone. The rotor was therefore removed from the machine. We placed a Hall effect probe in the center of the machine to measure the axial magnetic induction versus the current applied to the HTC coil. The comparison between the experimental measurement and the numerical model is shown in Fig. 14.33. Not surprisingly, there is a linear relationship between the two quantities. The results are very close; the error is due to the difficulty of placing the probe exactly in the center of the superconducting coil.

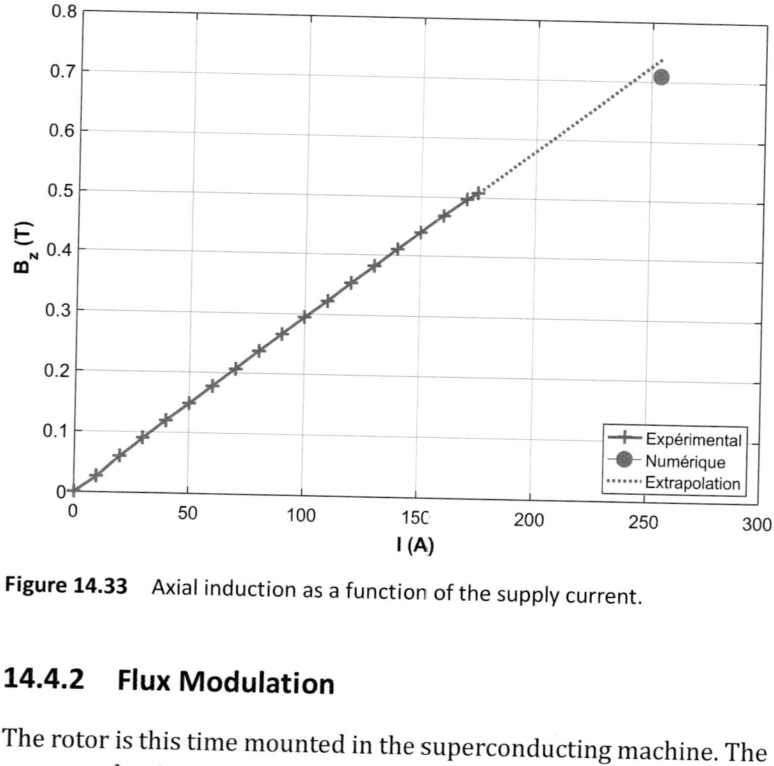

Figure 14.33 Axial induction as a function of the supply current.

14.4.2 Flux Modulation

The rotor is this time mounted in the superconducting machine. The superconducting machine is driven at a speed of 250 rpm and a Hall effect probe makes it possible to detect the variation in the magnetic field. The modulation of the magnetic field is shown in Fig. 14.34a–c at the mean radius and for 3 currents in the superconducting coil (50 A, 100 A, and 120 A).

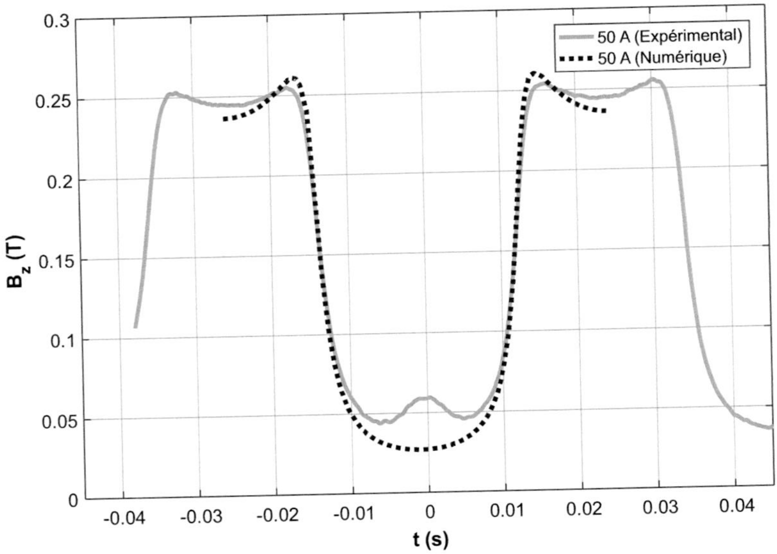

(a)

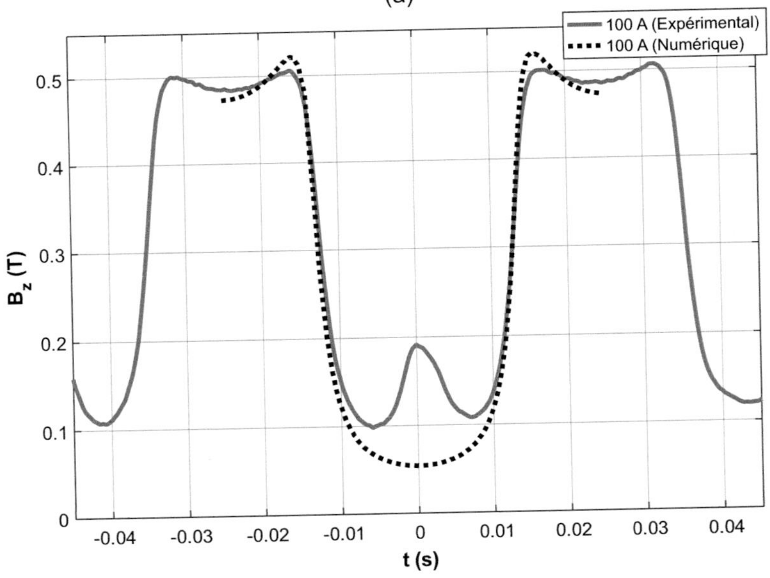

(b)

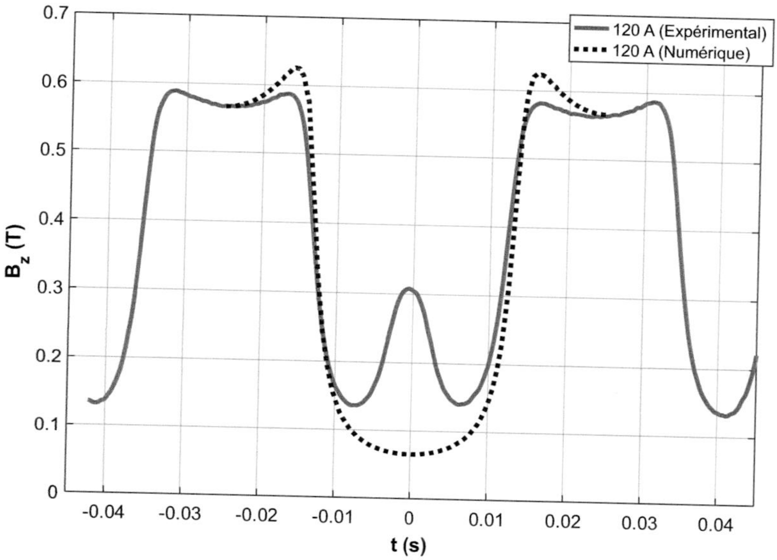

(c)

Figure 14.34 Modulation of the magnetic field for (a) 50 A, (b) 100 A, and (c) 120 A at the mean radius.

The results of a numerical calculation considering the hypothesis of the diamagnetic behavior of the screens are shown in these figures. First, we see that the numerical model is quite precise as to the variation of the magnetic field between the screens (maximum value). When approaching a screen, the magnetic field is slightly stronger with the numerical model. This comes from the perfect diamagnetic hypothesis of bulk superconductor. In this case, the magnetic field is brutally concentrated due to the change in permeability from $\mu = \mu_0$ to $\mu = 0$, while in reality the magnetic field penetrates the screen over a certain length, the variation is therefore smoother.

However, behind a screen, there appears an area where the magnetic field "crosses" the superconducting bulk. This is explained by the manufacture of the superconducting pellets which is made with 4 seeds, it appears at the growth border an area where the performance of the superconductor is less good. This is shown schematically in Fig. 14.35a. This behavior can be modeled by a numerical approach by considering a current density J_{c2} lower than the current density J_{c1} of the single crystal. A simulation result is shown in Fig. 14.35b on the penetration of the current and the magnetic field which explains this phenomenon. We note that by taking J_{c1} = 2250 A/mm^2 and J_{c2} = 30 A/mm^2, the numerical results are in adequacy with the experimental results.

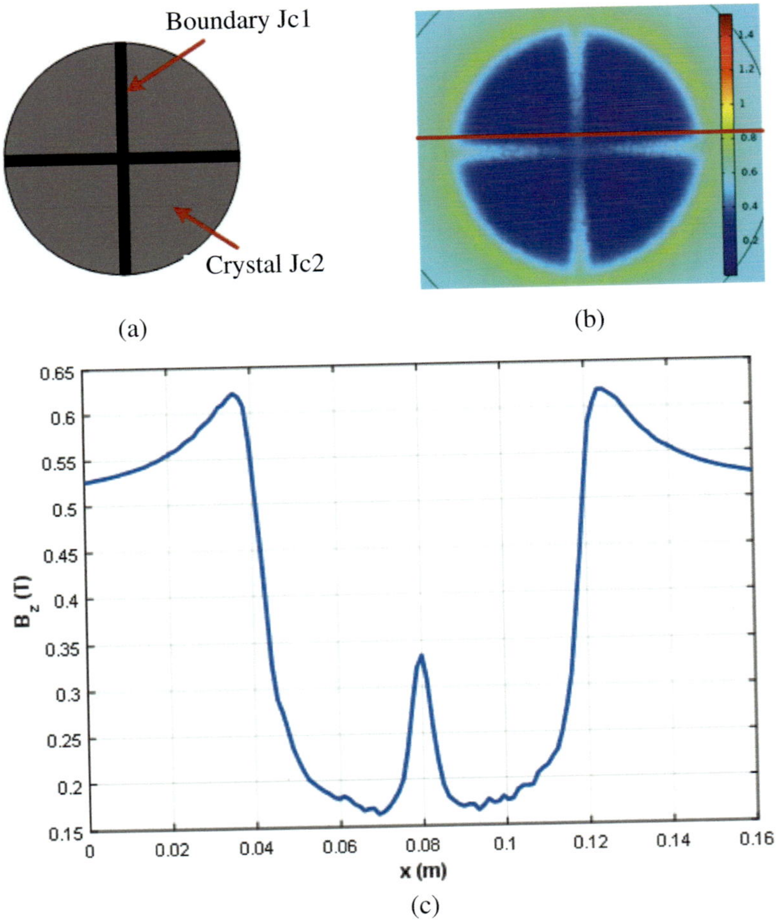

Figure 14.35 (a) Diagram of a superconductive solid with several seeds (b) field mapping of the multi-seeds bulk (c) simulation result with J_{c1} = 2250 A/mm^2 and J_{c2} = 30 A/mm^2.

14.4.3 No Load Tests

Figure 14.36 shows the phase to phase voltages found at the terminals of one of the armatures when the superconducting coil is supplied with a current of 120 A and the machine rotates at 250 rpm. We can notice the sinusoidal shape of these voltages. A slight imbalance appears between the different voltages, this is due to the bonding of the armature coils in the second cryostat in G10 which is not perfectly flat (slight variation in the air gap).

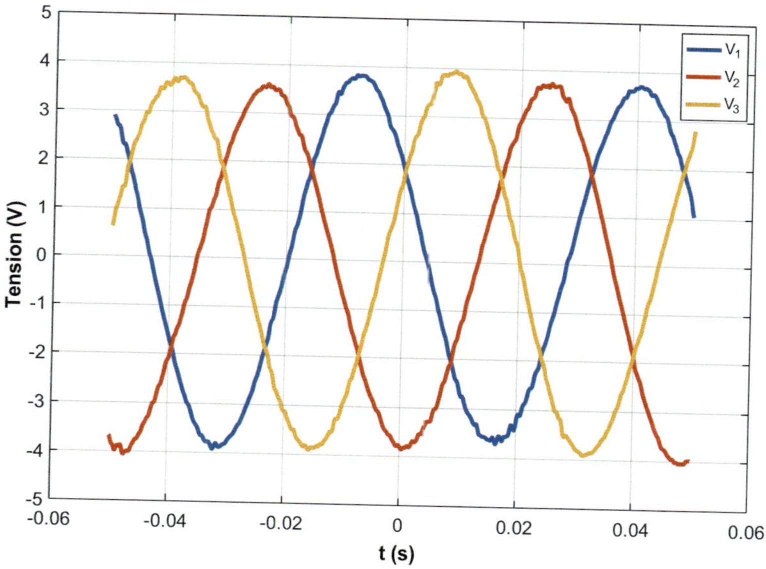

Figure 14.36 Voltages measured at the terminals of one of the phases for I = 120 A and N = 250 rpm.

By extrapolation, when we multiply by 20 (nominal speed 5000 rpm), then by 2 (critical current of the HTC coil) and finally by 2 (the 2 induced in series), we obtain a maximum voltage between 296 V and 316 V.

14.4.4 Cool Down

Figure 14.37 shows the cooling of the superconducting coil. We need approximately 4 h to reach a temperature of 33 K. It also appears that the temperature of the helium at the input and at the exit of the exchanger with the coil. When the inlet and outlet are at the same temperature, the gas no longer heats up and the minimum temperature is therefore reached.

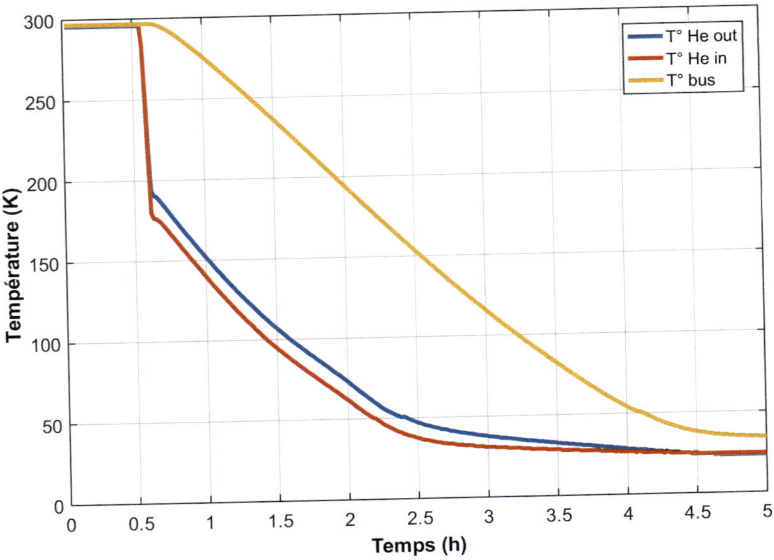

Figure 14.37 Cooling of the superconducting coil.

14.5 Conclusion

Following this chapter, it appears that many superconducting motors have been produced—not only laboratory prototypes but also industrial demonstrators. They have proven their efficiency and even their reliability.

The obstacle to marketing is their prohibitive price linked to the cost of drivers, despite their undeniable qualities. Indeed, these motors are about four to five times more expensive than a conventional permanent magnet motor.

A 50 kW prototype of superconducting aeronautic motor has been designed and manufactured, and the first tests have been successful. In the rest of the work, load tests and a new design for a 500 kW machine are planned.

Acknowledgments

The author would like to thank the Direction Générale de l'Armement (DGA) for its support and interest.

References

[AIN15]: M. D. Ainslie, H. Fujishiro, Modelling of bulk superconductor magnetization, *Superconductor Science and Technology.*, vol. 28, no. 5, p. 053002, 2015.

[ALH16]: R. Alhasan, T. Lubin, B. Douine, Z. M. Adilov, J. Lévêque, Test of an original superconducting synchronous machine based on magnetic shielding, *IEEE Transactions on Applied Superconductivity*, vol. 26, no. 4, pp. 1–5, 2016.

[BAR05]: P. N. Barnes, M. D. Sumption, G. L. Rhoads, Review of high power density superconducting generators: Present state and prospects for incorporating YBCO windings, *Cryogenics*, vol. 45, no. 10–11, pp. 670–686, 2005.

[BER16]: K. Berger, B. Gony, B. Douine, J. Lévêque, Magnetization and demagnetization studies of an HTS bulk in an iron core, *IEEE Transactions on Applied Superconductivity*, vol. 26, no. 4, pp. 1–7, 2016.

[BLA12]: R. Blaugher, Superconductivity at Westinghouse, *IEEE/CSC & ESAS European Superconductivity News Forum*, No. 20, pp. 4–8, April 2012.

[BRA02]: H. J. M. ter Brake, G. F. M. Wiegerinck, Low-power cryocooler survey, *Cryogenics*, vol. 42, no. 11, p. 705 718, 2002.

[EVO]: https://avidtp.com/product/evo-motors/

[FEL11]: J. L. B. Felder, Turboelectric Distributed Propulsion in a Hybrid Wing Body Aircraft, présenté à 20th International Society for Airbreathing Engines, Gothenburg, Sweden, 2011.

[HIR03]: M. Hirakawa, S. Inadama, K. Kikukawa, E. Suzuki, H. Nakasima, Developments of superconducting motor with YBCO bulk magnets, *Physica C: Superconductivity*, vol. 392, pp. 773–776, 2003.

[ITO95]: Y. Itoh, et al., High-temperature superconducting motor using Y-Ba-Cu-O bulk magnets, *Japanese Journal of Applied Physics*, vol. 34, no. 10R, p. 5574, 1995.

[LAM15]: M. Lampérth, A. Malloy, A. Mlot, M. Cordner, Assessment of axial flux motor technology for hybrid powertrain integration, *World Electric Vehicle Journal*, vol. 7, no 2, pp. 187–194, 2015.

[LEV02]: J. Lévêque, A. Rezzoug, A new kind of superconducting current limiter, *Superconductor Science and Technology*, vol. 15, no. 4, p. 630, 2002.

[LEV09]: J. Leveque, B. Douine, A. Rezzoug, The design of an original structure of current limiter, *Superconductor Science and Technology*, vol. 22, no. 12, p. 125021, 2009.

[MAT05]: H. Matsuzaki, et al., An axial gap-type HTS bulk synchronous motor excited by pulsed-field magnetization with vortex-type armature copper windings, *IEEE Transactions on Applied Superconductivity*, vol. 15, no. 2, pp. 2222–2225, 2005.

[MIK06]: M. Miki, et al., Development of a synchronous motor with Gd–Ba–Cu–O bulk superconductors as pole-field magnets for propulsion system, *Superconductor Science and Technology*, vol. 19, no. 7, pp. S494, 2006.

[MIK10]: M. Miki, et al., Applied HTS bulks and wires to rotating machines for marine propulsion, in *AIP Conference Proceedings*, 2010, vol. 1218, pp. 254–260.

[NAP16]: *Emissions, Commercial Aircraft Propulsion and Energy Systems Research: Reducing Global Carbon Emissions*. National Academies Press, 2016.

[NET07]: D. Netter, J. Lévêque, B. Douine, P. J. Masson, A. Rezzoug, Design and testing of a superconducting rotating machine, *IEEE Transactions on Applied Superconductivity*, vol. 17, no. 1, pp. 27–33, 2007.

[NIT18]: Nitta, Superconducting rotating machines: A review of the past 30 years and future perspectives, *Journal of Physics: Conference Series*, vol. 1054, p. 012081, 2018.

[PAR05]: A. Parviainen, M. Niemela, J. Pyrhonen, J. Mantere, Performance comparison between low-speed axial-flux and radial-flux permanent-magnet machines including mechanical constraints, in *IEEE International Conference on Electric Machines and Drives*, pp. 1695–1702, 2005.

[PHI]: http://www.phi-power.com/en/home/.

[PRI18]: L. Prigozhin V. Sokolovsky, 3D simulation of superconducting magnetic shields and lenses using the fast Fourier transform, *Journal of Applied Physics*, vol. 123, no. 23, p. 233901, 2018.

[PYR13]: J. Pyrhonen, T. Jokinen, V. Hrabovcova, *Design of Rotating Electrical Machines*. John Wiley & Sons, 2013.

[ROS07]: J. A. Rosero, J. A. Ortega, E. Aldabas, L. Romeral, Moving towards a more electric aircraft, *IEEE Aerospace and Electronic Systems Magazine*, vol. 22, no. 3, pp. 3–9, 2007.

[TER18]: Y. Terao, W. Kong, H. Ohsaki, H. Oyori, N. Morioka, Electromagnetic design of superconducting synchronous motors for electric aircraft propulsion, *IEEE Transactions on Applied Superconductivity*, vol. 28, no. 4, pp. 1–5, 2018.

[TIX91]: P. Tixador, Y. Brunet, P. Vedrine, Y. Laumond, J. L. Sabrie, Electrical tests on a fully superconducting synchronous machine, *IEEE Transactions on Magnetics*, vol. 27, no. 2, pp. 2256–2259, 1991.

[TRO14]: T. Trollier, J. Tanchon, Y. Icart, A. Ravex, High capacity 30 K remote helium cooling loop, *AIP Conference Proceedings*, vol. 1573, pp. 1461–1466, 2014.

[SHA00] E. Shaanika, et al., Core loss of a bulk HTS synchronous machine at 2 and 3 T rotor magnetisation, *IEEE Transactions on Applied Superconductivity*, vol. 30, no. 1, pp. 1-6, Jan. 2020.

[WAN04]: Rong-Jie Wang, M.J. Kamper, Calculation of eddy current loss in axial field permanent-magnet machine with coreless stator, *IEEE Transactions on Energy Conversion*, vol. 19, issue 3, 2004.

[XIA11]: W. Xian, Y. Yan, W. Yuan, R. Pei, T. A. Coombs, Pulsed field magnetization of a high temperature superconducting motor, *IEEE Transactions on Applied Superconductivity*, vol. 21, no. 3, pp. 1171–1174, 2011.

[YAM99]: K. Yamaguchi, et al., 70 MW class superconducting generator test, *IEEE transactions on Applied Superconductivity*, vol. 9, no. 2, pp. 1209–1212, 1999.

[ZHA16]: X. Zhang, K. S. Haran, High-specific-power electric machines for electrified transportation applications-technology options, in *2016 IEEE Energy Conversion Congress and Exposition (ECCE)*, pp. 1–8, 2016.

Chapter 15

Superconducting Fault Current Limiter

Quan Li
School of Engineering, University of Edinburgh,
Edinburgh, EH9 3JL, UK
Quan.Li@ed.ac.uk

Until now, operators of public and industrial electrical networks could only have limited protection from high short-circuit currents, by either the use of complicated protection equipment or overrated components. The vast development of distributed power generation, such as wind power and solar energy, and the ever-increasing demands for electric power have pushed medium voltage power grids up to their operating limits. Consequently, short circuits would occur more regularly and it is more likely to cause very high and uncontrollable fault currents that lead to power failures with physical damage to the power networks and associated electrical equipment.

Superconducting fault current limiters (SFCL) have been successfully demonstrated to effectively curb electric faults. These devices limit short-circuit currents below a safe value to ensure circuit breakers to function, and they can be adapted into the existing

High-T_c Superconducting Technology: Towards Sustainable Development Goals
Edited by Muralidhar Miryala

ISBN 978-981-4877-65-7 (Hardcover), 978-1-003-16468-5 (eBook)
www.jennystanford.com

protection configurations without additional investment. The fault current is not completely cut but only limited, which allows the effective fault detection and appropriate operation for the equipment downstream. The functionality of SFCL purely relies on the physical nature of superconductors, which combines superconducting state with perfect conductivity under normal operating conditions and non-superconducting state with high resistance when currents exceed their critical values. This nature enables automatic fault detection and self-controlled fault limiting and isolation. The transition from superconducting to non-superconducting state takes only milliseconds, hence fault current can be limited nearly immediately. After the fault is cleared, the superconductors return to superconducting state spontaneously. This makes reclosing operations possible without any servicing requirements.

For power networks, the application of SFCL allows new grid layouts, and the choice of locations for electricity generation would be more flexible. This is particularly beneficial to the forthcoming renewable electricity networks with distributed renewable power generation.

This chapter provides a comprehensive overview of the existing SFCL technology. All major types of SFCLs will be introduced, including the most popular resistive and saturated-core SFCL, as well as magnetic-shielded, coreless, transformer, flux-lock, bridge, resonance, and novel hybrid SFCLs. Some specially designed three-phase SFCLs will also be included in this chapter. The physical principles, specifications, and operations of these SFCLs will be analyzed and discussed in great detail.

15.1 Resistive SFCL

Resistive SFCL, along with saturated-core SFCL, is the most popular type of SFCL, which has been demonstrated at various voltage levels as in Table 15.1.

The principle of resistive SFCLs is to commutate the fault current by turning a superconducting element into a shunt resistor, which curbs fault currents to drop down below an acceptable level. This allows circuit breakers to be able to function effectively. This operation can be accomplished through switching the superconducting element from zero resistance (superconducting state) into its resistive state (non-superconducting state). Such switching is automatic once fault currents exceed the critical current of the superconducting element and force it to quench. Superconductors exhibit high resistivity after quench, and almost all fault currents can be diverted into the shunt resistor branch.

Table 15.1 Resistive SFCL demonstration at various voltage levels

Voltage	Year	Developer	Country
4.2 kV	2000	Siemens	Germany
8 kV	2001	ABB	Switzerland
10 kV	2004	Nexans	Germany
13.2 kV	2007	KEPRI	Korea
63.5 kV	2008	Nexans	Germany
110 kV	2009	Nexans	Germany
115 kV	2011	AMSC/Siemens	USA
138 kV	2013	SuperPower	USA
154 kV	2016	KEPCO	Korea
220 kV	2018	Zhongtian Co. Ltd	China
160 kV DC	2020	Southern Power Grid	China

An example shown in Fig. 15.1 explains the operation mechanism of a resistive SFCL (Gray, 1978). R_S represents a shunt resistor with a pre-determined amount of resistance designed to match the fault level of power networks. The other resistor highlighted by the dashed-line box is a superconducting element. Its resistance is 0 Ω during normal operations and the resistive SFCLs are invisible to power networks. When a fault occurs, the resistance drastically increases far above the shunt resistor and consequently diverts the fault current. In this way, the fault current is curbed.

The resistive SFCLs are simple in concept and easy to fabricate. However, a large amount of superconducting material is required to achieve high resistance at non-superconducting state. Besides, superconductors quench at each fault and this may cause potential damage and degradation of properties in long term.

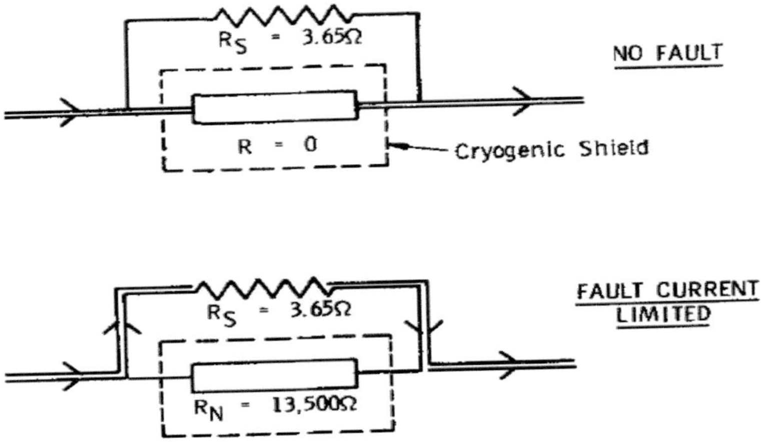

Figure 15.1 Schematic representation of a resistive SFCL, with the fault current diverted through the shunt resistor, R_S, by the superconducting element when faults occur (Gray, 1978).

15.2 Flux-Flow Resistive SFCL

Resistive SFCLs have simple structure and they respond to faults automatically. However, the superconducting components quench over faults. This results in substantial joule loss that turns into heat, due to which the temperature of superconductors increases and exceeds their critical value. When faults are cleared, it takes relatively long time for these superconducting components to recover and to be cooled back to superconducting state. This thermal cooling process could be time-consuming, while SFCLs are required to recover to normal operating state as quickly as possible for rapid reclosing.

The concept of flux-flow resistive SFCL was proposed to solve this issue. Such SFCLs limit fault currents only by the flux-flow resistance of superconductors, so that their temperature do not rise above the critical point. The flux-flow resistance appears in superconductors when the transport currents exceed their critical currents but their temperatures remains below the critical point. Since the flux-flow resistance disappears immediately when transport currents drop back below the critical current, such SFCLs have almost zero recovery time.

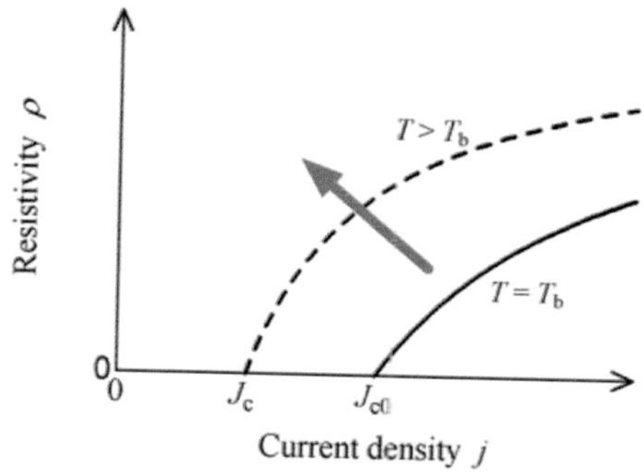

Figure 15.2 Resistivity of superconductors against current density at different temperatures. T_b is liquid nitrogen temperature, and J_{c0} is the current density at liquid nitrogen temperature (Shimizu, 2002, 2003).

The flux-flow resistive SFCLs are working within the critical temperature of superconductors, so that their resistance is relatively lower than resistive SFCLs under quench. Figure 15.2 illustrates the relationship between superconductor resistivity and current density at different temperatures (Shimizu, 2002, 2003). The resistance at liquid nitrogen temperature T_b (flux-flow resistive SFCLs without quench) is apparently lower than that at temperatures above T_b (resistive SFCL at quench). Therefore, a precise design is necessary to identify the required resistance, and apply sufficient superconducting material. Besides, since flux-flow resistive SFCLs do not quench, compared with resistive SFCLs, more superconductors are required to achieve the required resistance level and curb the same fault current level.

15.3 Saturated-Core SFCL

Saturated-core SFCL is another popular type of SFCL that has been successfully demonstrated at various voltage levels mainly in China and USA, as shown in Table 15.2.

Saturated-core SFCL is inductive fault current limiter, which introduces very high inductive impedance into power networks when fault occurs. The general concept of a saturated-core SFCL is shown in Fig. 15.3 (Moriconi, 2011). This type of SFCLs consists of iron cores, superconducting DC windings with their own power supplies, and AC coils connected to power transmission lines. The iron cores are made of ferromagnetic material that saturates above certain magnetizing ampere turns. These iron cores are surrounded by the superconducting DC windings energized by an individual DC power supply. The AC coils from power transmission lines share the magnetic path with the superconducting DC windings through the iron cores.

Table 15.2 Saturated-core SFCL demonstration at various voltage levels

Voltage	Year	Developer	Country
11 kV	2008	InnoPower	China
15 kV	2009	Zenergy Power	USA
138 kV	2011	Zenergy Power	USA
220 kV	2012	InnoPower	China
500 kV	2020	State Grid	China

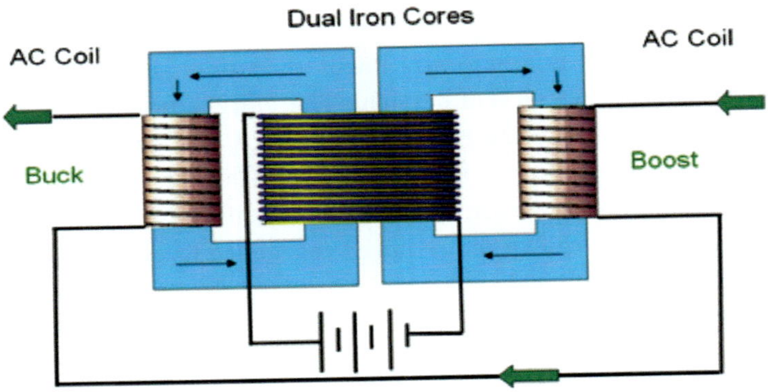

Figure 15.3 The basic concept and structure of saturated-core SFCLs. The part connected to a DC power supply is a superconducting winding (Moriconi, 2011).

During normal operations, the iron cores are highly saturated with a low relative permeability. The required magnetizing ampere turns are produced by the superconducting DC windings. Since superconductors are capable of carrying extremely large electric currents, the superconducting DC windings are very compact in space and efficient with almost zero loss, compared with conventional normal conducting windings. The relationship of magnetic flux Φ and magnetizing ampere turns Σ*NI* (similar as *B*-*H* curve) is illustrated by Fig. 15.4 (Parton, 1977). Normally the superconducting DC windings maintain the cores to work in a saturated condition centered by the point O, which is above the knee of the *Φ*- Σ*NI* curve. The AC current from power networks introduces alternating flux change between point A and B, however, such change is very small under the saturation condition. This guarantees that the saturated-core SFCLs are almost invisible to the power networks.

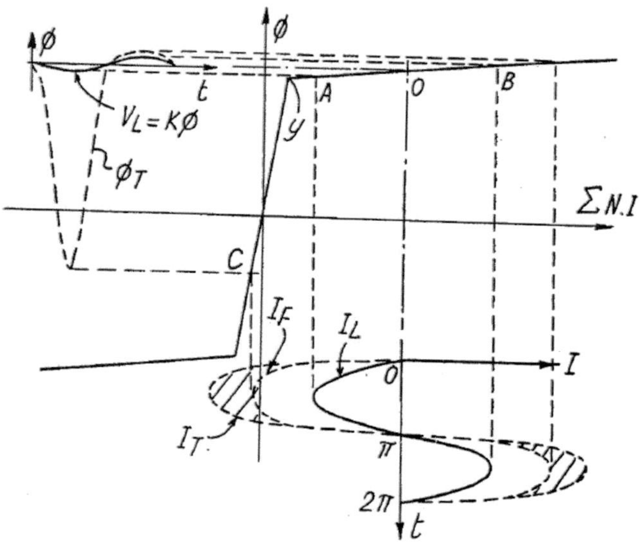

Figure 15.4 Change of magnetic flux against magnetizing ampere turns in saturated-core SFCLs. The flux and current waveforms (Φ_t and I_t) illustrate normal status (solid line) and fault-limiting operations (dashed line) (Parton, 1977).

When fault occurs, as indicated by the dashed line in Fig. 15.3, the fault current IF is much higher than the normal current. It drives the iron cores out of saturation state (below the keen point y) and forces the flux Φ_T to change drastically. This causes a large flux swing and induces electromotive force in the AC coils, which is in the opposite direction to the supply voltage. The SFCL exhibits high impedance and thereby limits the fault current.

Unlike resistive SFCLs, superconductors do not quench in saturated-core SFCLs. The superconducting windings are used for high magnetic flux, hence only a relatively small amount of superconducting material is required compared with the resistive SFCLs. However, the iron cores make such devices very heavy and large in size.

15.4 Magnetic-Shielded SFCL

Similar to saturated-core SFCLs, magnetic-shielded SFCLs exhibit high inductance when fault occurs and thus limit fault current. Hence these two types are also known as inductive SFCLs.

The schematic structure of magnetic-shielded SFCLs is shown in Fig. 15.5, which consists of an iron core, superconducting windings, and a normal conductor coil (Xin, 2018). These parts are assembled coaxially, so that the superconducting windings form a shielding layer between the iron core and the coil. The iron core is made of ferromagnetic material and the superconducting windings can be a solenoid or a set of separate coils subject to the performance requirements. The normal conductor coil is series-connected to the power networks.

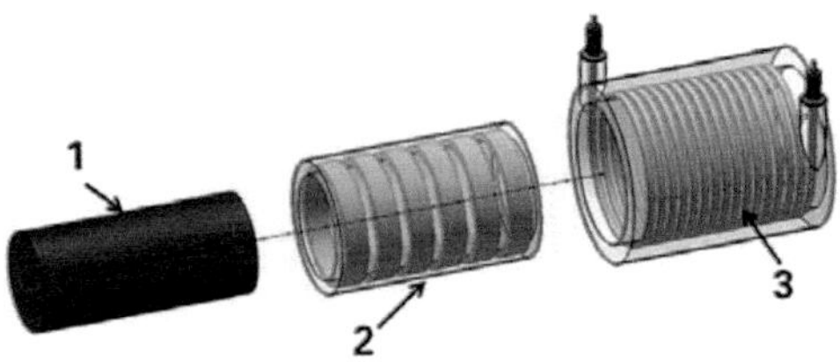

Figure 15.5 Structure of magnetic-shielded SFCLs, with (1) an iron core, (2) superconducting windings, and (3) coils connected to power lines (Xin, 2018).

The iron core is shielded by the superconducting windings during normal operations, and no magnetic flux flows through it. Therefore, the magnetic-shielded SFCL acts as an air-core solenoid with low inductance that depends on the leakage flux in the windings. Under fault conditions, the current in the normal conductor coil rises rapidly and generates high magnetic field. This field quickly penetrates the superconducting shield layer, and magnetic flux starts to flow in the iron core. Suddenly, the SFCL turns into an iron-core inductor and exhibits high impedance. The fault current is thus limited by the increased impedance. An example demonstrating the relationship between the SFCL impedance and network current is presented in Fig. 15.6 (Bashkirov, 1991). A rapid increase of impedance is triggered by the network current above the fault level.

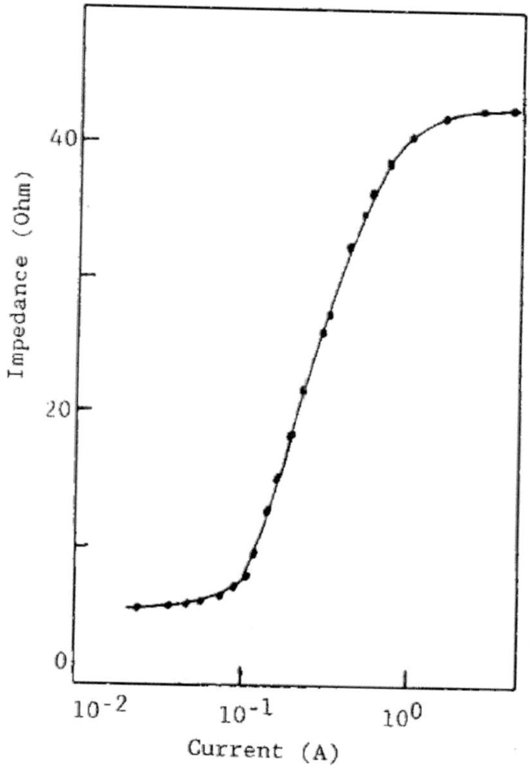

Figure 15.6 Relationship between the impedance of a shielded SFCL and electric current (Bashkirov, 1991).

The magnetic-shielded SFCLs require sufficient large iron cores to limit current effectively. Therefore, the same as saturated-core SFCLs, magnetic-shielded SFCLs are very heavy. However, since superconductor is needed for the shielding layer only, much less superconducting material is required compared with resistive SFCLs.

15.5 Coreless SFCL

Iron cores are essential for saturated-core and magnetic-shielded SFCLs, which enable quick response to fault with a significant increase of impedance. However, they also make these SFCLs heavy and bulky. To overcome this problem, coreless SFCLs were invented based on the structure of magnetic-shielded SFCLs with iron cores removed.

A structural illustration of coreless SFCLs is shown in Fig. 15.7a, consisting of primary copper windings, primary superconducting windings, and secondary superconducting windings (Kozak, 2012). Its structure is very similar to magnetic-shielded SFCLs, with iron cores replaced by air. The primary copper and superconducting windings are mechanically linked in parallel through interconnections, and then connected to power networks. The secondary superconducting windings are short-circuited and interact with the primary windings through electromagnetic coupling, similar as transformers. All these windings form a current-limiting unit, as shown in Fig. 15.7b, and a limiter contains several such units, the number of which is determined upon fault current level and circuit breaker requirements.

During normal operation, the primary superconducting windings are in superconducting state, and their resistance is nearly zero. With the primary copper windings connected in parallel, current almost entirely flows through the superconducting windings. The magnetic flux is compensated by the shorted secondary superconducting windings, and the whole SFCL exhibit very low impedance to power networks.

At fault conditions, current increases suddenly and exceeds the critical current of the primary superconducting windings. Superconductors quench and the fault current is diverted to the primary copper windings. The overall impedance increases rapidly due to the resistance jump of the primary superconducting windings and high inductance due to the exposure of the primary copper windings.

Coreless SFCLs are much lighter than saturated-core and magnetic-shielded SFCLs, since iron cores are no longer required. However, due to the low magnetic permeability of air, the impedance of coreless SFCLs over fault is much lower. Besides, the same as magnetic-shielded SFCLs, the superconductors in coreless SFCLs also quench and release significant heat when fault occurs. This results in a longer recovery time and higher cooling requirements.

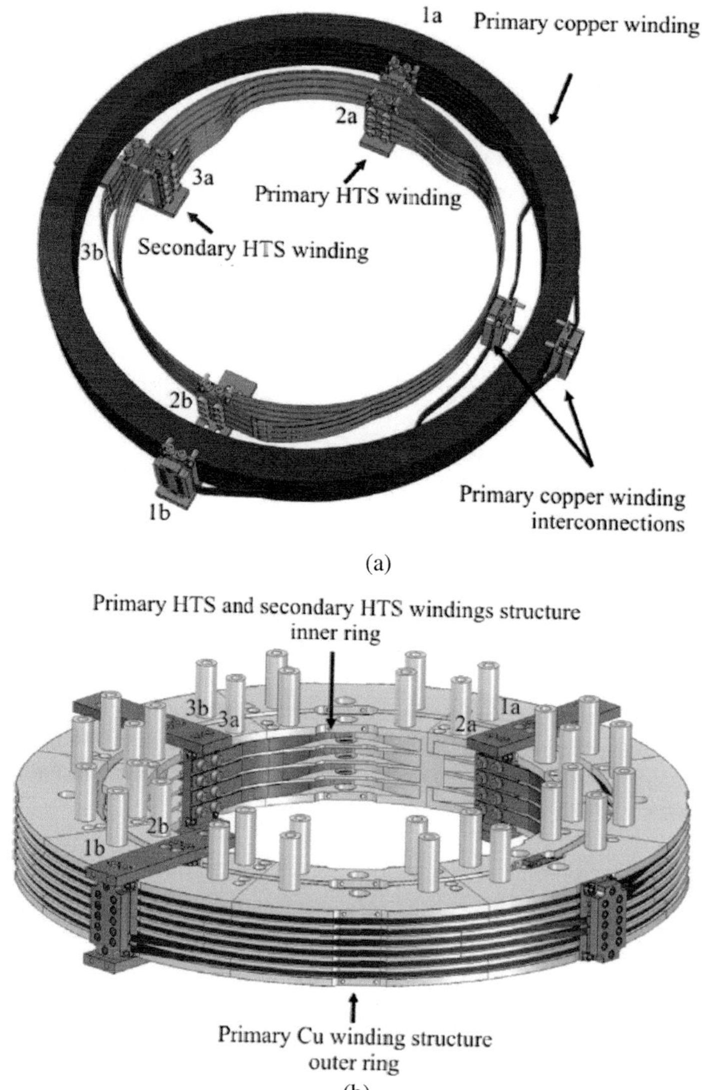

Figure 15.7 (a) Structure of coreless SFCLs with primary copper windings (connected to 1a and 1b), primary superconducting windings (2a and 2b) and secondary superconducting windings (3a and 3b). (b) A current-limiting unit. HTS, high-temperature superconducting (Kozak, 2012).

15.6 Transformer SFCL

Through a transformer, circuit impedance of the secondary winding can exhibit a very different property to the primary winding. Based on this concept, transformer SFCLs have been developed. Superconductors are connected to a transformer secondary winding and used as a variable resistor, shown as Fig. 15.8 (Yamaguchi, 2004). The transformer primary winding is connected to the power transmission lines, with *Z* representing the customer load. This structure is similar to resistive SFCLs, which have superconductors connected to the power lines directly.

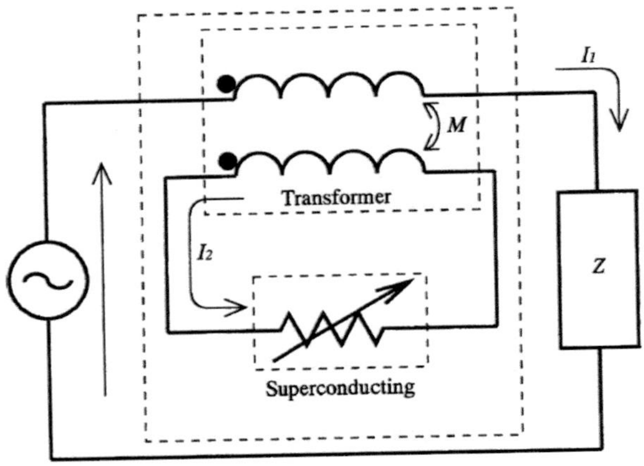

Figure 15.8 Schematic structure of transformer SFCL with load, Z, and superconducting component connected to the transformer secondary winding (Yamaguchi, 2004).

During normal operations, the superconductors are in superconducting state with zero resistance, therefore the total impedance of the transformer is very low. When fault occurs, large fault current induces high current in the secondary winding and forces the superconductors to quench. The secondary winding exhibits high impedance and this is converted to the primary side to curb the fault current.

Transformer SFCLs adopted the advantage of power transformers that physically isolate superconductors from the main power lines. Besides, high turns ratio can enlarge the limiting effect of the superconductors, hence less superconducting material is required.

However, transformer SFCLs contain iron cores that are heavy and superconductors quench during faults.

15.7 Flux-Lock SFCL

The concept of flux-lock SFCL was introduced in the 1990s by T. Matsumura. Its structure is show in Fig. 15.9 (Matsumura, 1997). Three coils are magnetically coupled through a shared iron core. Coil 1 and 2 are in parallel and then connected to the power lines. A superconducting element is inserted into the coil 2 branch, which is placed under the magnetic field coil powered by coil 3. The flux-lock SFCLs are similar to resistive SFCLs, with coil 1 and 3 removed. The superconductors function as a variable resistor, adjusting the total impedance based on fault conditions.

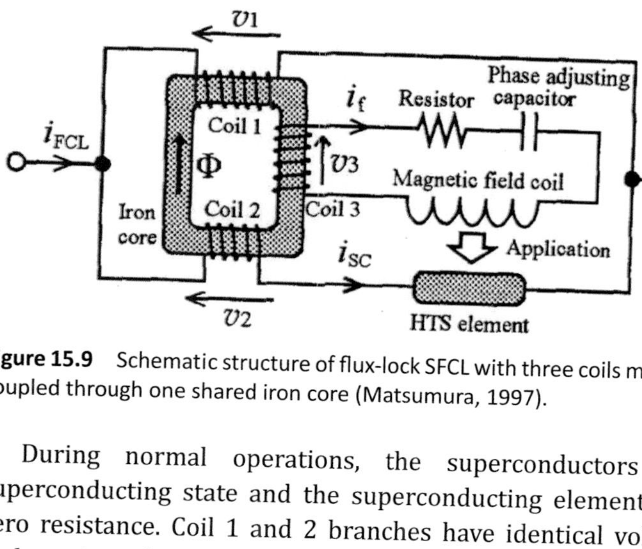

Figure 15.9 Schematic structure of flux-lock SFCL with three coils magnetically coupled through one shared iron core (Matsumura, 1997).

During normal operations, the superconductors are in superconducting state and the superconducting element exhibits zero resistance. Coil 1 and 2 branches have identical voltages, v_1 and v_2, since there is no voltage drop across the superconducting element. Equivalently, coil 1 and 2 are directly connected in parallel. The magnetic flux produced by coil 1 cancels that from coil 2 and there is no flux change in the iron core. Coil 3 is not activated and v_3 is zero.

At fault conditions, high fault current drives the superconductors out of superconducting state. The voltages across coil 1 and 2 are no longer balanced. Time-varying flux appears in the iron core, thereby inducing voltage in coil 3. Driven by this voltage, current flows through the magnetic field coil. The magnetic field produced is applied on the superconducting element that enhances quench and further increases its resistance. The SFCL exhibits high impedance overall and limits fault currents.

The flux-lock SFCLs have shorter response time to faults than the above SFCLs, since in addition to fault current magnetic field is applied to accelerate the quench of superconductors. However, the structure of such SFCLs is more complex than the SFCLs above, and they are heavy due to the iron cores.

15.8 Bridge SFCL

The concept of bridge SFCLs is to combine superconductors with solid-state power electronic devices, which offer advantages in simplicity over the conventional FCLs. The bridge SFCL consists of, as shown in Fig. 15.10a, a diode bridge and a bias superconducting coil (Boenig, 1983). A DC power supply is connected to the superconducting coil in series, and provides a constant bias voltage. The bias voltage is high enough to overcome the voltage drop of diodes, so that the diode bridge is always open and conducting current.

During normal operations, the DC power supply produces a bias current Io, which is designed to be higher than the regular peak current from the power network Imax. This current, in addition to the network current, flows through all diodes and their current waveforms are shown in Fig. 15.10b. As long as I_0 is higher than I_{max}, all currents in the diodes are positive and the diodes stay open. In this case there is only DC current flowing through the superconducting coil. The whole SFCL is invisible to the power network due to the zero resistance of the superconducting coil.

When fault occurs, the network current rises quickly above the bias current. Once I_{max} is higher than I_0, one branch of diodes, either D1/D4 or D2/D3, become reversed biased and ceases to conduct. Meanwhile, the superconducting coil is automatically switched into the power network and exhibits inductance. The inductance is predesigned to match fault levels and curbs current increase during faults.

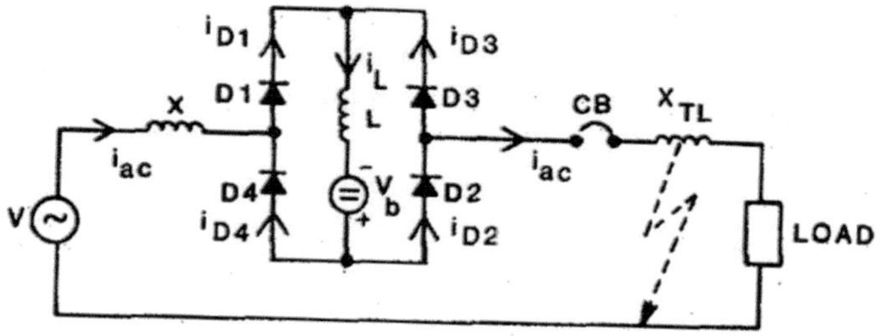

(a)

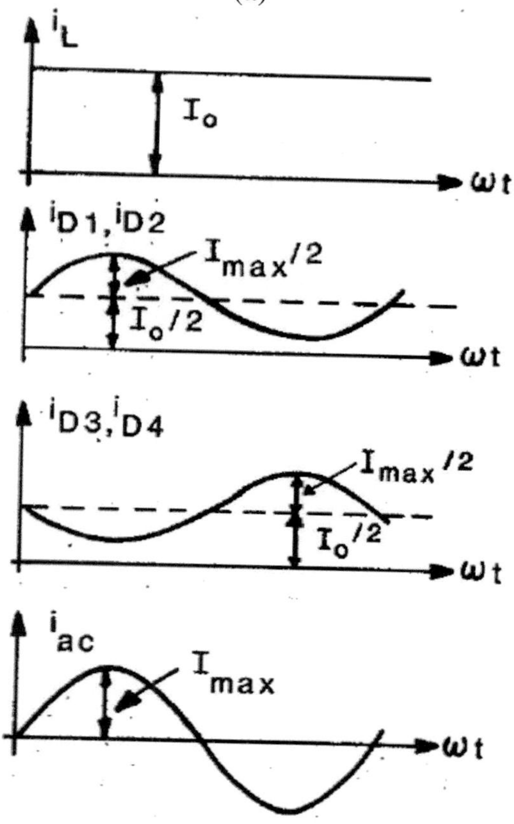

(b)

Figure 15.10 Schematic structure of a bridge SFCL (a) and the steady state currents of each circuit branch (b). The bridge SFCL consists of a diode bridge (D1 to D4), a bias superconducting coil (L) and a circuit breaker (CB) (Boenig, 1983).

The superconductors in bridge SFCLs only carry DC currents during normal operations, which significantly reduce loss. Besides, superconductor do not quench when fault occurs. These advantages simplify cooling system associated and greatly benefit the practical operation of such SFCLs. However, the application of power electronic diodes introduces extra loss and heat that require additional cooling. Besides, large network currents require high current level diodes, which could be expensive and large in size. A DC power supply is required to provide the bias current, which makes the current limiting circuit complicated and difficult to implement.

15.9 Resonance SFCL

The principle scheme of resonance SFCLs is to create resonance between an inductor and a capacitor so that the overall voltage drop is zero. As shown in Fig. 15.11 (Tosato, 2001), a resonance SFCL consists of a superconducting coil L and a capacitor C. R represents their resistance, and R_{n} and L_{n} are the power transmission line resistance and inductance.

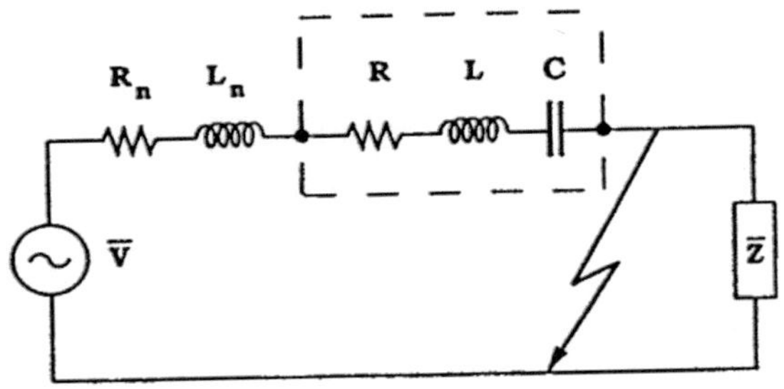

Figure 15.11 Schematic structure of resonance SFCLs (Tosato, 2001).

During normal operations, the series L-C circuit is designed to be resonance at the network frequency, which is usually around 50–60 Hz. The voltage drop is zero across the reactive components, since ωL is equal to $1/\omega C$.

When fault occurs, current increases suddenly and breaks the balanced voltage between inductance and capacitance. The L-C circuit exhibits impedance to the network and forces the current to increase gradually. This slows down the fault and allows circuit breakers to operator.

Resonance SFCLs have very simple structure, and they respond to faults automatically. However, the resonance may introduce very high voltage during operation. Besides, to achieve appropriate impedance, large capacitance is required, which is heavy and expensive, associated with a bulky oil tank for high voltage applications.

15.10 Hybrid SFCL

Hybrid SFCLs are developed based on the combination of hybrid fault current limiters and superconductors. Hybrid fault current limiters consist of fast fault current sensing relays, arc quenching devices, and fast switches, which are operated by electromagnetic repulsion force. These fault current limiters show rather viable performance. However, they lack accuracy in terms of fault evaluation and judgment and stability for multiple operations. Hybrid SFCLs were proposed to solve these issues by combining the sensitive superconductors and effective operations of fast mechanical switches and circuit breakers.

Hybrid SFCLs consist of superconductors, semiconductors, conventional fuses or circuit breakers, and fast switches. Figure 15.12 shows the fundamental operation concept of hybrid SFCLs (Lee, 2008: p. 624). During normal operations, currents flow through the superconductor and interrupter. Due to the nearly-zero resistance of the superconductor, the whole SFCL exhibits very low impedance. When fault occurs, high fault current forces the superconductor to quench and exhibit high resistance. The fault current is diverted to the drive coil, and the interrupter is forced to open due to the strong electromagnetic force produced by the drive coil.

Based on this concept, a hybrid SFCL design is proposed in Fig. 15.13 (Lee, 2008: p. 620). The interrupter is the main contact at the bottom of the fast switch. The drive coil is in the middle connected to current limiting unit with fuse. When fault occurs, the auxiliary contact (at the top of the fast switch) closes quickly, since high fault current flows through the driving coil and produces enough electromagnetic repulsion force. Meanwhile, the main contact of the fast switch opens and all current is transferred from the superconductor to the current limiting unit.

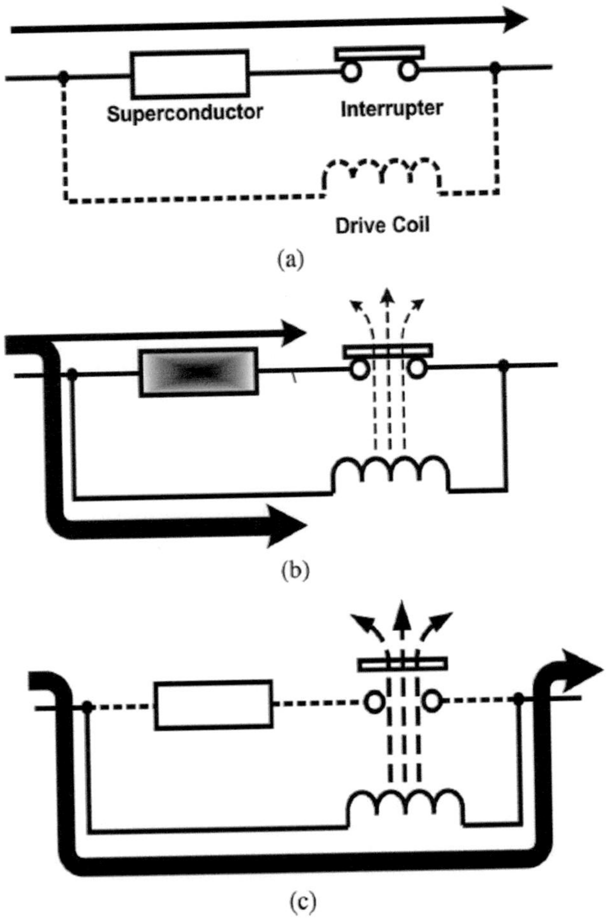

Figure 15.12 Concept of Hybrid SFCLs using superconductors to divert fault currents and open switch (Lee, 2008: p. 624).

Unlike other SFCLs, the superconductors in hybrid SFCLs do not limit fault current. Instead, they are acting as very sensitive sensors to trigger switch-off operations, when fault occurs. Hence, the required superconducting material can be significantly reduced with substantial benefits, such as low cost, low losses, and simpler cryogenic requirements. In additional, there is not high voltage stress and recovery time is relatively short. However, the fuse needs to be replaced after each operation.

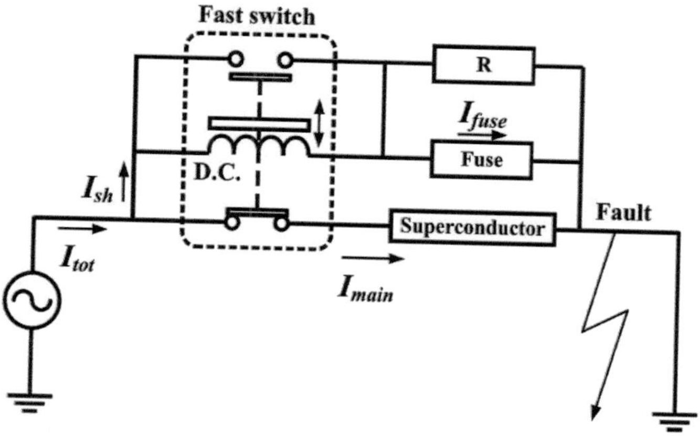

Figure 15.13 Hybrid SFCL proposed by Korea Electric Power Research Institute and LS Industrial Systems (Lee, 2008: pp. 620). After the fault is cleared, the current in the driving coil reduces and the auxiliary contact will open due to the low electromagnetic force remaining. The main contact of the fast switch closes again, and current is diverted back to the superconductor. This switch-off operation will repeat if the fault is not 100% cleared and reclosing is required.

To solve this problem, two designs were proposed, as shown in Fig. 15.14 (Hyun, 2009; Hyun, 2009). Both designs simplified the structure and replaced fuse with SFCL or other tapes of reactors to limit fault current, so that maintenance after each fault operation can be avoided. The fast switches in these two designs are identical with a main interrupter, a driving coil and an auxiliary contact bar. However, they are connected to the superconductor and current limiting unit in different ways, and different limiting functions can be realized.

In Fig. 15.14a, the main interrupter is connected in series with the superconductor, and then the whole branch is in parallel with the driving coil and the current limiting unit. When fault occurs, the superconductor quenches and current reduces simultaneously in the superconductor and in the main interrupter. The fault current is diverted to the driving coil and current limiting unit, which can be immediately controlled and curbed. Large current in the driving coil forces the main interrupter to open and the auxiliary contact to close. Current in the superconductor drops to zero within the first half cycle, and the current limiting unit functions quickly meanwhile. Because of this, such hybrid SFCL realizes first half cycle limiting operation.

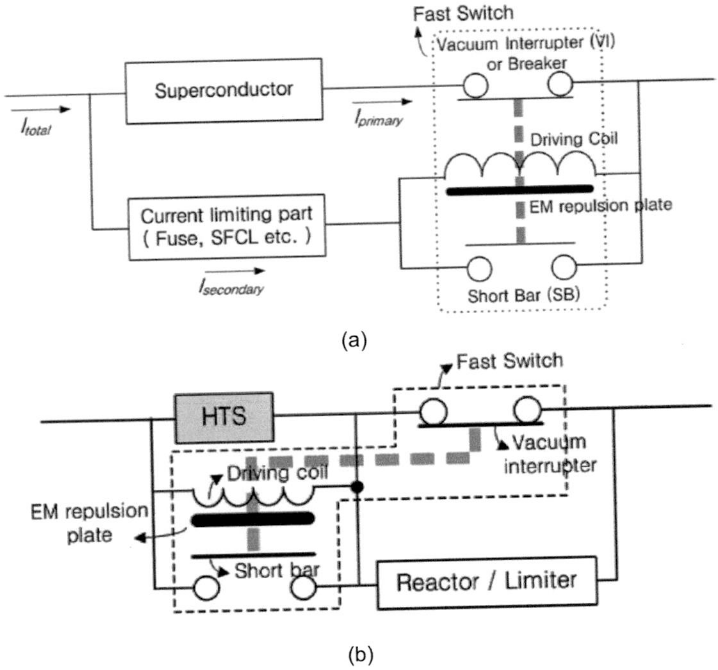

Figure 15.14 Hybrid SFCL with (a) the first half cycle limiting operation, and (b) the first half cycle non-limiting operation (Hyun, 2009; Hyun 2009).

In Fig. 15.14b, the main interrupter is connected in parallel with the current limiting unit, which as a whole connects to the superconductor in series. This is very different from the structure in Fig. 15.17a. When fault occurs, the superconductor quenches and diverts current to the driving coil. However, the fault current still flows through the main interrupter, and the current limiting unit in parallel is not active. At the end of the first half cycle, current in the main interrupter drops to zero, and the driving coil forces it to open and the auxiliary contact to close. From now on, the current limiting unit starts to perform, and fault current is limited. This all happens in the second half cycle, so such hybrid SFCL realizes first half cycle non-limiting operation.

Another design applies two different superconductors to realize fault current sensing. This design is shown in Fig. 15.15, with BSCCO 2223/Ag and YBCO-coated superconductors (Park, 2009). These two superconductors are wound into two coils

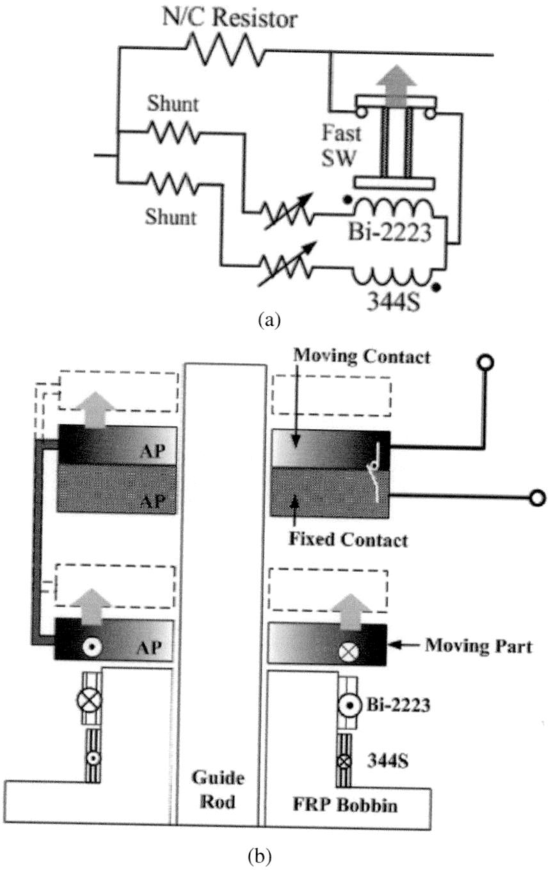

Figure 15.15 (a) Hybrid SFCL with various superconductor materials to form a non-inductive coil, a fast switch and current limiting resistor, and (b) an example of such SFCL (Park, 2009).

and assembled in opposite direction to form a non-inductive coil. During normal operations, both superconductors exhibit nearly-zero resistance and current is equally shared between those two coils. The whole coil stays non-inductive, and the fast switch does not act. When fault occurs, due to different quench characteristics, the two superconductors exhibit different resistance and conductor different currents. This will turn the coil into inductive and force the fast switch to act. An example of the coil structure is shown in Fig. 15.15b. The moving contact is a fast switch. With currents equally distributed in the two superconductors, the whole coil is non-inductive, and the moving contact is closed. When fault occurs, this balance is broken and the electromagnetic force will lift the moving contact to make it open. After the fault is cleared, due to gravity, the moving contact will automatically return to its previous closed position. This enables reclosing operation and avoids services after each fault operation.

In most cases, SFCLs are used to protect a certain range of fault current. For the networks experiencing multiple levels of fault currents, a new design is proposed in Fig. 15.16 (Ebrahimpour, 2013). Solid-state power electronic switches are integrated into hybrid SFCLs to enable control of different fault current levels.

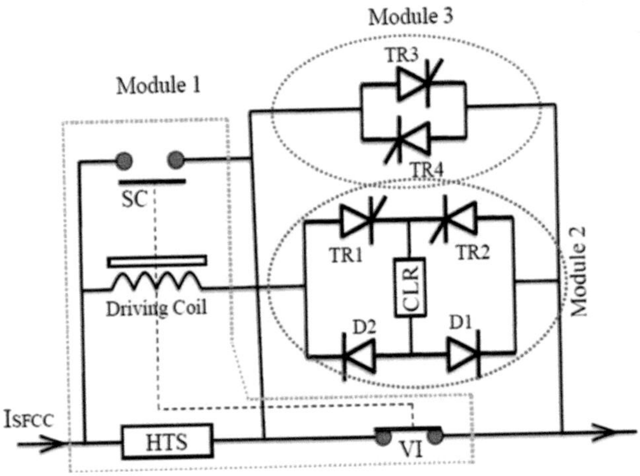

Figure 15.16 Hybrid SFCL with solid-state power electronic switches. HTS: high-temperature superconductor. CLR, current limiting resistor; SC, short contact; VI, vacuum interrupter (Ebrahimpour, 2013).

Module 1 is the fast switch. The operation of Module 1 is similar to the hybrid SFCLs above. The superconductor is still used for fault current sensing and current commutation. Unlike the other hybrid SFCLs, this design consists two modules: Module 2 is for fault current limiting and controlling, and Module 3 for fault current bypassing.

Module 2 consists of a current limiting resistor and a half-controlled bridge rectifier with thyristors and diodes. This still keeps the control function while avoiding additional cost of a full-controlled bridge rectifier. During normal operations, current flows through the superconductor and the main interrupter. There is no current in this module, and all the thyristors are off. When fault occurs, the superconductor quenches and diverts current to the driving coil. Meanwhile, the thyristors open to allow fault current to flow through the current limiting resistor. The two thyristors will open alternatively for half cycle each. For relatively low fault currents, they open at 0° trigger angle to allow the current limiting resistor to fully function. For very high fault currents, their trigger angles are between 0° and 180°, which are determined by the RMS value of current in the nearest half cycle. The current sensing will continue to monitor this value in order to provide the most accurate trigger angles.

For very low fault current levels, for instance faults in small networks or island networks, fault current limiting should be avoided, as long as it may postpone the overcurrent relays operation. In such cases, Module 3 works instead of Module 2. The two thyristors open alternatively for half cycle each to conduct the fault current. They open at 0° trigger angle and function as diodes.

15.11 Three-Phase SFCL

All the SFCLs presented in the sessions above are designed for single-phase systems or one phase in three-phase systems. To protect all three phases, three SFCL units are required and each unit protects one phase individually. These SFCLs can be operated, serviced, and repaired individually. There are also SFCLs specifically designed for three-phase systems. These SFCLs protect all three phases simultaneously, and such SFCLs are mainly applied to resolve both single-line-to-ground and three-line-to-ground faults.

15.11.1 Transformer Three-Phase SFCL

Figure 15.17 shows a three-phase SFCL based on the concept of transformer SFCLs (Zhang, 2002). Three single-phase transformers are connected to a three-phase power supply. Their primary windings are conducting line currents, while the secondary windings are connected to a single superconducting coil. All the three transformers are connected in parallel and they share the superconducting coil.

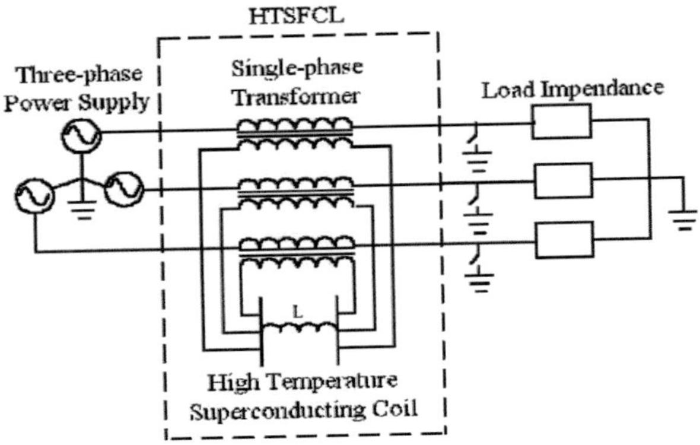

Figure 15.17 Transformer three-phase SFCL with superconducting coil (Zhang, 2002).

During normal operations, the total magnetic flux in the transformer core is zero, since the three-phase currents are balanced and they cancel each other. Therefore there is no induced voltages in the secondary windings, and the current in the superconducting coil is zero. Even though the three phases are not perfectly balanced, the total flux would be considerably small and the induced current is far below the critical current of the superconducting coil. In this case, the SFCL exhibits very low impedance consisting of the leakage reactance and winding resistance.

When a single-line-to-ground fault occurs, the current increases suddenly in one phase, and the three-phase currents become significantly unbalanced. This results in large induced current in the superconducting coil that leads to quench. The superconducting coil now exhibits resistance and inductance, and accordingly the transformer impedance increases that limits the fault current. Such SFCLs are very sensitive to single-line-to-ground faults. However, if fault occurs in all three phases simultaneously, induced currents from each phase cancel each other and the transformers will not exhibit high impedance to curb the fault current. This is the disadvantage of the transformer three-phase SFCLs, but such three-phase fault happens very rarely.

15.11.2 SFCL Three-Phase Reactor

Figure 15.18 shows another design for the three-phase SFCL (Shimizu, 1992, 1993). This new type SFCL reactor has three superconducting windings with the same numbers of turns, of which each is connected to one phase. These superconducting windings are wound on a single iron core, hence they share the same magnetic flux path.

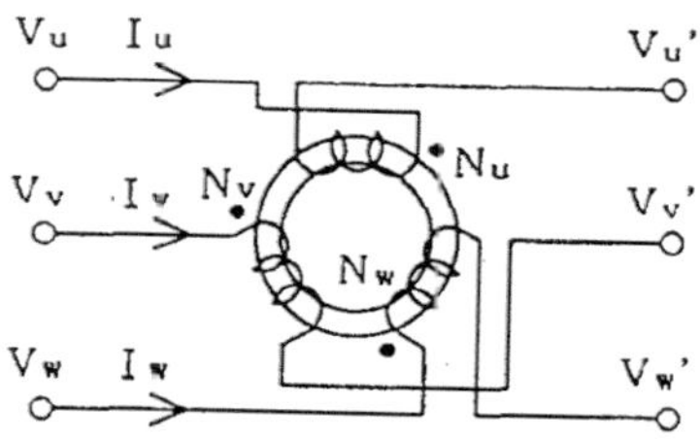

Figure 15.18 SFCL reactor with three superconducting windings, N_v, N_u and N_w, wound on a single iron core (Shimizu, 1992, 1993).

In a normal operating condition, the SFCL reactor exhibits very small impedance due to flux leakage, since the three-phase currents produce zero total flux. When a single-line-to-ground fault occurs, the reactance becomes very large because the balance between three phases currents is broken. In this case, fault current is limited by the large zero-phase-sequence reactance. With a proper design of the fault limiting level, superconductors wound quench during such faults.

When two-phase or three-phase faults occur, the reactance of such SFCL reactor does not increase. This is a similar issue as the transformer three-phase SFCLs. However, since the superconductors are directly connected to the transmission lines, when fault currents exceed their critical current level, quench happens and fault currents are limited by the increasing resistance of these superconductors. Therefore, such SFCL reactors work for not only single-phase faults, but also the other tapes of faults.

15.11.3 Three-Phase Winding and Magnetic Shield Combined SFCL

Due to the existence of different types of three-phase faults, a new type three-phase SFCL was designed to solve both symmetrical faults (three-phase) as well as unsymmetrical faults (single-phase). As shown in Fig. 15.19, the three-phase winding and magnetic shield combined SFCL is composed of a three-leg iron core and five identical superconducting coils (Cai, 2010). Three of the coils, a1, b1 and c1, are set to the left leg of the iron core. These coils form a three-phase reactor. The other two coils, b2 and c2, are set to the central leg together with a bulk superconductor or a short-circuited superconducting coil. Coil a1 is connected to phase a transmission lines in series, with coil b1/b2 to phase b, and coil c1/c2 to phase c. All the coils are identical, with only c2 in reversed winding.

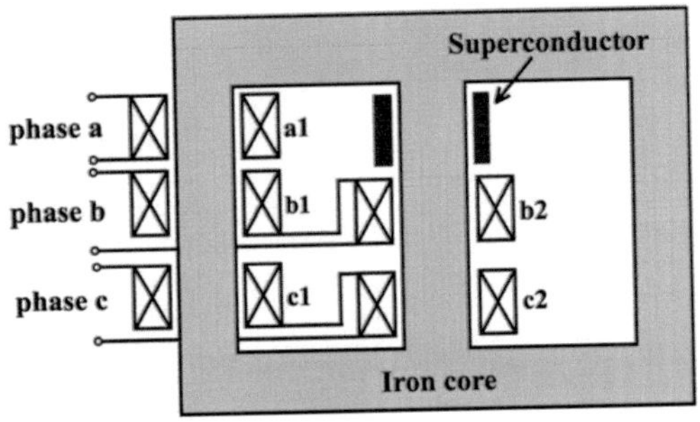

Figure 15.19 The structure of a three-phase winding and magnetic shield combined SFCL, with five superconducting coils in phase a, b, and c. The superconductor in solid black is a short-circuited superconducting coil or bulk superconductor (Cai, 2010).

During normal operations, three-phase currents are balanced and the total magnetic flux in the left leg of the iron core is zero. The magnetic flux in the central leg is generated by b2 and c2, and it is shielded by the bulk superconductor (or the short-circuited superconducting coil). The entire SFCL exhibits only the resistance of these superconducting coils, which is nearly zero.

In unsymmetrical faults such as single-line-to-ground faults, the three-phase balance is lost, and the magnetic flux in the iron core is not zero any more. The superconducting coils together with the iron core exhibit high inductance that limit the fault currents. When symmetrical faults occur, such as three-phase faults, the current limiting effect from the three-phase windings is very limited. However, coils b2 and c2 generate very high magnetic flux, and the bulk superconductor (or the short-circuited superconducting coil) would quench due to which large inductance appears and limit the fault current.

In the power networks, above 90% faults are single-phase, and such SFCLs limit fault current without quench, which greatly improves their recovery time.

References

Bashkirov, Y. A., Fleishman, L. S., Patsayeva, T. Y., Sobolev, A. N., and Vdovin, A. B. (1991). Current-limiting reactor based on high-T_c superconductors, *IEEE Trans. Magn.*, **27**(2), pp. 1089–1092.

Boenig, H. J., and Paice, D. A. (1983). Fault current limiter using a superconducting coil, *IEEE Trans. Magn.*, **19**(3), pp. 1051–1053.

Cai, Y., Okuda, S., Odake, T., Yagai, T., Tsuda, M., and Hamajima, T. (2010). Study on three-phase superconducting fault current limiter, *IEEE Trans. Appl. Supercond.*, **20**(3), pp. 1127–1130

Ebrahimpour, M., Vahidi, B., and Hosseinian, S. H. (2013). A hybrid superconducting fault current controller for DG networks and microgrids, *IEEE Trans. Appl. Supercond.*, **23**(5), 5604306.

Gray, K. E., and Fowler, D. E. (1978). A superconducting fault-current limiter, *J. Appl. Phys.*, **49**(4), pp. 2546–2550.

Hyun, O. B., Sim, J., Kim, H. R., Park, K. B., Yim, S. W., and Oh, I. S. (2009). Reliability enhancement of the fast switch in a hybrid superconducting fault current limiter by using power electronics switches, *IEEE Trans. Appl. Supercond.*, **19**(3), pp. 1843–1846.

Hyun, O. B., Park, K. B., Sim, J., Kim, H. R., Yim, S. W., and Oh, I. S. (2009). Introduction of a hybrid SFCL in KEPCO grid and local points at issue, *IEEE Trans. Appl. Supercond.*, **19**(3), pp. 1946–1949.

Kozak, J., Majka, M., Kozak, S., and Janowski, T. (2012). Design and tests of coreless inductive superconducting fault current limiter, *IEEE Trans. Appl. Supercond.*, **22**(3), 5601804.

Lee, B. W., Park, K. B, Sim, J., Oh, I. S., Lee, H. G., Kim, H. R., and Hyun, O. B. (2008). Design and experiment of novel hybrid type superconducting fault current limiters, *IEEE Trans. Appl. Supercond.*, **18**(2), pp. 624–627.

Lee, B. W., Sim, J., Park, K. B., and Oh, I. S. (2008). Practical application issues of superconducting fault current limiters for electric power systems, *IEEE Trans. Appl. Supercond.*, **18**(2), pp. 620–623.

Matsumura, T., Uchii, T., and Yokomizu, Y. (1997). Development of flux-lock-type fault current limiter with high-T_c superconducting element, *IEEE Trans. Appl. Supercond.*, **7**(2), pp. 1001–1004.

Moriconi, F., De La Rosa, F., Darmann, F., Nelson, A., and Masur, L. (2011). Development and deployment of saturated-core fault current limiters in distribution and transmission substations, *IEEE Trans. Appl. Supercond.*, **21**(3), pp. 1288–1293.

Park, D. K., Chang, K. S., Yang, S. E., Kim, Y. J., Ahn, M. C., Yoon, Y. S., Kim, H. M., Park, J. W., and Ko, T. K. (2009). Analytical and experimental studies on the hybrid fault current limiter employing asymmetric non-inductive coil and fast switch, *IEEE Trans. Appl. Supercond.*, **19**(3), pp. 1896–1899.

Parton, K. C. (1977). Current Limiting devices for alternating current systems, *U.S. Patent*, 4,045,823.

Shimizu, H., Yokomizu, Y., Goto, M., Matsumura, T., and Murayama, N. (2003). A study on required volume of superconducting element for flux flow resistance type fault current limiter, *IEEE Trans. Appl. Supercond.*, **13**(2), pp. 2052–2055.

Shimizu, S., Kado, H., Uriu, Y., and Ishigohka, T. (1992). Single-line-to-ground fault test of a 3-phase superconducting fault current limiting reactor, *IEEE Trans. Magn.*, **28**(1), pp. 442–445.

Shimizu, S., Tsukamoto, O., Ishigohka, T., Uriu, Y., and Ninomiya, A. (1993). Equivalent circuit and leakage reactances of superconducting 3-phase fault current limiter, *IEEE Trans. Appl. Supercond.*, **3**(1), pp. 578–581.

Shimizu, H., Yokomizu, Y., Matsumura, T., and Murayama, N. (2002). Proposal of flux flow resistance type fault current limiter using Bi2223 high T_c superconducting bulk, *IEEE Trans. Appl. Supercond.*, **12**(1), pp. 876–879.

Tosato, F., and Quaia, S. (2001). Reducing voltage sags through fault current limitation, *IEEE Trans. Power Del.*, **16**(1), pp. 12–17.

Xin, Y. (2018). A novel superconducting DC fault current limiter, *China Patent*, CN107819326A.

Yamaguchi, H., Kataoka, T., Yaguchi, K., Fujita, S., Yoshikawa, K., and Kaiho, K. (2004). Characteristics analysis of transformer type superconducting fault current limiter, *IEEE Trans. Appl. Supercond.*, **14**(2), pp. 815–818.

Zhang, J., Gao, Z., Song, N., Wen, H., Xiao, L., Lin, L., and Zuo, Z. (2002). Dynamic simulation and tests of a three-phase high T_c superconducting fault current limiter, *IEEE Trans. Appl. Supercond.*, **12**(1), pp. 896899.

Chapter 16

Mechanical Properties and Fracture Behaviors of Superconducting Bulk Materials

Akira Murakami
National Institute of Technology, Ichinoseki College, Takanashi, Hagisho, Ichinoseki, Iwate 021-8511, Japan
amura@ichinoseki.ac.jp

Superconducting devices equipped with superconducting bulk materials are promising for contribution to the establishment of sustainable society. Superconducting bulk materials are subjected to electromagnetic force and thermal stress in the devices. Since the electromagnetic force and thermal stress increase as the bulk size and critical current become larger, improvements of the mechanical properties of superconducting bulk materials contribute the development of high-performance superconducting devices. Clear understanding of the mechanical properties is also important for the design of superconducting devices. In this review, mechanical properties of superconducting bulk materials are summarized in association with the evaluation methods. Fracture surface

High-T_c Superconducting Technology: Towards Sustainable Development Goals
Edited by Muralidhar Miryala

ISBN 978-981-4877-65-7 (Hardcover), 978-1-003-16468-5 (eBook)
www.jennystanford.com

morphologies of conventional superconducting bulk materials that contain pores and modified bulk materials that contain few pores are also mentioned.

16.1 Introduction

Superconducting devices equipped with superconducting bulk materials, such as NMR and flywheel energy storage system, are promising for contribution to the establishment of sustainable society. It is well known that superconducting bulk materials can trap strong magnetic field in compact space [1–4]. Magnetic levitation ability is also interesting feature of superconducting bulk materials besides large current transport ability. Superconducting bulk materials are subjected to electromagnetic force and thermal stress in the superconducting devices [5–14]. Electromagnetic force applies to bulk materials during the magnetization process as schematically shown in Fig. 16.1. It has been reported that self-destruction occurred due to the electromagnetic force [6–8,12].

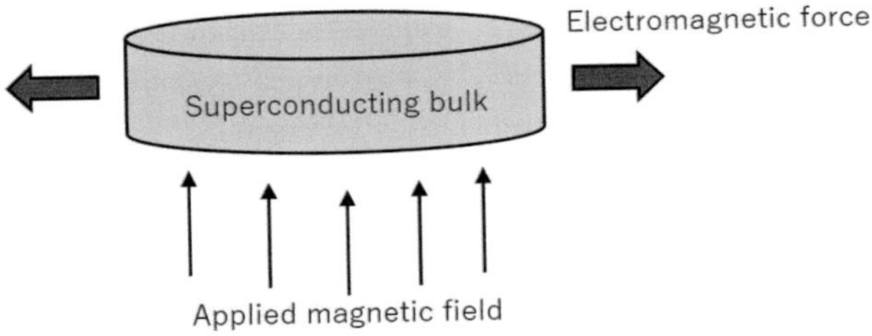

Figure 16.1 Schematic illustration of electromagnetic force during magnetization process.

REBaCuO (RE: Y or rare-earth elements) and MgB_2 are promising for liquid-helium-free superconducting bulk materials. REBaCuO bulk materials have to be a single-grain to trap strong magnetic field. Superconducting transition temperature (Critical temperature) of MgB_2 is lower than that of REBaCuO. Although the maximum trapped magnetic field values of MgB_2 bulk materials are also lower than those of REBaCuO bulk materials, MgB_2 bulk materials do not have to be a single-grain. While REBaCuO bulk materials are fabricated by melt-processing using a seed crystal, MgB_2 bulk materials are fabricated by just sintering. Enlargement of the bulk size contributes to the development of high-performance superconducting devices.

However, due to the undesirable nucleation, it is difficult to obtain large REBaCuO bulk piece. On the other hand, it is not so difficult to obtain large MgB_2 bulk piece. Thus, it is considered that not only REBaCuO bulk materials but also MgB_2 bulk materials are promising for practical application.

REBaCuO bulk materials contain micro-cracks perpendicular to the thickness direction. Such pre-existing micro-cracks are caused by phase transformation from tetragonal to orthorhombic during oxygen annealing process that is carried out after the melt-processing. Oxygen annealing is indispensable for the excellent superconducting properties. Such defects commonly cause degradation of mechanical properties of materials due to the reduction of net cross-sectional area and stress concentration around the defects. Since both the directions of electromagnetic force and pre-existing micro-cracks are perpendicular to the thickness direction of bulk materials, it is deduced that self-destruction of REBaCuO bulk materials does not largely depend on the existence of pre-existing micro-cracks.

Electromagnetic force and thermal stress increase as the bulk size and critical current become larger. Improvements of the mechanical properties, such as fracture strength and fracture toughness, are important for the development of high-performance superconducting devices equipped with superconducting bulk materials. Clear understanding of the mechanical properties is also important for the design of superconducting devices. Thus, evaluations of the mechanical properties of REBaCuO [15–91] and MgB_2 [92–98] bulk materials have been carried out extensively. However, the mechanical properties, especially fracture strength, depend on the evaluation methods. In this review, mechanical properties of superconducting bulk materials are summarized in association with the evaluation methods.

16.2 Evaluation Methods of Mechanical Properties

While tensile tests are commonly carried out to evaluate mechanical properties of metallic materials that have ductility, bending tests are commonly carried out for superconducting bulk materials that are brittle. It is difficult to prepare tensile test specimens that have taper for brittle materials. Also, it is difficult to do tensile tests without the fracture around chucking part. Schematic illustrations of bending tests are shown in Fig. 16.2, together with the bending moment diagrams (BMD) that illustrate the bending moment distribution caused by loading. 3- or 4-point bending load is applied to the specimen cut from superconducting bulk materials. Bending stresses caused by 3-point bending load σ_3 and 4-point bending load σ_4 are calculated by using Eqs. 16.1 and 16.2, respectively, where M_{max} is the maximum moment, I is moment of inertia of area, P is applied load, L is outer supporting span, l is inner loading span, and w and t are width and thickness of the specimen, respectively.

$$\sigma_3 = \frac{M_{max}}{I} \cdot \frac{t}{2} = \frac{3PL}{2wt^2} \tag{16.1}$$

$$\sigma_4 = \frac{M_{max}}{I} \cdot \frac{t}{2} = \frac{3P(L-l)}{2wt^2} \tag{16.2}$$

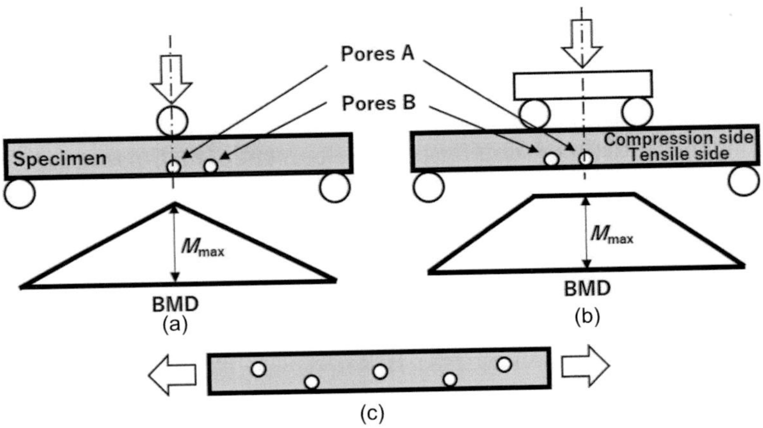

Figure 16.2 Schematic illustrations and bending moment diagrams BMD of (a) 3-point bending test, (b) 4-point bending test, and (c) tensile test.

In the 3-point bending test, the maximum moment causes only at the center of tensile side surface of the specimen. Conventional superconducting bulk materials contain pores and their fracture strength largely depends on the distribution of pores. While Pore A that exists at the center of tensile side surface affects the fracture strength in the 3-point bending test as schematically shown in Fig. 16.2a, Pore B does not affect the fracture strength. On the other hand, both Pores A and B affect the fracture of specimen in the 4-point bending test as schematically shown in Fig. 16.2b. Thus, the fracture strength evaluated through 3-point bending test is commonly higher than that evaluated through 4-point bending test. In the tensile test, all pores affect the fracture strength as schematically shown in Fig. 16.2c. Thus, the fracture strength values obtained through bending tests are overestimated.

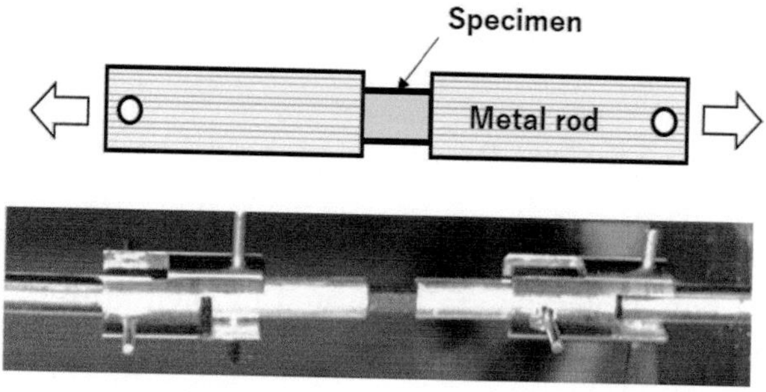

Figure 16.3 Schematic illustration and photograph of tensile test for REBaCuO bulk materials.

Tensile test method for REBaCuO bulk materials has been developed as reported in Ref. [24,25]. Figure 16.3 presents schematic illustration and photograph of the tensile test for REBaCuO bulk materials. Both ends of the specimen are glued to metal rods by using epoxy resin glue. After the metal rods are connected to the universal joints, tensile load is applied. Since the electromagnetic force causes in the direction perpendicular to the thickness of superconducting bulk materials as mentioned above, both the bending and the tensile test specimens are commonly cut from superconducting bulk materials such that the longitudinal direction corresponds to the direction perpendicular to the thickness of superconducting bulk materials. It has been reported that the bending strength of a conventional MgB_2 bulk material with the packing ratio of around 50% is lower than those of REBaCuO bulk materials [93]. Thus, the mechanical properties of conventional MgB_2 bulk materials also can be evaluated by this tensile test method without detachment between the specimen and metal rods. MgB_2 bulk materials are commonly fabricated by capsule method that is a sintering process performed under ambient pressure. In order to obtain high packing ratio MgB_2 bulk materials, hot isostatic pressing, and spark plasma sintering are used for the fabrication of MgB_2 bulk materials. Packing ratio and bending strength of a hot isostatic pressed MgB_2 bulk material are around 92% and 220 MPa [93] and those of a spark plasma sintered MgB_2 bulk material are around 98% and 315 MPa [95], respectively. These bending strength values are quite high in comparison with those of REBaCuO bulk materials that are mentioned in the followings. Thus, it is difficult to perform tensile test for such high packing ratio MgB_2 bulk materials without detachment between the specimen and metal rods.

16.3 Mechanical Properties and Fracture Behaviors

REBaCuO bulk materials are fabricated by melt-processing using a precursor that mainly consists of $REBa_2Cu_3O_x$ (RE123) and RE_2BaCuO_5 (RE211). After the melt-processing, RE211 secondary phase particles are dispersed into RE123 superconducting matrix. Small amount of Pt or CeO_2 addition is effective in inhibiting the coarsening of RE211 particles, that is indispensable for the excellent superconducting properties. Ag addition is effective in improving the mechanical properties [65–68]. Effects of RE211 and Ag contents on the mechanical properties are summarized in Ref. [69].

In order to obtain large REBaCuO bulk piece, fabrication process using a precursor with compositional gradient has been developed. GdBaCuO single-grain bulk piece with 150 mm in diameter can be obtained by that fabrication method. The precursor is prepared such that the peritectic decomposition temperature decreases with the distance from the center of it. Dy or Y is added to the precursor to decrease the peritectic decomposition temperature. It has been reported that mechanical properties of such super-large single-grain bulk materials are comparable to those of smaller REBaCuO bulk materials [70–75].

Conventional REBaCuO bulk materials fabricated by melt-processing in air or low O_2 atmosphere contain pores that cause degradation of the mechanical properties. It has been reported that melt-processing in O_2 atmosphere is effective in eliminating pores from REBaCuO bulk materials and also in improving the mechanical properties [24,25,76–88]. Figure 16.4 presents schematic illustration about atmosphere in the melt-processing for REBaCuO bulk materials. Figure 16.5 presents optical micrographs of REBaCuO bulk materials melt-processed in different O_2 atmosphere. Precursors of these bulk materials were heated in 50, 75, and 100% O_2 atmosphere, respectively. After the heating, one seed crystal was placed on the top of them and they were gradually cooled down in air for the crystal growth. Number of pore was decreased with increasing O_2 pressure. The porosities evaluated through image analysis were 4 and 14% for the bulk materials melt-processed in 75 and 50% O_2 atmosphere, respectively [79]. Few pores are observed for the bulk material melt-processed in 100% O_2 atmosphere.

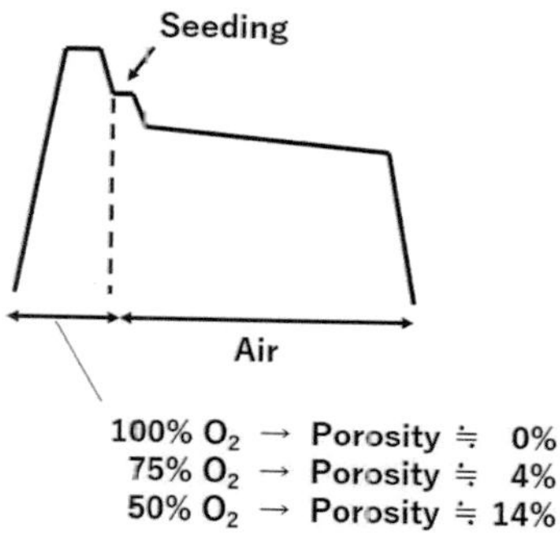

Figure 16.4 Schematic illustration about atmosphere in the melt-processing for REBaCuO bulk materials.

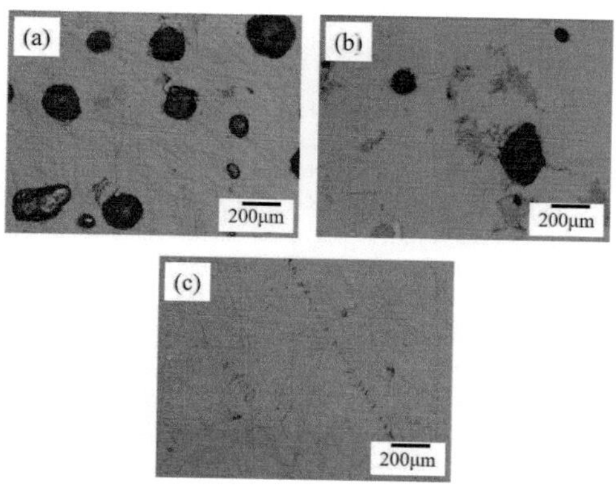

Figure 16.5 Optical micrographs of REBaCuO bulk materials melt-processed in (a) 50%, (b) 75%, and (c) 100% O_2 atmosphere.

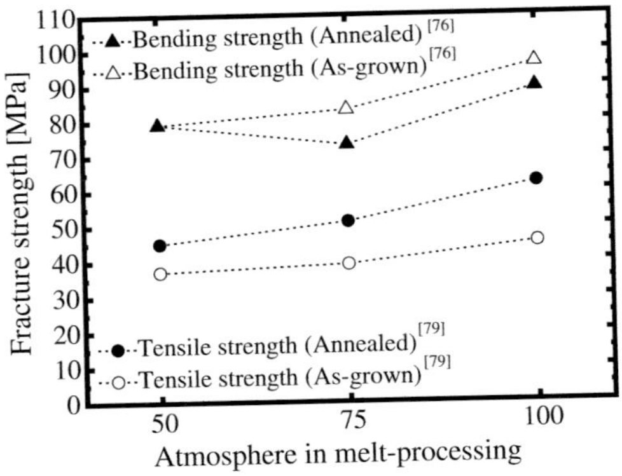

Figure 16.6 Average fracture strength values obtained through 3-point bending test and tensile test at room temperature of as-grown and oxygen annealed specimens cut from DyBaCuO bulk materials melt-processed in 50, 75, and 100% O_2 atmosphere [76,79].

Figure 16.6 presents the average fracture strength values obtained through 3-point bending test and tensile test at room temperature of the as-grown (non-annealed) and the oxygen annealed specimens cut from DyBaCuO bulk materials melt-processed in 50, 75, and 100% O_2 atmosphere [76,79]. Figure 16.7 presents approximation curves of the relationship between fracture strength and porosity. It has been reported that the relationship between fracture strength σ_f and porosity p of brittle materials can be approximated by using the following equation [99].

$$\sigma_f = \sigma_0 = \exp(-bp) \tag{16.3}$$

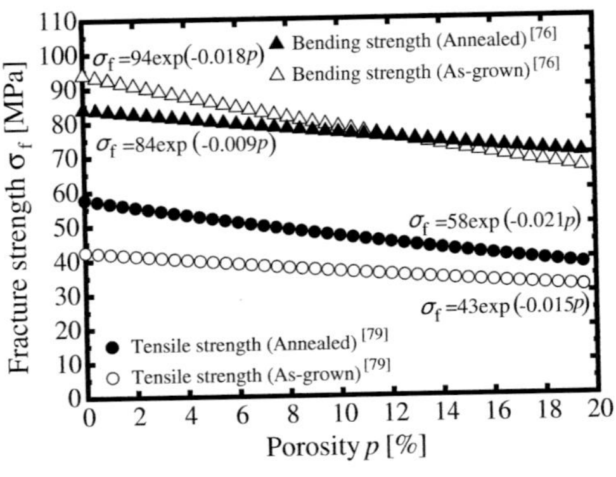

Figure 16.7 Relationship between average fracture strength values and porosity at room temperature of DyBaCuO bulk materials melt-processed in 50, 75, and 100% O_2 atmosphere [76,79].

Where σ_0 is the fracture strength at the porosity of 0% and b is constant. The relationship between Young's modulus and porosity of brittle materials also can be approximated by using an exponential equation [100]. The bending strength and the tensile strength are improved with increasing O_2 pressure in the melt-processing. Such an improvement of the fracture strength is attributable to the increase of net cross-sectional area and decrease of stress concentration region with decreasing the porosity. The tensile strength values are lower than the bending strength values. Such a difference between the bending strength and tensile strength is caused by the limited maximum moment region in the bending test specimens as schematically shown in Fig. 16.2. In order to obtain low-porosity REBaCuO bulk materials, infiltration growth technique that uses a precursor schematically shown in Fig. 16.8 is also useful. It has been reported that the tensile strength of a low-porosity REBaCuO bulk material fabricated by the infiltration growth technique is comparable to that of a low-porosity bulk material melt-processed in 100% O_2 atmosphere [91].

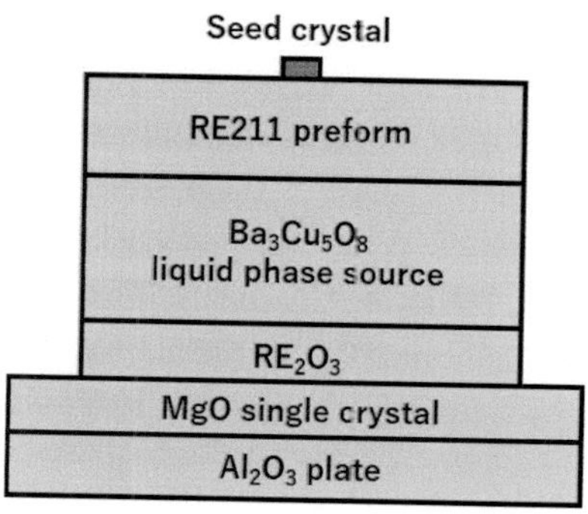

Figure 16.8 Schematic illustration of stacking of a seed, RE_2BaCuO_5, and liquid pre-form for the growth of single-grain bulk.

Figure 16.9 presents tensile fracture surfaces of as-grown specimens cut from DyBaCuO bulk materials melt-processed in 50% and 100% O_2 atmosphere, respectively. Flow-like patterns formed by the crack propagations are expanded from the left top part of the photographs. Through the observation for the flow-like patterns, the fatal crack initiation site can be initiated as marked by arrows. Large pores are observed around the crack initiation site of specimens cut from REBaCuO bulk material that contain pores [79,91]. On the other hand, an exceptional pore, some inclusion that consists of Pt and segregated RE211 secondary phase particles are observed around the crack initiation site of specimens cut from low-porosity REBaCuO bulk materials fabricated by the melt-processing in 100% O_2 atmosphere and infiltration growth technique [79,91].

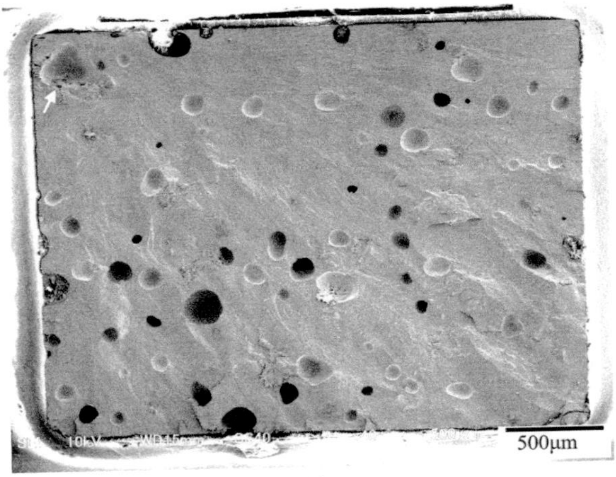

(a)

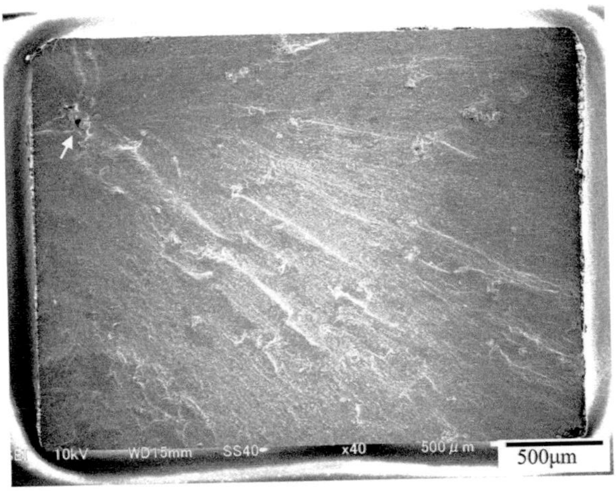

(b)

Figure 16.9 Tensile fracture surfaces of DyBaCuO bulk materials melt-processed in (a) 50% and (b) 100% O_2 atmosphere.

16.4 Conclusion

Mechanical properties of superconducting bulk materials REBaCuO and MgB_2 are summarized in association with the evaluation methods. While tensile tests are commonly carried out to evaluate mechanical properties of metallic materials that have ductility, bending tests are commonly carried out for superconducting bulk materials that are brittle. However, the fracture strength values obtained through bending tests are overestimated. In order to overcome the problem, tensile test method for superconducting bulk materials has been developed. While conventional REBaCuO bulk materials fabricated in air or low O_2 atmosphere contain pores, REBaCuO bulk materials fabricated by melt-processing in 100% O_2 atmosphere or infiltration growth technique contain few pores. The tensile strength values, that are lower than the bending strength values, increase with decreasing the porosity. While large pores are observed around the crack initiation site of specimens cut from conventional REBaCuO bulk materials that contain pores, some inclusion that consists of Pt and segregated RE211 secondary phase particles are observed around the crack initiation site of modified REBaCuO bulk materials that contain few pores.

References

1. Durrell, J. H., Dennis A. R., Jaroszynski J., Ainslie M. D., Palmer K. G. B., Shi Y. H., Campbell A. M., Hull J., Strasik, M., and Hellstrom, E. E. (2014). A trapped field of 17.6 T in melt-processed, bulk Gd-Ba-Cu-O reinforced with shrink-fit steel, *Supercond. Sci. Technol.* **27**, no. 082001.

2. Tomita, M., and Murakami, M. (2003). High-temperature superconductor bulk magnets that can trap magnetic fields of over 17 tesla at 29 K, *Nature* **421**, pp. 517–520.

3. Krabbes, G., Fuchs, G., Verges, P., Diko, P., Stover, G., and Gruss, S. (2002). 16 T trapped fields in modified YBaCuO: materials aspects, *Physica C* **378–381**, pp. 636–640.

4. Krabbes, G., Fuchs, G., Schatzle, P., Gruss, S., Park, J. W., Hardinghaus, F., Hayn, R., and Drechsler, S. L. (2000). YBCO-monoliths with trapped fields more than 14 T and peak effect, *Physica C* **341–348**, pp. 2289–2292.

5. Ren, Y., Weinstein, R., Liu, J., Sawh, R. P., and Foster, C. (1995). Damage caused by magnetic pressure at high trapped field in quasi-permanent magnets composed of melt-textured Y-Ba-Cu-O superconductor, *Physica C* **251**, pp. 15–26.
6. Miyamoto, T., Katagiri, J., Nagashima, K., and Murakami, M. (1999). Effect of Ag addition on the mechanical properties of bulk superconductors, *IEEE Trans. Appl. Supercond.* **9**, pp. 2066–2069.
7. Miyamoto, T., Nagashima, K., Sakai, N., and Murakami, M. (2000). Mechanical properties of bulk superconductors, *Supercond. Sci. Technol.* **13**, pp. 816–819.
8. Miyamoto, T., Nagashima, K., Sakai, N., and Murakami, M. (2000). Direct measurements of mechanical properties for large-grain bulk superconductors, *Physica C* **340**, pp. 41–50.
9. Tsuchimoto, M., and Takashima, H. (2000). Numerical evaluation of stresses of a bulk HTS in field-cooled magnetization, *Physica C* **341–348**, pp. 2467–2468.
10. Miyamoto, T., Nagashima, K., Sakai, N., and Murakami, M. (2001). Measurements of the thermal stresses in large-grain Y-Ba-Cu-O superconductors, *Physica C* **349**, pp. 69–74.
11. Tsuchimoto, M., Takashima, H., Tomita, M., and Murakami, M. (2002). Numerical stress evaluations of bulk superconductors with resin impregnation, *Physica C* **378–381**, pp. 718–721.
12. Nariki, S., Sakai, N., Matsui, M., and Murakami, M. (2002). Effect of silver addition on the mechanical and field trapping properties of Gd-Ba-Cu-O bulk superconductors, J. Japan Inst. Metals **66**, pp. 3096–314 (in Japanese).
13. Tsuchimoto, M., and Murata, K. (2003). Numerical evaluations of maximum stresses in bulk superconductors, Physica C **392–396**, pp. 575–578.
14. Namba, S., Fujishiro, H., Naito, T., Ainslie, M. D., and Huang, K. Y. (2019). Electromagnetic strain measurements and two-directional mechanical stress estimation for a REBaCuO ring bulk reinforced by a metal ring during field-cooled magnetization, *Supercond. Sci. Technol.* **32**, no. 125011.
15. Goyal, A., Oliver, W. C., Funkenbusch, P. D., Kroeger, D. M., and Burns, S. J. (1991). Mechanical properties of highly aligned $YBa_2Cu_3O_{7-\delta}$, effect of Y_2BaCuO_x particles, *Physica C* **183**, pp. 221–233.
16. Fujimoto, H., Murakami, M., and., and Koshizuka, N. (1992). Effect of Y_2BaCuO_5 on fracture toughness of YBCO prepared by a MPMG process, *Physica C* **203**, pp. 103–110.

17. Yu, F., White, K. W., and Meng, R. (1997). Mechanical characterization of top-seeded melt-textured $YBa_2Cu_3O_{7-\delta}$ single crystal, *Physica C* **276**, pp. 295–308.

18. Leenders, A., Ullrich, M., and Freyhardt, H. C. (1997). Influence of thermal cycling on the mechanical properties of VGF melt-textured YBCO, *Physica C* **279**, pp. 173–180.

19. Leenders, A., Ullrich, M., and Freyhardt, H. C. (1999). Mechanical properties of TSMG-YBCO, *IEEE Trans. Appl. Supercond.* **9**, pp. 2074–2077.

20. Okamura, T., Inoue, Y., Sakai, N., Takebayashi, S., Murakami, M., Munakata, F., and Akimune, Y. (1997). Anisotropy in Knoop hardness of MPMG-processed YBCO bulk superconductor, *Adv. Supercond.* **9**, pp. 729–732.

21. Schatzle, P., Krabbes, G., Gruβ, S., and Fuchs, G. (1999). YBCO/Ag bulk material by melt crystallization for cryomagnetic applications, *IEEE Trans. Appl. Supercond.* **9**, pp. 2022–2025.

22. Goretta, K. C., Diko, P., Jiang, M., Cuber, M. M., Xu, M., Ostenson, J. E., and Sengupta, S. (1999). Annealing and mechanical properties of bulk Y-Ba-Cu-O, *IEEE Trans. Appl. Supercond.* **9**, pp. 2081–2084.

23. Sakai, N., Seo, S. J., Inoue, K., Miyamoto, T., and Murakami, M. (2000). Mechanical properties of RE-Ba-Cu-O bulk superconductors, *Physica C* **335**, pp. 107–111.

24. Sakai, N., Yoneda, K., Seo, S. J., and Murakami, M. (1999). Mechanical properties of Sm-Ba-Cu-O bulk superconductors, *TEION KOGAKU (J. Cryo. Assoc. Jpn)* **34**, pp. 609–615 (in Japanese).

25. Sakai, N., Mase, A., Ikuta, H., Seo, S. J., Mizutani, U., and Murakami, M. (2000). Mechanical properties of Sm-Ba-Cu-O/Ag bulk superconductors, *Supercond. Sci. Technol.* **13**, pp. 770–773.

26. Yoshino, Y., Iwabuchi, A., Noto, K., Sakai, N., and Murakami, M. (2001). Vickers hardness properties of YBCO bulk superconductor at cryogenic temperature, *Physica C* **357–360**, pp. 796–798.

27. Yoshino, Y., Iwabuchi, A., Katagiri, K., Noto, K., Sakai, N., and Murakami, M. (2002). An evaluation of mechanical properties of YBaCuO and (Sm,Gd)BaCuO bulk superconductors using Vickers hardness test at cryogenic temperatures, *IEEE Trans. Appl. Supercond.* **12**, pp. 1755–1758.

28. Yoshino, Y., Iwabuchi, A., Takahashi, N., Katagiri, K., Noto, K., Sakai, N., and Murakami, M. (2004) Fracture toughness of SmBaCuO bulk superconductors at cryogenic temperatures, *IEEE Trans. Appl. Supercond.* **14**, pp. 1118–1121.

29. Yoshino, Y., Iwabuchi, A., Takahashi, N., Katagiri, K., and Noto, K. (2005). Elastic recovery of ReBaCuO bulk superconductors during indentation process at cryogenic temperatures, *Physica C* **426–431**, pp. 714–719.

30. Tomita, M., and Murakami, M. (2002). Mechanical properties of bulk superconductors with resin impregnation, *Supercond. Sci. Technol.* **15**, pp. 808–812.

31. Tomita, M., Murakami, M., and Katagiri, K. (2002). Reliability of mechanical properties for bulk superconductors with resin impregnation, *Physica C* **378–381**, pp. 783–787.

32. Tomita, M., and Murakami, M. (2002). Durability of bulk superconductors against the vibration load experiment, *Physica C* **378–381**, pp. 842–847.

33. Tomita, M., and Murakami, M. (2003). Effect of resin impregnation on the prevention of crack propagation for bulk Y-Ba-Cu-O superconductors, *Physica C* **392–396**, pp. 562–566.

34. Murakami, A., Katagiri, K., Kasaba, K., Shoji, Y., Noto, K., Sakai, N., and Murakami, M. (2002). Tensile strength and fracture surface topography of Sm-Ba-Cu-O bulk superconductors, *Supercond. Sci. Technol.* **15**, pp. 1099–1104.

35. Katagiri, K., Murakami, A., Sato, T., Okudera, T., Sakai, N., Muralidhar, M., and Murakami, M. (2002). Stress-strain characteristics and fracture surface morphology of (Sm,Gd)-Ba-Cu-O bulk superconductor, *Physica C* **378–381**, pp. 722–726.

36. Matsui, M., Sakai, N., Murakami, M., and Osamura, K. (2003). Mechanical properties of Nd-Ba-Cu-O bulk superconductors, *Supercond. Sci. Technol.* **16**, pp. 1086–1091.

37. Okudera, T., Murakami, A., Katagiri, K., Kasaba, K., Shoji, Y., Noto, K., Sakai, N., and Murakami, M. (2003). Fracture toughness evaluation of YBCO bulk superconductor, *Physica C* **392–396**, pp. 628–633.

38. Kaneko, T., Iida, K., Yoshioka, J., Sakai, N., Murakami, M., Murakami, A., and Katagiri, K. (2003). Mechanical properties of Y-Ba-Cu-O blocks welded by Er-Ba-Cu-O solder, *Physica C* **392–396**, pp. 673–676.

39. Kaneko, T., Iida, K., Kono, T., Sakai, N., Murakami, M., Koshizuka, N., Murakami, A., and Katagiri, K. (2004). Mechanical properties of Y-Ba-Cu-O blocks welded by silver added Y-Ba-Cu-O solder, *Physica C* **412–414**, pp. 678–682.

40. Murakami, A., Katagiri, K., Kasaba, K., Shoji, Y., Noto, K., Teshima, H., Sawamura, M., and Murakami, M. (2004). Mechanical properties of Gd123 bulk superconductor at liquid nitrogen temperature, *IEEE Trans. Appl. Supercond.* **14**, pp. 1050–1053.

41. Katagiri, K., Murakami, A., Shoji, Y., Teshima, H., Sawamura, M., Iwamoto, A., Mito, T., and Murakami, M. (2004). Tensile and bending mechanical properties of bulk superconductors at room temperature, *Physica C* **412–414**, pp. 633–637.

42. Murakami, A., Katagiri, K., Kasaba, K. Noto, K., Teshima, H., Sawamura, M., Sakai, N., and Murakami, M. (2004). Low temperature mechanical properties of Y123 bulk superconductor fabricated by the modified QMG process, *Physica C* **412–414**, pp. 673–677.

43. Kasaba, K., Katagiri, K., Murakami, A., Sato, G., Sato, T., Murakami, M., Sakai, N., Teshima, H., and Sawamura, M. (2005). Tensile testing method for rare earth based bulk superconductors at liquid nitrogen temperature, *Physica C* **426–431**, pp. 639–643.

44. Kasaba, K., Teshima, H., Hokari, T. Sato, T., Katagiri, K., Shoji, Y., Murakami, A., and Hirano, H. (2006). Advanced tensile testing methods for bulk superconductors at cryogenic temperatures, *Physica C* **445–448**, pp. 427–430.

45. Katagiri, K., Sato, T., Murakami, A., Kasaba, K., Shoji, Y., Noto, K., Teshima, H., and Sawamura, M. (2005). Fracture toughness of REBaCuO bulk superconductor at liquid nitrogen temperature, *Physica C* **426–431**, pp. 709–713.

46. Katagiri, K., Nyilas, A., Sato, T., Hatakeyama, Y., Hokari, T., Teshima, H., Iwamoto, A., and Mito, T. (2006). Temperature dependence of the mechanical properties of melt-processed Dy-Ba-Cu-O bulk superconductors evaluated by three point bending tests, *Supercond. Sci. Technol.* **19**, pp. S545–S549.

47. Murakami, A., Katagiri, K., Miyata, H., Kasaba, K., and Shoji, Y. (2006). Distribution of mechanical properties in a single-grain Sm123 bulk superconductor at liquid nitrogen temperature, *IEEE Trans. Appl. Supercond.* **16**, pp. 1003–1006.

48. Fujimoto, H. (2006). Mechanical properties of Y123/Ag superconducting bulks, *IEEE Trans. Appl. Supercond.* **16**, pp. 1043–1046.

49. Murakami, A., Katagiri, K., Kasaba, K., Miyata, H., and Shoji, Y. (2006). Bending mechanical properties of a single-grain Y123 bulk superconductor at liquid nitrogen temperature, *Physica C* **445–448**, pp. 361–365.

50. Yoneda, K., and Ye, J. (2006). Crack propagation and acoustic emission behavior of silver-added Dy123 bulk superconductor, *Physica C* **445–448**, pp. 371–374.

51. Yoneda, K., Ye, J., and Tomita, M. (2005). Acoustic emission behavior under bending deformation of YBCO bulk superconductor, *Physica C* **426–431**, pp. 705–708.

52. Sato, T., Katagiri, K., Hokari, T., Hatakeyama, Y., Murakami, A., Nyilas, A., Kasaba, K., Teshima, H., and Hirano, H. (2006). Evaluation of Young's modulus of RE123 bulk superconductors by three point bending tests, *Physica C* **445–448**, pp. 422–426.

53. Katagiri, K., Hatakeyama, Y., Sato, T., Kasaba, K., Shoji, Y., Murakami, A., Teshima, H., and Hirano, H. (2006). Evaluation of mechanical properties of Dy123 bulk superconductors by 3-point Bending Tests, *Physica C* **445–448**, pp. 431–435.

54. Roa, J. J., Capdevila, X. G., Martinez, M., Espiell, F., and Segarra, M. (2007). Nanohardness and Young's modulus of YBCO samples textured by the Bridgman technique, *Nanotechnology* **18**, no. 385701.

55. Roa, J. J., Jimenez-Pique, E., Capdevila, X. G., Martinez, M., and Segarra, M. (2008). Yield strength, shear stress and toughness of YBCO samples textured by Bridgman technique, *J. Phy.: Conf. Ser.* **97**, no. 012116.

56. Roa, J. J., Bartolome, E., Bozzo, B., Capdevila, X. G., Granados, X., and Segarra, M. (2010). Nano-mechanical properties of silver-welded YBCO bulks, *J. Phy.: Conf. Ser.* **234**, no. 012034.

57. Bartolome, E., Roa, J. J., Bozzo, B., Segarra, M., and Granados, X. (2010). Effective silver-assisted welding of YBCO blocks: mechanical versus electrical properties, *Supercond. Sci. Technol.* **23**, no. 045013.

58. Oka, T., Itoh, Y., Yanagi, Y., Tanaka, H., Takashima, S., Yamada, Y., and Mizutani, U. (1992). Critical current density and mechanical strength of $YBa_2Cu_3O_{7-\delta}$ superconducting composites containing Zr, Ag and Y_2BaCuO_5 dispersions by melt-processing, *Physica C* **200**, pp. 55–64.

59. Nariki, S., Sakai, N., Murakami, M., and Hirabayashi, I. (2004). Effects of Dy_2BaCuO_5 contents on microstructure and mechanical strength of Ag-added Dy-Ba-Cu-O bulk superconductors, *Physica C* **412–414**, pp. 651–656.

60. Murakami, A., Katagiri, K., Kasaba, K., Shoji, Y., Noto, K., Teshima, H., Sawamura, M., and Murakami, M. (2003). Mechanical properties of Gd123 bulk superconductors at room temperature, *Cryogenics* **43**, pp. 345–350.

61. Murakami, A., Miyata, H., Hashimoto, R., and Katagiri, K. (2007). Effects of Dy211 content on bending mechanical properties of Dy123 bulks at room temperature, *IEEE Trans. Appl. Supercond.* **17**, pp. 3059–3062.

62. Murakami, A., Katagiri, K., Noto, K., Kasaba, K., Shoji, Muralidhar, M., Sakai, N., and Murakami, M. (2002). Tensile mechanical properties of

(Nd,Eu,Gd)-Ba-Cu-O bulk superconductors at room and liquid nitrogen temperatures, *Physica C* **378–381**, pp. 794–797.

63. Matsui, M., Sakai, N., and Murakami, M. (2003). Effect of Nd422 addition on the mechanical properties of Nd-Ba-Cu-O bulk superconductors, *Physica C* **384**, pp. 391–398.

64. Murakami, A., Katagiri, K., Kasaba, K., Shoji, Y., Noto, K., Sakai, N., and Murakami, M. (2003). Mechanical properties of Sm-Ba-Cu-O bulk superconductors at room temperature, *Physica C* **392–396**, pp. 557–561.

65. Nariki, S., Sakai, N., Tomita, M., and Murakami, M. (2002). Mechanical properties of melt-textured Gd-Ba-Cu-O bulk with silver addition, *Physica C* **378–381**, pp. 779–782.

66. Fujimoto, H. (2003). Flexural strength of melt-processed Y-Ba-Cu-O bulk superconductors with Ag addition measured at 77 K, *Supercond. Sci. Technol.* **16**, pp. 1115–1119.

67. Matsui, M., Sakai, N., and Murakami, M. (2002). Effect of Ag_2O addition on the trapped fields and mechanical properties of Nd-Ba-Cu-O bulk superconductors, *Supercond. Sci. Technol.* **15**, 1092–1098.

68. Katagiri, K., Murakami, A., Kan, R., Kasaba, K., Noto, K., Muralidhar, M., Sakai, N., and Murakami, M. (2003). Effects of Ag content on the mechanical properties of (Nd,Eu,Gd)-Ba-Cu-O bulk superconductors, *Physica C* **392–396**, pp. 526–530.

69. Katagiri, K., Murakami, A., and Kasaba, K. (2008). Evaluation of mechanical properties of high-temperature superconducting bulks fabricated by melt-processing, *Cryogenics* **48**, pp. 87–94.

70. Murakami, A., Teshima, H., Morita, M., and Iwamoto, A. (2014). Distribution of mechanical properties in large single-grain RE-Ba-Cu-O bulk 150 mm in diameter fabricated using RE compositional gradient technique, *Physica C* **496,** pp. 44–48.

71. Murakami, A., Teshima, H., Morita, M., Kudo, T., and Iwamoto, A. (2014). Effects of RE compositional boundaries on fracture strength at 77 K in large single-grained RE-Ba-Cu-O bulk 150 mm in diameter, *J. Phy.: Conf. Ser.* **507**, no. 012034.

72. Murakami, A., and Iwamoto, A. (2019). Fracture strength properties of (Gd,Y)BaCuO large single-grain bulk at liquid nitrogen temperature, *J. Phy.: Conf. Ser.* **1293**, no. 012045.

73. Murakami, A., and Iwamoto, A. (2019). Mechanical properties of superconducting (Gd,Y)BaCuO large single-grain material fabricated by RE compositional gradient technique, *IEEE Trans. Appl. Supercond.* **29**, no. 8400406.

74. Murakami, A., Fujimoto, H., and Iwamoto, A. (2012). Fracture strength properties of Gd-Ba-Cu-O large single-grain bulk with CeO_2 addition, *Phy. Proc.* **27**, pp. 160–163.

75. Murakami, A., Fujimoto, H., and Iwamoto, A. (2013). Low temperature mechanical properties of RE-Ba-Cu-O large single-grain bulk 150 mm in diameter, *IEEE Trans. Appl. Supercond.* **23**, no. 6800505.

76. Murakami, A., Miyata, H, Hashimoto, R., Katagiri, K., and Iwamoto, A. (2009). Evaluations of mechanical properties in Dy123 single-grain bulk superconductors with low porosity, *IEEE Trans. Appl. Supercond.* **19**, pp. 2995–2998.

77. Fujimoto, H., and Murakami, A. (2011). Preparation and properties of high-quality melt growth Gd123 bulks with low void density - Flexural strength at 77 K -, *IEEE Trans. Appl. Supercond.* **21**, pp. 2718–2722.

78. Murakami, A., Otaka, K., Miura, T., and Iwamoto, A. (2011). Mechanical properties of Dy123 low porosity bulk superconductors at liquid nitrogen temperature, *IEEE Trans. Appl. Supercond.* **21**, pp. 3107–3110.

79. Murakami, A., and Iwamoto, A. (2020). Tensile properties of DyBaCuO low porosity bulk material melt-processed in oxygen atmosphere, *IEEE Trans. Appl. Supercond.* **30**, no. 6800105.

80. Fujimoto, H. (2009). Preparation and mechanical properties of large single domain GdBaCuO superconductor with low void density, *IEEE Trans. Appl. Supercond.* **19**, pp. 2933–2936.

81. Murakami, A., Miyata, H., Hashimoto, R., Otaka, K., Katagiri, K., and Iwamoto, A. (2009). Mechanical properties of low porosity Dy123 bulks with different Dy211 content, *Physica C* **469**, pp. 1207–1210.

82. Murakami, A., Otaka, K., and Iwamoto, A. (2010). Mechanical properties of Ag added Dy123 low porosity bulks, *Physica C* **470**, pp. 1185–1188.

83. Murakami, A., Otaka, K., and Iwamoto, A. (2010). Observations on fracture surfaces of Dy123 bulks with various porosities, *J. Phy.: Conf. Ser.* **234**, no. 012027.

84. Murakami, A., Otaka, K., Miura, T., and Iwamoto, A. (2011). Fracture toughness of Dy123 low porosity bulks at liquid nitrogen temperature, *Physica C* **471**, pp. 884–888.

85. Fujimoto, H., and Murakami, A. (2012). Mechanical properties of Gd123 superconducting bulks at 77 K, *Supercond. Sci. Technol.* **25**, no. 054017.

86. Fujimoto, H., and Murakami, A. (2012). Fracture toughness properties of Gd123 superconducting bulks, *Phy. Proc.* **36**, pp. 458–462.

87. Murakami, A., Otaka, K., Miura, T., and Iwamoto, A. (2012). Effects of Pt content on mechanical properties of Dy123 bulks, *Phy. Proc.* **36**, pp. 551–556.

88. Fujimoto, H., Murakami, A., Teshima, H., and Morita, M. (2013). Stress-strain behavior under static loading in Gd123 high-temperature superconductors at 77 K, *Cryogenics* **57**, pp. 6–11.

89. Murakami, A., and Iwamoto, A. (2017). Mechanical properties of EuBaCuO superconducting bulk material at liquid nitrogen temperature, *J. Phy.: Conf. Ser.* **871**, no. 012055.

90. Murakami, A., and Iwamoto, A. (2013). Microstructures and properties of Dy123 single-grain bulks fabricated from spark plasma sintered precursors, *Phy. Proc.* **45**, pp. 57–60.

91. Murakami, A., Muralidhar, M., and Iwamoto, A. (2020). Mechanical properties of REBaCuO single-grain bulk fabricated by infultration growth technique, *Supercond. Sci. Technol.* **33**, no. 024003.

92. Gorur, O., Nursoy, M., Terzioglu, C., Varilci, A., and Belenli, I. (2009). Influence of annealing temperature on the microstructure and mechanical properties of MgB_2, *J. Phy.: Conf. Ser.* **153**, no. 012012.

93. Murakami, A., Teshima, H., Naito, T., Fujishiro, H., and Kudo, T. (2014). Mechanical properties of MgB_2 bulks, *Phy. Proc.* **58**, pp. 98–101.

94. Murakami, A., Teshima, H., Naito, T., Fujishiro, H., Kudo, T., and Iwamoto, A. (2015). Low-temperature mechanical properties of bulk MgB_2 fabricated by hot isostatic pressing, *Phy. Proc.* **67**, pp. 982–986.

95. Murakami, A., Noudem, J., Guesmi, Z., Kudo, T., and Iwamoto, A. (2015). Mechanical properties of MgB_2 bulk fabricated by spark plasma sintering, *Phy. Proc.* **65**, pp. 77–80.

96. Murakami, A., Noudem, J., Guesmi, Z., and Iwamoto, A. (2016). Low temperature fracture strength of MgB_2 bulk processed by spark plasma sintering, *IEEE Trans. Appl. Supercond.* **26**, no. 8401705.

97. Murakami, A., Iwamoto, A., and Noudem, J. G. (2018). Effects of SPS pressure on the mechanical properties of high packing ratio bulk MgB_2 superconductor, *J. Phy.: Conf. Ser.* **1054**, no. 012051.

98. Murakami, A., Iwamoto, A., and Noudem, J. G. (2018). Mechanical properties of bulk MgB_2 superconductors processed by spark plasma sintering at various temperatures, *IEEE Trans. Appl. Supercond.* **28**, no. 8400204.

99. Duckworth, W. (1953). Discussion of Ryshkewitch paper by Winston Duckworth, *J. Amer. Ceram. Soc.* **36**, p. 68.

100. Spriggs, R. M. (1961). Expression for effect of porosity on elastic modulus of polycrystalline refractory materials, particularly aluminum oxide, *J. Am. Ceram. Soc.* **44**, pp. 628–629.

Index

For Product Safety Concerns and Information please contact our EU representative GPSR@taylorandfrancis.com Taylor & Francis Verlag GmbH, Kaufingerstraße 24, 80331 München, Germany